The Impact of Species Changes in African Lakes

CHAPMAN & HALL FISH AND FISHERIES SERIES

Amongst the fishes, a remarkably wide range of fascinating biological adaptations to diverse habitats has evolved. Moreover, fisheries are of considerable importance in providing human food and economic benefits. Rational exploitation and management of our global stocks of fishes must rely upon a detailed and precise insight of the interaction of fish biology with human activities.

The Chapman & Hall *Fish and Fisheries Series* aims to present authoritative and timely reviews which focus on important and specific aspects of the biology, ecology, taxonomy, physiology, behaviour, management and conservation of fish and fisheries. Each volume will cover a wide but unified field with themes in both pure and applied fish biology. Although volumes will outline and put in perspective current research frontiers, the intention is to provide a synthesis accessible and useful to both experts and non-specialists alike. Consequently, most volumes will be of interest to a broad spectrum of research workers in biology, zoology, ecology and physiology, with an additional aim of the books encompassing themes accessible to non-specialist readers, ranging from undergraduates and postgraduates to those with an interest in industrial and commercial aspects of fish and fisheries.

Applied topics will enhance synopses of fishery issues which will appeal to a wide audience of fishery scientists, aquaculturists, economists, geographers and managers in the fishing industry. The series will also contain practical guides to fishery and analysis methods and global reviews of particular types of fisheries.

Books already published and forthcoming are listed below. The Publisher and Series Editor would be glad to discuss ideas for new volumes in the series . . .

Available titles

1. **Ecology of Teleost Fishes**
   Robert J. Wootton
2. **Cichlid Fishes**
   Behaviour, ecology and evolution
   Edited by Miles A. Keenlyside
3. **Cyprinid Fishes**
   Systematics, biology and exploitation
   Edited by Ian J. Winfield and Joseph S. Nelson
4. **Early Life History of Fish**
   An energetics approach
   Ewa Kamler

5. **Fisheries Acoustics**
   David N. MacLennan and E. John Simmonds
6. **Fish Chemoreception**
   Edited by Toshiaki J. Hara
7. **Behaviour of Teleost Fishes**
   Second edition
   Edited by Tony J. Pitcher
8. **Genetics and Fish Breeding**
   Colin R. Purdom
9. **Fish Ecophysiology**
   J. Cliff Rankin and Frank B. Jensen
10. **Fish Swimming**
    John J. Videler
11. **On the Dynamics of Exploited Fish Populations**
    Raymond J.H. Beverton and Sidney J. Holt
    (Facsimile reprint)
12. **Sea Bass**
    Graham D. Pickett and Michael G. Pawson
13. **Fish Bioenergetics**
    Malcolm Jobling
14. **On the Sex of Fish and Gender of Scientists**
    Daniel Pauly
15. **Hake**
    Fisheries, ecology and markets
    Edited by Jürgen Alheit and Tony J. Pitcher
16. **Environmental Biology of Fishes**
    Malcolm Jobling
17. **Electric Fishes**
    History and behavior
    Peter Moller
18. **The Impact of Species Change in African Lakes**
    Edited by Tony J. Pitcher and Paul B. Hart
19. **Pike**
    Biology and exploitation
    Edited by John F. Craig
20. **Reef Fisheries**
    Edited by Nicholas V. C. Polunin and Callum M. Roberts

Forthcoming titles

Fisheries Ecology
Second edition
Edited by T.J. Pitcher and P. Hart

The Impact of Species Changes in African Lakes

Edited by

Tony J. Pitcher
Fisheries Centre
University of British Columbia
Vancouver
Canada

and

Paul J.B. Hart
Department of Zoology
Leicester University
Leicester
UK

CHAPMAN & HALL
London · Glasgow · Weinheim · New York · Tokyo · Melbourne · Madras

Published by Chapman & Hall, 2–6 Boundary Row, London SE1 8HN

Chapman & Hall, 2–6 Boundary Row, London SE1 8HN, UK

Blackie Academic & Professional, Wester Cleddens Road, Bishopbriggs, Glasgow G64 2NZ, UK

Chapman & Hall GmbH, Pappelallee 3, 69469 Weinheim, Germany

Chapman & Hall USA, 115 Fifth Avenue, New York NY 10003, USA

Chapman & Hall Japan, ITP-Japan, Kyowa Building, 3F, 2-2-1 Hirakawacho, Chiyoda-ku, Tokyo 102, Japan

Chapman & Hall Australia, 102 Dodds Street, South Melbourne, Victoria 3205, Australia

Chapman & Hall India, R. Seshadri, 32 Second Main Road, CIT East, Madras 600 035, India

First edition 1995

Typeset in 10/12pt Photina by AFS Image Setters Ltd, Glasgow

Printed in Great Britain by Cambridge University Press

ISBN 0 412 55050 4

A catalogue record for this book is available from the British Library

Library of Congress Catalog Card Number: 95-69068

∞ Printed on permanent acid-free text paper, manufactured in accordance with ANSI/NISO Z39.48-1992 and ANSI/NISO Z39.48-1984 (Permanence of Paper).

To the memory of Aggrey W. Kudhongania

Contents

List of contributors xix

Series foreword *Tony J. Pitcher* xxiii

Preface *Tony J. Pitcher and Paul J.B. Hart* xxv

General introduction *Paul J.B. Hart* xxvii
The impact of species changes in African lakes
References

1 Species changes and fisheries in African lakes: outline of the issues 1
Tony J. Pitcher
1.1 Introduction 1
1.2 Geographical scope of the book 2
1.3 Introduction of fish to African lakes 5
1.4 Establishment of introduced species 6
1.5 Impacts of fish introductions 9
1.6 Ecological models of fish introductions 10
1.7 Evaluation of the impacts of species changes 10
1.8 Mitigation of the impacts of species changes 13
1.9 General recommendations for development and assistance in African lake fisheries 14
Acknowledgements 15
References 15

Part One Introduced Nile Perch in Lake Victoria: Impacts on Biodiversity and Evaluation of the Fishery

2 Impact of environmental change, species introductions and ecological interactions on the fish stocks of Lake Victoria 19
Aggrey W. Kudhongania† and Deonatus B.R. Chitamwebwa

2.1 Introduction 19
2.2 Impact of exploitation 20
2.3 Impact of competition and hybridization 23
2.4 Impact of environmental changes 24
2.5 Impact of predation by *Lates* 25
2.6 Exploitation versus predation by *Lates* 27
2.7 The new fishery of Lake Victoria 28
2.8 Summary 29
Acknowledgements 30
References 30

3 Limnological changes in Lake Victoria since the Nile perch introduction 33
Peter B.O. Ochumba†
3.1 Introduction 33
3.2 Discussion 38
3.3 Conclusion 39
3.4 Summary 40
Acknowledgements 41
References 41

4 Impact of fish species introductions on the tilapias of Lakes Victoria and Kyoga 45
Timothy Twongo
4.1 Introduction 45
4.2 Ecological considerations 46
4.3 How introduced species affect indigenous fishes 51
Acknowledgements 55
References 55

5 Diversity and stability of fish stocks in Lakes Victoria, Kyoga and Nabugabo after establishment of introduced species 59
Richard Ogutu-Ohwayo
5.1 Introduction 59
5.2 Methods 60
5.3 Results 62
5.4 Discussion 76
5.5 Summary 79
Acknowledgements 79
References 80

6 Dynamics of the haplochromine cichlid fauna and other ecological changes in the Mwanza Gulf of Lake Victoria 83
Frans Witte, Tijs Goldschmidt and Jan H. Wanink
6.1 Introduction 83
6.2 Distribution and increase of the Nile perch 85
6.3 Decline of the haplochromines 88
6.4 Dynamics in morphology and ecology of haplochromines 91
6.5 Observations on other fish stocks and other organisms 99
6.6 Summary 104
Acknowledgements 106
References 106

7 An analysis of species changes in Lake Victoria: did the Nile perch act alone? 111
Alida Bundy and Tony J. Pitcher
7.1 Introduction 111
7.2 Methods 112
7.3 Results 117
7.4 Discussion 125
7.5 Conclusions 132
Acknowledgements 133
References 133

8 Analysis of species changes in Lake Victoria using ECOPATH, a multispecies trophic model 137
Jacques Moreau
8.1 Introduction 137
8.2 Lake Victoria: a severely damaged ecosystem 138
8.3 Material and methods 140
8.4 Results 148
8.5 Discussion 156
8.6 Conclusions 157
8.7 Summary 158
Acknowledgements 158
References 158

9 Assessment of the Nile perch fishery in Lake Victoria 163
Tony J. Pitcher and Alida Bundy
9.1 Introduction 163
9.2 Assembly of data 164
9.3 Analytical methods 167
9.4 Results 170
9.5 Discussion 175
9.6 Summary 179

Acknowledgements 179
References 179

10 Thirty years on: the development of the Nile perch fishery in Lake Victoria 181
J. Eric Reynolds, Dominique F. Gréboval and Piero Mannini
10.1 The making of an 'ichthyo-imbroglio' 181
10.2 Pre-perch, post-perch: the fisheries regimes compared 183
10.3 Impact assessment 199
10.4 Discussion: the outlook 205
10.5 Summary and conclusions 208
Acknowledgements 210
References 211

11 Socio-economic impacts of introduced species in Lake Victoria fisheries 215
Craig K. Harris, David S. Wiley and Douglas C. Wilson
11.1 Introduction 215
11.2 Individual and household impacts 217
11.3 Community and organizational impacts 227
11.4 Impacts at the national level 231
11.5 International and transnational impacts 234
11.6 Summary 237
Acknowledgements 238
References 238

Part Two Evaluation of Species Changes in Other African Lakes with Introduced Fish Species

12 Inshore fish population and species changes in Lake Kariba, Zimbabwe 245
Lawrence Karenge and Jeppe Kolding
12.1 Introduction 245
12.2 Background information 247
12.3 Brief history of the commercial fisheries 249
12.4 The experimental fishery 250
12.5 Changes in species composition 252
12.6 Trends in catch per unit effort 254
12.7 Discussion 263
12.8 Conclusion 270
12.9 Summary 271
Acknowledgements 272
References 272

13 The impact of an introduction of sardine into Lake Kivu 277
Hans H. de Iongh, Petronella C. Spliethoff and Fritz Roest
13.1 Introduction 277
13.2 Characteristics of Lake Kivu 278
13.3 Biology of *Limnothrissa* 281
13.4 Ecological impact of the sardines 288
13.5 The fishery 290
13.6 Socio-economic benefits of the sardine fishery 292
13.7 Summary 294
Acknowledgements 295
References 295

14 The persistence of two introduced tilapia species in Lake Naivasha, Kenya, in the face of environmental variability and fishing pressure 299
S. Mucai Muchiri, Paul J.B. Hart and David M. Harper
14.1 Introduction 299
14.2 Variability of the Lake Naivasha ecosystem 300
14.3 The fish community: species and their changes 301
14.4 The fishery for tilapia 302
14.5 Why have the tilapia species persisted in the face of heavy fishing pressure and habitat change? 307
14.6 Trophic ecology of the tilapias 307
14.7 Detritus as a buffer against habitat variability 314
14.8 Habitat variability and extinction rates in the fish community 316
14.9 Summary 317
Acknowledgements 318
References 318

15 Species changes in reservoir fisheries following impoundment: the case of Lake Itezhi-tezhi, Zambia 321
Ian G. Cowx and Cyprian K. Kapasa
15.1 Introduction 321
15.2 Materials and methods 322
15.3 Results 324
15.4 Discussion 329
15.5 Conclusions and recommendations 330
15.6 Summary 331
Acknowledgements 331
References 332

Part Three Evaluation of Species Changes in African Lakes without Introduced Species

16 Changes in species composition and abundance of fish populations in Lake Turkana, Kenya 335
Jeppe Kolding
16.1 Introduction 335
16.2 Brief history of the fishery 337
16.3 Experimental fishing surveys 340
16.4 Status of the major taxa 346
16.5 General trends in catch rates versus lake levels 353
16.6 Discussion 355
16.7 Conclusions 358
16.8 Summary 359
Acknowledgements 360
References 360

17 Management, conservation and species changes of exploited fish stocks in Lake Malaŵi 365
George F. Turner
17.1 Introduction 365
17.2 Fishing areas 366
17.3 Fisheries statistics 366
17.4 Stock identification and management 370
17.5 General trends in the fisheries 370
17.6 Tilapiine fisheries 372
17.7 Haplochromine trawl fisheries 380
17.8 Seine fisheries 386
17.9 Problems with fisheries assessment 388
17.10 Prospects 389
17.11 Summary 392
Acknowledgements 393
References 393

18 Changes in demersal cichlid communities as a result of trawling in southern Lake Malaŵi 397
George F. Turner, Denis Tweddle and Ramsey D. Makwinja
18.1 Introduction 397
18.2 Survey methods and analysis 398
18.3 Comparison of survey results 402
18.4 Discussion 408
18.5 Summary 411
Acknowledgements 412
References 412

19 Changes in species composition and abundance as a consequence of fishing in Lake Malombe, Malaŵi 413
Denis Tweddle, George F. Turner and Mohammed B.D. Seisay
19.1 Introduction 413
19.2 Methods 415
19.3 Results 416
19.4 Discussion 422
Acknowledgements 424

20 Effects of exploitation on the pelagic fish community in the south of Lake Tanganyika 425
Martin J. Pearce
20.1 Introduction 425
20.2 Summary of fish biology 426
20.3 On the relationships of the pelagic fisheries 428
20.4 Industrial catch and effort data from the SE Arm 429
20.5 Relationship between *S. tanganicae* and *L. stappersi* 434
20.6 Composition of the plankton 436
20.7 Food of *L. stappersi* 437
20.8 Discussion 437
20.9 Summary 440
Acknowledgements 440
References 441

21 Changes in the pelagic fisheries of northern Lake Tanganyika during the 1980s 443
Philippe Petit and Antoine Kiyuku
21.1 Introduction 443
21.2 Investigations into the fishery 443
21.3 Results 445
21.4 Conclusion 452
Acknowledgements 453
References 453

Part Four Overviews of Fish Introductions in African Lakes

22 Genetic impacts of fish introductions: a perspective on African lakes 457
Gary R. Carvalho and Lorenz Hauser
22.1 Introduction 457
22.2 Fish introductions in Africa: definitions, purposes and extent 459

22.3 Fish introductions in the African Greak Lakes: some special characteristics 462
22.4 Genetic impacts of fish introductions 463
22.5 Methods and their applications 478
22.6 Summary 484
Acknowledgements 485
References 485

23 Thinking the unthinkable: a candidate model for predicting sustainable yields of introduced fish species in African lakes 495
Tony J. Pitcher
23.1 Introduction 495
23.2 What factors might affect decisions to make new fish introductions? 497
23.3 Estimating sustainable yields from new introductions 499
23.4 Results 506
23.5 Discussion 509
23.6 Summary 522
Acknowledgements 522
References 523

24 Why is *Limnothrissa miodon* such a successful introduced species and is there anywhere else we should put it? 527
Brian E. Marshall
24.1 Introduction 527
24.2 Why is *L. miodon* such a successful species? 529
24.3 What are the ecological consequences of *Limnothrissa* introductions? 535
24.4 Where should we introduce *Limnothrissa?* 538
24.5 Summary 541
Acknowledgements 542
References 542

Appendix: Summary of characteristics of major Arican lakes 547
Lake Albert (Mobutu) 548
Lake Bengweulu 549
Lake Cahora Bassa 550
Lake Chad 551
Lake Chilwa 552
Lake Edward (Idi Amin) 553
Lake George 554
Lake Itezhi-tezhi (Kafue Gorge) 555
Lake Kainji 556

Lake Kariba 557
Lake Kivu 558
Lake Kyoga 559
Lake Malaŵi (Nyasa) 560
Lake Mweru 562
Lake Nasser/Nubia 563
Lake Tanganyika 564
Lake Turkana (Rudolf) 566
Lake Victoria 567
Lake Volta 569
References 570

Author index 571

Species index 581

Subject index 589

Contributors

Alida Bundy
Fisheries Centre, University of British Columbia, 2204 Main Mall, Vancouver, BC, Canada V6T 1Z4

Gary R. Carvalho
Marine and Fisheries Genetics Laboratory, School of Biological Sciences, University College of Swansea, Singleton Park, Swansea SA2 8PP, United Kingdom

Deonatus B.R. Chitamwebwa
Tanzania Fisheries Research Institute, Sota Station, P O Box 46, Shirati, Tanzania

Ian G. Cowx
Hull International Fisheries Institute, University of Hull, Hull HU6 7RX, United Kingdom

Tijs Goldschmidt
Research Group in Ecological Morphology, Zoölogisch Laboratorium, Rijksuniversiteit Leiden, Postbus 9516, 2300 RA Leiden, The Netherlands

Dominique F. Gréboval
Fishery Policy and Planning Division (FIPP), Fisheries Department, FAO/UN, Via delle Termi di Caracalla, 00100 Rome, Italy

David M. Harper
Department of Zoology, University of Leicester, University Road, Leicester LE1 7RH, United Kingdom

Craig K. Harris
Program on the Lakes of East Africa, African Studies Center, Michigan State University, East Lansing, Michigan 48824-1035, USA

Paul J.B. Hart
Department of Zoology, University of Leicester, University Road, Leicester LE1 7RH, United Kingdom

Lorenz Hauser
Marine and Fisheries Genetics Laboratory, School of Biological Sciences, University College of Swansea, Singleton Park, Swansea SA2 8PP, United Kingdom

Hans H. de Iongh
Centre for Environmental Studies, P O Box 9518, 2300 RA Leiden, The Netherlands

Cyprian K. Kapasa
Lake Mweru/Luapula Fisheries Research Station, P O Box 740005, Nchelenge, Zambia

Lawrence Karenge
Department of National Parks and Wildlife Management, Aquatic Research, P O Box 8365, Causeway, Harare, Zimbabwe

Antoine Kiyuku
Département des Pêches, B P 631, Bujumbura, Burundi

Jeppe Kolding
Department of Fisheries and Marine Biology, University of Bergen, High Technology Centre, N-5020 Bergen, Norway

Aggrey W. Kudhongania†
UFFRO, P O Box 343, Jinja, Uganda

Ramsey D. Makwinja
Department of Fisheries, Fisheries Research Station, P O Box 27, Monkey Bay, Malaŵi

Piero Mannini
Lake Tanganyika Research Project, GCP/RAF/271, FAO/UN, B P 1250, Bujumbura, Burundi

Brian E. Marshall
Department of Biological Sciences, University of Zimbabwe, P O Box MP 167, Mount Pleasant, Harare, Zimbabwe

Jacques Moreau
Dept of Inland Fisheries – INP/ENSAT, 145 Avenue de Muret, F-31076 Toulouse Cédex, France

S. Mucai Muchiri
Department of Fisheries, Moi University, P O Box 3900, Eldoret, Kenya

Peter B.O. Ochumba†
Kenya Marine and Fisheries Research Institute, Kisumu Laboratory, P O Box 1881, Kisumu, Kenya

Richard Ogutu-Ohwayo
Fisheries Research Institute, P O Box 343, Jinja, Uganda

Martin J. Pearce
Department of Fisheries, P O Box 9, Sinazeze, Zambia

Philippe Petit
Laboratoire d'Icthyologie, ENSAT/INP, 145 Avenue de Muret, F-31076 Toulouse Cédex, France

Tony J. Pitcher
Fisheries Centre, University of British Columbia, 2204 Main Mall, Vancouver, BC, Canada V6T 1Z4

J. Eric Reynolds
Fishery Policy and Planning Division (FIPP), Fisheries Deparment, FAO/UN, Via delle Termi di Caracalla, 00100 Rome, Italy

Fritz Roest
International Agricultural Centre, P O Box 88, 6700 AB Wageningen, The Netherlands

Mohammed B.D. Seisay
Department of Fisheries, Fisheries Research Station, P O Box 27, Monkey Bay, Malaŵi

Petronella C. Spliethoff
International Agricultural Centre, P O Box 88, 6700 AB Wageningen, The Netherlands

George F. Turner
Department of Zoology, University of Aberdeen, Aberdeen AB9 2TN, United Kingdom *and* SOAFD Marine Laboratory, P O Box 101, Victoria Road, Aberdeen AB9 8DB, United Kingdom

Denis Tweddle
Department of Fisheries, Fisheries Research Station, PO Box 27, Monkey Bay, Malaŵi

Timothy Twongo
Fisheries Research Institute, P O Box 343, Jinja, Uganda

Jan H. Wanink
Research Group in Ecological Morphology, Zoölogisch Laboratorium, Rijksuniversiteit Leiden, Postbus 9516, 2300 RA Leiden, The Netherlands

David S. Wiley
Program on the Lakes of East Africa, African Studies Center, Michigan State University, East Lansing, Michigan 48824-1036, USA

Douglas C. Wilson
Program on the Lakes of East Africa, African Studies Center, Michigan State University, East Lansing, Michigan 48824-1036, USA

Frans Witte
Research Group in Ecological Morphology, Zoölogisch Laboratorium, Rijksuniversiteit Leiden, Postbus 9516, 2300 RA Leiden, The Netherlands

Series foreword

Among the fishes, a remarkably wide range of biological adaptations to diverse habitats has evolved. As well as living in the conventional habitats of lakes, ponds, rivers, rock pools and the open sea, fish have solved the problems of life in deserts, in the deep sea, in the cold Antarctic, and in warm waters of high alkalinity or of low oxygen. Along with these adaptations, we find the most impressive specialization's of morphology, physiology and behaviour. For example, we can marvel at the high-speed swimming of the marlins, sailfish and warm-blooded tunas, air-breathing in catfish and lungfish, parental care in the mouth-brooding cichlids and viviparity in many sharks and toothcarps.

Moreover, fish are of considerable importance to the survival of the human species in the form of nutritious, delicious and diverse food. Rational exploitation and management of our global stocks of fishes must rely upon a detailed and precise insight of their biology.

The *Chapman and Hall Fish and Fisheries Series* aims to present timely volumes reviewing important aspects of fish biology. Most volumes will be of interest to research workers in biology, zoology, ecology and physiology but an additional aim is for the books to be accessible to a wide spectrum of non-specialist readers ranging from undergraduates and postgraduates to those with an interest in industrial and commercial aspects of fish and fisheries.

Since 1980, the biggest mass extinction event in modern times has taken place in Lake Victoria as a consequence of an introduced species. In the same time span, fisheries for introduced species in East Africa contributed more than half a million jobs and over half a billion dollars to local economies. This book contains 24 chapters by 36 authors that add to the debate about the impact of species change in 8 African lakes.

Two sets of species changes are considered, the first a consequence of heavy exploitation of endemic fish stocks in Rift Valley lakes like Lake Malawi, the second a result of fish introductions to lakes such as Lake Victoria and human-made lakes like Lake Kariba. Furthermore, the book compares and contrasts two kinds of introduced fish that have very different ecological roles: fish-eaters like the rapidly-growing Nile perch, and planktivores like the highly fecund freshwater sardines. The African lake sardines are not so well-known as the Nile perch and present an alternative perspective to the scenario in Lake Victoria: human-made lakes that have created highly successful fisheries without evident environmental harm.

This is an ambitious book because it addresses a major problem in biodiversity research — the intractable problem of introduced species that survive and successfully invade. It seems almost impossible to predict effects: in fact chaos theory suggests that such detail may be inherently unknowable. Moreover, we learn that the Nile perch story is not as simple as many have represented — many parallel changes have impacted the fish communities of Lake Victoria. Widespread anthropogenic factors are evidenced by major changes to the fauna of Lake Malawi, which has experienced no introductions, and where the most common fish in the present demersal trawl fishery was found to be an undescribed species.

The book encompasses an interdisciplinary view of many problems in African lakes: economics, sociology, limnology, community ecology, fish biology and assessment science. In many instances startling new perspectives are cast on what were thought to be familiar problems. Some new methods and models are presented. In common with all books in the *Chapman and Hall Fish and Fisheries Series*, all chapters have been refereed and edited to a common format. As Series Editor I hope that, as well as a compilation describing our current understanding, this work will become a useful source of ideas and focus for new research aimed at balancing necessary exploitation with a full understanding of the human role in aquatic ecosystem change

Professor Tony J. Pitcher
Editor, Chapman and Hall Fish and Fisheries Series
Director, Fisheries Centre, University of British Columbia, Vancouver, Canada

Preface

African lakes are important suppliers of fish protein to human populations in the region. In the past 90 years, major changes have been recorded in the fish fauna of many lakes in the East African Rift system: in Lake Victoria these include what has been called the largest mass extinction to occur in modern times. The aim of this book is to evaluate the impact of these species changes on the natural and human communities of the lakes. Some changes have been caused by increased fishing, but others have been brought about by exotic fish species introduced to create new fisheries. The problems of overfishing are well documented, but the consequences of introducing new species to a natural lake are less well understood. We hope that the material in this book provides analyses, case studies and concepts that will be of use in maximizing the benefits from introduced species, minimizing and mitigating any undesirable consequences, and evaluating the likely costs and benefits of proposed future fish introductions.

The book has its origins in a conference organized by Tony Pitcher and Alida Bundy at Imperial College, London, from 27 to 31 March 1992, as part of a project funded by the Overseas Development Administration (ODA) of the UK Government. This project aimed to examine the impact of changes in species composition on the biological, economic and social sustainability of fisheries in the African lakes. Fifty-seven scientists from 16 countries attended the London symposium, with sponsorship from the UK, Norwegian, Danish, Canadian, Japanese and Zimbabwean Governments and FAO supporting the travel and accommodation of delegates from Africa. The conference included discussion workshops on each of the major lakes, and round tables that focused on the major conceptual issues. The chapters in this book are based on papers given at the conference that have been reworked and revised in the light of discussions and of comments from anonymous referees and the editors.

The book like the ODA research project concentrates on introductions of Nile perch, *Lates niloticus*, to Lake Victoria and the freshwater sardine, *Limnothrissa miodon*, to the human-made Lake Kariba, and on changes to the Lake Malaŵi community caused by exploitation. Research on other Rift

Valley lakes, such as Lakes Tanganyika, Turkana, Kivu, Kyoga and Naivasha, and on the human-made Lake Itezhi-tezhi, has been included for comparison and to examine specific questions. Comparisons between the introduction of planktivores and piscivores, and between natural and human-made lakes, have been used to gain deeper understanding. Previously published work in this area has concentrated on ecology and conservation, while much of the relevant fisheries work has been scattered in the grey literature. In contrast, throughout this book an attempt has been made to give equal emphasis to both human and ecological costs and benefits of the species changes.

Following a brief outline of the issues in Chapter 1, the book is organized into four Parts. Part One contains ten chapters that focus exclusively on the Nile perch introduction to Lake Victoria. The four chapters in Part Two examine species changes in other lakes that have had fish introductions. Part Three contains six chapters on four lakes that have not had introduced species. Finally, Part Four comprises three perspectives on fish introductions that encompass a range of different lakes.

We would like to thank all the participants at the London symposium for freely contributing ideas and enthusiasm that has enhanced the material published here, our anonymous referees for constructive criticism, Professor John Beddington (MRAG, Imperial College) for generous support of the project and symposium, Dr John Tarbit and Mr Richard Beales of ODA for tireless action at the sharp end of government bureaucracy, our copy editor Dr Chuck Hollingworth for his contribution to clarity and accuracy by his unmatched attention to detail, and Nigel Balmforth and Martin Tribe at Chapman & Hall for their patience and support in producing the book.

This book is dedicated to the memory of Aggrey (Arthur) Kudhongania, who died during its preparation. Aggrey was the gentle and effective Director of the Jinja laboratory in Uganda for many years, the culmination of a distinguished career as a fisheries limnologist. Aggrey was calmly presenting a balanced view of the consequences of Nile perch introduction long before we thought of doing so in this book. We hope that the ideas presented here will do honour to his memory.

Tony J. Pitcher, Vancouver, Canada
Paul J.B. Hart, Leicester, UK

General introduction

Paul J.B. Hart

THE IMPACT OF SPECIES CHANGES IN AFRICAN LAKES

The preface explains the origins of this book in an aid project and conference examining the impact of changes in species composition on the biological, economic and social sustainability of fisheries in the African lakes.

The issues discussed by the authors of the twenty-four chapters are introduced by Pitcher in Chapter 1. This leads into Part One, which contains ten chapters on changes in Lake Victoria. This ecosystem has attracted attention because it highlights the conflict between the fishing requirements of the local population, the consequences of introductions and the need to preserve diversity. Before modern times, Lake Victoria had 200 or more species of haplochromine cichlids (Greenwood, 1974). Since the beginning of this century the number of species left has diminished. The introduction of the Nile perch, *Lates niloticus*, in the 1960s furnished a new source of predation which some say has been instrumental in reducing the diversity of the haplochromines. At the same time, people living around Lake Victoria need fish for food. They have increased the fishing pressure on the lake's fish stocks and this too may have contributed to the decline of the magnificent diversity that existed originally. Nile perch have been of great interest to the fishing community and some would say a great benefit.

The background to the changes in Lake Victoria is provided by Kudhongania and Chitamwebwa in Chapter 2. Over time the structure of the food web in the lake has been simplified by the reduction in cichlids and introduction of Nile perch and some *Tilapia* species. The lake prawn, *Caridina nilotica*, has increased in abundance and lately the water hyacinth, *Eichhornia crassipes*, has invaded. Apart from describing the changes in the fish community, Kudhongania and Chitamwebwa also give details of environmental fluctuations, such as rises in the lake's water level and an increasing incidence of deoxygenation in deeper water, that could also have contributed to changes in the fish community.

Six chapters follow, each analysing and interpreting the changes that have occurred in the lake. Ochumba (Chapter 3) provides limnological data from the lake with details of how the deoxygenation of deeper water has progressed and how fish make daily movements to avoid oxygen-poor layers. In Chapter 4, Twongo compares the changes in the tilapias of Lakes Victoria and Kyoga. The Nile tilapia, *Oreochromis niloticus*, has been very successful in both lakes and supports a valuable fishery. It coexists with Nile perch for reasons that are not yet clear. Oguto-Ohwayo observes in Chapter 5 that the decline in species diversity in Lake Victoria is not echoed in neighbouring lakes, such as Kyoga and Nabugabo, where Nile perch is also found. This finding supports those who suspect that the Nile perch is not the only cause of the decline in diversity in Lake Victoria.

Much of the publicity given to the changes in Lake Victoria has dwelt on the findings of the Dutch team that have surveyed the Mwanza Gulf in the southern part of the lake. The results of this work are summarized in Chapter 6 by Witte *et al.* Twenty years of work indicate that the loss of about 200 species of haplochromines coincided with the increase and spread of the Nile perch. It is proposed that the loss of detritivores has resulted in more microbial decomposition which has then led to lowered oxygen concentrations in water near the bottom. Algal blooms have been promoted by the lack of herbivores. The general conclusion is that the increased predation of Nile perch has reduced the haplochromines and led to a simplification of the ecosystem and to eutrophication of the lake. The new state is less stable, leading to the conclusion that the Nile perch fishery is unlikely to be sustainable. The wider issue of ecosystem change is analysed in Chapter 8 by Moreau, who uses the ECOPATH model as an analytical tool.

If the Nile perch was the main cause of decline in haplochromine species, then there should be good correlation between the increase in abundance of perch and decrease in the cichlids. In Chapter 7, Bundy and Pitcher analyse 20 years of catch data using multivariate statistical methods. The analysis shows that the Nile perch is unlikely to be the only cause of species changes in the lake. Evidence shows that the decline in haplochromines started before the Nile perch was abundant and is more likely to be associated with the increase in fishing activity. Nile perch have aggravated the decline that was already under way.

The fishery for Nile perch and its influence on the human community around the lake are discussed in three chapters. Pitcher and Bundy (Chapter 9) assess the Nile perch stock and estimate the sustainable yield, Reynolds *et al.* (Chapter 10) analyse the economics of the Nile perch fishery and Harris *et al.* (Chapter 11) provide a detailed discussion of the socio-economic changes induced by the Nile perch fishery. The stock assessment shows that by 1990 the species was overfished, with high capture of juveniles destablizing recruitment. If the current trend continues, the fishery is likely to collapse unless effort is cut back about 10% and size limits are enforced.

A collapse would be disastrous to the local communities. The fishery is estimated to have yielded a net gain of nearly US$ 300 million and has provided employment for at least 350 000 people directly, and perhaps over a million when peripheral activities are taken into account. These economic benefits have been accompanied by changes in the way the fishing fleet is owned and operated, with more differentiation of roles. Although wealth has increased, so have social tensions, there being an increasing amount of piracy and theft.

So much time and effort has been spent on Nile perch that there is a danger that the importance of *Rastrineobola* and lake prawn will be forgotten. *Rastrineobola* is an important commercial species, yet there is insufficient information about its biology. The potential yield may be in the region of a million tonnes a year. Lake prawn is an important species in the lake's food web, yet it too is now the subject of a fishery. The effects of this on the stability of the ecosystem are not known and there is a need for more publicly available information.

Part Two contains four chapters concerning other lakes in the region – Lake Kariba (Chapter 12, Karenge and Kolding), Lake Kivu (Chapter 13, de Iongh *et al.*), Lake Naivasha (Chapter 14, Muchiri *et al.*) and Lake Itezhi-tezhi (Chapter 15, Cowx and Kapasa) – that have all experienced species changes and provide interesting examples that help to understand better the changes occurring in the large lakes. The conditions in these lakes are very different, mainly because of their history. Lakes Kariba and Itezhi-tezhi are young reservoirs and have only recently stabilized after the first few years of expansion of diversity. Lake Naivasha has dried out completely in historical times and has had many unusual introductions such as the North American largemouth bass, *Micropterus salmoides*. Lake Kivu has unusual limnological conditions because of volcanic activity.

There have been no recorded introductions into Lakes Malaŵi, Tanganyika and Turkana, but the fish communities in these lakes have been variously influenced by fishing. The changes in Lake Turkana are discussed by Kolding in Chapter 16. The fish of the lake have been surveyed at long intervals and Kolding's work is the first modern study. He shows that the decline in the fishery is likely to have resulted from environmental changes rather than from increased fishing mortality.

Lake Malaŵi is often considered to be the jewel in the string of lakes which thread through the Rift Valley. It has spectacular species diversity with a high proportion of endemics. To aquarists in Europe, Lake Malaŵi is often the source of their most prized specimens. Despite this unique status, the lake is also a source of protein to many people who live in the hinterland. Fishing is active and is conducted by a number of methods. The fishery has an international component and chambo, *Oreochromis* spp., for example, can be purchased regularly on the fish market in Leicester, UK. The changes that have occurred in the lake are documented in three chapters (17, 18 and 19) of the six in Part Three.

In Chapter 17, Turner provides an overview of the lake's fisheries and assesses their status. The principal problem is the diversity of species caught in trawls and other gear, many of which have not been named. This makes proper stock assessment impossible and highlights the need for basic taxonomic work. In Chapter 18, Turner *et al.* look in detail at the commercial trawl fishery where there has been a shift in the proportions of each species making up the catch. An attempt is made to determine the sustainable multispecies catch. The final chapter on Lake Malaŵi (Chapter 19 by Tweddle *et al.*) analyses the noteworthy collapse of the artisanal fishery for chambo, which apart from appearing on the Leicester market also forms the basis for a national dish in Malaŵi. The decline is probably the result of increasing human population and of the introduction of new fishing methods. This chapter raises the ethical issue of trying to limit and regulate a fishery when people depending on it for food are starving.

Lake Tanganyika is longer and deeper than Lake Malaŵi, with a maximum depth in the south basin of 1470 m which is twice the maximum depth of Lake Malaŵi. The precipitous nature of Lake Tanganyika's shores make it an inhospitable environment for fishers. In the south the depth of 1470 m is reached only 4 km from the western shore (Beadle, 1974). It is only in the north and south of the lake that there are significant areas of shallow water. It is not surprising then that the two chapters (Chapter 20 by Pearce and Chapter 21 by Petit and Kiyuku) describe only the pelagic fisheries. In the north there have been problems with obtaining good data on landings, and piracy is a serious threat to the stability of the fishery.

Part Four presents three overviews. Making introductions of new species to lakes, together with heavy fishing are activities that have unexpected and often undesirable consequences for genetic diversity. None of the chapters discusssing changes in particular lakes addresses the genetic problems that might arise from introducing new species. The consequences of hybridization are discussed in a few, but only in so far as the processes undermined an attempt to introduce a new species. These deficiencies are made good by Carvalho and Hauser in Chapter 22. They discuss the genetic impacts of introductions, the implications for diversity, the likely effects of genetic bottlenecks and what happens when only a few individuals of an introduced species survive to form the basis of the stock (founder effect).

The final two chapters discuss the factors that might influence fisheries managers who are thinking about making an introduction into an African lake. The Nile perch in Lake Victoria has shown how risky introductions can be, with many potential consequences which are only realized in a particular set of circumstances. In Chapter 23, Pitcher describes a model that is designed to use data on the primary production of a lake to predict the yield that might be expected from an introduced species. Making predictions about introductions is always dependent on knowledge about the way aquatic ecosystems function. Even in the best-understood systems, unexpected events are always likely (see Kitchell, 1992, for an important case history that illustrates this). The lack of detailed knowledge of the lakes discussed in this book, makes the process of introduction an even more uncertain activity.

In the final chapter (Chapter 24), Marshall looks at the introduction of *Limnothrissa miodon* into Lake Kariba and analyses why it was successful. He then uses this to evaluate the information required before a new introduction is made. It goes without saying that information is required on the life history characteristics of the candidate species, on its growth and mortaility rates and on its reproductive characterstics. Knowing about these features still does not help in detail as there needs also to be an understanding of how the characteristics of the new species will interact with the biology of those species already in the lake. Introductions are often made on the assumption that there is an 'empty' niche. The trouble with this thinking is that the lake habitat is given a structure that it does not possess. The habitat is imagined to be like a set of pigeon-holes for letters, which exist even though there are no letters filling them. In ecosystems the niches are defined by the species that are present. In the pigeon-hole analogy, the letters define the number and sizes of the holes. Given a certain array of species in a habitat, the dynamics of the system will be determined. Return times and resilience of each species in the system will influence the response the ecosystem has to exploitation and to invasions (Pimm, 1991). All this means that ecological understanding has to be several orders of magnitude better than we have now before safe introductions can be made into unique ecosystems such as Lake Malaŵi.

Introducing new species into a reservoir, which has a short history and low diversity (such as Lake Kariba), can be contrasted with introductions into a complex ecosystem which contains a diverse fauna of endemic species (such as Lake Malaŵi). The potential loss in diversity and genetic resources from the reservoir is low if the introduced species increases its abundance beyond all expectations and eats up all the residents. If the same happened in the complex ecosystem, then the loss is irrecoverable. The key element in the decision to be made about whether or not an introduction should be made is the degree of uncertainty about the outcome. The uncertainty can be reduced by learning more about the systems, but the task is much harder for the complex ecosystem because there is much more to learn about and understand. As a result, we should agree not to introduce new species to complex lake ecosystems like Lake Malaŵi because the potential loss is too great to accept even though equally uncertain benefits could accrue.

REFERENCES

Beadle, L.C. (1974) *The Inland Waters of Tropical Africa. An Introduction to Tropical Limnology*, Longman, London, 365 pp.

Greenwood, P.H. (1974) *Cichlid Fishes of Lake Victoria, East Africa: The Biology and Evolution of a Species Flock*, British Museum (Natural History), London, 134 pp.

Kitchell, J.F. (ed.) (1992) *Food Web Management: A Case Study of Lake Mendota*, Springer-Verlag, Berlin, 553 pp.

Pimm, S.L. (1991) *The Balance of Nature? Ecological Issues in the Conservation of Species and Communities*, University of Chicago Press, Chicago, 434 pp.

Chapter one

Species changes and fisheries in African lakes: outline of the issues

Tony J. Pitcher

1.1 INTRODUCTION

The work presented in this book attempts to assess the impact of species changes in African lakes engendered by:

- introduction of fish species;
- impoundment;
- heavy exploitation.

All three of these factors have impinged recently on one or more African lakes and have led to large sustained changes in species composition: the objective of all authors in this book is to evaluate the impacts of these changes. Such evaluation needs a broad multidisciplinary perspective encompassing insight into both biological and human consequences of species change. On the biological side we evaluate faunal, ecological, habitat and genetic changes. The human aspects cover assessment, management, economics and socio-economics of fisheries for both introduced and endemic species. It is generally forgotten that the Rio Convention (UN Conference on Environment and Development) formulated two objectives: in addition to the often-quoted preservation of biodiversity, the convention advocated sustainable use. This

The Impact of Species Changes in African Lakes. Edited by Tony J. Pitcher and Paul J.B. Hart. Published in 1995 by Chapman & Hall, London. ISBN 0 412 55050 4.

book attempts to provide a balanced view of these two objectives for the fisheries in African lakes.

Reduction in biodiversity, conservation of unique endemic fauna, and the assessment of changes in habitat, species and genetic diversity are recurrent themes in evaluating impacts from the biological viewpoint. From the human side, recurrent themes are the evaluation of wealth generated by new fisheries, their sustainability and social equity, and comparative forecasts of the outcomes of a range of management scenarios. The attempt to reconcile these two perspectives on the impacts of species changes in African lakes represents at the same time both a formidable challenge and a productive, exciting opportunity for contemporary fishery scientists.

As we move towards a new century, fishery scientists are beginning to discover that there may be productive ways to accommodate conflicting opinions on the way that we humans change the world. This novel view is essential if we are to capture the synergy that may be generated by reconciliation of contrasting viewpoints, and avoid the ritual exchange of pejorative epithets that characterized examination of the consequences of introduction of Nile perch to Lake Victoria. The techniques of conflict resolution, recognition of acceptable trade-offs, and movement in directions where all players may gain, are some of the essential management tools needed. This chapter aims to introduce readers to the main issues surrounding species changes in the African Lakes.

1.2 GEOGRAPHICAL SCOPE OF THE BOOK

The main characteristics of the 19 major African lakes are summarized in the Appendix to this book (see also Craig, 1992) and their locations are shown in Fig. 1.1: six of these lakes have had fish species introduced to them. Unfortunately, many of these lakes are poorly documented and so the book concentrates on major lakes where there are substantial data, extensive ecological information, and contrasts in ecological and fishery status. Consequently, the emphasis is on two major rift valley lakes and one major human-made lake: Lakes Victoria, Malaŵi and Kariba.

Lake Victoria, the largest African lake and the third largest lake in the world (Table 1.1), has had several introductions of fish species, has undergone major changes in species composition and has suffered loss of biodiversity attributed to the introduced Nile perch. Moreover, it is subjected to heavy fishing pressure and has undergone substantial changes in limnology (Chapters 2 to 11).

Lake Malaŵi, the third largest lake in Africa and tenth in the world (Table 1.1), is the best known of the major rift lakes that has had no introductions, exhibits high biodiversity and endemicism, and is subjected to heavy fishing

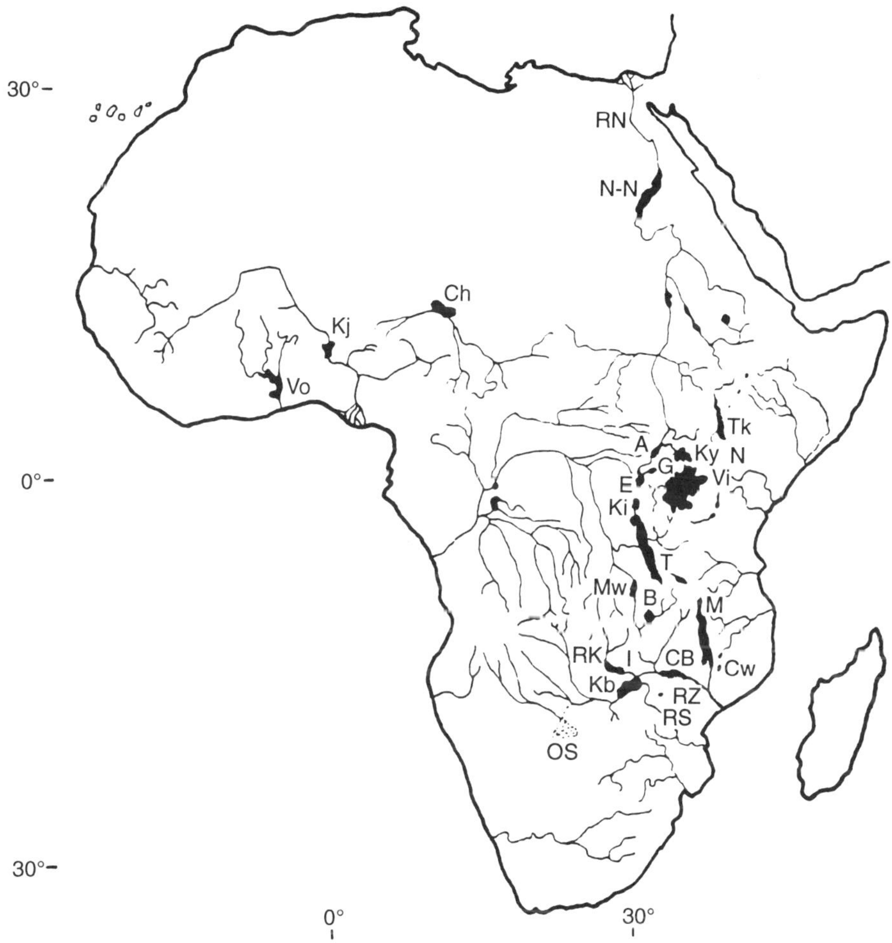

Fig. 1.1 Map showing the location of the major lakes of Africa (equal area projection: modified from Craig, 1992). Natural lakes = A, Albert; B, Bangweulu; Ch, Chad; Cw, Chilwa; E, Edward; G, George; Ki, Kivu; Ky, Kyoga; M, Malaŵi (and Malombe); Mw, Mweru; N, Naivasha; T, Tanganyika; Tk, Turkana; Vi, Victoria. Human-made lakes: CB, Cahora Bassa; I, Itezhi-tezhi; Kb, Kariba; Kj, Kainji; N-N, Nasser-Nubia; Vo, Volta. Rivers: OS, Okavango swamp; RK, Kafue; RN, Nile; RS, Sanyati; RZ, Zambezi.

Table 1.1 Approximate surface areas of great lakes

Region	Rank	Lake	Area (km^2)
World*	1	Caspian Sea	424 200
	2	Lake Superior	82 400
	3	*Lake Victoria*	69 500
	4	Aral Sea	63 800
	5	Lake Hurn	59 600
	6	Lake Michigan	58 000
	7	*Lake Tanganyika*	32 900
	8	Lake Baikal	31 500
	9	Great Bear Lake	31 100
	10	*Lake Malaŵi*	30 800
	11	Great Slave Lake	28 900
	12	Lake Erie	25 700
	13	Lake Winnipeg	24 300
	14	*Lake Chad*	20 700†
	...	...	
	21	*Lake Turkana*	9 100
Africa‡	1	Lake Victoria	69 500
	2	Lake Tanganyika	32 900
	3	Lake Malaŵi	30 800
	4	Lake Chad	20 700†
	5	Lake Turkana	9 100
	6	*Lake Volta*	8 845
	7	Lake Chilwa	7 500†
	8	Lake Albert	6 800
	9	*Lake Nasser*	6 216
	10	*Lake Kariba*	5 364
	11	Lake Mweru	5 200
	12	*Lake Itezhi-tezhi*	3 700
	13	Lake Kyoga	2 700
	14	Lake Kivu	2 700
	15	*Lake Cahora Bassa*	2 665
	16	Lake Edward	2 325
	17	Lake Bangweulu	2 072
	18	Lake Kainji	1 280

* African lakes in italics.

† Lakes Chad and Chilwa vary greatly according to rainfall: an average value for L. Chad might be 10 000 km^2.

‡ African lakes and impoundments: human-made lakes in italics include the largest human-made water bodies in the world.

pressure from both artisanal and industrial sectors that have caused significant species changes. Furthermore, it is important to consider it in detail because introductions of exotic fish species have been proposed (Chapters 17 to 19).

The human-made Lake Kariba, the tenth largest lake in Africa and the third largest human-made lake in the world, has an important fishery for the introduced Lake Tanganyika sardine, whereas the original Zambezi riverine fauna has undergone significant species changes as the lake has matured (Chapters 12, 23 and 24).

Lakes Kivu, Kyoga, Turkana, Itezhi-tezhi and Naivasha are also represented by chapters in the book. New work on the pelagic fisheries of Lake Tanganyika (the second largest lake in Africa) in the southern (Zambian) and northern (Burundian) areas is presented, but a recent synoptic book covers other aspects of fisheries ecology in Lake Tanganyika (Coulter, 1991).

1.3 INTRODUCTIONS OF FISH TO AFRICAN LAKES

Since the middle of the last century, 1354 introductions of 237 fish species into 140 countries have been documented worldwide (Welcomme, 1988). Africa has experienced only 147 introductions (11% of the total) of 50 fish species (21% of the total), 23 of these from outside Africa, although African fish have been exported widely elsewhere. The impact of introductions in Africa has perhaps received a high public profile on account of the concern about the effects of Nile perch on endemic fishes in Lake Victoria. The stated reasons for the introductions are listed in Table 1.2. The intention to create a fishery underlies only 16% of the introductions, most of which were associated with aquaculture. The most important successful introductions aimed at creating or improving fisheries are listed in Table 1.3, along with the rationale given at the time.

Contemporary justifications for the introduction of fish to improve or create

Table 1.2 Reasons given for fish introductions to African lakes

Reason	Number*	Percentage†
Aquaculture	72	49
Sport fishing	26	18
Create fishery	23	16
Control other fish	9	6
Control weed	6	4
Control mosquito	6	4
Control snail	2	1
Ornamental	1	1
Accidental	1	1
No reason given	1	1

*Number of introductions, total 147. Source: Welcomme (1988).
†Percentage of 147 introductions.

Table 1.3 Successful fish introductions carried out in Africa, with justifications made at the time

Species	Lake	Justification
Lates niloticus	Victoria	Convert trash fish (or sport fishery?)
	Kyoga	Convert trash fish
Oreochromis niloticus	Victoria	Supplement declining endemics
	Kyoga	Supplement declining endemics
Tilapia zillii	Victoria	Feed on macrophytes
	Kyoga	Feed on macrophytes
Limnothrissa miodon	Kariba	Fill vacant pelagic niche
	Kivu	Fill vacant pelagic niche
	Cahora Bassa	Fill vacant pelagic niche
	Itezhi-tezhi*	Fill vacant pelagic niche

* Establishment still to be verified.

fisheries have generally been formulated in ecological terms, although never on the basis of informed detailed knowledge of the environments. The overall principle seems to be that a component of the ecosystem is perceived to be in some way underutilized by the endemic fauna. In the case of Nile perch introduction to Lake Victoria in the 1960s, there is the suspicion that there was also a hidden agenda to create a sport fishery. These days, established protocols, as outlined in Chapter 23, need to be followed when considering an introduction, because there are biological, ecological, social and political factors that should be considered.

Two introduced species are given emphasis in the book:

- *Lates niloticus*, Nile perch, a piscivore when adult;
- *Limnothrissa miodon*, freshwater sardine, a pelagic zooplanktivore.

The success of these two species, and Nile tilapia, *Oreochromis niloticus*, introductions, in creating fisheries is evidenced by the observation that introduced fish have allowed per capita human fish consumption in the East African region to keep pace with a near doubling of the human population over a 17 year period (including Kenya, Malaŵi, Tanzania, Uganda, Zambia, Zimbabwe; in 1970, 9.0 kg per person per year, 49 million people; in 1987, 8.75 kg per person per year, 88 million people; J. Tarbit, pers. comm.).

1.4 ESTABLISHMENT OF INTRODUCED SPECIES

Other than the qualitative perception that a niche is unoccupied, or that some source of food is not fully used when compared with similar environments, a more precise insight of establishment of introduced species may be obtained by examining four ecological factors: habitat, food, competition and predation.

Physical habitat factors

Physical factors such as oxygen requirements, salinity, pH and temperature are indicators that inhibit introductions to obviously unsuitable habitats. These may easily be measured in the field; in many cases, records and information already exist. But for many introductions that have taken place, it is apparent that physical factors were not taken into account except in a superficial way. So there are no track records of detailed measurements that accompany the success or failure of these African introductions that would enable us to have better insight into precise conditions leading to establishment.

For putative introductions, in addition to the obvious physical factors, the presence and extent of suitable substrata for spawning would also be important. Some work of this kind was carried out on breeding sites in the Malaŵi chambo, *Oreochromis lidole*, as a candidate for introductions to human-made lakes in Africa (Turner *et al.*, 1991).

Food sources

The presence of suitable food has always been a major consideration in introductions that have been mooted. The perception that there is an excess of underutilized food is common to many introductions, but this is often unquantified and based on casual observations that ignore the reality of the aquatic ecosystem. For example, the presence of lake fly swarms in Lake Malaŵi compared with Lake Tanganyika suggested to J.L. Turner (1982) that they may be cropped more efficiently by clupeid planktivores than by the indigenous cyprinids or cichlids. If absence of lake fly from Lake Tanganyika was caused by sardine predation, to what would the sardines turn once they had eliminated lake fly from Lake Malaŵi? Doubtless, as in Lake Kariba where *Chaoborus* has been wiped out by sardines, they would find something to feed upon, but the point is that the logic behind the original suggestion was defective. Although Degnbol (1993), using an ECOPATH model, supports the view that most Lake Malaŵi production goes into flies, not fish, the work is based on old and uncertain data that did not properly cover the large planktonic fish community. The latest work on the pelagic system in Lake Malaŵi comprises a detailed survey of the whole of the planktonic community, and in fact shows that planktonic production is not underutilized by fish (A. Menz, pers. comm.; Chapters 17, 23).

Competition

It is difficult to predict the outcome of competition of an introduced species with residents, unless a niche is obviously vacant. Vacant niches occur in the pelagic zone of artificial lakes that only have fish of riverine origin, or in Lake Kivu where the volcanic birth of the lake restricted the founding fish species. Where competitive displacement occurs, it is often hard to measure and distinguish from possible environmental influences. In some cases, hybridization may complicate the picture, as in *O. esculentus* and *O. niloticus* in Lake Victoria (Chapter 4).

Predation

The evidence supports the contention that predators on successfully introduced species have little effect on persistence. Moreover, an introduced species that becomes successfully established and withstands heavy fishing pressure by humans is unlikely to be wiped out by natural predators. However, failures to establish might be a result of predation by endemic species, although there appears to be no documentation of an example.

Introduced sardines in Lake Kariba are eaten by tigerfish, birds, crocodiles and humans. There is no evidence that predation on introduced Nile perch or Nile tilapia in Lakes Victoria or Kyoga has been a significant factor affecting establishment, although some complex hypotheses about predation by haplochromine cichlids on juvenile Nile perch have been put forward to account for the long delay in the Nile perch explosion (Chapter 6).

Cannibalism is often cited as a symptom of the imminent collapse of fisheries for introduced species: this has been stated for sardines in Lake Kivu (Chapter 13) and Nile perch in Lake Victoria (Chapter 2). Fish are said to have been so excessively predaceous that they have to turn to eating their own young. A degree of cannibalism is widespread amongst fish (Smith and Reay, 1991), and so the view that this leads to instability is not based on ecological fact. Pike *Esox lucius* and trout *Salmo trutta* populations in temperate lakes and rivers persist with high levels of cannibalism. With high fecundity, high growth rate and juveniles that commence feeding on invertebrate food before switching to cannibalism, there seems no a priori reason why such a fish should not have self-sustaining populations. In fact, some of the most stable and resilient fisheries in the world are based on cannibalistic species. Many hake stocks, for example, have 40% cannibalism (Pitcher and Alheit, 1994).

In natural communities, of course, the role of predation remains controversial between those who advocate top-down control by predators and those who hold that predators exist merely on sufferance. There is some evidence that predation by *Lates* species on *Stolothrissa* in Lake Tanganyika can generate irregular predator/prey cycles (Coulter, 1991, and Chapter 20). In Lake

Milaŵi, natural predation by large predators has probably not been important in shaping the evolution of the fish community because there are none except for some large benthic catfish. Extensive harvesting, or predation, by humans has drastically altered the species composition in Lake Malaŵi over the past 20 years (Chapters 17, 18).

1.5 IMPACTS OF FISH INTRODUCTIONS

Introduced species clearly have an effect on the distribution and densities of their prey, but there is no unequivocal evidence that introductions can shift the whole nature of the ecosystem, although such an effect has been advocated for Lake Victoria (Chapter 6).

Impact of introducing piscivores

The effect of predator introductions in bringing about species changes has been well documented outside Africa (e.g. *Cichla ocellaris* exterminated 13 out of 17 endemic fish in Gatún Lake, Panama (Zaret and Paine, 1973; Zaret, 1984). The problem is that documentation of such effects has not led to any great predictive insight of what may happen. In most African situations, even the documentation is poor. Introduction of predators will reduce overall ecological production, but nevertheless may increase the amount of commercially harvestable fish (Chapter 24).

Introducing planktivores

The effect of introducing pelagic zooplanktivores may appear to be less dramatic, but the effect on the plankton community can be profound. In Lake Kivu, large species of plankton almost disappeared (Chapter 13). In Lake Kariba, there was a significant shift in size of the plankton species as the sardines became established in the early 1970s. Marshall (1993; Chapter 24) suggests that the absence of lake fly *Chaoborus* from Lake Tanganyika and its disappearance from Lake Kariba is due to sardine predation. Existing top predators may thrive on the new food source (e.g. tigerfish feeding on sardines in Lake Kariba, Chapter 12). The stability of systems where both sardines and their sympatric centropomid predators, for example from Lake Tanganyika, are introduced together, has not been explored (Chapter 23).

Introducing herbivores

Herbivores, such as tilapias or the Malaŵian chambo, should have less impact on the endemic fauna and environment and may provide larger harvestable biomass. Moreover, they may actually help by cropping excess algal production and reducing the incidence of algal blooms. Establishment of herbivores, which tend to be large and slow moving, may be compromised by predation from endemic predators.

1.6 ECOLOGICAL MODELS OF FISH INTRODUCTIONS

Ecologists would be delighted if they could develop a model of the process of successful establishment of introduced species, and if they could then use the model to explore how to mitigate undesirable consequences. All ecologists would agree that one prerequisite would be detailed knowledge of the ecology of the introduced species. But in the case of species that were introduced to African lakes, principally the Nile perch and freshwater sardine, there were no in-depth studies of the ecology prior to the introductions. Only in the mooted introduction of the herbivore *Oreochromis lidole* from Lake Malaŵi to Lake Nasser, has there been an attempt to describe basic trophic and reproductive ecology relevant to success of an introduction (Pitcher *et al.*, 1989).

But even where such ecological detail is known, it is evident that the success or failure of an introduction cannot be accurately predicted by existing knowledge of community ecology. It would require the invention of a general ecology of invasions (Townsend, 1991) and specific information about the community ecology of the African lakes to produce anything of use in this area. Furthermore, modelling the mitigation of an introduction is not likely to be successful in the absence of insight of the critical features of community structure. These are likely to be the resilience and stability of fluxes among predators, competitors for food, spawning sites and space, and the dynamics of prey populations that may be either primary foods, or food resources to which the target species switch through exigency.

The bottom line is that to model explicitly and convincingly the establishment and progress of a fish introduction, we would need an unprecedented degree of insight into the dynamic processes of aquatic ecology. At present, such an attempt would be so difficult that it would be doomed even in the most heavily studied northern temperate lakes such as Windermere, Kootenay Lake or Lake Ontario.

Nevertheless, despite these problems with explicit models, approximate methods may, with suitable expression of the uncertainties involved, provide helpful guidelines to the consequences of an introduction (Chapter 23).

1.7 EVALUATION OF THE IMPACTS OF SPECIES CHANGES

Of about 150 fish species introduced to African lakes, many have not established themselves, but comparison of the failures and successes is not especially illuminating. As outlined above, this area of applied ecology is not sufficiently well developed to let us make confident predictions about the success of future introductions on endemic faunas, or about the effects of fishing. We do not know which are the relevant features of the ecology and life history dynamics of the introduced or harvested species, or of the ecosystems into which they were placed. However, should full establishment of an introduced species take place or fishing produce measurable changes in community composition and structure, fisheries scientists and economists do have some tools with which to make reasonable predictions about the likely size of catches, to assess the associated uncertainties and to evaluate the impacts of the species changes.

Species changes from introductions

One can calculate that successful establishment of our two principal introduced species, Nile perch and freshwater sardine, in the twelve major natural lakes and the seven largest human-made lakes in Africa could provide yields in excess of 1.5 million tonnes of fish per year. A yield of this magnitude would be worth over a billion US dollars and would represent an approximate doubling of the existing recorded catch (Chapter 23). But work presented in this book shows clearly that potential yields from most of the natural water bodies, including existing introductions, could probably match this figure, in terms both of biomass and of economic worth if more effectively managed.

When one looks in detail at the issues that would determine new introductions in individual lakes (Chapter 23 reviews the factors that should be considered), there emerge a number of cases where the benefits in terms of fish yield would not be sufficient to outweigh the damage to endemic faunas that already sustain successful fisheries. The notable cases here are those of Lakes Malaŵi and Tanganyika, both lakes that have not been subjected to introductions. It is important that this case against introductions in Lakes Malaŵi and Tanganyika be made on grounds of ecological economics and not just of conservation *per se* (Chapter 23 gives model predictions). It is also important to realize that the conclusions about introductions are sensitive to fish prices. On a month-to-month basis, fish prices are fluid, but are likely to rise in relative terms in a world short of high-quality protein, with increasing demand for fish, and with a need for enhanced economic activity in developing countries. Evaluations of the benefits of fish introductions are therefore liable and subject to continuous review.

Species changes from fishing

Conventional single-species fisheries assessment ignores the impact of harvesting the target species on the rest of the fish community. But assessment of multispecies fisheries is a major methodological problem. One of the most commonly used methods requires, in its present form, vast amounts of information from fish stomach data (multispecies VPA: Magnusson, 1995) and such data are unlikely to be available for any African lake in the foreseeable future. Simple surplus-production approaches sometimes fit historical time series of total catch effort quite well (e.g. Tweddle and Magasa, 1989). Unfortunately, such techniques can be misleading because total catch may hold up over time, but the exploited fish community may meanwhile undergo major shifts in species composition (Chapter 16 gives an example from Lake Malaŵi).

Investigation through experiment

Latterly, the new disciplines of behavioural ecology and ecophysiology have had some spectacular successes in testing evolutionary hypotheses concerning the trade-offs of adaptation of behaviour and physiology within the natural environment. For example, large-scale field and enclosure experiments have established the general rules that fish (and other animals) use when they forage and how these rules affect habitat choice, life history dynamics and interaction with predators (review, Hart, 1993). But it is still hard to see how one could use such theoretical insight to make specific predictions about introduced species.

Rather than borrow theory, practical techniques from this discipline may be helpful. One useful development here might be to borrow the mesocosm experiment paradigm from contemporary ecology. Prior to an introduction, experiments covering a range of food, predator and competitive regimes might be set up in the kind of extensive pond facilities used by workers such as Werner, Mittelbach, Dill or Persson (see Hart, 1993, for references). Key parameters of the models used in foraging and habitat choice might then be established and some insight of the likely outcome of the introduction gained. The practical difficulties in doing this would be great for open water pelagic species, simply on account of the large size of pond required, but it might be more feasible for predators like Nile perch or herbivores like the Malaŵi chambo. A competition experiment between the Lake Victoria dagaa, *Rastrineobola*, or the Malaŵian usipa, *Engraulicypris* (= *Rastrineobola*), and the clupeid sardine, *Limnothrissa*, from Lakes Tanganyika or Kariba, would be of great interest. Such an experiment could be set up in such a controlled mesocosm if the practical husbandry problems could be solved. There are quite a few problems to overcome in the design of such work, for example short-term experiments of a few months might not match what might happen in real breeding populations.

Investigation through modelling

Because with introductions and human exploitation one is dealing with ecosystem effects through predation and competition, one might expect more theoretical help from community ecology. With a few exceptions, community ecology seems to have gone nowhere since energy flow modelling in the 1960s and 1970s. For example, cascade theory has not been very successful in predictions for African lakes. The study of diversity has been embroiled in methodological wrangles and esoteric theory, but as a practical science, has hardly lifted itself from the level of inventory. One encouraging exception is the simplified ecosystem model of ECOPATH (Christensen and Pauly, 1992). ECOPATH offers quantitative insight into major system flows, stabilities and cascade-type effects and has been applied to Lake Victoria in Chapter 8. (Further ECOPATH models of African lakes are given for Lakes Chad, George, Kariba, Malaŵi, Tanganyika and Turkana in Christensen and Pauly, 1993.)

Chapter 23 describes the development of approximate predictors of fish yields and potential yield. The availability and accuracy of data used to estimate the parameters required is often a constraint. Methods relying on the physical properties of the water bodies are easier to use and some have had reasonable success. For introduced sardines, the model developed here may soon be tested on Lake Itezhi-tezhi in Zambia.

The use of more biotic indicators, and attributes of lakes that can be measured from satellite, seem to be two profitable areas for developing approximate models. One can recommend two approaches: methodological studies, such as the verification of suitable frameworks for empirical and statistical testing, and more conceptual studies, such as the search for easily measured correlates of fish growth and production.

There is evidently much scope for innovative work here. For example, measuring colonies of fish-eating birds (e.g. in Lake Victoria, Wanink and Goudswaard, 1992) is easier than measuring fish, and yet might be used as an accurate indicator. Other possibilities might be the relative abundance and sizes of key plankters such as copepods, dipterans and cladocerans in relation to current primary production levels. Macrophyte growth may be related to survival of juvenile fish.

1.8 MITIGATION OF THE IMPACTS OF SPECIES CHANGES

Mitigation measures for threatened endemic species

Preservation of threatened species of endemic fish through establishment of sanctuaries or aquarium breeding programmes can evidently reduce the impact on biodiversity at small cost. In the African lakes with their valuable cichlid fauna, the world aquarium trade might provide financial support for such projects. Moreover, establishment of sanctuaries will benefit existing fisheries in addition to threatened ones. Habitat refuges may be constructed for species already known to have some elements of robustness in face of predation (e.g. rocky shore haplochromines in Lake Victoria, Chapter 6), or riverine breeding sanctuaries established for species living in human-made lakes (e.g. tigerfish in Lake Kariba, Chapter 12; nchila in Lake Malaŵi, Chapter 17).

Fish ranching

What of fish ranching? Salmon enhancement projects in Canada and the United States have been enormously expensive failures (cost/benefit ratio of 0.7, P. Pearse, pers. comm.; Hilborn, 1992), and moreover, by directing heavy fishing pressure on natural stocks, may threaten their survival and compromise genetic diversity. Unlike the situation in Africa, the salmon fisheries of the Pacific North West deal with relatively simple, well-understood ecosystems. Salmon are large, easily measured fish, their ecology has been subjected to unprecedented study, and they are caught in well-documented and regulated fisheries operating in closely defined geographical areas of the developed world. If simple schemes of breeding and enhancement do not work in such locales, they are unlikely to be robust enough to work in Africa.

1.9 GENERAL RECOMMENDATIONS FOR DEVELOPMENT AND ASSISTANCE IN AFRICAN LAKE FISHERIES

Bruton (1990) puts forward a balanced view of conservation of fishes in the African lakes. The material in this book supports his recommendation that higher priority should be placed on assisting African scientists to monitor and assess the fish resources of the lakes. There are many well-educated, innovative fishery scientists in the region (12 of whom are authors of chapters in this book), yet it is evident that their training focus is often driven by international funding that gives insufficient emphasis to applied fishery objectives. Donors, both in Europe and North America, need to be persuaded of the value of increasing African competence in commercial and artisanal fisheries assessment. Perhaps the emerging South Africa will be able to help in this regard.

In natural African lakes the sheer taxonomic difficulties in dealing with hundreds of endemic species can overwhelm many projects. Simple, easily used field guides to the main species pending formal taxonomy are vitally needed for lakes like Malaŵi and Tanganyika (Chapter 17).

Consequently, two sets of practical recommendations for work on species changes in the African lakes emerge from the work presented in this book. First, research programmes should support work that aims to:

- test candidate models for evaluating introductions;
- explore easily-measured indicators of fish population status;
- develop fish identification field guides;
- evaluate the feasibility of mesocosm experiments.

Second, development aid should attempt to facilitate:

- trans-national riparian management of African lake fisheries;
- suitable publication outlets for scientific work;
- training of African fisheries scientists.

ACKNOWLEDGEMENTS

This work was initiated under support from the Overseas Development Administration of the United Kingdom Government and completed with support from the Fisheries Centre at the University of British Columbia. I would like to thank Daniel Pauly and Alida Bundy for comments on the draft manuscript.

REFERENCES

Bruton, M.N. (1990) The conservation of the fishes of lake Victoria, Africa: an ecological perspective. *Env. Biol. Fishes*, **27**, 161–75.

Christensen, V. and Pauly, D. (1992) ECOPATH 2, a system for balancing steady state ecosystem models and calculating network characteristics. *Ecol. Modelling*, **61**, 169–85.

Christensen, V. and Pauly, D. (eds) (1993) *Trophic Models of Aquatic Ecosystems*, ICLARM, Manila, 390 pp.

Coulter, G.W. (ed.) (1991) *Lake Tanganyika and Its Life*, Oxford University Press, Oxford, 354 pp.

Craig, J.F. (1992) Human-induced changes in the composition of fish communities in the African Great Lakes. *Rev. Fish Biol. Fish.*, **2**, 93–124.

Degnbol, P. (1993) The pelagic zone of central Lake Malawi – a trophic box model, in *Trophic Models of Aquatic Ecosystems* (eds V. Christensen and D. Pauly), ICLARM, Manila, pp. 110–115.

Hart, P.J.B. (1993) Teleost foraging: facts and theories, *Behaviour of Teleost Fishes*, 2nd edn (ed. T.J. Pitcher), Chapman & Hall, London, pp. 253–84.

Hilborn, R. (1992) Hatcheries and the future of salmon in the Northwest. *Fisheries*, **17**(1), 5–8.

Magnusson, K.G. (1995) An overview of the multispecies VPA: theory and applications. *Rev. Fish Biol. Fish.* (in press)

Marshall, B.E. (1993) Biology of the African clupeid *Limnothrissa miodon* with reference to its small size in artificial lakes. *Rev. Fish Biol. Fish.*, **3**, 17–38.

Pitcher, T.J. and Alheit, J. (1994) What makes a hake? A review of the critical biological features that sustain global hake fisheries, in *Hake: Fisheries, ecology and markets* (eds T.J. Pitcher and J. Alheit), Chapman & Hall, London, pp. 1–14.

Pitcher, T.J., Turner, G.T. and Grimm, A.S. (1989) The cichlid fish *Oreochromis lidole* as a candidate for introduction to tropical impoundments. Report NRD/R4334 to Overseas Development Administration, London, UK, 103 pp.

Smith, C. and Reay, P. (1991) Cannibalism in teleost fish. *Rev. Fish Biol. Fish.*, **1**, 41–64.

Townsend, C.R. (1991) Exotic species management and the need for a theory of invasion ecology. *N.Z. J. Ecol.*, **15**, 1–3.

Turner, G.F., Witimani, J., Robinson, R.L., Grimm, A.S. and Pitcher, T.J. (1991) Reproductive isolation and the nest sites of Lake Malawi Chambo, *Oreochromis (Nyasalapia)* spp. *J. Fish Biol.*, **39**, 775–82.

Turner, J.L. (1982) Lake flies, water fleas and sardines. *Biological studies on the pelagic ecosystem of Lake Malawi*. Rome, FAO, FI: DP/MLW/75/019 Tech. Rep. 1, 165–73.

Tweddle, D. and Magasa, J. (1989) Assessment of multispecies cichlid fisheries of the SE arm of Lake Malawi. *J. Cons. int. Explor. Mer.*, **45**, 209–22.

Wanink, J.H. and Goudswaard, P.C. (1992) Effects of Nile perch introduction into Lake Victoria on the diet of pied kingfishers. *Hydrobiologia*, **190**, 1–10.

Welcomme, R.L. (1988) International introductions of inland fish species. *FAO Fish. tech. Pap.*, no. 294, 318 pp.

Zaret, T.M. (1984) Central American limnology and Gatun Lake, Panama, in *Ecosystems of the World* (ed. F.B. Taub), Elsevier, Amsterdam, pp. 447–65.

Zaret, T.M. and Paine, R.T. (1973) Species introduction in a tropical lake. *Science*, **182**, 449–55.

Part One

Introduced Nile Perch in Lake Victoria: Impacts on Biodiversity and Evaluation of the Fishery

Chapter two

Impact of environmental change, species introductions and ecological interactions on the fish stocks of Lake Victoria

Aggrey W. Kudhongania† and Deonatus B.R. Chitamwebwa

2.1 INTRODUCTION

Lake Victoria was traditionally regarded as a cichlid lake on account of the predominance of these species in the fish community (Greenwood, 1966). The cichlid fauna included two indigenous tilapiine species (Graham, 1929) and, through endemic explosive speciation (Greenwood, 1965), about 300 species of haplochromines (Van Oijen *et al.*, 1981). In addition, four exotic tilapiine species were introduced into the lake during the early 1950s (EAFFRO, 1964). There were only about 50 non-cichlid fish species in the lake (Lowe-McConnell, 1975).

The coexistence of nearly 350 fish species in one lake clearly illustrates the community diversity that some tropical systems can support. Despite the high diversity, the traditional commercial fishery depended on relatively few taxa (Table 2.1). As recently as the early 1980s, Lake Victoria has been experiencing a dramatic decline in species diversity, with a simultaneous

† Deceased.

The Impact of Species Changes in African Lakes. Edited by Tony J. Pitcher and Paul J.B. Hart. Published in 1995 by Chapman & Hall, London. ISBN 0 412 55050 4.

Table 2.1 Major commercal fish species (as % by weight of the catch) from Lake Victoria (1958)

Taxon	%
Oreochromis esculentus	23.2
Bagrus docmac	21.0
Haplochromines	18.0
Labeo victorianus	10.5
Oreochromis variabilis	8.0
Protopterus aethiopicus	6.0
Clarias mossambicus	3.5
Synodontis sp.	2.9
Schilbe mystus	2.7
Mormyrids	2.3
Barbus sp.	1.1
Alestes sp.	0.6
Other	0.2
Total	100.0

phenomenal increase in total annual fish yield. This decline demonstrates the fragility of the fish community to heavy fishing. Nile perch, *Lates niloticus* Linnaeus, which was introduced into Lake Victoria from Lakes Albert and Turkana (Gee, 1964), has been cast in the role of villain and proposed as the cause for the decline (Barel, 1985; Simmie, 1986; Payne, 1987; Chapters 6, 7). But *Lates*, which first appeared in the lake in 1960 (Hamblyn, 1960), could not have done more damage to the fishery than has been caused since time immemorial by man, a much more rapacious predator. This chapter compares the impact of exploitation, predation by Nile perch, competition and hybridization between introduced and native species and environmental change, on changes in the fish community in Lake Victoria.

2.2 IMPACT OF EXPLOITATION

The traditional Lake Victoria fishery was exploited by simple fishing gears, such as basket traps, hooks and seine nets of papyrus, each of which exerted little fishing effort. The most abundant and highly cherished fish species were *Oreochromis esculentus*, *Bagrus docmac* and *Labeo victorianus*, in decreasing order of significance.

The demand for fish from Lake Victoria was significantly stimulated by the development of urban centres along the lake shore and the arrival of the railway at the Nyanza Gulf in 1908 (Graham, 1929). Fishing was further intensified by the introduction of flax gill nets in 1905, and later by the

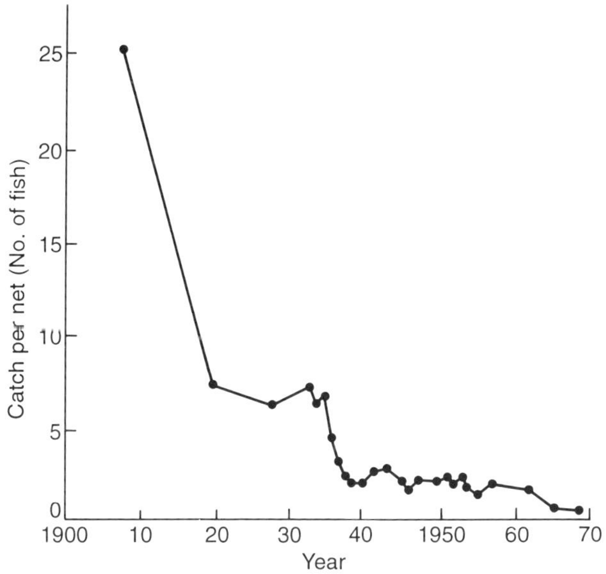

Fig. 2.1 Average catch per net of *Oreochromis esculentus*, partly based on data from Beverton (1959) and Garrod (1961).

introduction of non-selective beach seines in the early 1920s. Increased fishing pressure led to localized overfishing for the most popular fish, *O. esculentus*, as early as the late 1920s. The initial catch rate of more than 25 fish per net-night declined to about two fish by 1940 (Fig. 2.1). When the more efficient synthetic fibre gill nets were introduced in 1952, and the outboard engine in 1953 (Mann, 1969), the fishing effort increased by 100% but the yield went up by only 10% (Beauchamp, 1958). By 1955 the catch rates had become so low that it was no longer profitable to fish with the recommended 127 mm mesh nets. As a result, fishermen were tempted to fish with gill nets having smaller and smaller meshes, which led to the exploitation of immature fish and endangered the recruitment processes. Detailed studies by Beverton (1959) predicted that use of smaller-mesh gill nets would decrease both the immediate and long-term catches of *O. esculentus*, *Bagrus* and *Barbus*.

Labeo victorianus formed the most important commercial fish species along the affluent rivers of Lake Victoria (Fig. 2.2). This fishery deteriorated steadily following intensive gill netting of gravid individuals on their breeding migrations. Between 1955 and 1956, a marked decline in catch per net for *Labeo* was reported for Nzoia River, Kenya (Cadwalladr, 1965) and for the Kagera River (Garrod, 1961). In Lake Victoria *Labeo* catches declined from

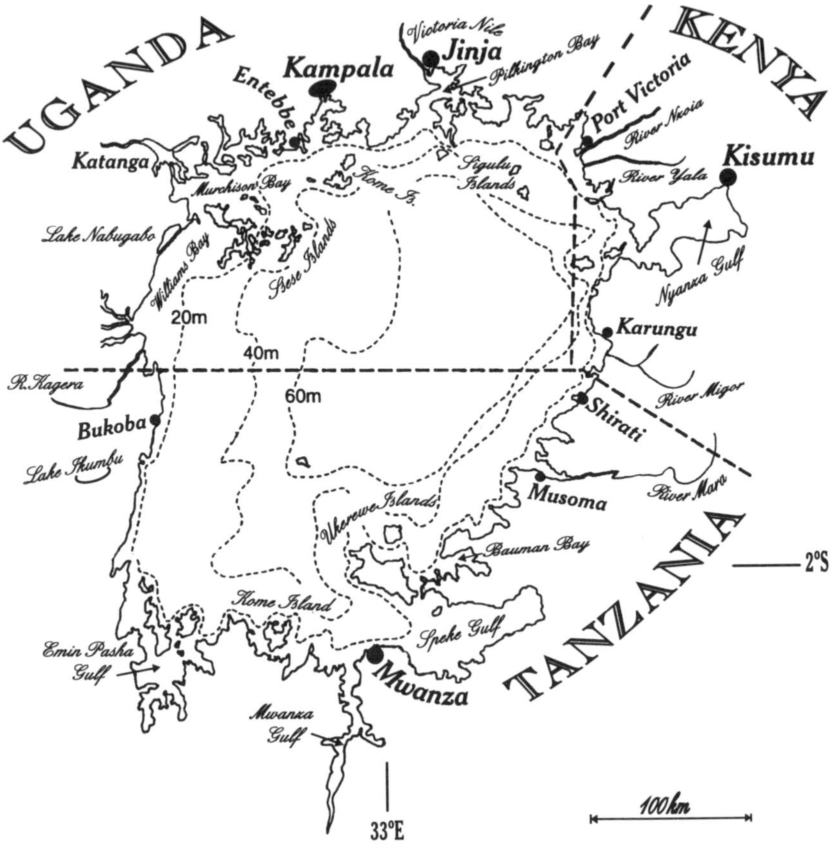

Fig. 2.2 Map of Lake Victoria (compare Chapter 6, Fig. 6.1), showing places mentioned in the text.

10.5% of the total catch in 1958 to less than 1% in 1970. Overfishing for *Labeo* affected 13 other anadromous or anadromous-like fish species (Whitehead, 1959).

The trend of overfishing for the wide range of species parallelled the continuously declining gill net mesh sizes, the increased use of beach seines and the virtual absence of management measures. The fishery had unlimited entry for the number of fishermen, fishing boats or fishing gears. Increased use of beach seines had damaging effects on some haplochromine and tilapiine stocks, especially on eggs and fry, and on breeding and nursery grounds (Welcomme, 1964). The beaches where seines are normally operated are also used by tilapiines and some haplochromine species as spawning and nursery grounds. Earlier studies had indicated that availability of suitable spawning sites was the most important factor limiting tilapiine populations in the lake (Beauchamp, 1958). It has also been shown that haplochromine stocks of Lake Victoria are not resilient to prolonged intensive exploitation pressure (Kudhongania and Twongo, 1985; Witte and Goudswaard, 1985). The sustained use of beach seines since the early 1920s (Mann, 1969) and the escalation of beach seining in many parts of the lake, particularly when littoral vegetation was reduced (section 2.5), eventually reduced both the stocks and recruitment capacities of many cichlid fishes in Lake Victoria.

A lake-wide bottom-trawl exploratory survey conducted between 1969 and 1971 (Kudhongania and Cordone, 1974) demonstrated that most of the traditional fish species were continuously declining to alarming levels. Data further showed that up to 1970 *Lates* was a significant component in neither the commercial catches nor the ecosystem. Lowe-McConnell (1975) observed that the Lake Victoria fishery was a classic example of a decline in catches following excessive fishing inshore.

2.3 IMPACT OF COMPETITION AND HYBRIDIZATION

To replenish the declining stocks of the two indigenous tilapiine fish species (*O. esculentus, O. variabilis*), four exotic tilapiines (*O. niloticus, O. leucostictus, Tilapia zillii* and *T. melanopleura*) were introduced into Lake Victoria during the early 1950s (EAFFRO, 1964; Welcomme, 1967). The establishment of four exotic tilapiines in the lake suddenly introduced interspecific competition with two indigenous species and enhanced the likelihood of genetic dilution through hybridization (Beauchamp, 1958; Lowe, 1958; Welcomme, 1967; Chapter 22).

Studies on the life history of indigenous tilapiines in Lake Victoria indicated that availability of suitable spawning sites was the most important factor limiting populations (Beauchamp, 1958). The native tilapiines were spatially segregated, with the smaller *O. variabilis* being more inshore than *O. esculentus*. When *T. zillii* and *O. leucostictus* became established, they occupied the same habitat as *O. variabilis*. Fryer (1961) and Welcomme (1966) observed direct competition for nursery grounds between *O. variabilis* and *T. zillii*. Adult *O. niloticus* competed for breeding grounds with *O. variabilis*. This competition was followed by a decline in *O. variabilis* catches (Lowe-McConnell, 1975). Although most other tilapiines showed considerable specialization regarding selection of spawning and nursery grounds, *O. niloticus* was more flexible.

Competition for food among the indigenous tilapiines was not thought to be significant (Beauchamp, 1958). However, *O. esculentus*, feeding only on planktonic diatoms (Welcomme, 1966), was at a competitive disadvantage to *O. niloticus*. The latter species ingests phytoplankton, planktonic deposits and epiphytic algae, and is capable of utilizing the bluegreen algae (Moriarty, 1973) which are abundant in Lake Victoria.

Hybridization also appears to have led to the restructuring of the tilapiine communities in Lake Victoria. Welcomme (1967) identified hybrids between *O. niloticus* and *O. variabilis*, and between *T. zillii* and *T. melanopleura*. Lowe (1958) had reported possible hybridization between *O. niloticus* and *O. esculentus*. In crosses involving *O. niloticus*, the morphological characters of the hybrids were apparently dominated by features of *O. niloticus*. It is probable that over time, the stocks lumped as *O. niloticus* could be a mixed population of such hybrids. The apparent dominance of this species could result from its abilities as a competitor. It grows to a larger size, has a faster growth rate, is more fecund, has a longer life span, has a wider food spectrum and is less habitat restricted than any of the other tilapiine species (Fryer and Iles, 1972).

2.4 IMPACT OF ENVIRONMENTAL CHANGES

Between the end of 1961 and into 1964, very heavy rains fell in East Africa. These were remembered by inhabitants as the 'Uhuru' rains because they coincided with the attainment of independence by East African countries. The most remarkable effect of the rains on Lake Victoria was to raise its level by about 2 m. Surface area also increased from 68 800 km^2 (Hutchinson, 1957) to about 75 000 km^2 (Beadle, 1974). Almost all the rooted vegetation was permanently submerged and eventually rotted away. Floating vegetation, such as papyrus swamps, was either washed ashore or disintegrated through wave action. The loss of the papyrus swamps in most bays and gulfs reduced protection against wave action, so producing a more exposed aspect. The outer fringe of lilies (*Nymphaea* spp.) disappeared (cf. Lake Naivasha, Chapter 14). Land vegetation in newly submerged areas also died, although stumps and branches took some time to rot away.

Some macrophytes, such as the ambatch shrub, *Aeschonomi* spp., and the bulrush *Typha australis*, re-established themselves quickly after the change in water level. Other plants, such as lilies and the aquatic weed *Vallisneria* spp., took longer to return. The process of macrophyte recolonization and re-establishment is still taking place and may require many more years before full recolonization is complete and the littoral has stabilized.

The change in water level and its consequences on the ecosystem occurred about 8–10 years after the introduction of the four exotic tilapiine species. The effects of the two must be confounded. Before the changes in water level it had been argued that the extensive papyrus swamps were limiting fry production in the tilapias as the swamps covered most of the potential spawning sites (Beauchamp, 1961). The loss of papyrus solved this putative problem favouring, in particular, *O. esculentus*. The dying stumps and branches of terrestrial vegetation also offered protection to fry on the nursery grounds, so boosting recruitment. Later developments counteracted this improvement. As the remnants of trees disappeared from shallow water, the nursery grounds were accessible to beach seines, a very unselective fishing method which also destroys tilapia nests. It is hypothesized that the combined effects of intensive gill netting, beach seining and loss of protected nursery areas combined to reduce *O. esculentus* recruitment, leading to further decline in the fishery. Confirmation of this hypothesis has not yet been achieved (Asila and Ogari, 1988; Getabu, 1988).

A further change observed is the gradual enrichment of the water (Bugenyi and Magumba, 1990). A comparison of data gathered before and after the development of the Nile perch fishery has shown that silicate concentrations have fallen by an order of magnitude while inshore values for chlorophyll concentration have risen from a range of 3–20 $\mu g\,l^{-1}$ to 13–71 $\mu g\,l^{-1}$ (Bugenyi and Magumba, 1990). This shift could be a result of the destabilization of the ecosystem through the large changes in the fish community (Barel *et al.*, 1991). Algal blooms are increasingly observed in the lake (Ochumba and Kibaara, 1989). These cannot be explained only in terms of run-off from increasingly fertilized land because blooms have also been observed in the dry season.

Fish kills are also becoming a regular phenomenon including large kills of the shrimp *Caridina nilotica* (pers. obs.). These kills are usually the result of anoxic conditions and are attributed to water low in oxygen being brought to the surface. The lack of oxygen in deeper water is possibly a result of an increased accumulation on the bottom of dead organic matter.

The invasion of the water hyacinth, *Eichhornia crassipes*, has further complicated the recovery of the macrophyte community around the lake. The weed is thought to have entered the lake through the River Kagera (E.F.B. Katunzi, pers. comm.). It is spreading fast and now covers most of the lake's shores. By December 1990 it was reported to have been seen in Tanzanian and Ugandan waters, and was sighted only 20 km south of the Kenya–Tanzania border in January 1991 (Shirati, pers. comm.). The presence of this plant could further complicate the changes occurring in the fishery.

2.5 IMPACT OF PREDATION BY *LATES*

Predation on fish caught for human consumption

Numerous predators existed in the traditional Lake Victoria fishery. The indigenous predators included *B. docmac, Clarias mossambicus, Protopterus*

Table 2.2 Combination of the lungfish, *Protopterus aethiopicus*, to total trawl catches of MV *Kiboko* (Tanzania) in 1986–1991*

Year	% (by weight) of catch
1986	1.5
1987	0.5
1988	1.2
1989	no data
1990	2.2
1991	1.8

*Source: MV *Kiboko* logbook (TAFIRI, Mwanza).

aethiopicus, *Schilbe* sp., *Barbus* sp. and a number of predatory haplochromine species (Graham, 1929; Corbet, 1961; Greenwood, 1966). Up until 1984, *Protopterus* contributed over 40% of trawl catches (Mosille, 1986), but since 1985 catches fell and have remained low (Table 2.2). The combined predation pressure exerted by these species on the stocks of the lake was not considered significant because most of them were mainly part-time predators.

Predation pressure became of potential importance with the establishment of *Lates* in the lake. But although *Lates* was introduced into the lake during the early 1960s, its catches became significant only from 1975, 1977 and 1978 on, in the Ugandan, Kenyan and Tanzanian sectors of the lake respectively (Fig. 2.2) (Ssentongo and Welcomme, 1985). In the Mwanza area (Tanzania) the *Lates* fishery developed mainly between 1982 and 1984 (Ligtvoet *et al.*, 1987). Determining when the fish first became a permanent member of the Lake ecosystem is harder than determining its established presence. What is known of first occurrences casts doubt on *Lates* as a cause of the changes that have occurred in the lake (Chapters 6, 7).

The role of predation by *Lates* on the decline of the food fish species in Lake Victoria appears minor, even though most fish species found in the lake have been identified in stomachs of piscivorous stages of the predator (Hamblyn, 1966; Gee, 1969; Okedi, 1970; Ogutu-Ohwayo, 1985). *Lates* is a generalist predator which will select the most abundant prey not greater than one-third of its length. Hopson (1972) found that in Lake Chad, *Lates* selected the most abundant prey. In Lake Kyoga, Ogutu-Ohwayo (1985) observed that *Lates* larger than 60 cm fed mainly on *O. niloticus* and young *Lates*, which were relatively the more abundant prey options. In Lake Victoria, *Lates* became fully established when the traditional food fish species had already declined to only 15% of the total ichthyomass (Kudhongania and Cordone, 1974). In the Kenyan part of Lake Victoria, for instance, fish landings had declined from 20 000 tons (18 200 t; 1 ton = 0.9091 tonne) in 1948 to

4200 tons in 1961. It is unlikely that in Lake Victoria, *Lates* which feeds on the most abundant prey, would have actively searched for the relatively rare food fish species (15%) in an environment where haplochromines were preponderant (85%).

Predation on haplochromines

There is evidence suggesting that predation by *Lates* had considerable impact on the haplochromine stocks of Lake Victoria. Okedi (1970) observed that haplochromines constituted 80% by weight of the food of *Lates*. In the north-western portion of Lake Victoria, Okaronon *et al.* (1985) noted a progressive decline in the experimental trawl catch rates of haplochromines following the firm establishment of *Lates* in the area. In the Nyanza Gulf (Fig. 2.2), a decline and eventual collapse of the haplochromine fishery between 1977 and 1983 was related by Ogari (1985) to continuous increases in the *Lates* population. Heavy predation pressure by *Lates* was considered more responsible for the decline in the haplochromine stocks in the southern part of Lake Victoria (Witte and Goudswaard, 1985). This evidence appears to be in line with the feeding habits of *Lates*, which selects the most abundant prey.

2.6 EXPLOITATION VERSUS PREDATION BY *LATES*

The evidence just outlined gives the impression that *Lates* predation has been the major cause of the decline in haplochromines. The information available does not distinguish between the effects of predation and the impact of fishing. Establishment of a trawl fishery for a moderate fish meal production plant in the Mwanza Gulf in 1973 provided an opportunity to separate out the impact on haplochromines of predation by *Lates* and of exploitation by humans. In 1978, 5 years after the introduction of industrial trawl fishing, the larger haplochromines had been depleted and the catch rates were very low (Witte and Goudswaard, 1985). Trawl catch data from RV *Mdiria* in Mwanza Gulf showed a decline in the haplochromine catch rates from 1750 $kg\,h^{-1}$ in 1976 to 680 $kg\,h^{-1}$ in 1985 (Goudswaard and Ligtvoet, 1987). Two years later, trawlable haplochromines had become extremely rare. Similar changes occurred in Lake Malaŵi after only 4 years of demersal trawling for haplochromines (Turner, 1977).

In the northern portion of the lake, where commercial trawl fishing was not established, experimental trawl catch rates for the haplochromines declined only slightly after a period of 10 years, from 670 $kg\,h^{-1}$ in 1969–1970 (Kudhongania and Cordone, 1974) to 540 $kg\,h^{-1}$ in 1981 (Okaronon *et al.*, 1985).

Given that *Lates* was fully established in the northern part of the lake in 1975 (Ssentongo and Welcomme, 1985) but much later in 1982 in Mwanza Gulf (Goudswaard and Ligtvoet, 1987), the above trawl catch data strongly suggest that exploitation by humans was more intensive and damaging than predation by *Lates*.

2.7 THE NEW FISHERY OF LAKE VICTORIA

Following the dramatic decline in species diversity, the new fishery is now dominated by three species: the introduced predator (*Lates*), the exotic herbivore (*O. niloticus*) and an indigenous zooplanktivore (*Rastrineobola argentea*). The stocks of *O. niloticus* and *R. argentea* increased in the presence of increasing populations of *Lates*. The decline in species diversity has been accompanied by a tremendous increase in annual landings. For the Uganda sector of Lake Victoria, for instance, the fish yield increased from 42 000 tons in 1970 to 120 000 tons in 1990. The major question is whether the highly increased catch levels would be sustained on a long-term basis and the following facts are relevant (see also Chapter 10).

- *Lates* is very fecund, being able to produce up to 16 million eggs in one breeding cycle (Ogutu-Ohwayo, 1988). It can feed at different trophic levels at each stage in its life history and can switch to the most abundant prey at any time. In its piscivorous stage, *R. argentea, O. niloticus* and juvenile *Lates* itself constitute the major prey species under the current situation. This regime could inherently restrain overpopulation of *Lates* beyond available prey sources.
- *O. niloticus* is fecund, has a fast growth rate, has a wide food spectrum and is flexible in its habitat selection (Fryer and Iles, 1972). This species is the only major fish in the lake now feeding on primary production and filling the trophic positions previously occupied by several original tilapiine and haplochromine herbivores and detritivores (Balirwa, 1989). *O. niloticus* thus has a sound ecological base to sustain its stocks in Lake Victoria.
- The rise of the *R. argentea* fishery was closely associated with the decline of the haplochromine stocks. The species now occupies a wide ecological range previously dominated by many haplochromine species and has become the main trophic link between secondary production (zooplankton and insects) and tertiary levels (*Lates*) (Wandera, 1989).

From the foregoing, if exploitation, predation, competition, environmental change and hybridization are considered in the context of the new fishery of Lake Victoria, it is most unlikely that the three species (*Lates, O. niloticus, R. argentea*) would be affected by competition (and hybridization) among themselves. Apart from environmental factors, exploitation and predation would be the main processes to influence the trend of the future Lake Victoria fishery.

Of the two factors, predation by Nile perch is unlikely to be the limiting factor. Stocks of *R. argentea* and *O. niloticus* increased in Lake Victoria despite an increasing population of *Lates*. This suggests that the two prey species are somehow resilient to predation by *Lates*. In Lake Albert, the native habitat for *Lates*, the indigenous fish species all still coexist with the predator (Ogutu-Ohwayo, pers. comm.). In Lake Tanganyika, over 90% of the annual fish harvest comes from four species of *Lates* and their two principal clupeid prey species (Bayona, 1988), although with cyclic variations in the relative abundance of the predator/prey components in the catch (Roest, 1988).

Exploitation by humans and major environmental change are likely to be the major factors in influencing the future trend of the Lake Victoria fisheries. Certain exploitation regimes appear to have been instrumental in the dramatic decline in species diversity of the indigenous fishery of Lake Victoria. Similar processes have been responsible for the disruption of fish stocks in other lakes (Turner, 1977; Robertson and Scavia, 1984).

Current fishing practices on Lake Victoria include gill nets, beach seines, cast nets, all with uncontrolled mesh sizes, and for *R. argentea*, mosquito seine nets with a similar lack of restriction. There are commercial trawling operations in some parts of the lake. In the total absence of any predator other than humans, there are already definite signals that the commercial landings of *Lates* are beginning to decline, in terms both of total weight and of individual size range. In Lake Kyoga, the annual fish landings dropped from a record high of 167 000 tons in 1977 to 88 000 tons in 1983 due to indiscriminate fishing practices (Twongo, 1986). It is therefore reasonable to assume that the mode and tempo of exploitation will greatly influence the future viability of the Lake Victoria fishery.

2.8 SUMMARY

Evidence is presented and discussed for the causes of the changes in the Lake Victoria fish community that have occurred since the beginning of the 20th century. The traditional fishery concentrated on *Oreochromis esculentus*, *Bagrus docmac* and *Labeo victorianus*, which were caught with primitive equipment causing low fishing mortality. Increases in demand, together with gear improvements, increased the fishing mortality and brought new species into the fishery. By the middle of the century, catch rates were well down on those of 50 years earlier. Serious depletion of important stocks of food fishes could have been caused by the introduction of four exotic tilapiine species in the late 1950s, drastic changes in water levels brought about by heavy rains in the early 1960s, combined with increased use of beach seines and fine-mesh nets. At about the same time the Nile perch, *Lates niloticus*, was introduced but made little contribution to the fishery for a decade or more after its introduction. Since the early 1970s the species has contributed an increasing proportion of the growing fish catch from the lake. At present the fishery is dominated by the Nile perch, with significant contributions from *Oreochromis niloticus* and *Rastrineobola argentea*, an endemic species. There is increasing evidence that the lake is becoming eutrophic, which has meant a higher incidence of algal blooms and fish kills from anoxic conditions. Experience from the south of the lake indicates that the decline of the haplochromine community is more likely to have resulted from overfishing than from heavy Nile perch predation.

ACKNOWLEDGEMENTS

AWK is very grateful to the International Development Research Centre (Canada) and to the Government of Uganda for financing the Nile perch project; to Drs F. Bugenyi and T. Twongo (UFFRO) for their useful comments on the draft manuscript; to Mr S.N. Sowobiu and Mrs F.K. Balirwa for the preparation of the figures and the manuscript, respectively.

REFERENCES

Asila, A.A. and Ogari, J. (1988) Growth parameters and mortality rates of Nile perch (*Lates niloticus*) estimated from length-frequency data in the Nyanza Gulf (Lake Victoria). In *Contributions to Tropical Fisheries Biology* (eds S.C. Venema, J.M. Christensen and D. Pauly), *FAO Fish Rep.*, no. 389, pp. 272–7.

Balirwa, J.S. (1989) In UFFRO Technical Report to IDRC on Nile perch (Uganda) Project, June 1989. (Mimeo)

Barel, C.D.N. (1985) *Cichlid Species Flock of Lake Victoria on the Verge of Extinction.* Pamphlet written for HEST, Kaiserstraat 63, 2311 GP Leiden.

Barel, C.D.N. and 12 Co-authors (1985) Destruction of fisheries in Africa's lakes. *Nature, Lond.*, **315**, 19–20.

Barel, C.D.N., Ligtvoet, W., Goldschmidt, T., Witte, F. and Goudswaard, P.C. (1991) The haplochlomine cichlids in Lake Victoria: an assessment of biological and fisheries interests, in *Cichlid Fishes, Behaviour, Ecology and Evolution* (ed. M.H.A. Keenleyside), Chapman and Hall, London, pp. 258–78.

Bayona, J.D.R. (1988) A review of the biology, productivity and fish predation in Lake Tanganyika. *CIFA Occ. Pap.*, no. 15, 1–17.

Beadle, L.C. (1974) *The Inland Waters of Tropical Africa*, Longman, London, 365 pp.

Beauchamp, R.S.A. (1958) Utilizing the natural resources of Lake Victoria for the benefit of fisheries and agriculture. *Nature, Lond.*, **18**, 1654–1656.

Beauchamp, R.S.A. (1961) East African fisheries. *Corona*, **13**(2), 57–60.

Beverton, R.J.H. (1959) Report on the state of the Lake Victoria fisheries. Fisheries Laboratory, Lowestoft. (Mimeo)

Bugenyi, F.B. and Magumba, K.M. (1990) The physico-chemistry of the northern waters of Lake Victoria, in *Prevailing Activities on Lake Victoria Basin (with particular reference to the fisheries of the lake)* (ed. P.O.J. Bwathondi), Tanzania Fisheries Research Institute, Dar es Salaam, pp. 43–55.

Cadwalladr, D.A. (1965) The decline in the *Labeo victorianus* Boulenger (Pisces: Cyprinidae) fishery of Lake Victoria and methods in the Nzoia river, Kenya. *E. Afr. Agr. For. J.*, **30**, 249–56.

Corbet, P.S. (1961) The food of non-cichlid fishes in the Lake Victoria basin, with remarks on their evolution and adaptation to lacustrine conditions. *Proc. Zool. Soc. Lond.*, **136**(I), 1–101.

EAFFRO (1964) East Africa Freshwater Fisheries Research Organisation. *Annual Report* 1964.

Fryer, G.E. (1961) Observations on the cichlid fish, *Tilapia variabilis* Boulenger in the Northern waters of Lake Victoria (East Africa). *Revue Zool. Bot. afr.*, **64**, 11–33.

Fryer, G.E. and Iles, T.D. (1972) *The Cichlid Fishes of the Great Lakes of Africa*, Oliver and Boyd, Edinburgh, 641 pp.

Garrod, D.J. (1961) The need for research on the population dynamics of tropical fish species with special reference to 'Ningu' (*Labeo victorianus*) from Lake Victoria, in *Hydrobiology and Inland Fisheries. Fourth Symposium*, pp. 54–9.

Gee, J.M. (1964) Nile perch investigations. *EAFFRO Ann. Rep.*, pp. 14–24.

Gee, J.M. (1969) A comparison of certain aspects of the biology of *Lates niloticus* (Linne) in some East African lakes. *Revue Zool. Bot. afr.*, **80**, 244–61.

Getabu, A. (1988) Aspects of the Lake Victoria fisheries with emphasis on *Oreochromis niloticus* and *Alestes sadleri* from the Nyanza Gulf, in *Contributions to Tropical Fisheries Biology* (eds S.C. Venema, J.M. Christensen and D. Pauly), *FAO Fish Rep.*, no. 389.

Goudswaard, P.C. and Ligtvoet, W. (1987) Recent developments in the fishery for haplochromines (Pisces: Cichlidae) and Nile perch, *Lates niloticus* (Pisces: Centropomidae) in Lake Victoria. Mwanza, Tanzania/Leiden, the Netherlands, *Report for HEST*, No. 35.

Graham, M. (1929) *The Victoria Nyanza and Its Fisheries. A report on the fishery survey of Lake Victoria, 1927–1928*. Crown Agents, London.

Greenwood, P.H. (1965) Explosive speciation in African lakes. *Proc. Roy. Inst.*, **40**(184), 256–69.

Greenwood, P.H. (1966) *The Fishes of Uganda*, 2nd edn, The Uganda Society, Kampala.

Hamblyn, E.L. (1960) The Nile perch project. *EAFFRO Ann. Rep.*, pp. 26–32.

Hamblyn, E.L. (1966) The food and feeding habits of the Nile perch, *Lates niloticus* (Linne) (Pisces: Centropomidae). *Revue Zool. Bot. afr.*, **74**, 1–2.

Hopson, A.J. (1972) A study of the Nile perch in Lake Chad. *Overseas Res. Publ.*, no. 19, HMSO, London.

Hutchinson, G.E. (1957) *A Treatise on Limnology, Vol. I, Geography, Physics and Chemistry*, John Wiley & Sons, New York, 1016 pp.

Kudhongania, A.W. and Cordone, A.J. (1974) Batho-spatial distribution patterns and biomass estimates of the major demersal fishes in Lake Victoria. *Afr. J. trop. Hydrobiol. Fish.*, **3**(2), 167–81.

Kudhongania, A.W. and Twongo, T. (1985) Some considerations for research and management of the Lake Victoria fisheries (Uganda). *FAO Fish. Rep.*, no. 335, pp. 139–42.

Ligtvoet, A., Chande, A.I. and Mosille, O.W.I. (1987) A preliminary description of the artisanal Nile perch (*Lates niloticus*) fishery in Southern Lake Victoria. (Mimeo)

Lowe, R.H. (1958) Observations on the biology of *Tilapia nilotica* Linne in East Africa waters. *Revue Zool. Bot. afr.*, **157** 129–70.

Lowe-McConnell, R.H. (1975) *Fish Communities in Tropical Freshwaters: Their Distribution, Ecology and Evolution*, Longman, London and New York.

Mann, M.J. (1969) A resume of the evolution of the *Tilapia* fisheries of Lake Victoria up to the year 1960. *EAFFRO Ann. Rep.*, pp. 21–27.

Moriarty, D.J.W. (1973) The physiology of digestion of blue-green algae in the cichlid fish *Tilapia nilotica*. *J. Zool. Lond.*, **171**, 25–39.

Mosille, O.I.I.W. (1986) Reproductive biology and some aspects of the production of the African lungfish (*Protopterus aethiopicus*) in Lake Victoria – Mwanza Gulf, MSc thesis, University of Dar es Salaam, 116 pp.

Ochumba, P.B.O. and Kibaara, D.I. (1989) Observations on bluegreen algal blooms in the open waters of Lake Victoria, Kenya. *Afr. J. Ecol.*, **27**, 23–34.

Ogari, J. (1985) Distribution, food and feeding habits of *Lates niloticus* in Nyanza Gulf of Lake Victoria (Kenya). *FAO Fish. Rep.*, no. 335, pp. 68–80.

Ogutu-Ohwayo, R. (1985) The effects of predation by Nile perch, *Lates niloticus* (Linne), introduced into Lake Kyoga (Uganda) in relation to the fisheries of Lake Kyoga and Lake Victoria. *FAO Fish. Rep.*, no. 335, pp. 18–41.

Ogutu-Ohwayo, R. (1988) Reproductive potential of Nile perch (*Lates niloticus* L.) and the establishment of the species in Lakes Kyoga and Victoria (East Africa). *Hydrobiologia*, **162**, 193–200.

Okaronon, J., Acere, T. and Ocenodongo, D. (1985) The current state of the fisheries in the northern portion of Lake Victoria. *FAO Fish. Rep.*, no. 335, pp. 89–98.

Okedi, J. (1970) Further observations on the ecology of Nile perch (*Lates niloticus* Linne) in Lakes Victoria and Kyoga. *EAFFRO Ann. Rep.*, pp. 42–55.

Payne, I. (1987) A lake perched on piscine peril. *New Scientist*, **115** (27 August), 50–54.

Robertson, A. and Scavia, D. (1984) North American Great Lakes, in *Ecosystems of the World 23: Lakes and Reservoirs* (ed. F.B. Taub), Elsevier, Amsterdam, pp. 135–76.

Roest, F.C. (1988) Predator–prey relations in Lake Tanganyika and fluctuation in the pelagic fish stocks. *CIFA Occ. Pap.*, **15**, 104–29.

Simmie, S. (1986) How Lake Victoria faces ecological disaster as Nile perch ruins other stocks. *Fishing News Int.* (July), p. 28.

Ssentongo, G.W. and Welcomme, R.L. (1985) Past history and current trends in the fisheries of Lake Victoria. *FAO Fish. Rep.*, no. 335, pp. 123–35.

Turner, J.L. (1977) Some effects of demersal trawling in Lake Malawi (Lake Nyasa) from 1968 to 1974. *J. Fish. Biol.*, **10**, 261–71.

Twongo, T. (1986) Current trends in the fisheries of Lake Kyoga, Uganda. (Mimeo)

Van Oijen, M.J.P., Witte, F. and Witte-Maas, E.L.M. (1981) An introduction to ecological and taxonomic investigations on the haplochromine cichlids from the Mwanza Gulf of Lake Victoria. *Neth. J. Zool.*, **31**, 149–74.

Wandera, S.B. (1989) Recent changes in the stocks of *Rastrineobola argentea* in Lake Victoria. *UFFRO tech. Rep. to IDRC*, June 1989.

Welcomme, R.L. (1964) The habitat and habitat preferences of the young of the Lake Victoria *Tilapia* (Pisces–Cichlidae). *Revue Zool. Bot. afr.*, **70**, 1–28.

Welcomme, R.L. (1966) Recent changes in the stocks of *Tilapia* in Lake Victoria. *Nature, Lond.*, **212**, 52–4.

Welcomme, R.L. (1967) Observations on the biology of the introduced species of *Tilapia* in Lake Victoria. *Revue Zool. Bot. afr.*, **76**, 249–79.

Whitehead, P.S.P. (1959) The anadromous fishes of Lake Victoria. *Revue Zool. Bot. afr.*, **59**, 329–63.

Witte, F. and Goudswaard, P.C. (1985) Aspects of the haplochromine fishery in Southern Lake Victoria. *FAO Fish. Rep.*, no. 335, pp. 81–8.

Chapter three

Limnological changes in Lake Victoria since the Nile perch introduction

Peter B.O. Ochumba†

3.1 INTRODUCTION

Lake Victoria, the third largest lake in the world (Chapter 1) and a source of livelihood for over thirty million people in Kenya, Uganda and Tanzania, is in a state of ecological transition. In the past 30 years, the Lake Victoria ecosystem has undergone one of the most profound ecological disruptions ever observed (Ochumba *et al.*, 1992).

The severity of the disturbance is the result of several interrelated forces, including the introduction of alien species, overfishing and increased nutrient inputs to the lake. The most decisive impact was attributable to the introdcution of Nile perch, *Lates niloticus*, and Nile tilapia, *Oreochromis niloticus*. Nile perch consumed many native fishes, eliminating them from the ecosystem and bringing them to commercial extinction (Barel *et al.*, 1985; Ogari and Dadzie, 1988; Ogutu-Ohwayo, 1990a; Baskin, 1992; Kaufman, 1992). The Nile tilapia outcompeted other tilapiine species, changing their feeding from phytoplankton to benthic organisms (Gophen *et al.*, 1992). These actions slightly altered the water quality and food chain connections in the lake. Changes in the fish community are described in Chapters 2 and 4–8.

† Deceased.

The Impact of Species Changes in African Lakes. Edited by Tony J. Pitcher and Paul J.B. Hart. Published in 1995 by Chapman & Hall, London. ISBN 0 412 55050 4.

Owing largely to the work of Fish (1957) and Talling (1965, 1966), Lake Victoria's limnology was better understood than its fishery biology during the time before the Nile perch introduction. Lake Victoria's shallow depth (< 100 m) and youth (< 750 000 years total, with a near total desiccation only 15 000 years ago; Kendall, 1969), set it apart from the much deeper, older lakes lying within the East African Rift Valley. Lake Victoria lacked a permanent hypolimnion. Instead, the lake underwent two cycles of stratification and mixing that corresponded to the long and short rainy seasons.

There is evidence that the nutrient chemistry of the lake also changed over the period of the transformation of the fishery (Hecky and Bugenyi, 1992). A comparison of Talling's (1966) stations in nearshore Ugandan waters between 1961 and 1988 showed no change in total phosphorus (Hecky and Bugenyi, 1992), but a tenfold increase in chlorophyll concentration. Large cyanobacteria are abundant in the open lake (Ochumba and Kibaara, 1989). Frequent localized fish kills have been recorded in nearshore shallow waters in association with temporary anoxic conditions (Ochumba, 1987, 1990). Increased algal biomass in Lake Victoria gives cause for concern as this evidence of eutrophication could lead to increased oxygen demand in the lake's deep waters and decrease the hypolimnetic volume habitable by fish during seasonal stratification.

A study summarized in this chapter was designed to test the hypothesis that anoxia is now persistent throughout the year, to examine the annual mixing and to determine the eutrophic status of the lake since the introduction of Nile perch. Limnology and fisheries data available in the literature before and after the Nile perch introduction were selected for this study, from Fish (1957), Newell (1960), Talling (1966), Kitaka (1971), Akiyama *et al.* (1977), Melack (1979a), LBDA (1984), Ochumba and Kibaara (1989), Calamari *et al.* (1992), Gophen *et al.* (1992) and Hecky and Bugenyi, (1992). In the 1950s, when stratification formed during the long and short rainy seasons, the oxycline sat at about 50 m depth, beneath which levels dropped on occasion to as low as 1 mg l^{-1}. Fish life flourished in the lake's deeper waters at all times of the year. In the 1980s and 1990s the situation for the June–August mixing seemed worse than that of the 1950s. Severe deoxygenation to shallow depths (Ochumba, 1990) was observed, indicating that most of the volume of Lake Victoria was incapable of sustaining aerobic life. A major factor may be the reduction of cichlid species that formerly helped prevent anoxic conditions within the lake by cropping phytoplankton. Therefore, the Nile perch and commercial fishing activity have possibly decimated the algae-grazing fish species and this has probably led to a significant increase in organic detritus into the hypolimnion (Kilham and Kilham, 1990a).

Between March 1990 and July 1992, a remote operated vehicle (Phantom 300) equipped with sonar, and a Hydrolab surveyor were used to gather fish location data and measure temperature, pH, dissolved oxygen and conductivity

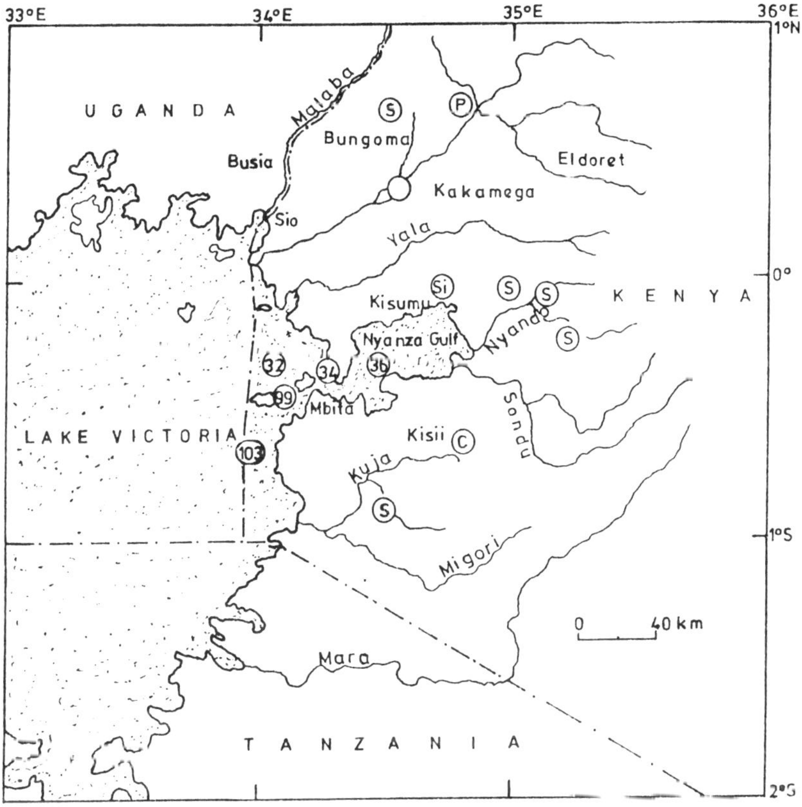

Fig. 3.1 The Kenyan portion of Lake Victoria, showing the drainage area, sampling stations 32, 34, 36, 99 and 103, and land sources of nutrients. Encircled numbers denote sampling stations; C, coffee processing factory; P, paper mill; S, sugar refinery; Si, sisal factories; —·—, international boundary.

at a series of sample stations (Fig. 3.1). Results from this work (reported in detail elsewhere) show that the lake is stratified for the entire year. Seasonality is still apparent, but mixing is restricted to periods during June, July and August when the oxycline, though persistent, has sunk at least to about 50 m. The region between 50 m and 20 m is subjected to year-round severe deoxygenation. During the study, extensive areas of Lake Victoria at the bottom were covered by anoxic waters which, under the influence of winds, currents and seiches, moved freely onshore and into the gulfs and bays. These events were associated with extensive fish kills.

Conductivity, Secchi depth, redox potential and pH had distinctively marked patterns at various sampling stations. There was a decrease in conductivity

Table 3.1 Catch composition (%) of major fish species and total landings (tonnes × 1000) in the Kenyan part of Lake Victoria*

Year	Tilapiines	*Clarias*	*Haplochromis* spp.	*Bagrus*	*R. argentea*	*Protopterus*	*L. niloticus*	Total (t × 1000)
1968	14.8	10.6	22.8	7.0	4.5	17.2	0.0	16.0
1969	26.6	7.6	36.8	5.5	2.9	9.3	0.1	17.0
1970	27.5	9.7	32.7	6.7	3.2	11.0	0.2	16.0
1971	21.1	12.5	32.0	7.1	5.1	12.8	0.3	15.0
1972	14.3	17.0	29.0	5.4	7.8	12.7	0.2	16.0
1973	10.1	15.7	33.2	8.6	10.5	13.0	0.9	17.0
1974	5.6	12.9	35.0	6.4	21.8	8.6	0.5	17.0
1975	3.9	15.6	27.9	8.4	27.4	1.1	0.1	17.0
1976	5.4	13.0	34.1	5.5	30.3	5.0	0.5	19.0
1977	7.4	9.1	32.4	6.0	34.7	4.0	1.1	19.0
1978	10.9	7.2	27.8	5.9	36.5	2.6	4.5	24.0
1979	9.0	10.0	21.6	5.8	30.5	1.5	14.0	31.0
1980	18.6	4.5	13.5	2.4	35.1	1.4	16.0	27.0
1981	10.2	2.6	2.1	1.1	20.4	0.5	59.7	46.0
1982	7.3	3.4	4.2	4.2	17.1	0.4	54.4	61.0
1983	5.5	2.7	0.8	3.1	21.3	0.3	67.7	77.0
1984	10.4	1.1	0.0	0.1	27.1	0.1	57.5	72.0
1985	10.7	0.6	0.0	0.1	29.2	0.2	56.5	90.0
1986	2.7	0.3	0.6	0.1	30.5	0.2	63.5	103.0
1987	2.8	0.6	0.3	0.0	24.5	0.1	69.1	113.0
1988	2.4	0.6	0.3	0.0	36.5	0.1	59.3	123.0
1989	2.3	1.4	1.5	0.1	38.5	0.1	54.3	225.0
1990	2.2	0.2	1.1	0.1	39.6	0.1	56.7	185.0
1991	2.4	0.2	0.5	0.0	39.2	0.0	57.3	175.0

*Sources: CIFA (1990); Fisheries Department – Annual Reports, 1990, 1991.

Table 3.2 Bottom trawl catches (kg ha^{-1}) of dominant fish species, 1969–1990*†

Species	Year (no. of hauls)				
	1969–70 (19)	1975 (69)	1977 (167)	1982–83 (54)	1989–90 (41)
Bagrus docmac	11.7	12.5	1.8	0.9	0.01
Clarias gariepinus	3.3	2.6	0.7	0.9	0.10
Haplochromis spp.	35.8	32.7	28.7	–	0.54
Labeo victorianus	0.1	0.1	0.1	0.1	0.10
Lates niloticus	0	0.8	2.8	29.0	32.7
Protopterus aethiopicus	3.7	10.7	0.3	–	0.01
Schilbe mystus	0.03	0.2	0.01	0	0.10
Synodontis spp.	2.10	0.20	0.50	0.04	0.10
Oreochromis variabilis	0.03	0.11	0.3	–	–
O. niloticus	0.01	0.20	0.70	1.40	0.83

*Source: Asilia and Ogari (1987).
†–, Not recorded; 0, catch densities less than 0.01 kg ha^{-1}.

from the shallow station 36 to the deep open lake stations. The measured Secchi depth and pH values were lower than in previous studies. Results of our study compared with others (Tables 3.1 and 3.2) show that Lake Victoria exhibits signs of eutrophication, including decreased transparency, increase in cyanobacterial biomass, elevated nutrient concentrations and hypolimnetic oxygen depletion. Sulphate concentrations are about one-hundredth of the lowest concentrations measured in the large lakes in the world. There has been a shift in the phytoplankton community towards cyanobacterial species, fewer green algae and diatoms. The zooplankton community is dominated by small-bodied species of copepods and cladocerans with low densities (Gophen *et al.*, 1992).

Remotely operated vehicle reconnaissance dives in the open lake revealed extensive areas littered with dead fishes and invertebrates, confirming the presence of severe oxygen depletion in the hypolimnion, as well as an oxycline that was shallower than before. Much of Lake Victoria may now be incapable of sustaining aerobic life. Systematic data from the portable fish finder indicated that Nile perch and bigger fish often moved into inshore channels and bays, leaving the open lake a virtual wasteland. During the brief mixing periods, the giant Nile perch streamed back out into the open lake. There were dense populations of shrimp *Caridina nilotica* below the oxycline. This may suggest that the bottom waters of Lake Victoria may be almost permanently deoxygenated and therefore cannot support a commercial fishery.

3.2 DISCUSSION

The main activities of people in the Lake Victoria catchment area are agriculture, fisheries and agro-based industries. These activities can be traced as far back as the 1920s when large forested areas were cleared for tea, coffee and sugar plantations. As the whole region lacks a unified agency for watershed management for Lake Victoria, increased soil loads, tree logging and other activities remain unchecked. In the absence of precise data on loads, the degree to which pollutants might be harmful cannot be accurately assessed. Many of these changes pre-date the Nile perch explosion, but records of environmental changes in the catchment area suggest a series of abrupt and severe changes in the aquatic environment of Lake Victoria during the 1980s (Hecky and Bugenyi, 1992). The above explanation is not the only explanation for the astonishing turn of events in Lake Victoria. A major factor is the elimination of cichlid species that formerly helped to safeguard conditions within the lake by cropping phytoplankton. The introduction of the Nile perch eliminated the major part of haplochromine populations (Witte *et al.*, 1992, Chapters 6, 7), resulting in the decline of grazing pressure on algae. It is possible therefore that algal matter is now incompletely removed, accumulating in the lake and enhancing anoxia in the open lake. Comparison of historical limnological observations with the modern condition indicates that the present situation is quite different in structure and function from that in the 1950s and 1960s (Hecky and Bugenyi, 1992). All these changes could be attributed to increased algal growth compared with 30 years ago, but intervening observations are lacking to confirm the causal link. However, dramatic changes in the 1960s were coincident with both the introduction of Nile perch and exotic tilapiines and with record high water levels on Lake Victoria (Welcomme, 1970). The increased nutrients may indicate that widespread disturbance, probably related to increasing land use, is causing eutrophication of Lake Victoria. It is possible that nutrient cycles in tropical lakes are controlled by organisms to a much greater extent than in temperate lakes (Kilham and Kilham, 1990b), which would reduce the significance of watershed changes. Our ability to explain the massive changes in Lake Victoria is still hampered by lack of information.

Hypolimnion water temperatures as low as 23.5 °C, which had been observed by Graham (1929), Worthington (1930), Fish (1957), Talling (1966) and Akiyama *et al.* (1977), were not seen during the recent studies on Lake Victoria. These data may suggest a response of the lake to a possible warming trend of climate in East Africa (Hastenrath and Kruss, 1992). Comparing data from Fish (1957), and Talling (1966), Ochumba and Kibaara (1989) with data from my recent study revealed that there has been a considerable oxygen deficiency in the lake. Oxygen depletion occurred in the lake earlier (Graham, 1929) when humans began physically to modify the watershed for large-scale agriculture, but the conditions responsible for oxygen depletion today are worse than in the past. Because of bathymetric changes, increased sedimentation and the modification which may have restricted water circulation, portions of Lake Victoria have exceeded the capacity to assimilate its oxygen demand; this has severely affected the biota (Ogutu-Ohwayo, 1990a; Kaufman, 1992). Hecky (1993) and Mugidde (1993) consider this development to be related to increased primary production associated with increased nutrient inputs from the watershed and atmosphere.

Recent changes in climate may have altered water column structure and caused mixing patterns to favour development of bluegreen algae and the loss of oxygen in the bottom waters of Lake Victoria (Kling, 1992). For Lake Victoria, the rainy seasons, floods, stratification, upwelling and rising lake levels may have direct effects on organisms and nutrient cycling (Melack, 1979b).

Modern limnologists argue about the relative importance of 'top-down' versus 'bottom-up' processes in regulating the structural dynamics of large lakes. For Lake Victoria, the hypothesis given greatest emphasis at present is 'top-down'. Predation by Nile perch altered the food web, destroying the detritivore and benthic insectivore fish assemblages, and thus terminating recycling from benthos to the water column (Witte *et al.*, 1992). The alternative or 'bottom-up' hypothesis for the lake is that nutrient inputs have increased, forcing algal blooms and a classical eutrophic deoxygenation cycle (Ochumba, 1990).

3.3 CONCLUSION

Lake Victoria has undergone dramatic alteration in recent years through a combination of eutrophication, species introductions and climate change. The effects of the introduced Nile perch alone could be large and further changes are likely. Exploitation is booming of an industrialized commercial fishery for Nile perch along with an international marketing and distribution network (Reynolds and Greboval, 1988; Chapter 11). The key question now is to what extent is the Nile perch fishery sustainable under the altered conditions in the lake? (See also Chapter 9.) This will depend on the availability of prey for the Nile perch and the way in which the prey species and human-induced changes in the environment are managed. Major food web changes in Lake Victoria are likely to continue. The population of Nile perch is heavily dependent on cannibalism at the moment (Ogari and Dadzie, 1988) and a consequent crash of the fishery has been predicted (Beadle, 1981; Ogutu-Ohwayo, 1990b). (But see Chapters 9 and 23 for a contrasting view of the role of cannibalism.) The interaction of trophic structure with water quality has been hypothesized to have led to a decrease in phytoplankton abundance through a trophic cascade effect in some large lakes (Scavia *et al.*, 1986; Lehman, 1988). Only further research will resolve the cause of the increase in algal biomass in Lake Victoria.

Urban, agricultural, agro-based industries, climate change and the introduction of Nile perch and competitive tilapiine species are the major factors implicated in the deterioration of Lake Victoria's ecosystem. *Oreochromis esculentus, O. variabilis* and over 200 haplochromine species are commercially extinct. Current fisheries legislation and operations aimed at reduction of soil erosion and sedimentation in rivers and the catchment area are insufficient to improve the lake's water quality. It is probably impossible to completely eliminate the Nile perch from Lake Victoria (Marten, 1979), but the reduction of algal matter is urgently required. Large-bodied cichlids are less preyed upon by Nile perch (Ogari, 1985; Ogutu-Ohwayo, 1985, 1990b). Therefore in the Kenyan sector we propose to develop techniques of rearing phytoplanktivorous and detritivorous large-bodied cichlids in protected bays or half-open lagoons and to release them into the entire lake when they are big enough not to be preyed upon by Nile perch.

3.4 SUMMARY

Lake Victoria exhibits traditional symptoms of eutrophication including decreased water transparency, cyanobacterial blooms, elevated nutrient concentrations and hypolimnetic deoxygenation. Changes in the phytoplankton community have altered available food sources to primary consumers, but these grazers have been considerably suppressed by higher trophic levels. Fish introduction modified the phytoplankton, zooplankton and fish assemblages in Lake Victoria and had the most intensive impact on water quality. Lake Victoria fish species are threatened by the worsening conditions in the lake itself and from the rivers in the catchment area. It is suggested that urban, agricultural and industrial pollution, climate change and the introduction of predatory Nile perch and competitive tilapiine species are the major factors in the deterioration of Lake Victoria's ecosystem. *Oreochromis esculentus* is commercially extinct and many more endemic species are endangered. Current fisheries legislation and operations aimed at reduction of soil erosion and sedimentation in flowing rivers in the catchment area are insufficient for significant improvement of the present deterioration of water quality. Localized manual harvest of papyrus, shoreline dredging and sandmining did not sufficiently improve lake conditions.

Stocking the lake with large-bodied cichlids is recommended to intensify pressure on algal and detrital matter. These grazers are less vulnerable to Nile perch predation and can probably reduce algal and detrital densities. Further long-term monitoring of the lake and its tributary rivers is required for administrative agencies to make decisions to protect Lake Victoria's resources.

ACKNOWLEDGEMENTS

I thank L. Kaufman, M. Gophen, W. Cooper, G. Kling, E. Okemwa, J. Ogari and A. Asila who made several contributions to the success of the work. Field work was supported by a grant from the USAID NO.DPE-5544-G-SS-7075-00 and NSF Grant BSR-9016552 to Les Kaufman. Simon Agembe, E. Odada and W. Oyieko assisted in field data collection.

REFERENCES

Akiyama, T., Kajumulo, A.A. and Olseen, S. (1977) Seasonal variation of plankton and physiochemical condition in Mwanza Gulf, Lake Victoria. *Bull. Freshwat. Fish. Res. Lab.*, **27**, 49–61.

Asila, A. and Ogari, J. (1987) Growth parameters and mortality rates of Nile perch *Lates niloticus* estimated from length frequency data in the Nyanza Gulf (Kenya) Lake Victoria. *FAO Fish. Rep.*, no. 389, pp. 271–87.

Barel, C.D.N., Dorit, R., Greenwood, P.H., Fryer, G., Hughes, N., Jackson, P.B.N., Kawanabe, H., Lowe-McConnell, R.H., Witte, F. and Yamaoka, K. (1985) Destruction of fisheries in Africa's lakes. *Nature, Lond.*, **315**, 19–20.

Baskin, Y. (1992) Africa's troubled waters. *Bioscience*, **42**, 476–81.

Beadle, L.C. (1981) *The Inland Waters of Tropical Africa: An Introduction to Tropical Limnology*, 2nd edn, Longman, London, 475 pp.

Calamari, D.A., Akech, M.O. and Ochumba, P.B.O. (1992) Conservation of the aquatic environment "Winam Gulf Basin Area" Preliminary Hazard Assessment. *FAO TCP.* Rep/Ken/158, 32 pp.

CIFA (Committee for the Inland Fisheries for Africa) (1990) Report of the fifth session of the subcommittee for the development and management of the fisheries in Lake Victoria. Mwanza, Tanzania, 12–14 September 1989. *FAO Fish. Rep.*, no. 430, 1–97.

Fish, G.R. (1957) A seiche movement and its effect on the hydrology of Lake Victoria. *Col. Off. Fish. Publ. London*, **10**, 1–68.

Gophen, M., Pollingher, U. and Ochumba, P.B.O. (1992) Ecological changes in Lake Victoria after the invasion of Nile perch (*Lates niloticus*): physico-chemical, plankton conditions and fish feeding. *USAID-CDR Report*, Washington, DC, 37 pp.

Graham, M. (1929) *The Victoria Nyanza and its Fisheries. A Report on the Fish Survey of Lake Victoria 1927–1928* and appendices. Crown Agents for the Colonies, London, 256 pp.

Hastenrath, S. and Kruss, P.D. (1992) Greenhouse indicators in Kenya. *Nature Lond.*, **355**, 503.

Hecky, R.E. (1993) The eutrophication of Lake Victoria. *Verh. Int. Verein. Limnol.*, **25** (in press).

Hecky, R.E. and Bugenyi, F.W.B. (1992) Hydrology and Chemistry of the African Great Lakes and water quality issues: problems and solutions. *Mitt. int. Verein. Limnol.*, **23**, 45–54.

Hecky, R.E. and Mngoma, S. (1990) The eutrophication of Lake Victoria. American Society of Limnology and Oceanography, 1990 Annual Meeting (Abstract).

Kaufman, L.S. (1992) Catastrophic change in species rich freshwater ecosystems: The lessons of Lake Victoria. *Bioscience*, **42**, 846–58.

Kendall, R.L. (1969) An ecological history of the Lake Victoria Basin. *Ecol. Monogr.*, **39**, 121–76.

Kilham, P. and Kilham, S.S. (1990b) Endless summer: internal loading processes dominate nutrient cycling in tropical lakes. *Freshwat. Biol.*, **24**, 379–389.

Kilham, S.S. and Kilham, P. (1990a) Typical limnology: do African lakes violate the first law of limnology? *Verh. Int. Verein. Limnol.*, **24**, 68–72.

Kitaka, G.E.B. (1971) An instance of cyclonic upwelling in the southern offshore waters of Lake Victoria. *Afr. J. Trop. Hydrobiol. Fish.*, **1**, 85–92.

Kling, G. (1992) Lake Victoria: structure and function of a tropical ecosystem. *Ecosystem Program Report*, University of Michigan, Ann Arbor, 48 pp.

LBDA (Lake Basin Development Authority) (1984) Technical report on current and future implications of development to the aquatic environment of Lake Victoria. *Winam Gulf Baseline Study Report*, LBDA, Kisumu, Kenya, 55 pp.

Lehman, J.T. (1988) Algal biomass unaltered by food web changes in Lake Michigan. *Nature, Lond.*, **332**, 537–8.

Marten, G.G. (1979) Impact of fishing on the inshore fishery of Lake Victoria (East Africa). *J. Fish. Res. Bd Can.*, **36**, 891–900.

Melack, J.M. (1979a) Photosynthetic rates in four tropical African Freshwaters. *Freshwat. Biol.*, **9**, 555–71.

Melack, J.M. (1979b) Temporal variability of phytoplankton in trophical Lakes. *Oceologia*, **44**, 1–7.

Mugidde, R. (1993) The increase in phytoplankton primary productivity and biomass in Lake Victoria (Uganda). *Verh. int. Verein. Limnol.*, **25**, 846–9.

Newell, B.S. (1960) Hydrology of Lake Victoria. *Hydrobiologia*, **15**, 363–83.

Ochumba, P.B.O. (1987) Periodic massive fish kills in the Kenyan part of Lake Victoria. *Wat. Qual. Bull.*, **12**, 119–22, 130.

Ochumba, P.B.O. (1990) Massive fish kills within the Nyanza Gulf of Lake Victoria. *Hydrobiologia*, **208**, 93–9.

Ochumba, P.B.O. and Kibaara, D. (1989) Observation on blue-green algal blooms in the open waters of Lake Victoria, Kenya. *Afr. J. Ecol.*, **27**, 23–34.

Ochumba, P.B.O., Gophen, M. and Pollingher, U. (1992) Ecological changes in Lake Victoria after the invasion of Nile perch (*Lates niloticus*): the catchment, water quality and fisheries management, in *Proceedings from the Aquaculture/Schistosomiasis Network Meeting in Manila, Philippines*. Washington, DC.

Ogari, J. (1985) Distribution, food and feeding habits of *Lates niloticus* in the Nyanza Gulf of Lake Victoria (Kenya). *FAO Fish. Rep.*, no. 335, pp. 68–80.

Ogari, J. and Dadzie, S. (1988) The food of the Nile perch *Lates niloticus* after the disappearance of the haplochromine cichlids in the Nyanza Gulf of Lake Victoria. *J. Fish. Biol.*, **32**, 571–7.

Ogutu-Ohwayo, R. (1985) The effects of predation by Nile perch *Lates niloticus* (Linne), introduced into Lake Kyoga (Uganda) in relation to the fisheries of Lake Kyoga and Lake Victoria. *FAO Fish. Rep.*, no. 335, pp. 18–41.

Ogutu-Ohwayo, R. (1990a) The decline of native fishes of Lake Victoria and Kyoga (East Africa) and the impact of introduced species, especially Nile perch (*Lates niloticus*) and the Nile tilapia *Oreochromis niloticus*. *Env. Biol. Fishes*, **27**, 81–96.

Ogutu-Ohwayo, R. (1990b) Changes in the prey ingested and the variations in the Nile perch and other fish stocks in Lake Kyoga and the northern waters of Lake Victoria (Uganda). *J. Fish. Biol.*, **37**, 55–63.

Reynolds, J.E. and Greboval, D.F. (1988) Socio-economic effects of the evolution on Nile perch fisheries in Lake Victoria: a review. *CIFA tech. Pap.*, no. 17, 148 pp.

Scavia, D., Fahenensteil, G.L., Evans, M.S., Jude, D.J. and Lehman, J.T. (1986) Influence of salmonine predation and weather on long-term water quality trends in Lake Michigan. *Can. J. Fish. aquat. Sci.*, **43**, 435–43.

Talling, J.F. (1965) The photosynthetic activity of phytoplankton growth in Lake Victoria (East Africa). *Int. Revue ges. Hydrobiol.*, **50**, 1–32.

Talling, J.F. (1966) The annual cycle of stratification and phytoplankton growth in Lake Victoria (East Africa). *Int. Revue ges. Hydrobiol.*, **51**, 545–621.

Witte, F., Goldschmidt, T., Wannik, J., van Oijen, M., Goudswaard, K., Witte-Mass, E. and Bouton, N. (1992) The destruction of an endemic species flock. Quantitative data on the decline of the haplochromine cichlids of Lake Victoria. *Env. Biol. Fishes*, **34**, 1–28.

Welcomme, R.L. (1970) Studies on the effect of abnormally high water levels on the ecology of fish in certain shallow regions of Lake Victoria. *J. Zool., Lond.*, **160**, 405–36.

Worthington, E.B. (1930) Observations on the temperature, hydrogen ion concentrations and other physical conditions of the Victoria and Albert Nyanzas. *Int. Rev. ges. Hydrobiol.*, **25**, 328–57.

Chapter four

Impact of fish species introductions on the tilapias of Lakes Victoria and Kyoga

Timothy Twongo

4.1 INTRODUCTION

Early in the twentieth century, *Oreochromis esculentus* and *O. variabilis* were the most important fish species in a diverse multispecies fishery in Lakes Victoria and Kyoga, exploited by traditional fishing gears such as spears, fish weirs and fish baskets (Graham, 1929). Stocks of the two tilapias were abundant; excellent catches were obtained with little effort. However, yield to the fisherman of *O. esculentus* and *O. variabilis*, as well as that of other popular fishes such as *Bagrus docmac* and *Labeo victorianus*, declined steadily over several decades under increasing exploitation pressure using mainly flax gill nets, which were later replaced by synthetic fibre nets (Beverton, 1959; Garrod, 1960; Chapter 2). This steady decline largely contributed to the decision to introduce several tilapia species into Lake Victoria and the lakes of the Kyoga Basin (Kyoga lakes), ostensibly to boost fishery production. The introduction of non-indigenous fishes into the two lake systems as a management measure later included predatory Nile perch, *Lates niloticus*.

Interaction between the indigenous fish species and those introduced into the environments of the two lake systems has produced some unanticipated results. And, although the impact of those interactions on the lake ecosystems has yet to be fully evaluated, early observations imply that there are complex

The Impact of Species Changes in African Lakes. Edited by Tony J. Pitcher and Paul J.B. Hart. Published in 1995 by Chapman & Hall, London. ISBN 0 412 55050 4.

and profound consequences that will possibly influence the entire biodiversity resources and the aquatic environment of Lake Victoria and the Kyoga lakes. This Chapter comprises a brief discussion of the likely impact of the interactions on the biology and ecology of tilapias in the two lake systems, based on the rather limited data available.

4.2 ECOLOGICAL CONSIDERATIONS

Distribution and abundance of a species in a given environment will be greatly influenced by its biological characteristics and ecological requirements. Aspects of these are outlined below as a basis for the discussion on the impact of introduced fish species on the tilapias of Lakes Victoria and Kyoga.

Distribution and fishery of indigenous tilapias

In Uganda, *O. esculentus* and *O. variabilis* are indigenous and endemic to Lake Victoria (Graham, 1929; Fig. 2.2) and the small associated lakes such as Nabugabo; to the Kyoga lakes (Fig. 4.1) and to the Victoria Nile above the Murchison Falls (Greenwood, 1966). Resident populations of *O. variabilis* occurred also in affluent streams of Lakes Victoria and Kyoga, which they entered in even larger numbers early in the rainy season, presumably to breed (Graham, 1929). The above distribution was, however, widened when *O. variabilis* was extensively stocked in dams and ponds throughout Uganda as well as in Kenya and Tanzania (Greenwood, 1966).

The spatial distribution of the above fishes in Lake Victoria was generally inshore (Graham, 1929), but largely segregated such that *O. variabilis* occurred closer to the shoreline in shallow (typically <6 m), more exposed environments, whereas *O. esculentus* preferred gulfs and bays (generally <20 m deep) sheltered from strong waves to favour muddy bottoms (Graham, 1929; Lowe, 1956; Fryer, 1961). In shallow Lake Kyoga (the largest of the Kyoga Lakes; average depth 3 m) the two fishes were similarly but less strongly segregated such that *O. esculentus* was more common in open waters and *O. variabilis* occurred mainly close to the shore, particulary among water lilies (Worthington, 1928; Fisheries Department, 1951).

The two indigenous tilapias of Lake Victoria and the Kyoga lakes were the most exploited fish species in the early fishery of the two lake systems, and their generally inshore distribution in Lake Victoria contributed to this significance. The introduction of the more efficient flax gill net into the fishery in 1906 and its subsequent widespread use by 1916 greatly increased fishing pressure on the larger and more commercially preferable *O. esculentus*, so that by 1920 localized overfishing of the species had set in (Graham, 1929). The stocks of *O. esculentus* subsequently declined steadily through the 1930s

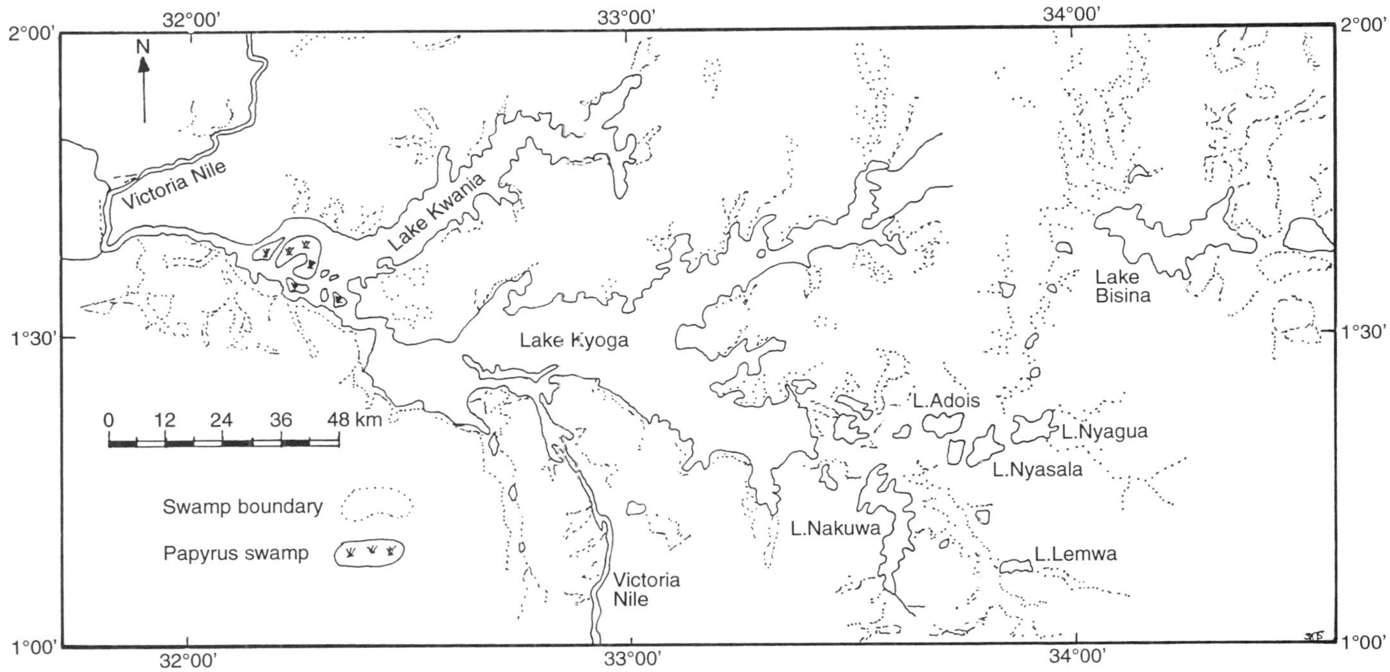

Fig. 4.1 Map of the Kyoga lakes and associated swamp.

and 1940s as fishing pressure intensified in direct response to increasing demand for fish and to the use of more efficient gill nets (Beverton, 1959; Garrod, 1961). Attempts to regulate gill net mesh sizes so as to control overfishing were difficult to implement, mainly because of the tri-national nature of Lake Victoria. There was a brief resurgence in the catches of *O. esculentus* early in the 1960s (Kudhongania and Cordone, 1974b) due to the rise in lake level following exceptionally high rainfall, which created more favourable ecological conditions (Temple, 1964; Welcomme, 1966, 1969). The fishermen had turned to the more preferred *O. esculentus*, the catches of which had temporarily improved. However, soon the landed catch of this species and that of *O. variabilis* in Lake Victoria resumed the earlier decline and by 1984 *O. esculentus* had disappeared from the experimental trawl catches in the Northern portion of the lake. To date, this species is extremely rare in Lake Victoria as well as in the major lakes in the Kyoga Basin (Lakes Kyoga and Kwania). However, isolated stocks of *O. esculentus* and *O. variabilis* still occur in some small lakes associated with Lake Victoria such as Lakes Kayugi, Kayanja and Manywa (Ogutu-Ohwayo, 1993), as well as in some of the minor Kyoga lakes such as Adois and Nyasala (Fig. 4.1) where *O. variabilis* is apparently more common.

Introduced fish species, their distribution and fishery

Four fish species (three tilapias and Nile perch) introduced into Lake Victoria and the Kyoga lakes in the 1950s and early 1960s (Gee, 1964; Welcomme, 1966, 1967; Stoneman and Rogers, 1970) are well established in various habitats of these lake environments.

Oreochromis leucostictus and Tilapia zillii

These species were introduced at various sites around Lake Victoria between 1951 and 1960 and into the Kyoga Lakes between 1954 and 1957. Both fishes originated from Lake Albert, where they are native, but some of these fish had been multiplied in ponds and dams in Uganda and Kenya. Their native range includes Lakes Edward and George (for *O. leucostictus*) as well as Lakes Turkana and Chad, the River Nile (below Murchison Falls), West African rivers and Palestine (for *T. zillii*). *Tilapia rendalli*, which closely resembles *T. zillii*, was also introduced into Lake Victoria, but failed to establish.

Establishment of *O. leucostictus* and *T. zillii* in the two lakes was rapid and the two species appeared in the commercial catches from 1959. Habitat preference of *O. leucostictus* in the new environments compared quite closely with that in the native Lakes George and Edward (Lowe, 1957) and its habitat preference in Lakes Victoria and Kyoga has not changed significantly. *O.*

leucostictus frequents shallow muddy environments close to the marginal swamp fringe, particularly in bays, inlets, lagoons and affluent rivers in Lake Victoria and the Kyoga lakes. In northern Lake Victoria, adult specimens of this fish have been caught in deeper water, particularly in large bays with muddy bottoms. On the other hand, the spatial distribution of *T. zillii* appears to have shifted from the initially preferred shallow, more turbulent as well as sheltered waters along the lake shore, to rather restricted environments. In Lake Kyoga, for instance, *T. zillii* occurs mainly under cover of submerged and floating macrophytes such as *Ceratophyllum, Mylliophyllum, Potamogeton, Nymphaea* and *Pistia*, and in sheltered bays, often close to the papyrus fringe. Although this fish appears to be similarly distributed in Lake Victoria, more information is required.

The commercial fishery of *O. leucostictus* and *T. zillii* attained considerable significance in the 1960s and 1970s as *O. esculentus* and *O. variabilis* continued to decline (Welcomme, 1966). However, this fishery appears to have been quickly overshadowed by the upsurge in the commercial catches of introduced *O. niloticus*. In Lake Kyoga, *O. leucostictus* is caught mainly in basket traps and undersized gill nets (< 3 inch (7.6 cm) mesh, stretched) set along the marginal papyrus mats or among aquatic macrophytes where considerable stocks still occur (Twongo, unpubl. data). The same gear is used to harvest *T. zillii* in similar environments but the stocks of this species in formerly preferred habitats appear to be declining.

Oreochromis niloticus

Oreochromis niloticus together with the predator *L. niloticus* are clearly the most successful so far of the fishes introduced into Lake Victoria and the Kyoga lakes. *O. niloticus* was introduced into Lake Victoria between 1954 and 1962 and into the lakes of the Kyoga Basin in 1954 to 1957. The fish rapidly established itself into the shallow and comparatively small Lake Kyoga, appearing in the commercial catches in 1959, and by the end of the 1960s *O. niloticus* constituted over 30% of commercial landings (Fig. 4.2). Distribution was already virtually lakewide. By early 1970, *O. niloticus* in Lake Victoria was distributed largely inshore, thriving best within the 10 m depth contour (Kudhongania and Cordone, 1974a). In both Lake Victoria and the Kyoga lakes, *O. niloticus* shows an extremely wide habitat tolerance inshore. The species occurs on hard (rocky, gravelly and sandy) bottoms as well as on soft (muddy) ones; among aquatic macrophytes and in macrophyte-free environments; in sheltered conditions and in turbulent waters. In western Lake Kyoga, where a huge papyrus swamp was partially submerged (Fig. 4.1), *O. niloticus* supports a very active fishery among 'floating islands' of papyrus. Under these shallow (less tham 1 m in places), decomposing conditions, oxygen tensions are likely to drop considerably at night. However,

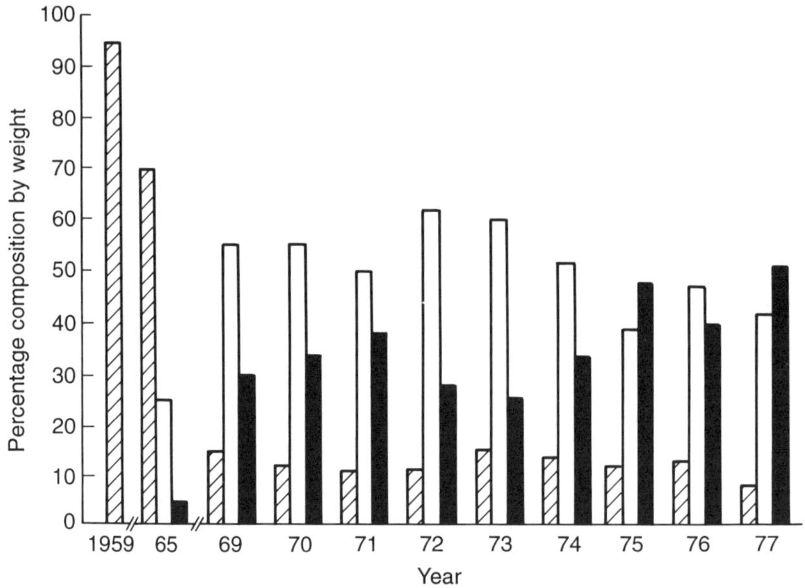

Fig. 4.2 Trends in relative yield to the fisherman of major indigenous and introduced fishes in the commercial fishery of the Kyoga lakes, 1959 to 1977. Hatched bars, indigenous fishes; solid bars, Nile tilapia, *Oreochromis niloticus*; open bars, Nile perch, *Lates niloticus*.

O. niloticus has been reported to avoid very low oxygen tensions (Welcomme, 1967).

The commercial fishery of *O. niloticus* in the Kyoga lakes attained special prominence in the late 1960s when it overtook the combined contribution of the indigenous fish species to rank second to that of *L. niloticus* (Fig. 4.2). This ranking was maintained until the late 1970s, when the catches of *O. niloticus* equalled and eventually overtook commercial landings of *L. niloticus*. Through the 1980s *O. niloticus* has been, and still is the most important commercial species in a basically two-species commercial fishery. Overall assessment of the contribution of *O. niloticus* to the commercial landings of Lake Victoria is rather difficult in view of the unavailability of data and the early tendency to lump the tilapia catches together. However, *O. niloticus* steadily overshadowed the commercial fishery of the indigenous fishes, particulary the tilapias, because of its large size and ecological success and eventually replaced them. For instance, in Ugandan waters the fishery of *O. niloticus* ranked second to that of *L. niloticus* until the late 1980s when there was a surge in landings of *Rastrineobola argentea*.

Lates niloticus

L. niloticus was introduced into the larger lakes in the Kyoga Basin (Lakes Kyoga, Kwania and Bisina) between 1954 and 1957. This piscivorous fish established rapidly and built up robust stocks which by the close of the 1960s contributed almost 50% of the commercial fishery (Fig. 4.2). On the other hand, the spread and full establishment of *L. niloticus* in Lake Victoria, where it was introduced in several locations around the lake in the early 1960s, took more than 18 years. Commercial catches of *L. niloticus* became significant from 1977, 1981 and 1983 in the Kenyan, Ugandan and Tanzanian sectors of Lake Victoria respectively (Okaronon *et al.*, 1985; Hughes, 1986; Goudswaard and Witte, 1985). In each case the commercial fishery of *L. niloticus* soon overtook all the other fishes to rank highest and this position has been maintained (Chapter 7).

L. niloticus, like *O. niloticus*, is distributed in almost all habitats in Lake Victoria, that are not deficient in oxygen. The larger fish (>10 kg) prefer waters deeper than 3 m but are more widely distributed in areas less than 30 m deep, although they can thrive in deeper waters (Goudswaard and Witte, 1985). However, the anoxic conditions observed in some parts of the lake (Hecky, 1993) are likely to influence the bathymetric distribution of *L. niloticus* in this lake (Chapter 3).

4.3 HOW INTRODUCED SPECIES AFFECT INDIGENOUS FISHES

The above review indicates that the drastic decline in the stocks of indigenous *O. esculentus* and *O. variabilis* in Lakes Victoria and Kyoga until the mid 1960s, when introduced fishes attained commercial significance, was due to destructive fishing practices and the absence of effective management (Ogutu-Ohwayo, 1990; Kudhongania *et al.*, 1992; Chapter 7). However, even after introduced fishes were firmly established and contributed a more preferred commercial fishery, the stocks of the native tilapias continued to decline when exploitation of introduced tilapias should have relieved them. Secondly, most native fishes, including the two endemic tilapias, have virtually disappeared from Lake Victoria and the major Kyoga lakes, where introduced fishes are abundant. But stable stocks of *O. esculentus* and *O. variabilis* still occur in the small lakes of the Victoria and Kyoga basins that are largely free from introduced fishes, particularly Nile perch. Overall, it would appear that fish species introductions have had a negative impact on the diversity, distribution and abundance of indigenous fishes, notably the tilapias, in Lake Victoria and the Kyoga basin lakes.

A review of environmental preferences indicates that the preferred habits of introduced *T. zillii* are similar to those of *O. variabilis*, whereas introduced

O. niloticus, a habitat generalist like *L. niloticus*, invaded the preferred habitats of both indigenous tilapias. It is interesting to note that *O. niloticus* coexists with *O. leucostictus* in less-deoxygenated habitats. However, it appears that reduced oxygen limits the spread of *O. niloticus* into all of the habitats of *O. leucostictus* because the latter species can tolerate considerable deoxygenation. On the other hand, the distribution and abundance of *T. zillii*, particularly in Lake Kyoga, is declining from that observed as late as 1978. This fish species appears to be confined to patches of aquatic macrophytes, which are now highly restricted in Lake Kyoga due to the operation of seine nets offshore and, moreover, the species competes for habitats with *O. leucostictus* in sheltered papyrus-fringed bays. Reduction in macrophyte cover does not adequately account for its decline in Lake Kyoga. The reasons may be that this fish is not an obligate macrophyte feeder and it is not known how tolerant *T. zillii* is to deoxygenation.

If the observed habitat overlap between species accounts for the observed changes in the distribution and abundance of tilapias in Lakes Victoria and Kyoga, it would be essential to establish the processes and mechanisms of any detrimental interactions. The impact on the tilapias appears to have been mediated through competitive advantage in various ecological processes. Therefore, detailed studies on ecological features and processes of the tilapias such as those outlined below may reveal interactive situations and mechanisms that facilitated the impact of introduced fish species. Unfortunately, data on many of the aspects outlined are limited or lacking altogether.

Reproductive potential/success

In this discussion the terms reproductive potential and success refer to those features and processes that influence the production of young and their survival to recruitment and sexual maturity. Factors may include sex ratio, fecundity, competition for spawning and nursery sites, spawning behaviour and parental care, all of which facilitate survival of young through the vulnerable early phase.

Breeding status

Data on spawning status of introduced tilapias in Lake Victoria between 1965 and 1966 (Welcomme, 1967) are combined in Fig. 4.3 with similar data for *O. niloticus* collected from Lake Kyoga between 1978 and 1985 by the author.

All data sets indicate spawning activity throughout the year with two minor peaks during the rainy season. The ratio of ripe males to ripe females in *O. niloticus* almost consistently favours males (i.e. exceeds 1.0), especially in Lake Kyoga, but the reverse is generally true for *O. leucostictus* and *T. zillii*. Maintaining excess ripe males may enhance spawning activity, may facilitate

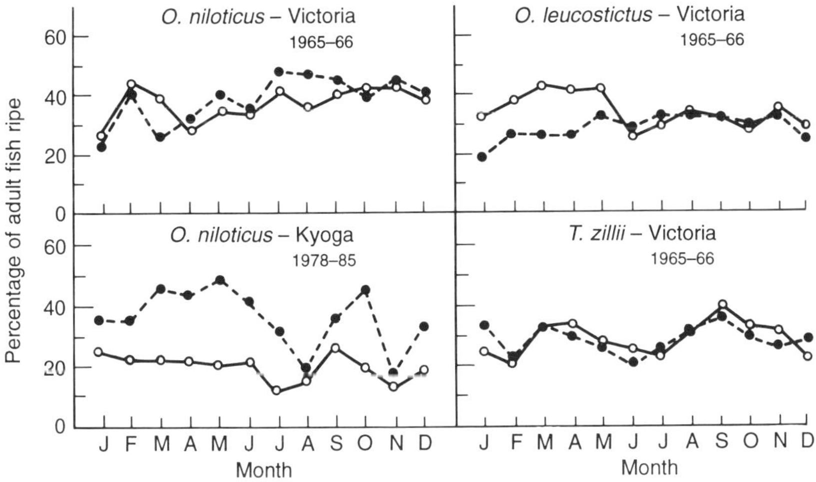

Fig. 4.3 Breeding status of *Oreochromis leucostictus*, *Oreochromis niloticus* and *Tilapia zillii* as indicated by numbers of mature females (○) and males (●) throughout the year. Data for *O. niloticus* in Lake Kyoga were collected in 1978–1985 by the author, other data are for 1965–1966 and are taken from Welcomme (1967).

more successful fertilization and is consistent with maternal brooding whereby female *O. niloticus* retire to a quiet brooding site alone. On the other hand, the evolution of paired guarding of eggs at the nest in *T. zillii* may be viewed as consistent with the observed more or less equal numbers of ripe members of each sex which reduces disruptive aggressive encounters on brooding grounds.

Competition for spawning sites and nurseries

Earlier studies (Lowe, 1956; Fryer, 1961; Welcomme, 1964, 1967) demonstrated likely competition for spawning, brooding and nursery sites between native *O. variabilis* and introduced *T. zillii* in Lake Victoria. Both species require exposed, sandy environments to spawn, and perhaps for brooding they need shallow rocky shores as nurseries. The same studies also indicated possible competition for breeding and nursery grounds between native *O. esculentus*, which frequented sheltered bays, and introduced *O. leucostictus*, the preferred habitat of which was shallow environments with muddy bottoms. However, *O. esculentus* tended to avoid deoxygenated environments.

O. niloticus is said to breed on relatively shallow, sandy or gravelly shores. In Lake Kyoga, where it was introduced, this species has been observed by the author to spawn in environments with a sandy or muddy bottom. The fry of *O. niloticus* frequent shallow rocky, gravelly and muddy shores as nursery grounds. Welcomme (1967) supposed that *O. niloticus* used similar environments to breed and shelter their young in Lake Victoria.

Control of spawning and nursery sites in a competitive situation is likely to ensure reproductive success of the victor and reproductive failure of the displaced species. There are indications that *O. niloticus* has a higher reproductive success than any of the other tilapias in view of its wider habitat tolerance for spawning and nursery purposes, its higher fecundity and an apparently more aggressive posture in competitive situations. These attributes would enhance its ecological success over other tilapias.

The vast reproductive potential and success of *O. niloticus* (Ogutu-Ohwayo, 1988) seems to be the key to the proliferation of this species in the introduced habits of Lakes Victoria and Kyoga. *Lates* has inherited from its centropomid ancestors the normal teleost strategy of high fecundity (up to 17 million eggs), and in common with most fishes, has no parental care. The ability of young *Lates* to survive in a wide range of nursery environments (cf. *O. niloticus*) brings them into competition with most tilapias, the only possible exception being *O. leucostictus*, the young of which may frequent low-oxygen environments.

Competition for food, and food habits

Competition for food among the tilapias is likely to be more important at the nursery stage when similar items such as encrusted algae and macroinvertebrates are the main food. Competitive ability to take over shared nursery sites is, therefore, likely to influence ecological success of a species. However, ability to eat a wide range of items increases the chances of obtaining adequate food, especially under competition. Therefore *O. esculentus*, which was reported to utilize only planktonic diatoms for food (Welcomme, 1966), would have been at a competitive disadvantage to *O. niloticus*; the latter species not only ingests phytoplankton, planktonic detritus and epiphytic algae, but is capable of utilizing bluegreen algae (Moriarty, 1973), which are abundant in Lakes Victoria and Kyoga. The flexibility in food habits of *O. niloticus* has been further demonstrated by inclusion of a large proportion of invertebrates in its adult diet in Lake Victoria (Balirwa, 1990).

On the other hand, *T. zillii*, which can feed on macrophytes as well as on phytoplankton, does not seem to have converted this attribute into improved stocks in Lakes Victoria and Kyoga. This may be partly because of the continued decline of macrophyte cover in these lakes. But *T. zillii* could also be losing the struggle at nursery grounds, where the young fish depend entirely on phytoplankton and macroinvertebrates. The species originally shared nursery grounds with *O. variabilis* and the more successful *O. niloticus*.

Growth

Comparative growth data for the tilapias in Lakes Victoria and Kyoga are scanty. The growth of *T. zillii* in Lake Victoria was found to be similar to that of native *T. esculentus* (Garrod, 1959) but faster than that of *T. variabilis* (Fryer, 1961). This factor would have given *T. zillii* early competitive advantage over *O. variabilis* at their shared nurseries. However, the growth of *O. niloticus* is generally believed to be faster than that of most tilapias (Fryer and Iles, 1972) and this trait largely accounts for the species' notable success in aquaculture and possibly also in natural habitats.

Hybridization

Although hybridization in the tilapias of Lakes Victoria and Kyoga is often suspected and has occasionally been observed, neither the viability of specific crosses, nor the reproductive viability of the progeny is certain. Welcomme (1967) identified hybrids between *O. variabilis* and *O. niloticus* and between *T. zillii* and *T. rendalli*. Hybridization involving *O. esculentus* and introduced tilapia has been cited as possibly responsible for the drastic decline of this species through genetic dilution. Suspicion has often been expressed that the very large Nile tilapia found in Lakes Victoria and Kyoga may actually be hybrids, although equally large specimens of *O. niloticus* (over 6 kg) have been reported in Lake Turkana (Greenwood, 1966). It is clear, therefore, that hybridization in the tilapias of Lake Victoria and the Kyoga Basin requires more detailed study to evaluate its impact on these fishes (Chapter 22).

ACKNOWLEDGEMENTS

I am grateful to Richard Ogutu-Ohwayo for reading through the manuscript and to the Renewable Resources Assessment Group at Imperial College, London, for inviting me and for meeting my travel costs to the workshop.

REFERENCES

Balirwa, J.S. (1990) The effect of ecological changes in Lake Victoria on present trophic characteristics of *Oreochromis niloticus* in relation to the species role as a stabilizing factor of biomanipulation. *Int. Agric. Centre, Wageningen, Occ. Pap.*, no. 3, 58–66.

Beverton, R.J.H. (1959) Report on the state of the Lake Victoria fisheries. Fisheries Laboratory, Lowestoft. (Mimeo)

Fisheries Department (1951) *Annual Report, Uganda Fish. Depart., Dec. 1951, Appendices 4 & 5*. Government Printers.

Fryer, G. (1961) Observations on the biology of the cichlid fish *Tilapia variabilis* Boulenger in the northern waters of Lake Victoria (East Africa). *Revue Zool. Bot. afr.*, **64**, 1–33.

Fryer, G. and Iles, T.D. (1972) *The Cichlid Fishes of the Great Lakes of Africa: Their Biology and Evolution*. Oliver and Boyd, London, 641 pp.

Garrod, D.J. (1959) The growth of *Tilapia esculenta* Graham in Lake Victoria. *Hydrobiologia*, **12**, 268–98.

Garrod, D.J. (1960) The fisheries of Lake Victoria, 1954–1959. *E. Afr. Agric. For. J.*, **26**, 42–8.

Garrod, D.J. (1961) The rational exploitation of the *Tilapia esculenta* stocks of the Northern Buvuma Island area of Lake Victoria. *E. Afr. Agric. For. J.*, **27**, 69–76.

Gee, J.M. (1964) Nile perch investigations. *EAFFRO Ann. Rep. 1962/1963*, pp. 14–24.

Goudswaard, P.C. and Witte, F. (1985) Observations on Nile perch, *Lates niloticus* (L), 1758, in Tanzanian waters of Lake Victoria. *FAO Fish. Rep.*, no. 335, pp. 62–5.

Graham, M. (1929) A report on the fishing survey of Lake Victoria 1927–1928, and appendices. Crown Agents. London, 255 pp.

Greenwood, P.H. (1966) *The Fishes of Uganda*. The Uganda Society, Kampala, 131 pp.

Hecky, R.E. (1993) Peter Kilham Memorial Lecture. The eutrophication of Lake Victoria. *Int. Verein. Limnol.*, **25**, 39–48.

Hughes, N.F. (1986) Changes in feeding biology of the Nile perch, *Lates niloticus* L. (Pisces: Centropomidae) in Lake Victoria, East Africa, since its introduction in 1960, and its impact on the native fish community in the Nyanza Gulf. *J. Fish Biol.*, **29**, 541–8.

Kudhongania, A.W. and Cordone, A.J. (1974a) Batho-spatial distribution patterns and biomass estimate of the major demersal fishes in Lake Victoria. *Afr. J. trop. Hydrobiol. Fish.*, **3**, 15–31.

Kudhongania, A.W. and Cordone, A.J. (1974b) Past trends, present stocks and possible future state of the fisheries of the Tanzania part of Lake Victoria. *Afr. J. trop. Hydrobiol. Fish.*, **3**, 167–81.

Kudhongania, A.W., Twongo, T. and Ogutu-Ohwayo, R. (1992) Impact of the Nile perch on the fisheries of Lakes Victoria and Kyoga. *Hydrobiologia*, **232**, 1–10.

Lowe (McConnell), R.H. (1956) Observations on the biology of *Tilapia* (Pisces–Cichlidae) in Lake Victoria, East Africa. *E. Afr. Fish. Res. Org. supp. Pub.*, **1**, 1–72.

Lowe, R.H. (1957) Observations on the diagnosis and biology of *Tilapia leucostictus* Trewavas in East Africa. (Pisces: Cichlidae). *Revue Zool. Bot. afr.*, **55**, 353–73.

Moriarty, D.J.W. (1973) The physiology of digestion of blue-green algae in the cichlid fish *Tilapia nilotica*. *J. Zool., Lond.*, **171**, 25–39.

Ogutu-Ohwayo, R. (1988) Reproductive potential of the Nile perch, *Lates niloticus L.* and the establishment of the species in lakes Kyoga and Victoria (East Africa). *Hydrobiologia*, **162**, 193–200.

Ogutu-Ohwayo, R. (1990) The decline of the native fishes of lakes Victoria and Kyoga (East Africa) and the impact of introduced species, especially the Nile perch *Lates niloticus* and the Nile tilapia *Oreochromis niloticus*. *Env. Biol. Fishes*, **27**, 81–96.

Ogutu-Ohwayo, R. (1993) The effects of predation by Nile perch, *Lates niloticus* L., on the fish of Lake Nabugabo, with suggestions for conservation of endangered endemic cichlids. *Conserv. Biol.*, **7**, 701–11.

Okaronon, J., Acere, T. and OcenOdongo, D. (1985) The current state of the fisheries in the northern portion of Lake Victoria (Uganda). *FAO Fish. Rep.*, no. 335, pp. 89–98.

Stoneman, J. and Rogers, J,F. (1970) Increase in fish production achieved by stocking exotic species (Lake Kyoga, Uganda). *Uganda Fish. Dept, Occ. Pap.*, no. 3, pp. 16–19.

Temple, P.H. (1964) Lake Victoria levels. *Proc. E. Afr. Acad.*, **2**, 50–58.

Welcomme, R.L. (1964) The habitat and habitat preferences of the young of the lake Victoria *Tilapia* (Pisces Cichlidae). *Revue Zool. Bot. afr.*, **70**, 1–28.

Welcomme, R.L. (1966) Recent changes in the stocks of *Tilapia* in Lake Victoria. *Nature, Lond.*, **212**, 52–4.

Welcomme, R.L. (1967) Observations on the biology of the introduced species of *Tilapia* in Lake Victoria. *Revue Zool. Bot. afr.*, **76**, 249–79.

Welcomme, R.L. (1969) The effect of rapidly changing water level in Lake Victoria upon the commercial catches of *Tilapia* (Pisces: Cichlidae) in *Int. Symp. Man-made Lakes*, Ghana Academy of Science, Achimota (ed. L. Obeng), Ghana Univ. Press, Accra.

Worthington, E.B. (1928) *A Report on the Fishing Survey of Lakes Albert and Kyoga.* London, Crown Agents for the Colonies, 136 pp.

Chapter five

Diversity and stability of fish stocks in Lakes Victoria, Kyoga and Nabugabo after establishment of introduced species

Richard Ogutu-Ohwayo

5.1 INTRODUCTION

In the first half of this century, Lakes Victoria, Kyoga and Nabugabo had a diverse and similar native fish fauna (Graham, 1929; Worthington, 1929; Cambridge Nabugabo Biological Expedition, 1962). The fisheries of these lakes were based on about 14 fish taxa of which two were tilapiine cichlids; *Oreochromis esculentus* and *O. variabilis* were the most important. The haplochromine cichlids were represented by more than 300 species in Lake Victoria, more than 99% of them endemic (Greenwood, 1974; Witte *et al.*, 1992a,b). These formed at least 80% of the demersal fish stocks in Lake Victoria up to the 1980s (Kudhongania and Cordone, 1974; Okaronon *et al.*, 1985). Haplochromine cichlids were also abundant in Lakes Kyoga and Nabugabo (Worthington, 1929; Cambridge Nabugabo Biological Expedition, 1962).

By the 1960s, stocks of the native tilapiines and other large species had

The Impact of Species Changes in African Lakes. Edited by Tony J. Pitcher and Paul J.B. Hart. Published in 1995 by Chapman & Hall, London. ISBN 0 412 55050 4.

been reduced by overfishing (Jackson, 1971; Ogutu-Ohwayo, 1990a). The fishing emphasis then shifted to the smaller species, mainly *Rastrineobola argentea* and the haplochromines.

The predatory Nile perch, *Lates niloticus*, was introduced into Lakes Victoria, Kyoga and Nabugabo to feed on haplochromines and convert them into a larger fish of greater food and recreational value (Gee, 1964; Ogutu-Ohwayo, 1985). This was done amidst the controversy that the Nile perch would deplete stocks of other fishes and reduce fish species diversity in these lakes (Fryer, 1960; Anderson, 1961). Four tilapiine species, *Oreochromis niloticus* (Nile tilapia), *O. leucostictus, Tilapia zillii*, and *T. rendalli* were also introduced to these lakes at about the same time (Welcomme, 1967; Chapter 4).

There have been considerable changes in the fish species composition in Lakes Victoria, Kyoga and Nabugabo since the Nile perch and other fishes were introduced. Populations of most of the native species have declined and many species have become extinct (Witte *et al.*, 1992a,b). The original decline in the fish stocks was due to overfishing (Jackson, 1971; Ogutu-Ohwayo, 1990a) but the recent and more drastic decline has been due to predation by the introduced Nile perch (Ogari and Dadzie, 1988; Ligtvoet and Mkumbo, 1990; Ogutu-Ohwayo, 1990b; Witte *et al.*, 1992a,b; but see Chapter 7). Nile perch feed on invertebrates, changing to a piscivorous diet with size (Hamblyn, 1966; Gee, 1969; Okedi, 1971; Ogutu-Ohwayo, 1990a,b). As the suitable types of prey were depleted in the new habitats, Nile perch switched to other prey types, to the extent of feeding even on its own young. There are fears that the Nile perch will overshoot its food supply, resulting in reduction of its own population, leading to a collapse in the fishery.

The changes in fishery yield and in fish species diversity in Lakes Victoria, Kyoga and Nabugabo were examined to see how the composition and relative abundance of the fish species had changed in the presence of the Nile perch and if there was any evidence that fish species composition and diversity had stabilized or would do so. It was assumed that fish species diversity had declined and stabilized or would continue to decline and finally collapse. Stability would occur at a level unique to Lakes Victoria, Kyoga and Nabugabo or similar to that in Lake Albert, the original and native habitat of Nile perch stocks which were introduced into the three lakes. The hypothesis examined was that fish species composition and diversity had declined and stabilized. In a stable state, there would be no further drastic decline in (1) fish species composition and (2) fish species diversity.

5.2 METHODS

The changes in the types of fish taxa and in fish species diversity were examined by analysing commercial catch records and experimental fishing data collected from: (1) Lake Kyoga, where Nile perch were introduced in 1955 and were well established by 1965; (2) Lake Nabugabo, where Nile perch were introduced in 1960 and are well established; (3) Lake Victoria, where Nile perch were introduced in the early 1960s and became well established between 1977 and 1983; (4) Lake Albert, the native habitat of Nile perch stocks introduced into the other three lakes (Fig. 5.1). Analyses were based on: (1) commercial catch records collected by the Fisheries Departments of Uganda, Kenya and Tanzania just before and after the rapid increases in Nile perch stocks; these included commercial catch records of

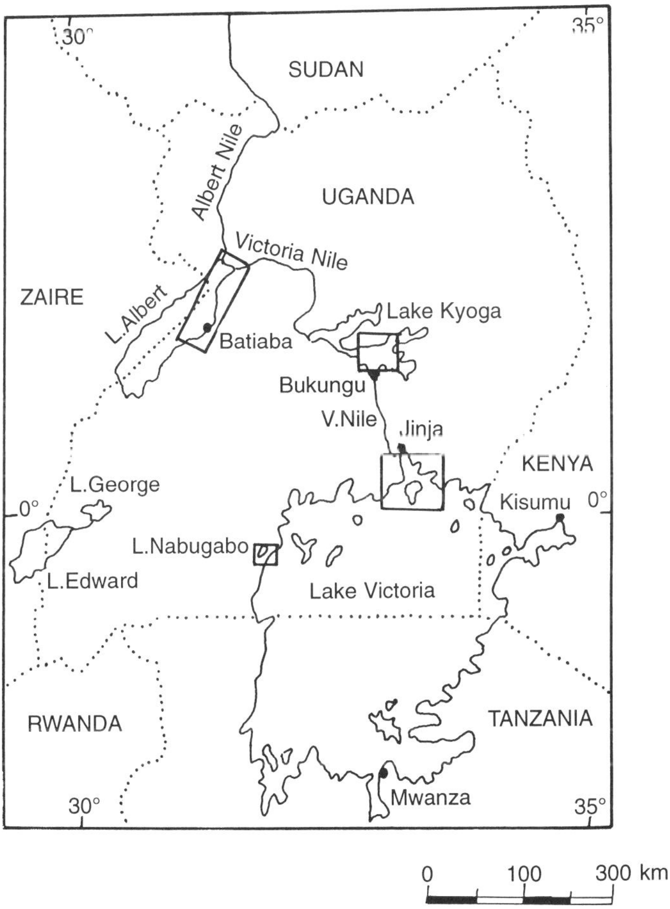

Fig. 5.1 Map of Uganda showing the location of Lakes Victoria, Nabugabo, Kyoga and Albert. The main sampling zone on each lake is indicated by a rectangle.

Lake Kyoga from 1963 to 1989 and of the Kenyan, Ugandan and Tanzanian regions of Lake Victoria and Lake Albert between 1970 and 1990; (2) the composition of experimental trawl catches collected from Lake Victoria between 1970 and 1992; (3) the composition of experimental catches collected from Lakes Victoria, Kyoga, Nabugabo and Albert between 1988 and 1993.

Experimental fishing was conducted using gill nets, beach seines and trawl nets. Two seine nets were used. A fine-mesh seine net of 10 mm stretched mesh and 20 m in length was used to sample small fish such as *R. argentea* and juveniles of larger species. This was the least selective gear and was capable of catching the smallest taxa. The second seine net of 100 m length with wings of 102 mm mesh and a bag of 51 mm mesh was used to catch larger fish. Trawling was performed only on Lake Victoria, using a cod end mesh size of 19 mm. The data from trawling were compared with earlier surveys of Kudhongania and Cordone (1974) and Okaronon *et al.* (1985). The gill nets used were of 25.4 mm to 305 mm stretched mesh. On each lake, an effort was made to sample different habitat types in respect of location (inshore or offshore), water depth, substrate type, and presence or absence of aquatic macrophytes and other refugia.

Because experimental fishing on Lake Albert was difficult, commercial fishermen's catches were sampled. These were representative because the size ranges of gill nets and seine nets used by commercial fishermen were, with the exception of the fine 10 mm seine net, similar in mesh size ranges to those used in experimental fishing on Lakes Victoria, Kyoga and Nabugabo. On Lake Albert, fishermen's catches were sampled to represent the gill nets and the 51 mm seine net catches and only the 10 mm mesh seine net was operated.

The total number and weight of each fish taxon captured was recorded to give fish species composition. For Lake Albert, the whole catch from a single boat was treated as above.

Changes in fish species diversity were examined using the Shannon–Weaver index of diversity H' (Pielou, 1969). The diversity index H' was estimated using the formula:

$$H' = -\sum \{p_i \ln p_i\} \qquad (5.1)$$

where p_i is the proportion of each taxon i in the sample.

5.3 RESULTS

Changes in commercial catch composition in Lake Victoria

The changes in yield of different taxa in the Kenyan, Ugandan and Tanzanian regions of Lake Victoria between 1970 and 1990 are illustrated in Figs 5.2,

5.3 and 5.4. (These data are used on a whole-lake basis in Chapter 7 – *Eds.*) Before the stocks of the introduced species started to increase rapidly, twelve to fourteen taxa occurred regularly among commercial catches. Two tilapiines (*O. esculentus* and *O. variabilis*), *Bagrus docmac, Protopterus aethiopicus, Clarias gariepinus* and haplochromine cichlids were the major commercial species. Others included *Labeo victorianus, R. argentea, Schilbe intermedius*, mormyrids, *Synodontis* spp., *Barbus* spp., and *Brycinus* spp.

In the Kenyan region (Fig. 5.2), the total quantity of fish landed had remained between 14 918 t (tonnes) and 18 677 t between 1970 and 1976. As the introduced species became established, total yield increased rapidly from 19 332 t in 1977 to 165 802 t in 1990. This was due to an increase in the yield of the two introduced species: that of *L. niloticus* from 94 t in 1976 to 71 414 t in 1990 and that of *O. niloticus* from 421 t in 1976 to 38 305 t in 1990. The only native species the yield of which persistently increased during this period was *R. argentea*. The yield of all other taxa declined and some like *Brycinus* spp., *Synodontis* spp., *L. victorianus* and haplochromines disappeared from the catch. There are, however, indications (as confirmed by experimental gear catch rates below) (Fig. 5.2) that by 1989, haplochromine stocks in the Kenyan region of Lake Victoria had started to recover.

In the Ugandan region (Fig. 5.3), total yield had decreased from 41 670 t in 1970 to 9999 t in 1980. As the stocks of the introduced species increased, total yield increased from 17 000 t in 1981 to 132 382 t by 1989. This was again due to increases in the contribution of the Nile perch from 13 980 t in 1983 to 101 257 t in 1989. The yield of tilapiines, which had decreased to very low levels in the early 1980s, increased from 382 t in 1983 to 20 218 t in 1989 following establishment of *O. niloticus*. The yield of *R. argentea* also increased from less than 100 t before 1984 to 7052 t by 1989. The landings of all the other taxa declined and some like *Synodontis* spp., *L. victorianus* and haplochromines were virtually absent from commercial catches by 1983.

Detailed commercial catch records for the Tanzanian region of Lake Victoria were available only up to 1987. Total fish yield increased only slightly between 1970 and 1982 (Fig. 5.4). When the stocks of the introduced species, especially Nile perch, increased from 1983 onwards, the total quantity of fish landed increased rapidly from 72 586 t in 1983 to 216 403 t in 1986. This was again due to increases in landings of Nile perch from 274 t in 1981 to 123 879 t in 1986. Although catch statistics for tilapiines were not separated into their composite species, the Nile tilapia was, as in the Kenyan and the Ugandan regions, the source of the increased landings of tilapiines from 6359 t in 1970 to 17 630 t in 1986 (Bwathondi, 1988). The only native species the yield of which increased was again *R. argentea*, which increased from 252 t in 1976 to 9825 t in 1986. Unlike in the Kenyan and Ugandan regions, only the yield of haplochromines had started to decline by 1987 and

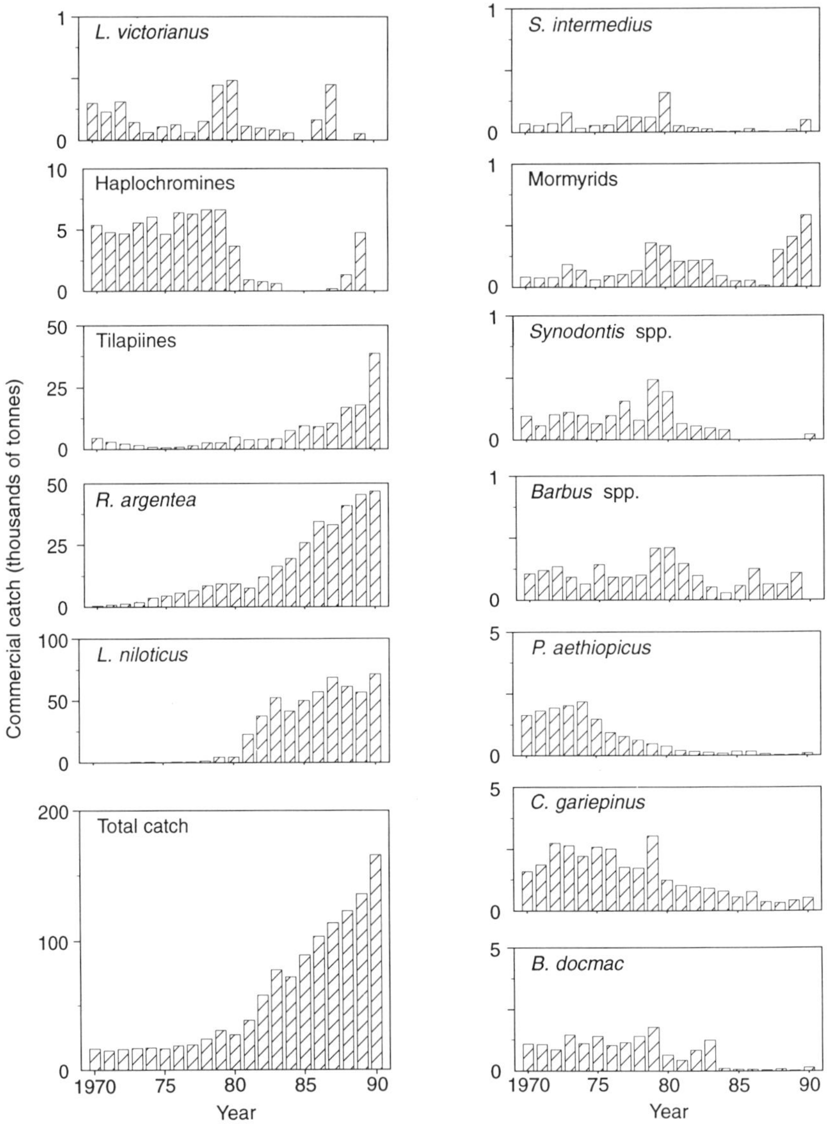

Fig. 5.2 Changes in commercial fishery yield from the Kenyan region of Lake Victoria between 1970 and 1990. Note differing scales.

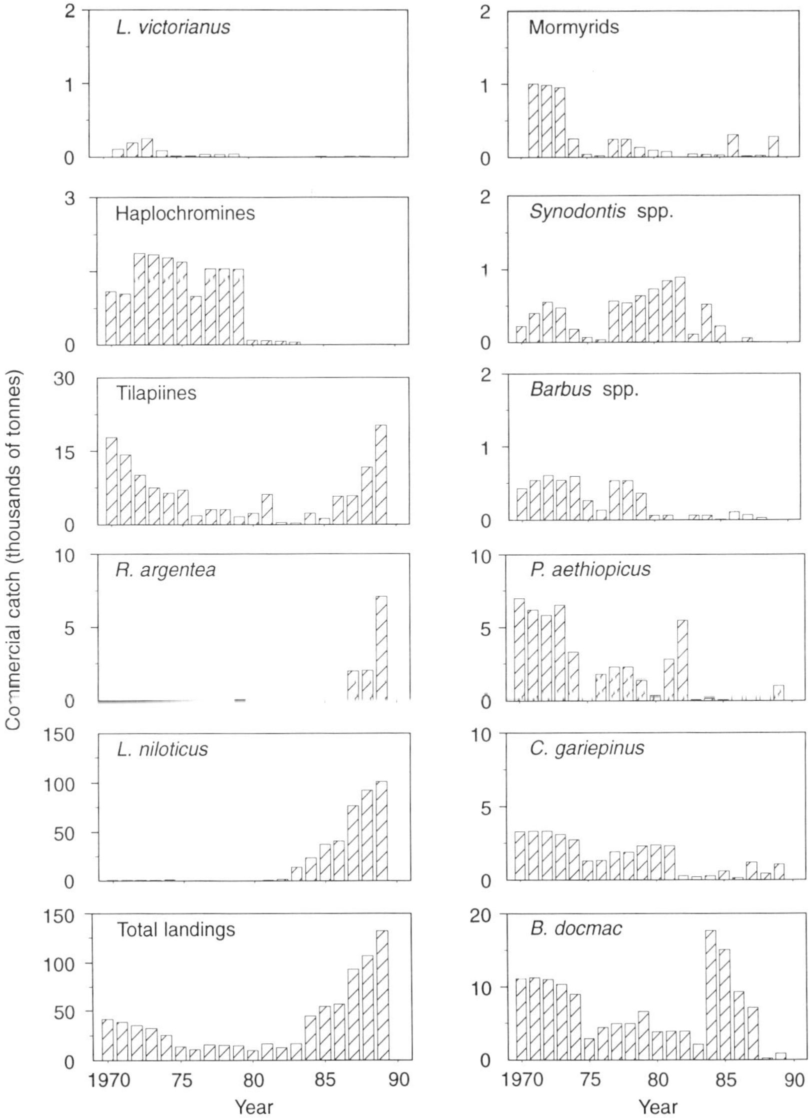

Fig. 5.3 Changes in commercial fishery yield from the Ugandan region of Lake Victoria between 1970 and 1990. Note differing scales.

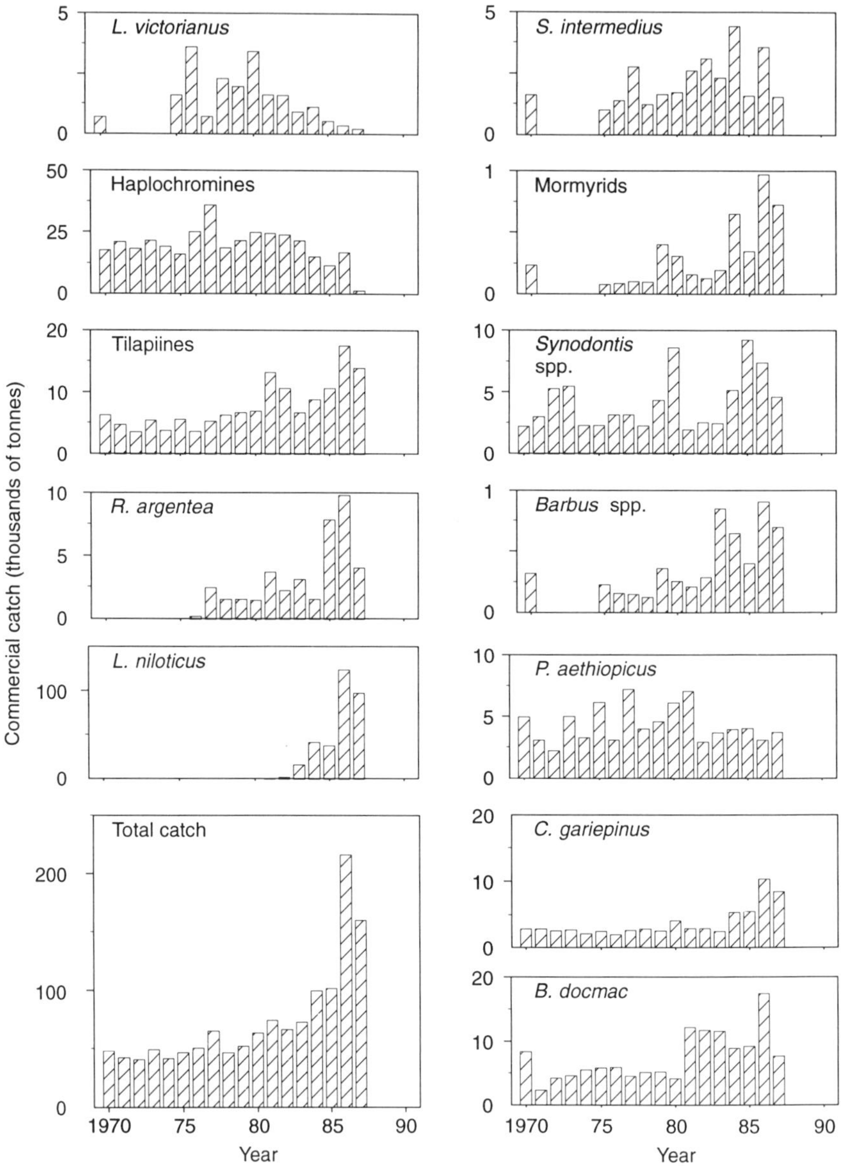

Fig. 5.4 Changes in commercial fishery yield from the Tanzanian region of Lake Victoria between 1970 and 1990. Data for 1988 and 1990 were not available. Note differing scales.

that of other species had not changed much. However, populations of the other species in the Tanzanian region of Lake Victoria later declined, as had been the case in the Ugandan and Kenyan regions (Witte *et al.*, 1992a,b).

Changes in trawl catch composition in Lake Victoria

The changes in composition and relative abundance of fish species in Lake Victoria were also manifested in experimental trawl catch composition (Fig. 5.5). Haplochromines constituted up to 80% of the demersal fish stocks in Lake Victoria around 1970 (Kudhongania and Cordone, 1974) and remained the most abundant taxa in the northern part of the lake near Jinja up to 1982 (Okaronon *et al.*, 1985). After 1982, the proportion of haplochromines declined from 76.2% in 1983 to 6.9% in 1985. During the same period, the proportion of Nile perch increased from 16.8% in 1983 to 90.4% in 1985 and that of the Nile tilapia from 0.4% to 23.9% in 1988 with only 2% being made up by the other species. By 1985, haplochromines were virtually absent from trawl catches. Other species, namely *P. aethiopicus*, *B. docmac*, *T. zillii*, *O. leucostictus* and *Synodontis afrofischeri*, were caught in very small quantities. *R. argentea*, although abundant, was not recorded because it is too small to be retained by the mesh size of the cod end used.

Catch composition of seine nets and gill nets in Lake Victoria

Data obtained from experimental beach seines and gill nets in Lake Victoria provide additional evidence of similarities in changes in species composition between commercial and experimental catches in the lake (Table 5.1). The catch from the 10 mm seine net, which was the only fishing gear that could retain *R. argentea*, was dominated by *R. argentea*, Nile tilapia and Nile perch. Haplochromines, which were originally very abundant, formed only 5.8% of the biomass of fish caught in the 10 mm seine net. Nile perch contributed 97% of both the gill net catch and that in the 51 mm mesh seine net. Nile tilapia was next in abundance. Other species, which comprised haplochromines, mormyrids, *O. leucostictus* and *T. zillii*, were caught in very small quantities.

Changes in commercial catch composition of Lake Kyoga

Fish species composition of Lake Kyoga prior to the establishment of introduced species was similar to that of Lake Victoria (Worthington, 1929). *O. esculentus*, *O. variabilis*, *P. aethiopicus*, *B. docmac*, *C. gariepinus*, *S. intermedius* and *Barbus* spp. were the major commercial species. Haplochromines were also abundant. These remained the dominant species up to the time of the introductions (Ogutu-Ohwayo, 1985).

Nile perch and the non-native tilapiine species were introduced earlier and

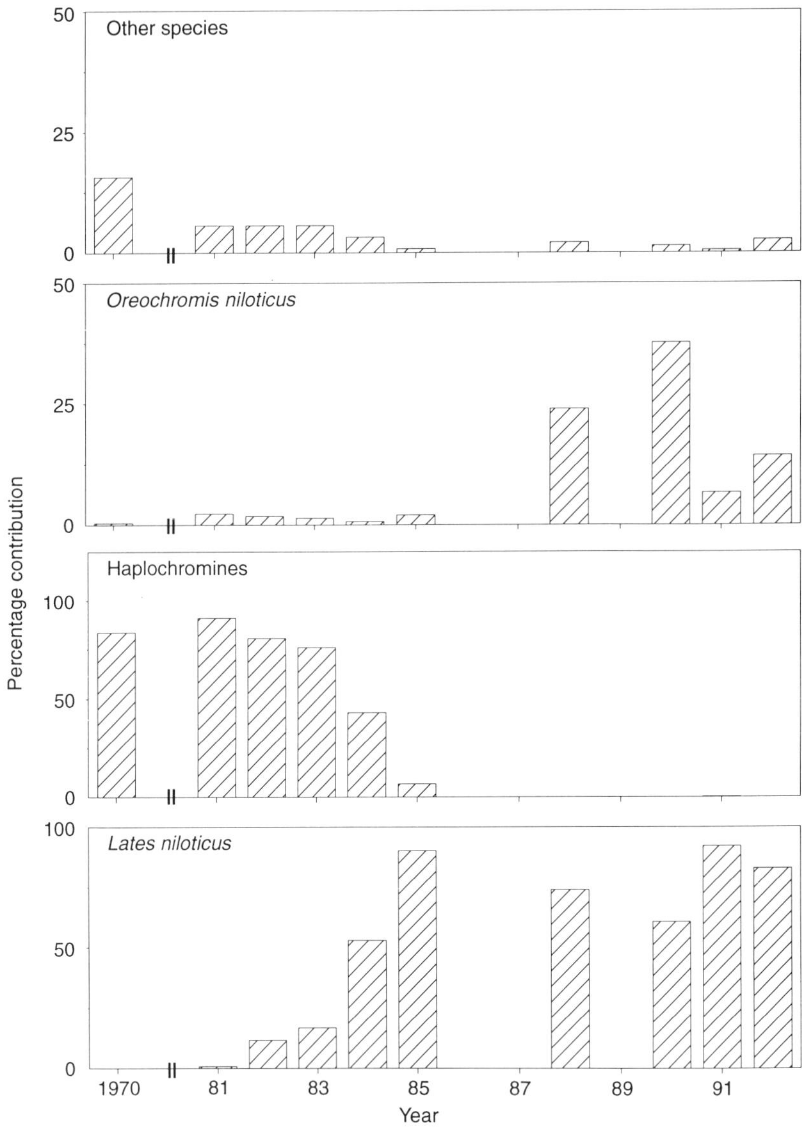

Fig. 5.5 Changes in trawl catch composition in the northern (Ugandan) region of Lake Victoria between 1970 and 1992. Note differing scales.

Table 5.1 Perentage contribution (by weight) of different fish taxa caught in experimental sampling in Lake Victoria from 1988 to 1993

Fish taxa	10 mm Seine	51 mm Seine	Gill nets	Trawl 19 mm
Lates niloticus	13.07	97.89	97.17	80.01
Oreochromis niloticus	22.76	1.93	1.01	14.43
Rastrineobola argentea	39.10			
Haplochromines	5.77	0.06	0.03	0.09
Mormyrids			1.59	
Bagrus docmac				0.74
Clarias gariepinus				0.05
Oreochromis leucostictus			0.01	0.00
Protopterus aethiopicus				0.92
Synodontis afrofischeri				0.09
Tilapia zillii		0.13	0.02	0.12
Unidentified tilapiines	14.25			3.47
Total weight sampled (kg)	213	4847	1836	5580

spread faster in Lake Kyoga than in Lake Victoria (Gee, 1964). Unfortunately, unlike those for Lake Victoria, the catch statistics of Lake Kyoga were not split into their constituent species. Stocks of the introduced species in Lake Kyoga increased rapidly from 1963 onwards (Fig. 5.6). This resulted in increases in total catch from 18 261 t in 1964 to a peak of 167 200 t in 1978. These increases were due to the contribution of Nile perch, which increased from 657 t in 1964 to 71 000 t in 1978, and that of the introduced tilapiines, especially the Nile tilapia which increased from 589 t in 1964 to 80 960 t in 1985. The yield of the native tilapiines, which had dominated the fishery since its development, and that of all the other native species declined. However, total yield declined from 165 000 t recorded at peak production in 1977 to 54 706 t by 1989. During this period, the yield of the Nile perch decreased from 71 000 t in 1977 to 15 000 t in 1989.

Experimental catch composition from Lake Kyoga

As in Lake Victoria, the 10 mm seine net catches were dominated by Nile perch, Nile tilapia and *R. argentea* (Table 5.2). Nile perch and Nile tilapia were again the dominant species in the gill net and the 51 mm seine. However, the proportion of Nile tilapia was higher in Lake Kyoga than in Lake Victoria whereas that of the Nile perch was higher in Lake Victoria than in Lake Kyoga. Haplochromines were more commonly encountered in Lake Kyoga than in Lake Victoria. Other taxa which included *T. zillii*, *O. leucostictus*, *S. mystus*,

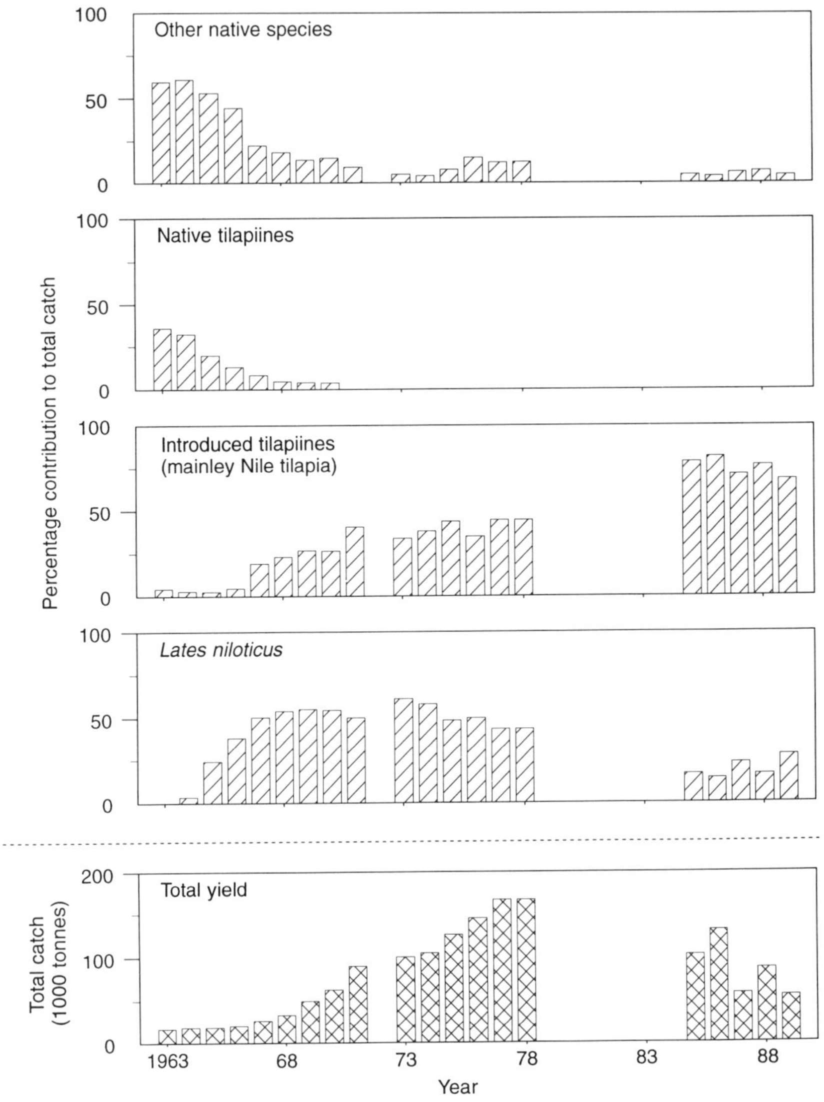

Fig. 5.6 Changes in the contribution of different fish taxa to total yield since establishment of introduced species in Lake Kyoga. Data for 1979–1984 were not available.

Table 5.2 Percentage contribution (by weight) of different fish taxa caught in experimental sampling in Lake Kyoga between 1988 and 1993

Fish taxa	10 mm Seine		51 mm Seine		Gill nets	
	1988/90	1991/93	1988/90	1991/93	1988/90	1991/93
Lates niloticus	59.23	61.60	25.75	74.48	65.55	79.50
Oreochromis niloticus	34.63	13.03	68.25	23.31	32.54	18.85
Rastrineobola argentea	5.17	13.17				
Haplochromines	0.98	0.78	0.00	0.04	0.03	0.26
Tilapia zillii		5.55	5.29	1.57		0.02
Oreochromis leucostictus		0.19	0.38	0.37		0.01
Barbus spp.		0.04	0.01	0.01	0.39	0.92
Schilbe intermedius			0.02		1.31	0.02
Clarias gariepinus					0.18	0.41
Mormyrids						0.02
Protopterus aethiopicus			0.29	0.13		
Synodontis afrofischeri			0.00	0.04		0.00
Afromastacembelus frenatus			0.02	0.06		0.00
Labeo victorianus		0.12				0.00
Unidentifiable tilapiines		5.52				
Total weight sampled (kg)	38.8	51.5	2737	903	1855	5027

Table 5.3 Catch rate (fish per night) of haplochromines in 25.4 mm and 38.1 mm stretched mesh gill nets of 90 m length. Each boat in the fishing fleet had one net of each mesh

	1988/90	1991	1992	1993
Number of fishing nights	40	91	93	73
Maximum no. of fish in a night	30	48	348	213
Mean fish per night	1.58	2.14	13.52	21.27
SE	0.83	0.85	4.19	4.97

P. aethiopicus, *C. gariepinus*, *Barbus* spp., *S. afrofischeri*, *L. victorianus*, mormyrids and *Afromastacembelus frenatus*, were caught in very small quantities. The two originally most abundant tilapiines and the catfish *B. docmac* were not caught at all.

From 1991, haplochromine stocks in Lake Kyoga started to increase as shown by the experimental catch rates (Table 5.3) and as prey of Nile perch (Ogutu-Ohwayo, pers. obs.). However, these haplochromines consisted of only one species, *Astatotilapia latifasciata* (Wandera and Namulemo, pers.

comm.). This increase may not, therefore, represent an overall recovery in the haplochromine species flock but rather an improvement in stocks of one species that may have the capacity to persist in the presence of the Nile perch.

Experimental catches from Lake Nabugabo

There were no commercial catch statistics from Lake Nabugabo. When the Cambridge expedition surveyed the lake in 1962, the most important fish taxa among commercial catches were similar to those of Lakes Victoria and Kyoga: *P. aethiopicus, B. docmac, S. intermedius, C. gariepinus, O. esculentus, O. variabilis* and haplochromines. The expedition also recorded thirteen non-cichlids and eight haplochromine species from experimental samples (Cambridge Nabugabo Biological Expedition, 1962; Ogutu-Ohwayo, 1993).

The relative importance of the fish taxa recorded from Lake Nabugabo between 1991 and 1992 is shown in Table 5.4. The 10 mm seine net was dominated by Nile perch, Nile tilapia, *R. argentea* and *T. rendalli*. Nile perch and Nile tilapia were again the most abundant species in the gill nets and the 51 mm seine net. However, *S. intermedius* was caught in large numbers in gill nets in the lake and contributed 6.96% to the gill net catches. Also *T. rendalli, B. jacksonii* and *T. zillii* made significant contribution to gill net catches. As in Lakes Victoria and Kyoga, other taxa, which included *C. gariepinus*, mormyrids, *O. leucostictus* and *S. afrofischeri*, were caught only in very small quantities.

Table 5.4 Percentage contribution (by weight) of different fish taxa caught in experimental sampling in Lake Nabugabo between 1991 and 1992

Fish taxa	10 mm Seine	51 mm Seine	Gill nets
Lates niloticus	78.72	86.92	67.73
Oreochromis niloticus	8.92	8.28	15.17
Rastrineobola argentea	5.10		
Haplochromines	0.28	0.59	1.66
Schilbe intermedius		0.77	6.96
Mormyrids		0.01	0.05
Clarias gariepinus			0.25
Oreochromis leucostictus			0.19
Synodontis afrofischeri	0.00	0.01	0.08
Brycinus jacksonii	0.07	0.14	2.77
Tilapia zillii	0.00	1.31	2.23
Tilapia rendalli	6.91	1.97	2.92
Barbus spp.	0.00		
Total weight sampled (kg)	196	280	317

Changes in commercial catch composition of Lake Albert

The original state of the fishery of Lake Albert, the original and native habitat of Nile perch, Nile tilapia, *T. zillii* and *O. leucostictus* which were introduced to Lakes Victoria, Kyoga and Nabugabo, is given by Worthington (1929). The most important commercial taxa, in order of importance, were *Citharinus citharus, Lates* spp., Nile tilapia, *Bagrus* spp., *Alestes baremose, Hydrocynus forskalii, Synodontis shall* and *Mormyrus caschive.*

The contribution of different taxa to total yield in Lake Albert for the period 1971–1989 is illustrated in Fig. 5.7. The fish species composition of the lake has not changed much from that recorded by Worthington (1929). Fourteen taxa still occur regularly among commercial catches. *Lates* spp., tilapiines (mainly Nile tilapia), and two pelagic characid taxa, *Hydrocynus* spp. and *Alestes* spp., were still among the most important commercial species. The only originally important species which was no longer abundant among commercial catches was *C. citharus.* This species was depleted due to overfishing during the first part of the 20th century (Cadwalladr and Stoneman, 1966).

Experimental sampling carried out on Lake Albert in 1990–1992 (Table 5.5) showed that the commercial catch records gave an accurate picture of what was present in the lake. *Lates* spp., *Hydrocynus* spp., *Alestes* spp. and tilapiines were the most abundant taxa in the samples examined. Even *C. citharus*, which had been feared extinct from the lake, were encountered in very small numbers. Unlike Lakes Victoria, Kyoga and Nabugabo, where many fish taxa have disappeared, no fish taxa have vanished from Lake Albert since the first fishery survey of the lake.

Changes in fish species diversity

The changes in diversity of fish taxa in the Kenyan, Ugandan and Tanzanian regions of Lake Victoria between 1970 and 1990 are compared with that in Lake Albert in Fig. 5.8. In the Kenyan and Ugandan regions, there were two distinct levels of fish species diversity corresponding to the periods before and after establishment of introduced species. In the Kenyan and Ugandan regions, there was no significant change in H' (at 95% confidence level) before the rapid decline, while in the Tanzanian region, H' increased significantly during this period. After the rapid decline in H' in the Kenyan and Ugandan regions, fish species diversity did not again change significantly. Also, fish species diversity in Lake Albert increased significantly during the entire period under review. There seem to be no further reduction in fish species diversity in the Kenyan and Ugandan regions of Lake Victoria, indicating that it has stabilized. However, fish species diversity in the Kenyan and Ugandan regions after the rapid fall was significantly lower than that of Lake Albert for the same period.

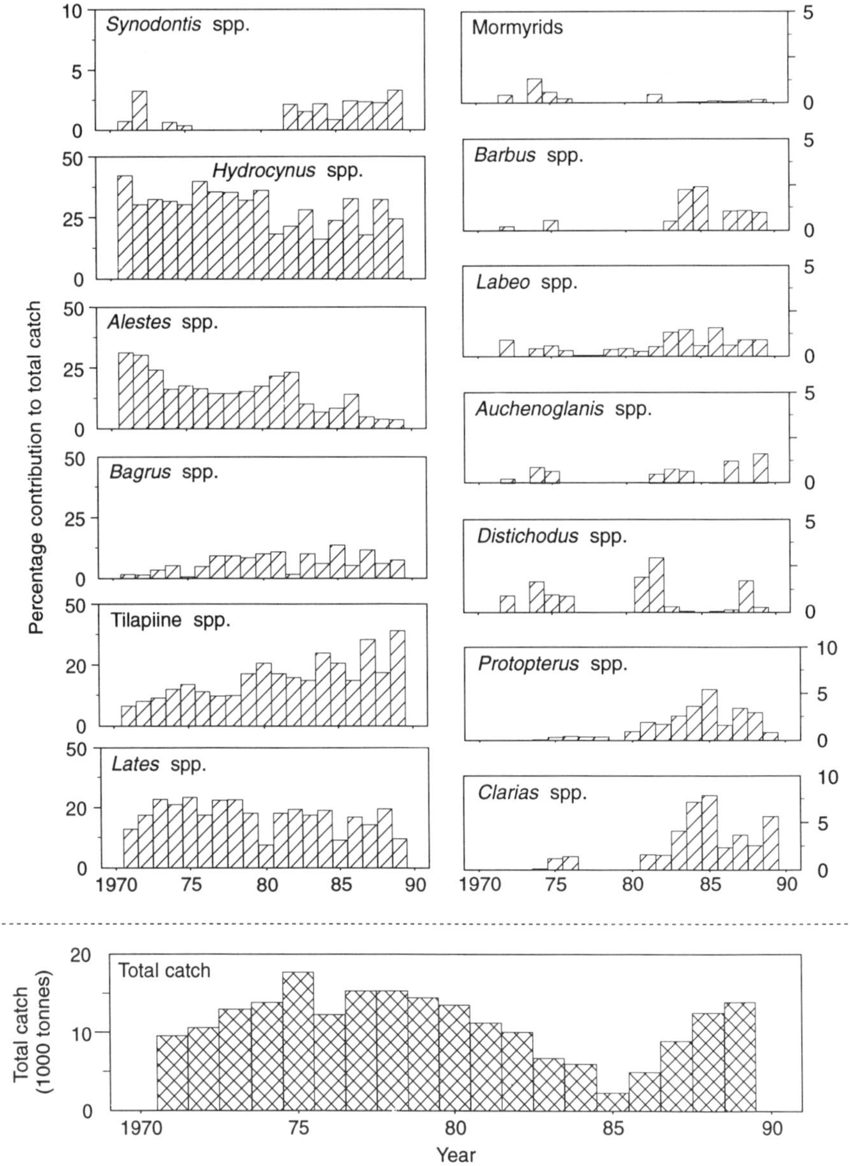

Fig. 5.7 Contribution of different fish taxa to total yield in Lake Albert from 1971 to 1989. Note differing scales.

Table 5.5 Percentage contribution (by weight) of different fish taxa to experimental samples examined in Lake Albert during the period 1989–1992

Fish taxa	10 mm Seine	51 mm Seine	Gill nets
Alestes baremose		1.96	25.27
Alestes dentex			0.65
Brycinus nurse	21.03	1.76	0.01
Lates niloticus	24.66	22.65	21.96
Lates macrophthalmus		8.42	1.03
Barbus bynni		2.34	14.95
Hydrocynus spp.	15.08	0.78	14.66
Distichodus niloticus		0.18	8.63
Oreochromis niloticus		32.54	4.76
Oreochromis leucostictus		0.31	0.03
Tilapia zillii	14.90	0.56	0.08
Mormyrops spp.		1.35	2.05
Mormyrus niloticus			0.36
Mormyrus kannume		1.66	0.34
Mormyrus victorie			0.03
Gnathonemus spp.		0.09	0.19
Clarias gariepinus	0.13	2.68	1.31
Bagrus bayad	3.05	14.55	1.30
Bagrus docmac		0.36	0.39
Synodontis spp.		0.26	0.61
Labeo horie		0.75	0.45
Labeo coubie		0.03	0.20
Auchenoglanis occidentalis	1.91	5.75	0.44
Malapterurus electricus		0.90	0.14
Citharinus citharus			0.11
Haplochromines	4.63	0.13	0.03
Schilbe mystus		0.01	0.02
Neobola bredoi	0.12		
Barilus niloticus	14.50		
Total weight sampled (kg)	69	1145	4830

This suggests that fish species diversity in Lake Victoria has stabilized at a lower level than that of the original habitats of the introduced species.

There were no detailed catch statistics for Lakes Kyoga and Nabugabo to facilitate similar comparisons. However, comparisons based on gill net samples (Tables 5.1, 5.2, 5.4 and 5.5) support the conclusion that fish species diversity in Lakes Victoria, Kyoga and Nabugabo after establishment of Nile perch was lower than that of Lake Albert ($H' = 0.14$ for Lake Victoria, 0.58 for Lake Kyoga, 1.13 for Lake Nabugabo and 2.07 for Lake Albert).

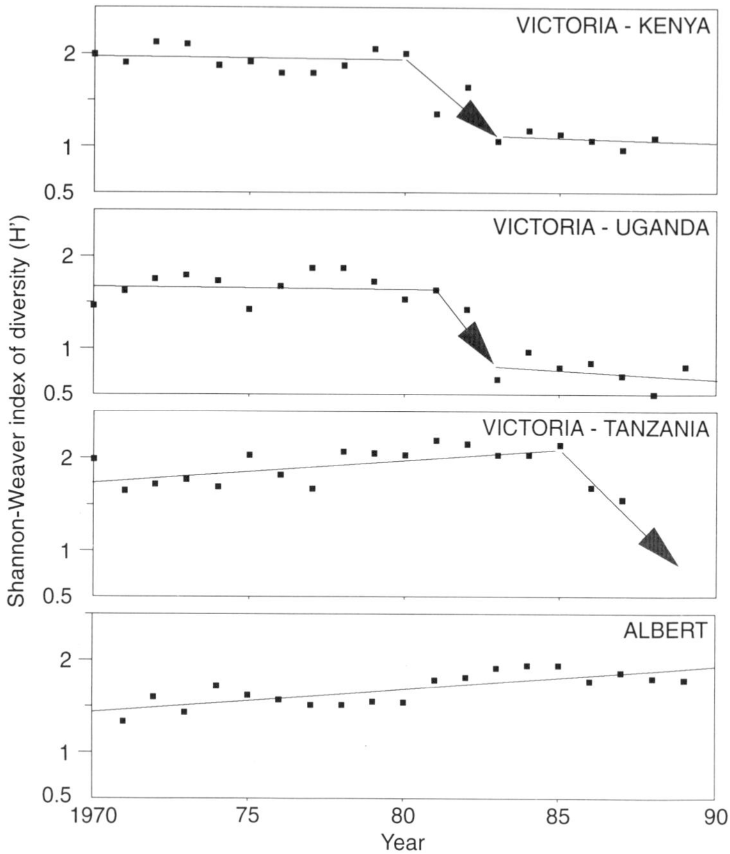

Fig. 5.8 Changes in diversity (H') of fish in Lakes Victoria and Albert between 1970 and 1990. Arrowheads show the periods of rapid fall in diversity. Line segments are fitted by least-squares regression.

5.4 DISCUSSION

The composition of fish taxa in Lakes Victoria, Kyoga and Nabugabo has changed from a fishery comprising about 14 taxa to one dominated by three species, *L. niloticus, O. niloticus* and *R. argentea*. The fish species diversity is similar in Lakes Victoria, Kyoga and Nabugabo, but lower than that of Lake Albert, the original habitat of Nile perch and the tilapiines. However, fish species composition and fish species diversity in Lakes Victoria and Kyoga is considered to be stable. *L. niloticus, O. niloticus* and *R. argentea* have remained the dominant species in Lake Victoria for over 20 years and for probably a longer period in Lakes Kyoga and Nabugabo. There are also similarities between the current fisheries of Lakes Victoria, Kyoga and Nabugabo and those of Lake Albert. Nile perch and Nile tilapia, which are the dominant species in Lakes Victoria, Kyoga and Nabugabo, are also the dominant species in Lake Albert. The two remaining major components of the catch in Lake Albert, *Alestes* spp. and *Hydrocynus* spp., are both pelagic and can be regarded as ecologically analogous to *R. argentea* of Lakes Victoria, Kyoga and Nabugabo, at least in respect to resilience and susceptibility to predation by the Nile perch.

Total fish yield of Lakes Victoria, Kyoga and probably Nabugabo increased following establishment of introduced species. It is possible that these fisheries will not sustain the peak yields recorded in these lakes (Chapter 9). Total yield from Lake Kyoga, where Nile perch was established earlier than in Lake Victoria, has declined (Fig. 5.6). Although this decline was attributed to intensive use of small-mesh beach seines (Twongo, pers. comm.) which cropped immature Nile perch and Nile tilapia, disrupted courtship of tilapiines on breeding grounds and destroyed nests, the declines were more prominent among Nile perch populations and could have been partly due to the reduction in haplochromines.

The large quantities of fish landed in Lakes Victoria and Kyoga soon after the introduced species became established were initially supported by large quantities of vulnerable haplochromine cichlids, which were then abundant in the lakes (Gee, 1964, 1969; Hamblyn, 1966; Okedi, 1971). These haplochromines, which formed the main prey of Nile perch, became depleted and are no longer abundant. The depletion of haplochromine prey could have contributed to the decline in the Nile perch yield recorded in Lake Kyoga. This notion is supported by the fact that the proportion of Nile perch in experimental catches has increased since the stocks of haplochromines in the lake started to increase (Tables 5.2, 5.3).

The factors that have facilitated improvement of haplochromine stocks in Lake Kyoga would be useful in enhancement of some of the fish species that have been depleted from Lakes Victoria, Kyoga and Nabugabo due to the presence of Nile perch. The increase in haplochromine stocks in Lake Kyoga has coincided with the invasion and spread of the water hyacinth, *Eichhornia crassipes*, which invaded Lake Kyoga in 1988 and became well established between 1991 and 1992 (Twongo, 1991). Its floating beds might be providing cover and allowing the haplochromines, the remnants of which had taken cover in marginal macrophytes, to recolonize the lake and to multiply. Although the hyacinth is a dangerous weed, Goeltenboth and Kristyanto (1987) have noted that it can provide a valuable habitat as long as it is kept under control. Willoughby *et al.* (1993) have observed that the hyacinth can support a large and diverse community. There are plans to control the spread of the water hyacinth in Lakes Victoria and Kyoga. If the above hypothesis is correct, this may re-expose the haplochromines in Lake Kyoga to predation which might again result in a reduction in haplochromine stocks. Also, even if the hyacinth has invaded Lake Victoria, its distribution in the lake may not become as extensive as in Lake Kyoga, because Lake Victoria is deeper and more open. The water hyacinth thrives in shallow sheltered areas. The haplochromine cichlids may, therefore, never recover to the same extent in Lake Victoria as in Lake Kyoga.

Even if haplochromine stocks in Lakes Victoria, Kyoga and Nabugabo improve, it appears that these will consist of a few specific types that can persist in the presence of the Nile perch. As described above, only one species is so far recovering in Lake Kyoga. Generally the lakes to which Nile perch is native have very few haplochromine species; Lake Albert has four and Lake Turkana only three (Greenwood, 1974). The only exception is Lake Tanganyika, which has haplochromine species-flocks, but these occur in the presence of different species of *Lates* (Chapter 23). Also the Lake Tanganyika haplochromines are confined to rocky inshore areas where they can take shelter from *Lates* spp. (Fryer and Iles, 1972). In Lake Victoria, it is also the rock-dwelling haplochromines that have been least depleted (Ogutu-Ohwayo, 1990c; Witte *et al.*, 1992a,b). Therefore, protection of refugia such as rocky outcrops and marginal vegetation will be beneficial to the preservation and recovery of haplochromines and other species in Lakes Victoria, Kyoga and Nabugabo.

Apart from the introduced species, especially Nile perch, other factors such as overexploitation and environmental degradation have affected, and are likely to continue to affect, stability and sustainability of fish stocks in Lakes Victoria, Kyoga and Nabugabo. Overexploitation and the use of destructive fishing practices contributed greatly to depletion of fish stocks in Lakes Victoria and Kyoga prior to establishment of the Nile perch and the tilapiines (Jackson, 1971; Ogutu-Ohwayo, 1990a). Whereas *Brycinus nurse* and the prawn, *Caridina nilotica*, which are major prey of Nile perch in Lake Albert, are not commercially exploited in that lake, *R. argentea* is heavily exploited in Lake Victoria and even *Caridina* is being exploited now (Ogari and Asila, 1992). Juvenile Nile perch, which are major prey of subadult and adult Nile perch, are also heavily exploited and this almost led to a total collapse of the Nile perch fishery in Lake Kyoga (Twongo, pers. comm.). There is already a decline in the size of *R. argentea* in Lake Victoria due to the combined effect of predation and human exploitation (Wandera, 1992). Control of exploitation of *R. argentea*, juvenile Nile perch and *C. nilotica* will contribute greatly to the stability of fish stocks in Lakes Victoria, Kyoga and Nabugabo.

The decrease in fish species diversity in Lake Victoria has been concurrent with major environmental changes (Chapter 7). Algal biomass in Lake Victoria has increased two- to threefold (Hecky, 1993; Mugidde, 1993; Chapter 3). Accumulation of excess dead organic matter has increased oxygen demand and led to development of a deep anoxic layer the volume of which has been increasing. Spread of the low-oxygen water to the surface has often resulted in mass fish kills (Ochumba and Kibaara, 1989). The increases in algal production are likely to be stimulated by increased nutrient inputs into the lake from the catchment area and from the atmosphere. Control of activities that increase such inputs will also benefit the stability of fish stocks.

5.5 SUMMARY

Lakes Victoria, Kyoga and Nabugabo had a similar native fish fauna of high species diversity. Since the introduction and establishment of Nile perch and several tilapiine species, total yield increased but fish species diversity declined drastically. Before the increase in the stocks of the introduced species, fish species diversity in Lake Victoria was higher than that of Lake Albert, the original habitat of the introduced species. After establishment of the introduced species, fish species diversity in Lake Victoria decreased to a significantly lower level than that of Lake Albert and similar declines occurred in Lakes Kyoga and Nabugabo. Lakes Victoria, Kyoga and Nabugabo are now dominated by two introduced species, *Lates niloticus* and *Oreochromis niloticus*, and one native species, *Rastrineobola argentea*. Fish species composition of these lakes has not decreased further since the three species became dominant, suggesting that it has stabilized although total yield has started declining in some of the lakes. Haplochromines are also recovering in some of the lakes, especially in areas with cover which may provide refuge from predation. However, the three dominant species are the main commercial species as well as the main prey of Nile perch. Also, the environment of these lakes, especially Lake Victoria, is becoming eutrophic which further threatens aquatic life. Therefore, the current fish species diversity can only be sustained through management measures that will control fishing and predation pressure, and degradation of the environment.

ACKNOWLEDGEMENTS

I am grateful to S.B. Wandera for useful comments and criticisms on the manuscript and to R. Amina, L. Kibirige and G. Malyonga for assisting with collection of the data. I am also grateful to the Renewable Resources Assessment Group at Imperial College for inviting me and meeting my travel expenses to the workshop. This work was done with the aid of a grant from the International Development Research Centre (IDRC), Ottawa, Canada.

REFERENCES

Anderson, A.M. (1961) Further observations concerning the proposed introduction of Nile perch into Lake Victoria. *E. Afr. Agric. J.*, **26**, 195–201.

Bwathondi, P.O.J. (1988) The state of Lake Victoria fisheries, Tanzania sector. *FAO Fish. Rep.*, no. 388, pp. 29–35.

Cadwalladr, D.A. and Stoneman, A.J. (1966) *A Review of the Fisheries of the Ugandan Waters of Lake Albert, East Africa, 1928–1965/66, with Data Mainly from 1953* (East African Freshwater Fisheries Research Organization Supplementary Publication), Government Printers, Entebbe, Uganda.

Cambridge Nabugabo Biological Expedition (1962) *Preliminary Report*, Makerere University, Kampala, 21 pp. (Mimeo)

Fryer, G. (1960) Concerning the proposed introduction of Nile perch into Lake Victoria. *E. Afr. Agric. J.*, **25**, 267–70.

Fryer, G. and Iles, T.D. (1972) *The Cichlid Fishes of the Great Lakes of Africa: Their Biology and Evolution*. Oliver and Boyd, Edinburgh, 641 pp.

Gee, J.M. (1964) Nile perch investigations. *Ann. Rep. E. Afr. Freshwat. Res. Org.* (1962/63): 14–24.

Gee, J.M. (1969) A comparison of certain aspects of the biology of *Lates niloticus* (Linne) in some East African lakes. *Revue Zool. Bot. afr.*, **80**, 244–62.

Goeltenboth, F. and Kristyanto, A. (1987) Ecological studies on the Rawa Pening Lake (Central Java): some biotic parameters of the semi-natural lake. *Trop. Ecol.*, **28**, 101–16.

Graham, M. (1929) *The Victoria Nyanza and its Fisheries. A Report on the Survey of Lake Victoria 1927–1928 and Appendices*, Crown Agents for Colonies, London, 255 pp.

Greenwood, P.H. (1974) The *Haplochromis* species (Pisces: Cichlidae) of Lake Rudolf, East Africa. *Bull. Br. Mus. (Nat. Hist.), Zool.*, **27**, 141–75.

Hamblyn, E.L. (1966) The food and feeding habits of the Nile perch, *Lates niloticus* (Linne) (Pisces: Centropomidae). *Revue Zool. Bot. afr.*, **74**, 1–28.

Hecky, R.E. (1993) Peter Kilham Memorial Lecture. The eutrophication of Lake Victoria. *Int. Verein. Limnol.*, **25**, 39–48.

Jackson, P.B.N. (1971) The African Great Lakes fisheries: past, present and future. *Afr. J. trop. Hydrobiol. Fish.*, **1**, 35–49.

Kudhongania, A.W. and Cordone, A.J. (1974) Batho-spatial distribution patterns and biomass estimates of the major demersal fishes in Lake Victoria. *Afr. J. trop. Hydrobiol. Fish.*, **3**(1), 15–31.

Ligtvoet, W. and Mkumbo, O.C. (1990) Stock assessment of Nile perch in Lake Victoria. *FAO Fish, Rep.*, no. 430, pp. 35–74.

Mugidde, R. (1993) The increase in phytoplankton production and biomass in Lake Victoria (Uganda). *Int. Verein. Limnol.*, **25**, 846–9.

Ochumba, P.B.O. and Kibaara, D.I. (1989) Observations on the blue-green algal blooms in the open waters of Lake Victoria, Kenya. *Afr. J. Ecol.*, **27**, 23–34.

Ogari, J. and Asila, A. (1992) Status of fisheries of Lake Victoria Kenya Sector. *FAO Fish. Rep.*, no. 475, 15–23.

Ogari, J. and Dadzie, S. (1988) The food of the Nile perch, *Lates niloticus* (L.), after disappearance of the haplochromine cichlids in the Nyanza Gulf of Lake Victoria (Kenya). *J. Fish Biol.*, **32**, 571–7.

Ogutu-Ohwayo, R. (1985) The effects of predation by the Nile perch, *Lates niloticus* (Linne) introduced into Lake Kyoga (Uganda) in relation to the fisheries of Lake Kyoga and Lake Victoria. *FAO Fish. Rep.*, no. 335, pp. 18–41.

Ogutu-Ohwayo, R. (1990a) The decline of the native fishes of Lakes Victoria and

Kyoga (East Africa) and the impact of the introduced species, especially the Nile perch, *Lates niloticus* and the Nile tilapia, *Oreochromis niloticus*. *Env. Biol. Fishes*, **27**, 81–96.

Ogutu-Ohwayo, R. (1990b) Changes in the prey ingested and the variations in Nile perch and other fish stocks in Lake Kyoga and the northern waters of Lake Victoria (Uganda). *J. Fish Biol.*, **37**, 55–63.

Ogutu-Ohwayo, R. (1990c) The reduction in fish species diversity in Lakes Victoria and Kyoga (East Africa) following human exploitation and introduction of non-native fishes. *J. Fish Biol.*, **37** (Suppl. A), 207–8.

Ogutu-Ohwayo, R. (1993) The effects of predation by the Nile perch, *Lates niloticus* L., on the fish of Lake Nabugabo, with suggestions for conservation of endangered endemic cichlids. *Conservation Biol.*, **7**, 701–11.

Okaronon, J., Acere, T. and Ocenodongo, D. (1985) The current state of the fisheries in the northern portion of Lake Victoria (Uganda). *FAO Fish. Rep.*, no 335, pp. 89–98.

Okedi, J.Y. (1971) Further observations on the ecology of the Nile perch (*Lates niloticus* Linne) in Lake Victoria and Lake Kyoga. *Ann. Rep. E. Afr. Freshwat. Fish. Res. Org. (1970)*, 42–55.

Pielou, E.C. (1969) *An Introduction to Mathematical Ecology*, Wiley Interscience, New York.

Twongo, T. (1991) Implications of the water hyacinth infestation for fisheries, with particular reference to Lake Kyoga, in *The Water Hyacinth in Uganda* (ed. K. Thompson), FAO F1:TCP/UGA/9153/A, pp. 19–23.

Wandera, S.B. (1992) The effect of Nile perch predation and human exploitation on the population structure of *Rastrineobola argentea* in Lake Victoria, Kyoga and Nabugabo. Report of the sixth session of the Sub-Committee for the development and management of the fisheries of Lake Victoria. Jinja, Uganda, 10–13 February, 1992. *FAO Fish. Rep.*, no. 475, 48 pp.

Welcomme, R.L. (1967) Observations on the biology of the introduced *Tilapia* in Lake Victoria. *Revue Zool. Bot. afr.*, **76**, 249–79.

Willoughby, N.G., Watson, I.G., Lauer, S. and Grant, I.F. (1993) An investigation into the effects of water hyacinth on the biodiversity and abundance of fish and invertebrates in Lake Victoria, Uganda. Technical Report. Natural Resources Institute. Overseas Development Administration, UK, 26 pp.

Witte, F., Goldschmidt, T., Wanink, J., van Oijen, M.J.P., Goudswaard, K., Witte-Maas, E. and Bouton, N. (1992a) The destruction of an endemic species flock: quantitative data on the decline of haplochromine cichlids of Lake Victoria. *Env. Biol. Fishes*, **34**, 1–28.

Witte, F., Goldschmidt, T., Goudswaard, P.C., Ligtvoet, W., van Oijen, M.J.P. and Wanink, J.H. (1992b) Species extinction and concomitant ecological changes in Lake Victoria. *Neth. J. Zool.*, **42**, 214–32.

Worthington, E.B. (1929) *A Report of the Fishing Survey of Lakes Albert and Kyoga, March–July 1929*. Crown Agents for Colonies, London, 136 pp.

Chapter six

Dynamics of the haplochromine cichlid fauna and other ecological changes in the Mwanza Gulf of Lake Victoria

Frans Witte, Tijs Goldschmidt and Jan H. Wanink

6.1 INTRODUCTION

Lake Victoria, the largest tropical lake in the world (Fig. 6.1), was until recently a typical cichlid lake. Eighty percent of the demersal ichthyomass of this East African lake consisted of haplochromine cichlids (Kudhongania and Cordone, 1974a,b). The more than 300 haplochromine species, of which 99% were endemic, exploited virtually all food sources in the lake (Greenwood, 1974, 1981; van Oijen *et al.*, 1981; Witte and van Oijen, 1990). Each species had its own unique combination of food and habitat preferences (e.g. van Oijen, 1982; Goldschmidt *et al.*, 1990). The haplochromines were only locally important to fisheries (Kudhongania and Cordone, 1974a). In the 1970s a small-scale trawl fishery on these fishes was developed in the southern part of the lake near Mwanza. At that time the mean catch rate was more than $1000\,kg\,h^{-1}$ (Goudswaard and Ligtvoet, 1988; Barel *et al.*, 1991; Witte *et al.*, 1992b).

The Impact of Species Changes in African Lakes. Edited by Tony J. Pitcher and Paul J.B. Hart. Published in 1995 by Chapman & Hall, London. ISBN 0 412 55050 4.

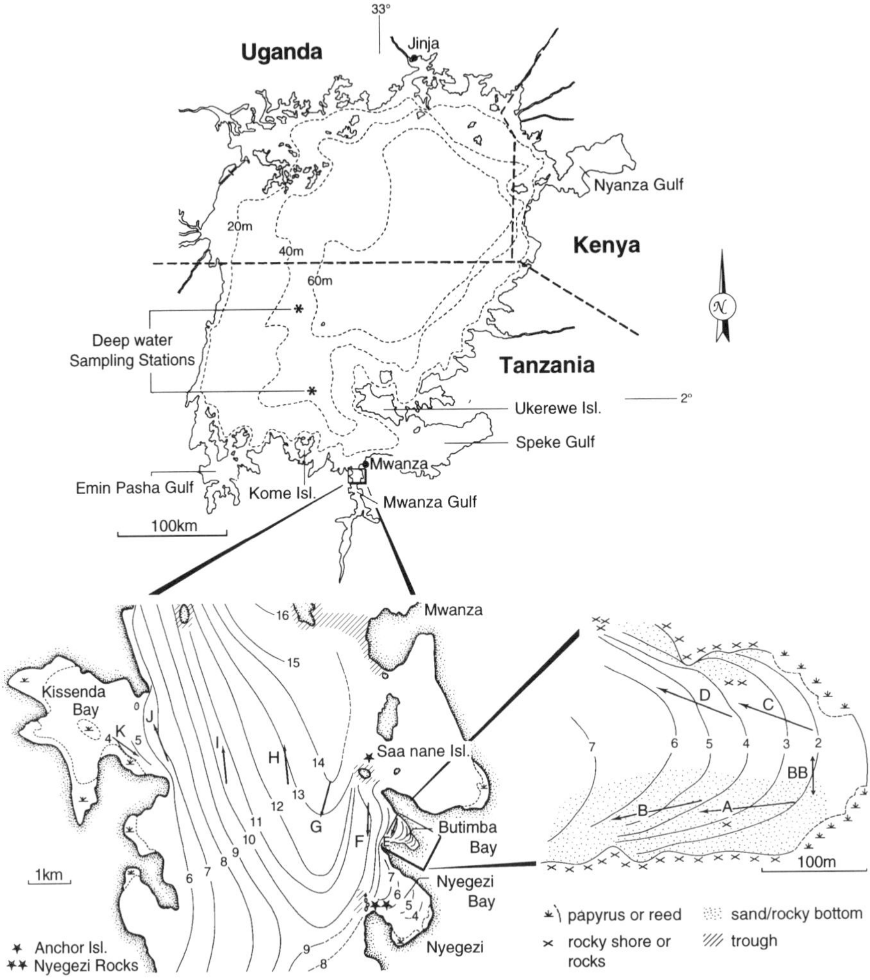

Fig. 6.1 Map of Lake Victoria and the sampling areas of the Haplochromis Ecology Survey Team (HEST). Depths in metres. Letters denote stations on transects. Nyanza Gulf is also called Kavirondo or Winam Gulf.

Besides haplochromine cichlids, about ten other genera, belonging to several families, contributed to the fisheries (Chapters 2, 4, 5, 7). To boost the fisheries, several fish species were introduced into the lake in the 1950s (Lowe-McConnell, 1987; Welcomme, 1988). Among these was the Nile perch, a large piscivore. Harrison (1991) reports that, using morphological characters,

Nile perch in Lake Victoria cannot be unambiguously identified to one of the three species (*Lates niloticus, L. macrophthalmus* and *L. longispinis*) that occur in the lakes of origin, Lakes Albert and Turkana. A genetic analysis to solve this problem is required (and is in progress – *Eds*).

During the first 25 years after the introduction, Nile perch catches were insignificant. However, in the early 1980s an explosive increase was observed. Simultaneously, a decline of the stocks of many other fish species was reported in several areas of the lake (e.g. Arunga, 1981; Okemwa, 1981; Barel *et al.*, 1985; Hughes, 1986; Acere, 1988a; Ogutu-Ohwayo, 1990a; Witte *et al.*, 1992a,b). Most striking was the decline of the haplochromines, which were the dominant prey of the Nile perch.

In the second half of the 1980s, Nile perch constituted more than 90% of the total weight caught in demersal trawls (Ligtvoet and Mkumbo, 1990). Simultaneously, other great changes in the ecosystem were observed. In this chapter we survey the changes which occurred in the lake during and after the Nile perch upsurge with an emphasis on the haplochromine stocks in the Mwanza Gulf. We demonstrate: (1) the role of Nile perch in the decline of haplochromines; (2) the differences in susceptibility between haplochromine species to disturbance by predation and fisheries; (3) ecological and morphological changes in some surviving haplochromine species; (4) other changes in the ecosytem that may be related to the disappearance of the haplochromines.

6.2 DISTRIBUTION AND INCREASE OF THE NILE PERCH

Nile perch has been recorded from experimental trawl catches in the Mwanza Gulf since 1972, the year in which a research vessel became operational in this area. However, the mean catch rates were initially low ($<10\,\mathrm{kg\,h^{-1}}$). After 1982 a sharp increase in Nile perch catches was observed (Fig. 6.2; Barel *et al.*, 1991; Witte *et al.*, 1992b). Initially, the Nile perch catches in the Mwanza Gulf comprised mainly adult individuals of more than 10 kg. In the middle of 1984, suddenly large numbers of subadult Nile perch (15–39 cm SL; weighing about 0.5–2 kg) were caught (Goudswaard and Witte, 1985), and in 1985, for the first time, large numbers of juveniles smaller than 10 cm SL were observed in the trawls (Goudswaard, 1988). This indicates that the initial increase of the Nile perch in the Mwanza Gulf was mainly due to migration of adults and subadults from other areas. A similar observation was made by Zaret and Paine (1973) on *Cichla ocellaris* in Gatún Lake (Panama). According to them, this predator, introduced into a tributary of the Chagres Rivers, "moved as a wave, with a leading edge composed of subadults". We suppose that the wave-like distribution may be due to local overpopulation in areas where the fishes bred, subsequently followed by migration of a part of the population. It is still unknown why

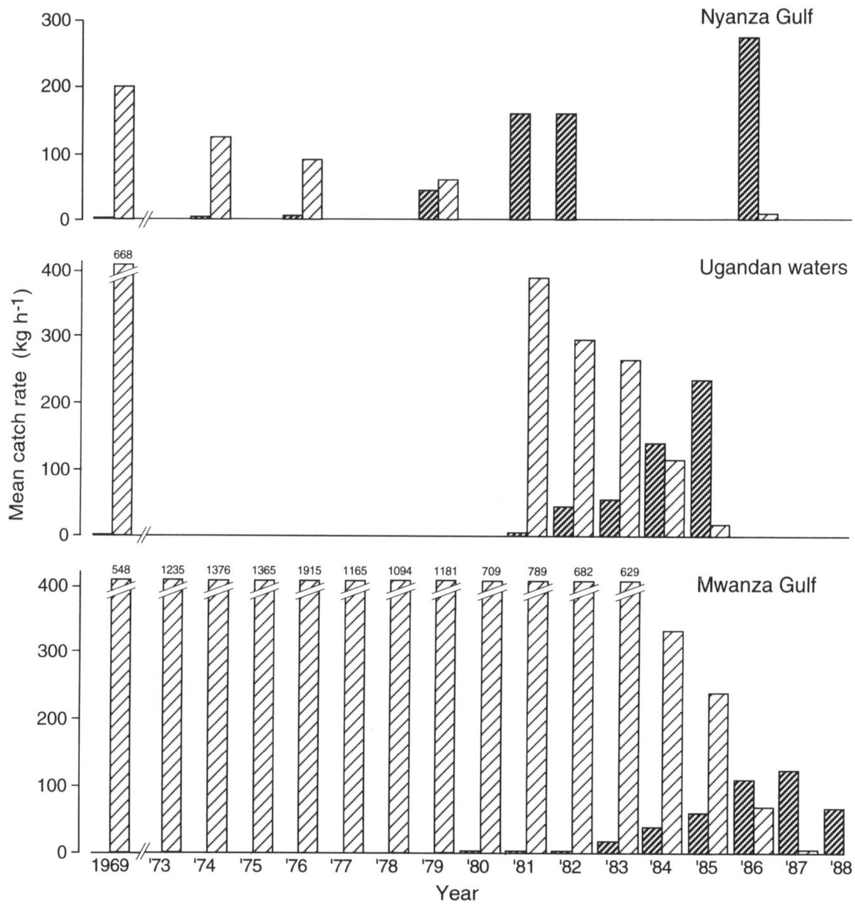

Fig. 6.2 Mean catch rates of haplochromines (light hatching) and Nile perch (bold hatching) in bottom trawl catches made in the Nyanza Gulf (Kenya), Ugandan waters and the Mwanza Gulf (Tanzania). All 1969 data are from Kudhongania and Cordone (1974a). Nyanza Gulf data from: Benda (1981), Okemwa (1981), Fisheries Department Nairobi, Kenya (1988), Ogari and Asila (1990). Data for Ugandan waters are from Acere (1988a). Mwanza Gulf data were collected by the Nyegezi Freshwater Fisheries Training Institute and by P.C. Goudswaard (HEST/TAFIRI). Tows in the Ugandan and Tanzanian waters were made with trawlers ranging from 105 to 180 hp utilizing conventional bottom trawls with 18–25 m headropes and 20 mm cod-end mesh. In the Nyanza Gulf, part of the catches were made with an 85 hp trawler, towing a net with a 13.7 m headrope. In this area mainly 38 mm cod-end mesh was used.

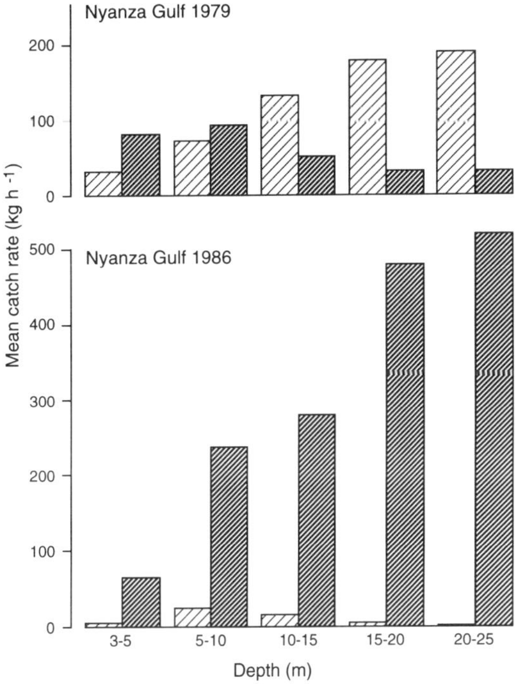

Fig. 6.3 Mean catch rates of haplochromines (light hatching) and Nile perch (dark hatching) in bottom trawl catches at different depths in the Nyanza Gulf in 1979 and 1986. After Fisheries Department Nairobi, Kenya (1988).

the Nile perch boom started in the shallow waters of the relatively isolated Nyanza Gulf in Kenya.

A lake-wide trawl survey in 1969–1971 revealed that Nile perch was restricted to waters less than 29 m deep, with the highest catch rates ($2\,kg\,h^{-1}$) in waters 4–9 m deep (Kudhongania and Cordone, 1974a). In 1979, when the Nile perch upsurge started in the Nyanza Gulf (Fig. 6.2), the highest catch rates were also obtained from waters less than 10 m deep. However, from 1982 onwards the majority of the Nile perch in the Nyanza Gulf were caught at depths of 10–25 m (Fig. 6.3; Fisheries Department Nairobi, 1988). In the Tanzanian part of the lake, Nile perch have been caught over a depth range from 1 m to 60 m (Goudswaard and Witte, 1985; Ligtvoet and Mkumbo, 1990). The highest catch rates in 1984/85 were obtained from water between 16 and 50 m deep. With decreasing depth, catches became smaller (Fig. 6.4; Goudswaard and Ligtvoet, 1988). Although few catches were made at depths exceeding 50 m, these scanty data also suggest a decrease. Juveniles smaller than 5 cm TL were mainly found in

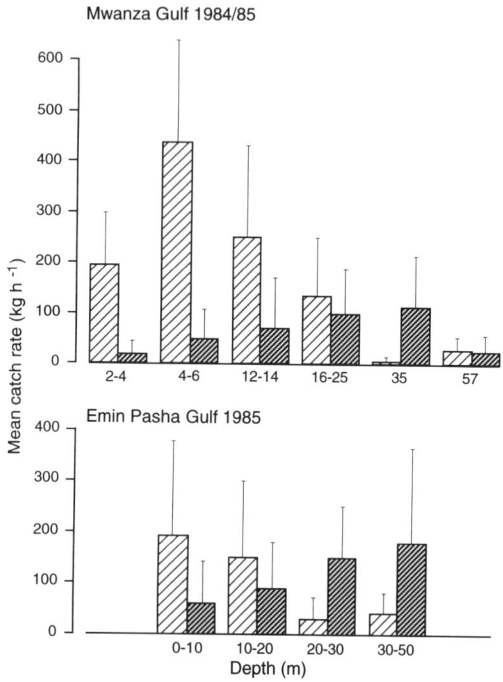

Fig. 6.4 Mean catch rates and one standard deviation (thin bar) of haplochromines (light hatching) and Nile perch (dark hatching) in bottom trawl catches at different depths in the Mwanza Gulf and the Emin Pasha Gulf. After Goudswaard and Ligtvoet (1988).

littoral and sublittoral areas (E.F.B. Katunzi, pers. comm.). Generally, Nile perch is a bottom dweller, but it may migrate into the water column when oxygen concentrations near the bottom become critical (Ligtvoet and Mkumbo, 1990). Rocky areas represent one of the few habitats in Lake Victoria not to have been invaded by the Nile perch. Although Nile perch occurs near these areas, it was rarely observed between boulders, and never in the narrow crevices where rock-dwelling cichlids are common (N. Bouton and O. Seehausen, pers. comm.).

6.3 DECLINE OF THE HAPLOCHROMINES

Since 1979, when monitoring was started, the catches of haplochromines at station G (depth 13–15 m) along a transect across the Mwanza Gulf (Fig. 6.1) decreased strongly in numbers (Table 6.1). Within 10 years the mean bottom-trawl catches per 10 min dropped from over 1300 individuals to almost zero. The total number of species caught per sampling season decreased from 37 to one in 1987/88. In 1990/92 the number of individuals increased slightly and the number of species increased from one to four. This may be indicative of a slight revival. Other data (unpublished) indicate that, except for the shallow littoral areas (see below), the situation at this station is representative for the whole Mwanza Gulf and probably for most areas outside the Mwanza Gulf. Indeed, decreases in biomass of haplochromines were observed lake-wide (Fig. 6.2; Arunga, 1981; Okemwa, 1981; Acere, 1988a; Ogutu-Ohwayo, 1990a).

There are indications that intensive fishing has had negative effects on the haplochromines in Lake Victoria, especially on the larger species (Marten, 1979; Witte and Goudswaard, 1985; Goudswaard, 1988). According to several authors (Acere, 1988b; Harrison *et al.*, 1989; Kudhongania *et al.*, 1992) fishing was more harmful to the haplochromines than Nile perch predation (or just as much; Chapter 7 – *Eds*). However, the following observations demonstrate that the impact of the Nile perch on the haplochromine stocks was much greater than that of fishery.

- Until the haplochromines had virtually disappeared they were the main food items of Nile perch (Ligtvoet and Mkumbo, 1990; Mkumbo and Ligtvoet, 1992).
- Haplochromines also vanished from large areas without (high) fishing pressure (Witte *et al.*, 1992a,b).
- The Nile perch upsurge in different areas of Lake Victoria occurred in different years (Fig. 6.2). Each time the boom of the Nile perch population corresponded with a strong decline of the haplochromines.
- Before the sublittoral and offshore haplochromines had virtually disappeared from the lake, the mean catch rates by depth of Nile perch and haplochromines showed inverse patterns (Figs 6.3, 6.4; Fisheries Department Nairobi, 1988; Goudswaard and Ligtvoet, 1988). In the Nyanza Gulf a shift of this pattern was observed. In 1979, catch rates of Nile perch were lowest in deep water (15–25 m), while in the same areas haplochromine catches were highest. In 1986 Nile perch catches were highest at these depths and haplochromine catches lowest (Fig. 6.3).
- The species richness in the Mwanza Gulf did not decrease to the same degree at all depths. Species that were restricted to littoral habitats near submerged vegetation and between rocks were least affected (Witte *et al.*, 1992b). In these shallow areas the Nile perch catches were relatively low (Fig. 6.4; Goudswaard and Ligtvoet, 1988).

Of the more than 120 haplochromine species which had been caught along the transect and the neighbouring rocky shores, at least 80 had vanished by 1990 (Witte *et al.*, 1992b). As well as those species of the transect, the

Table 6.1 Mean number of haplochromines and total number of species per sampling season (between parentheses) per trophic group, in bottom trawls of 10 minutes duration at station G. No samples are available for 1989

Trophic group	1979/80 ($n = 13, 18$)*		1981/82 ($n = 10, 11$)		1983/84 ($n = 17, 17$)		1985/86 ($n = 4, 4$)		1987/88 ($n = 9, 9$)		1990/92 ($n = 4, 4$)	
	Mean	SD	Mean	SD	Mean	SD	Mean	SD	Mean	SD	Mean	SD
Detritivores/ phytoplanktivores	1101.2 (8)	±814.0	791.2 (9)	±569.1	216.3 (7)	±232.7	18.8 (3)	±16.1	0		1 (2)	±1.4
Zooplanktivores	153.1 (10)	±312.7	187.9 (8)	±240.3	45.7 (10)	±62.1	62.8 (3)	±43.4	0.3 (1)	±0.7	18.3 (2)	±35.2
Insectivores	25.6 (3)	±20.4	15.5 (4)	±12.5	1.1 (3)	±1.6	0		0		0	
Molluscivores	5.4 (2)	±8.4	0.5 (2)	±0.8	0.1 (2)	±0.3	0		0		0	
Piscivores	3.5 (11)	±8.1	2.0 (6)	±1.9	0.1 (1)	±0.2	0		0		0	
Rest	32.8 (3)	±21.3	7.3 (2)	±5.8	0.1 (2)	±0.2	0		0		0	
Total	1321.5 (37)	±932.2	1004.4 (31)	±745.6	263.5 (25)	±279.0	81.5 (6)	±45.5	0.3 (1)	±0.7	19.3 (4)	±34.6

* n, Number of catches used for calculating mean number of individuals and total number of species, respectively.

total number (302+) of haplochromine species counted by HEST included deep-water species and geographically restricted species in littoral and sublittoral areas other than the Mwanza Gulf. Virtually all deep-water species seem to have vanished. If the decline in other littoral and sublittoral areas is comparable to that in the Mwanza Gulf, we estimate that about 200 of the more than 300 original species have disappeared or are threatened with extinction.

6.4 DYNAMICS IN MORPHOLOGY AND ECOLOGY OF HAPLOCHROMINES

Size reduction

Before the haplochromines had vanished from the sublittoral part of the Mwanza Gulf, a dramatic change in the size structure of the haplochromine stock had been observed (Fig. 6.5). This size reduction was partly due to eradication of species with a large adult size. These species were most

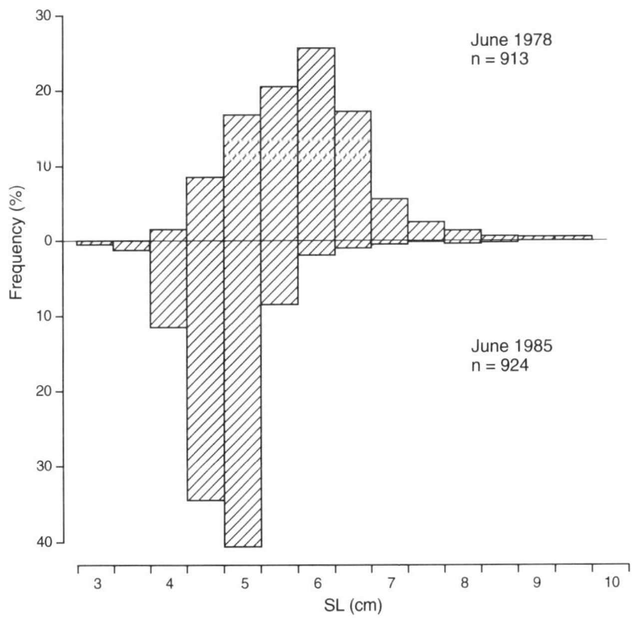

Fig. 6.5 Length–frequency distribution of haplochromines in the northern part of the Mwanza Gulf in June 1978 and June 1985. After Witte and Goudswaard (1987).

susceptible to trawling with small meshes, and possibly to size-selective predation by Nile perch (Witte and Goudswaard, 1985; Goudswaard, 1988; Witte *et al.*, 1992a, b). A decrease in abundance of large species due to a trawl fishery was also observed in the SE Arm of Lake Malaŵi (Turner, 1977a,b). However, with the exception of perhaps one species, none of these larger species was eradicated (Chapter 18): after the initial changes the catch composition in the Lake Malaŵi trawl fishery remained stable for more than 15 years.

Besides a decrease of large species, a decrease in mean (and maximum) adult size of many small zooplanktivorous and detritivorous species was also observed (Fig. 6.6; Witte and Goudswaard, 1985; Witte and Witte-Maas, 1987; Goldschmidt and Witte, 1990; Witte *et al.*, 1990). We observed a decrease of the size at first maturity in all the zooplanktivores and detritivores investigated (e.g. Table 6.2), while age at first maturity (about 1 year; Witte, 1981; Goldschmidt *et al.*, 1993; R.O. Pompier, pers. comm.) did not seem to change. The size reduction appeared to be reversible. Individuals from the dwarfed population of *H. piceatus* which were caught at a size of 3–5 cm SL and subsequently kept in tanks, grew to sizes that were larger than those of normal individuals, i.e. individuals caught in the period 1977–1980 (Fig. 6.7). This suggests that growth is a phenotypically plastic trait (Stearns and Koella, 1986; Witte *et al.*, 1990). According to Stearns and Koella (1986) and Stearns (1992), age and size at maturity depend on growth–mortality relationships. We have no data on juvenile and adult mortality in relation to growth rate and consequently cannot determine whether this theory holds for the Lake Victoria haplochromine species.

The dwarfed individuals of *H. piceatus* differed from normal individuals in more characters than size alone. Differences were found for measures of body depth, caudal peduncle depth, eye diameter and lower jaw width (Witte and Witte-Maas, 1987). The body depth was the most striking difference. At standard lengths less than 40 mm, the body depths of normal and dwarfed individuals were similar. At larger sizes an increasing divergence was observed, with the dwarfed specimens having a smaller body depth (Fig. 6.7). Wild-caught juveniles of the dwarfed population, raised in the laboratory, obtained body depths that can be regarded as extrapolations of those of the normal individuals (Fig. 6.7).

Both before and after the Nile perch upsurge, *H. piceatus* migrated during ontogeny from the shallow littoral habitat in Butimba Bay to the deeper sublittoral habitat in the Mwanza Gulf (R.O. Pompier, pers. comm.; F.W., pers. obs.). Trawl fishery and Nile perch were more common in the deeper habitat than in the shallow. The divergence in body depth coincided with this habitat shift (Fig. 6.7). Witte *et al.* (1990) tentatively suggested that increased swimming activity in the sublittoral habitat, as a result of increased predation pressure, might explain the decreased body depth. (A reverse

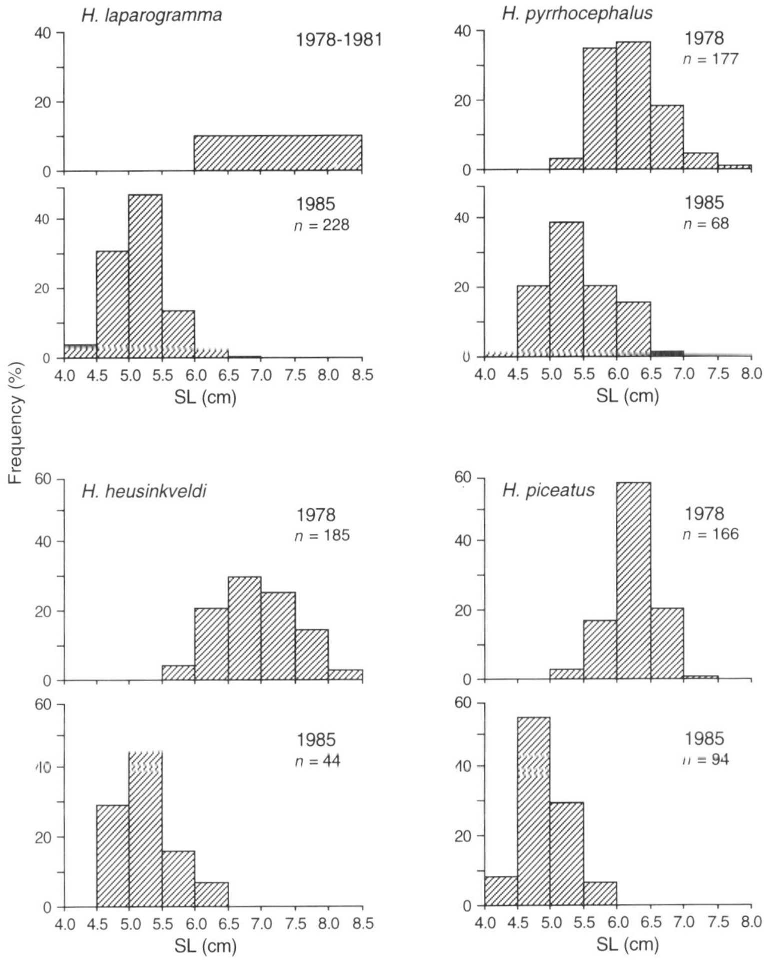

Fig. 6.6 Length–frequency distribution of four zooplanktivorous haplochromine species from the northern part of the Mwanza Gulf in 1978–1981 and 1985. As only few individuals of *H. laparogramma* were caught in this area before 1981, only the length range is given. After Witte and Witte-Maas (1987).

morphological change in response to predation pressure has been demonstrated experimentally by Brönmark and Miner (1992). Crucian carp, *Carassius carassius*, develop deeper bodies when exposed to predation by pike, *Esox lucius*, reducing predator effectiveness. Possibly two types of predator avoidance exist: (1) increased swimming speed due to shallow body depth; (2) decreased chance to be swallowed due to high body depth – *Eds.*)

Table 6.2 Size (SL, mm) of ripe and brooding females of *H. piceatus* caught in 1977–1980 and in 1985 in Butimba Bay. Differences in mean standard length are significant (*t*-test, $P < 0.005$)

Year	Smallest ♀	Largest ♀	Mean ± SD	*n*
1977–1980	54	74	63.2 ± 4.2	106
1985	46	53	49.0 ± 1.8	21

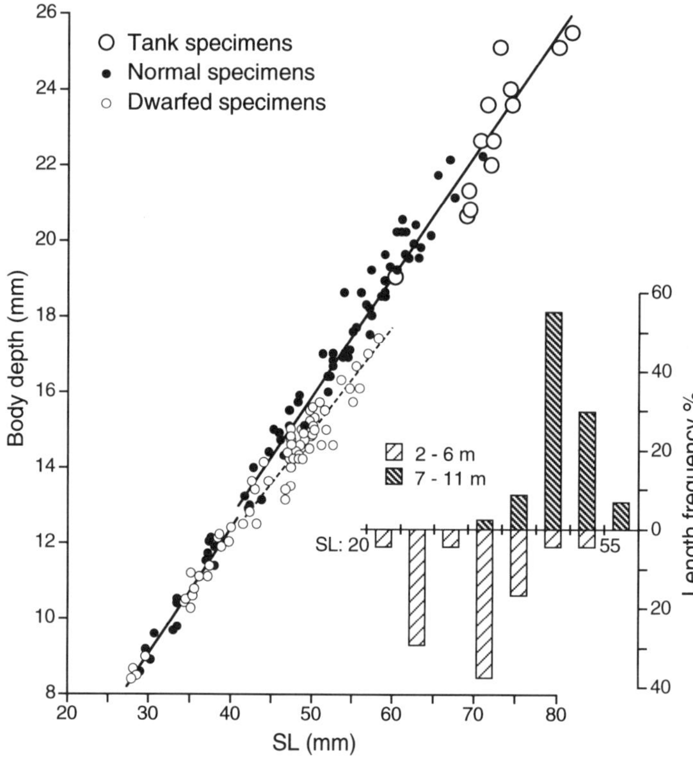

Fig. 6.7 Relations between body depth and standard length of normal, dwarfed and tank specimens of *H. piceatus*. Normal specimens are from the period 1977–1980, dwarfed specimens were collected in 1985, tank specimens in 1984 and 1985. Slope differences for body depth of normal and dwarfed specimens smaller than 40 mm SL are not significant; for specimens larger than 40 mm SL the difference is significant (F-test, $P < 0.01$). Bars represent length–frequency distributions of dwarfed individuals in the littoral and sublittoral habitat. After Witte *et al.* (1990).

Surviving species

Witte *et al.* (1992a,b) demonstrated that haplochromine species disappeared at different rates. Large and rare species were more susceptible to extinction than small and abundant species. The smaller the habitat overlap of a haplochromine species with Nile perch, the lesser the destructive impact of the predator. Both detritivores and zooplanktivores comprise relatively small and abundant species (Witte and van Oijen, 1990). Moreover, zooplanktivores have a reduced habitat overlap with Nile perch because they are partly pelagic (Goldschmidt *et al.*, 1990). Thus, in the sublittoral areas of the Mwanza Gulf it was zooplanktivorous and detritivorous species that showed the slowest rate of decline (Witte *et al.*, 1992b) and some species are still present.

Other factors that may have contributed to differential survival can be deduced from a comparison of four zooplanktivorous species, studied both prior to and during the Nile upsurge (Witte, 1984; Goldschmidt and Witte, 1990; Goldschmidt *et al.*, 1990; Wanink, 1991). The three morphologically similar species of the *laparogramma* group (*H. laparogramma, H. pyrrhocephalus* and *H. heusinkveldi*), and the broadly similar *H. piceatus* (Witte and Witte-Maas, 1987), differed in abundance and distribution pattern in the Mwanza Gulf (Witte, 1984; Goldschmidt *et al.*, 1990). *H. piceatus* occurred predominantly at the entrances of bays at a depth of 7–10 m, the rare *H. laparogramma* was mainly restricted to water of 20–35 m, and the other two species were found in open water with depths of 7–20 m (Fig. 6.8; Goldschmidt *et al.*, 1990).

From the beginning of the 1980s, the four species showed a strong decline in size with time (see above). Simultaneously with the Nile perch upsurge, the populations of *H. piceatus, H. heusinkveldi* and *H. pyrrhocephalus* collapsed (Figs 6.8, 6.9). However, the densities of *H. laparogramma*, which had always been a rare species within the Mwanza Gulf, initially increased at station G (Fig. 6.9) and in other parts of the Mwanza Gulf. Finally, when Nile perch densities continued to increase, *H. laparogramma* collapsed as well at station G. The initial increase of *H. laparogramma* in the Mwanza Gulf might have been the result of a migration to shallower areas where the predation pressure of Nile perch was lower and/or of competitive release when the other species of the *laparogramma* group were decimated.

Of the four zooplanktivorous species discussed here, only *H. laparogramma* and *H. pyrrhocephalus* have survived the Nile perch boom. After 1987 the highest numbers of these two species were caught at the shallow station J and for *H. laparogramma* the densities at this station were even higher than before the Nile perch boom (Fig. 6.8).

It is still unclear why the potential of the ecologically and morphologically similar zooplanktivorous species to co-occur with Nile perch was so different. *H. piceatus* lived closer to the bottom than the other three species (Goldschmidt *et al.*, 1990) and may thus have had a greater potential habitat overlap with

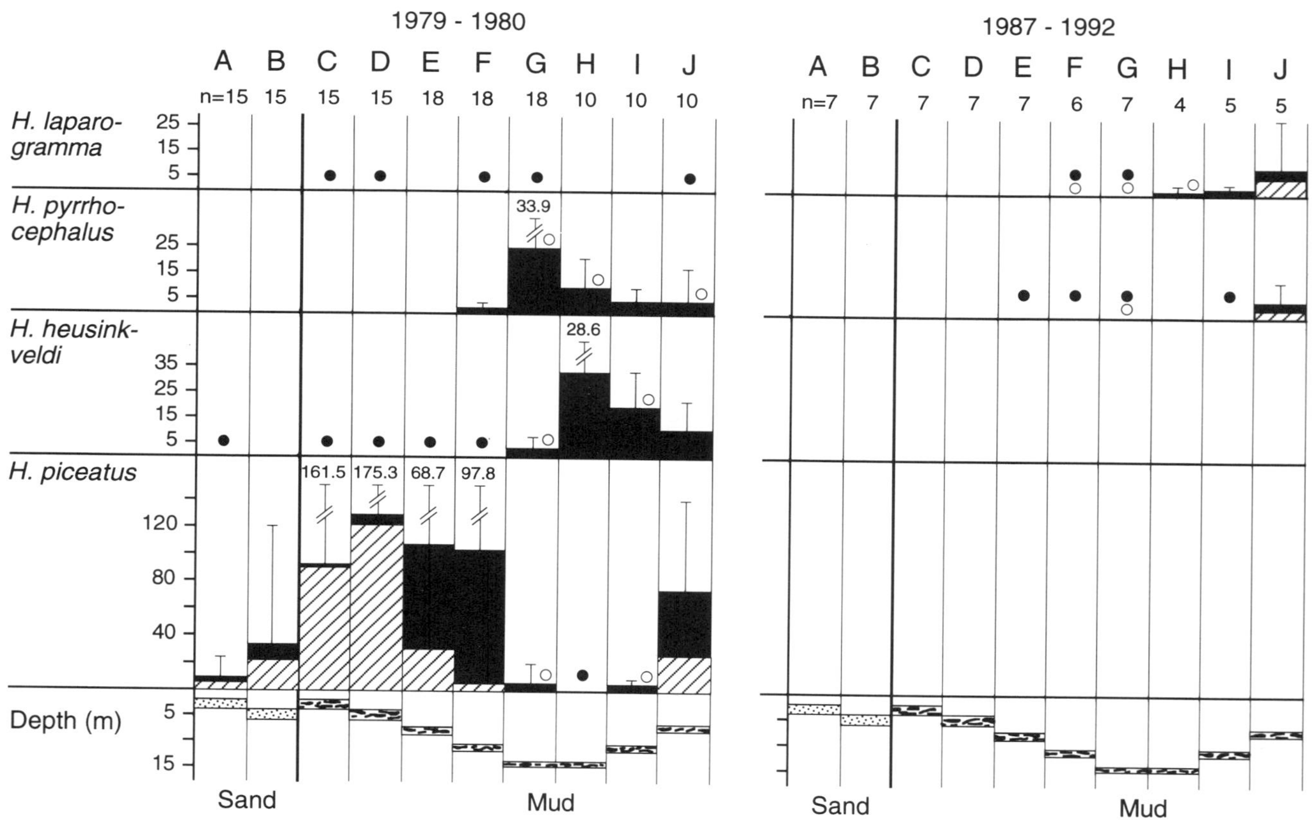

Fig. 6.8 Relative densities (mean number in samples of 2–2.5 kg) and one standard deviation (thin bar) of zooplanktivores in bottom trawl catches along the transect across the Mwanza Gulf in 1979–1980 and 1987–1992. *n*, number of catches; solid bars, adults; hatched bars, juveniles; filled circles, mean number of adults, less than one; open circles, mean number of juveniles, less than one. Note differing scales.

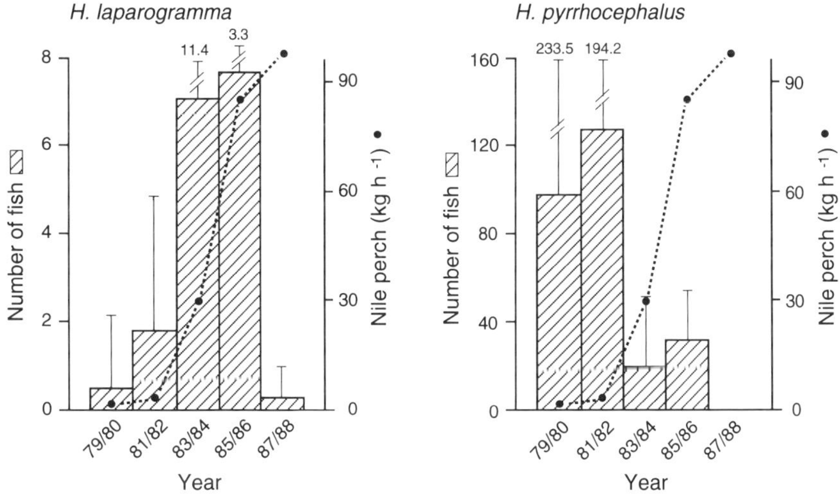

Fig. 6.9 Numbers of *H. laparogramma* and *H. pyrrhocephalus* individuals and one standard deviation (thin bars) in bottom trawl catches of 10 min duration at station G during the Nile perch upsurge. Note differing scales.

the bottom-dwelling Nile perch. *H. laparogramma*, *H. pyrrhocephalus* and *H. heusinkveldi* were more pelagic, and the latter two showed very little segregation in distribution.

These three species differed in reproductive characteristics (Goldschmidt and Witte, 1990). A preliminary study of the reproductive biology of *H. laparogramma* and *H. pyrrhocephalus* after the Nile perch boom showed an increase in both absolute and relative fecundity (Fig. 6.10; Wanink, 1991). The relatively large body size of the few ripe females collected in 1988 partly accounts for the rise in absolute fecundity. Even the maximum fecundity of *H. laparogramma* and *H. pyrrhocephalus* (18 and 40 eggs respectively) in 1983 was considerably lower than the minimum in 1988 (31 and 59 respectively), while the ranges of standard length overlapped. The increase in relative fecundity also indicates that the high clutch sizes in 1988 are not only due to the relatively large size of the females. Some indications have been found that the fraction of body weight devoted to ovaries increased. Perhaps *H. piceatus* and *H. heusinkveldi* were already producing the maximum eggs possible because mouth size constrains fecundity in these cichlids (C.D.N. Barel, pers. comm.; Witte and Witte-Maas, 1987; Goldschmidt and Witte, 1990). Alternatively, before the Nile perch upsurge, *H. pyrrhocephalus* and *H. laparogramma* had the lowest gonadosomatic index among the zooplanktivores

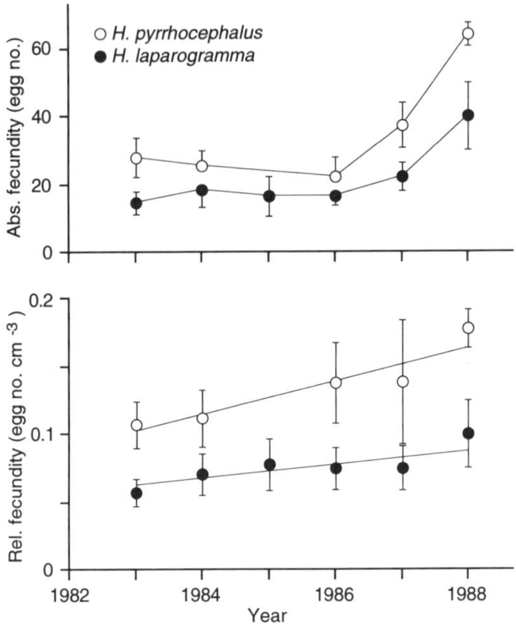

Fig. 6.10 Changes in mean absolute and relative fecundity of *H. laparogramma* and *H. pyrrhocephalus*. Successive sample sizes: *H. laparogramma*: 9, 11, 27, 25, 14, 3; *H. pyrrhocephalus*: 18, 37, 18, 5, 6. Thin bars indicate standard deviations. Both regressions are significant, $P < 0.005$. After Wanink (1991); note that in Wanink (1991) the relative fecundity of *H. pyrrhocephalus* in 1987 was presented wrongly.

(Goldschmidt and Witte, 1990) and these species may have had a potential to increase the gonad volume.

Another difference between species that survived and disappeared is that *H. laparogramma* and *H. pyrrhocephalus* were all-year-round spawners, whereas in contrast, *H. piceatus* and *H. heusinkveldi* were seasonal spawners (Goldschmidt and Witte, 1990). This may have led to higher lifetime fecundity in the former two species by the ability to spawn more frequently once mature.

In June 1991, 84 detritivores were caught along the transect, mainly at station E. Although probably four 'species' were involved (O. Seehausen, pers. comm.; F.W., pers. obs.), they could not be identified unambiguously to any of the species that had been caught along the transect during intensive sampling over the 1979–1982 period (Witte *et al.*, 1992b). There are four possible explanations for these observations. (1) The 'species' come from areas that had thus far not been sampled by HEST and, like *H. laparogramma*, had changed their distribution pattern. (2) In response to the recent environmental changes (see below), phenotypic changes may have occurred, such plasticity being common in haplochromine cichlids (Greenwood, 1965; Hoogerhoud, 1986; Witte *et al.*, 1990). (3) The 'species' are hybrids of other species that originally occurred at the transect. Based on laboratory experiments, Crapon de Caprona and Fritzsch (1984) suggested that accidental hybridization between haplochromine species could occur if under naturally 'forced' conditions one sex of a species is absent or rare. The extremely low population densities of the haplochromine species at the transect after 1984, must also have reduced the chances that individuals would meet conspecific mates and may have been conducive to hybridization. (4) They are formerly rare haplochromine species that have been favoured by the new limnological and biotic status of the lake.

6.5 OBSERVATIONS ON OTHER FISH STOCKS AND OTHER ORGANISMS

Changes in the abundance of other fish stocks

During the past decades, many fish species other than haplochromines showed a strong decline in Lake Victoria. In some cases the decline was caused by overfishing, and in other cases the Nile perch may be blamed (Table 6.3; Goudswaard, 1988; Ogutu-Ohwayo, 1990a; Witte *et al.*, 1992a). However, not all fish species decreased after the increase of the Nile perch. The abundance (measured as catch per unit effort) of the pelagic zooplanktivorous cyprinid, *Rastrineobola argentea*, which has a small habitat overlap with the benthic Nile perch, increased strongly in the period between 1982 and 1987 (Wanink, 1991; Witte *et al.*, 1992a). As in the haplochromines, a considerable decrease in size with time was observed in *R. argentea* (Wanink, 1991). The striking difference between zooplanktivorous haplochromines and *R. argentea* in ability to coexist with Nile perch may be due to their difference in reproductive strategies (Bruton, 1990; Wanink, 1991). *R. argentea* has a mean fecundity which is higher (1073 ± 313, $n = 22$; Wanink, 1991) than that in haplochromines. However, other explanations for the increase of *R. argentea* cannot be ruled out, e.g. if predation pressure of Nile perch on zooplanktivorous haplochromines was stronger than on *R. argentea*, reduced competition for food might have benefited the latter species.

According to official landing statistics, the catch rates of the introduced tilapiine *Oreochromis niloticus* also increased (CIFA, 1990; Ogutu-Ohwayo, 1990b). This species occurs in littoral and sublittoral waters. The juveniles which live in very shallow areas (among others along rocky shores), have only a narrow habitat overlap with adult Nile perch. In areas where they are indigenous (e.g. Lake Albert), both species coexist. Some of the

Table 6.3 Changes in the fauna and flora of Lake Victoria during the past decades

Organism	Change	Possible cause	References
MAMMALS			
Lutra maculicollis (spotted-necked otter)	Feeding behaviour in 1980s	Shift in densities of haplochromines and *Oreochromis niloticus*	Kruuk and Goudswaard (1990)
BIRDS			
Ceryle rudis (pied kingfisher)	Feeding behaviour in 1980s	Shift in densities of haplochromines and *Rastrineobola argentea.*	Wanink *et al.* (1993), Wanink and *Goudswaard (1994)*
Phalacrocorax carbo and *P. africanus* (cormorants)	Feeding behaviour in 1980s	Shift in densities of haplochromines and *Rastrineobola*	Goudswaard and Wanink (1993)
FISH			
Lates niloticus? (Nile perch)	Introduction in 1950s, strong increase in 1980s	?	Arunga (1981), Okemwa (1984), Barel *et al.* (1985, 1991), Goudswaard and Witte (1985), Okaronon *et al.* (1985), Hughes (1986), Goudswaard and Ligtvoet (1988), Ogutu-Ohwayo (1990a), Witte *et al.* (1992b)
Bagrus docmac and *Clarias gariepinus*	Decrease since 1980s	Competition for food with Nile perch and predation by Nile perch?	Arunga (1981), Hughes (1986), Goudswaard (1988), Ogutu-Ohwayo (1990a)
Synodontis victoriae	Decrease since 1980s	Predation by Nile perch?	Arunga (1981), Hughes (1986), P.C. Goudswaard, unpubl. data
Protopterus aethiopicus	Decrease since 1970s, 1980s,	Disappearance of breeding habitat (papyrus swamps)?	Hughes (1986), Ogutu-Ohwayo (1990a), P.C. Goudswaard, unpubl. data
Rastrineobola argentea	Increase since 1980s	Competitive release for food, due to disappearance of zooplanktivorous haplochromines?	Bwathondi (1990), Ogutu-Ohwayo (1990a), Wanink (1991), Witte *et al.* (1992a)

Table 6.3 (Continued)

Organism	Change	Possible cause	References
Labeo victorianus	Decrease since 1950s	Overfishing	Cadwalladr (1965), Ogutu-Ohwayo (1990a)
Oreochromis esculentus and *O. variabilis*	Decrease since beginning of century	Overfishing and competition with introduced *O. niloticus*. Hybridization with *O. niloticus*?	Fryer (1961), Fryer and Iles (1972), Welcomme (1967), Lowe-McConnell (1987), Ogutu-Ohwayo (1990a)
Oreochromis niloticus	Introduction in 1950s, strong increase in 1980s	Increase due to reduced fishing pressure with small-meshed nets after Nile perch upsurge?	CIFA (1990), Ogutu-Ohwayo (1990a), P.C. Goudswaard and W. Ligtvoet, unpubl. data
Haplochromines	Strong decrease since 1970s	Locally to some extent overfishing, but mainly Nile perch predation	Marten (1979), Okemwa (1984), Barel *et al.* (1985, 1991), Okaronon *et al.* (1985), Hughes (1986), Goudswaard and Ligtvoet (1988), Ogutu-Ohwayo (1990a,b), Witte *et al.* (1992a,b)
INVERTEBRATES			
Caridina nilotica	Increase in 1980s	Diminished predation pressure on juveniles due to decline of haplochromines?	Witte *et al.* (1992a), Goldschmidt *et al.* (1993)
Oligochaetes	Increase? in 1980s	?	Kaufman (1992), Witte *et al.* (1992a)
Chaoborid larvae and chironomid larvae	Increase in 1980s Increase in 1980s	Diminished predation pressure due to decline of haplochromines and other insectivorous fishes	Kaufman (1992), Mbahinzireki (1992), Witte *et al.* (1992a)
Molluscs	Increase? in 1980s	Diminished predation pressure due to decline of haplochromines, *Synodontis victoriae* and *Protopterus aethiopicus*	Kaufman (1992), Mbahinzireki (1992), Witte *et al.* (1992a)

Table 6.3 (Continued)

Organism	Change	Possible cause	References
PHYTOPLANKTON			
Microcystis and other bluegreen algae	Increase in 1980s	Increasing eutrofication and eradication of the considerable biomass of haplochromines that used to feed on these algae	Ochumba and Kibaara (1989), Ochumba (Chapter 3, this volume), Witte *et al.* (1992b), Goldschmidt *et al.* (1993)
MACROPHYTES			
Nymphaea spp. and *Ceratophyllum demersum*	Increase? at end of 1980s	Replacement of the introduced plant-eating *Tilapia zillii* by the phytoplanktivorous *Oreochromis niloticus* and decrease of the bottom-digging haplochromines?	Witte *et al.* (1992a), Kudhongania and Chitamwebwa (Chapter 2, this volume)
Eichhornia crassipes	Invasion and increase at end of 1980s	?	Njuguna (1991), Witte *et al.* (1992a), Kudhongania and Chitamwebwa (Chapter 2, this volume)

above-described changes in the fish stocks have also been observed in Lake Kyoga, which had been stocked with Nile perch a few years earlier than Lake Victoria (Chapter 5). In Lake Kyoga the introduced Nile perch and *O. niloticus* as well as the native *R. argentea* also became dominant while other native species declined (Ogutu-Ohwayo, 1990a,b).

Changes in the abundance of other organisms

Concurrently with the changes in the fish stocks, the abundance of several other organisms in Lake Victoria increased (Table 6.3).

The few quantitative data which are available on the densities of the formerly rare atyid prawn *Caridina nilotica* indicate a strong increase in the Mwanza Gulf transect (Witte *et al.*, 1992a; Goldschmidt *et al.*, 1993). The cause of the increase of prawns is not yet understood. *C. nilotica* feeds on bottom debris (Fryer, 1960a) which is unlikely to have been a limiting resource in the lake even when eaten by haplochromines. Possibly, the predation pressure on juvenile prawns has diminished because formerly widespread haplochromines have not been replaced by small Nile perch, which dwell inshore (E.F.B. Katunzi, pers. comm.). At least part of the prawn population is known to breed in deeper water (>15 m; E.H.F.M. Straetmans and W. Ligtvoet, pers. comm.).

An increased occurrence of bluegreen alga *Microcystis* sp. was observed in the Kenyan part of the lake (Ochumba and Kibaara, 1989). In the Mwanza Gulf, Secchi disc readings in the sublittoral part of the transect were much lower in 1987 than in 1980 and algal blooms seemed more intense (Witte *et al.*, 1992a). Although increased nutrient input to the lake is likely to have contributed to the algal blooms (Hecky, 1993) the disappearance of the haplochromines cannot, a priori, be ignored as a possible cause (Chapter 3). Bluegreen algae were one of the food sources of the detritivorous/phytoplanktivorous haplochromines and the major food source of the pelagic phytoplanktivores. Together these two trophic groups constituted about 50% of the total haplochromine biomass in the sublittoral areas of the lake (Goldschmidt *et al.*, 1993).

Macrophytes were formerly scarce in Butimba Bay, the main littoral study area of HEST (Fig. 6.1) but, since the end of the 1980s, the shallow inshore area of this bay (3 m depth) has been covered with *Nymphaea* spp., *Ceratophyllum demersum* and (more recently) with the alien *Eichhornia crassipes*. Witte *et al.* (1992a) have suggested that the original high densities of haplochromines, through their continuous digging in the bottom, might have prevented the settling of rooted plants (e.g. *Nymphaea* spp.). The recent decrease of the introduced plant-eating *Tilapia zillii* may also have contributed to the increase of macrophytes.

There are also indications of increased densities of oligochaetes, molluscs and chaoborid and chironomid larvae in the lake (Kaufman, 1992; Mbahinzireki, 1992; Witte *et al.*, 1992a). *Chaoborus* larvae were an important food source for insectivorous haplochromines (Hoogerhoud *et al.*, 1983), zooplanktivorous haplochromines (Goldschmidt *et al.*, 1990) and other insectivorous fish species that were common in the pre-Nile perch system (MacDonald, 1956; Corbet, 1961). In the present system, predation on larval and pupal stages of *Chaoborus* has probably been reduced drastically. It should be noted that *Chaoborus* larvae are not an important food source for *R. argentea*, which replaced the zooplanktivorous haplochromines (J.H.W., pers. obs.). Chironomid larvae were included in the diets of insectivorous, detritivorous and molluscivorous haplochromines (Hoogerhoud *et al.*, 1983; Witte and van Oijen, 1990; Goldschmidt *et al.*, 1993). In the present system they form a part of the diet of Nile perch smaller than 20 cm SL (Ogari and Dadzie, 1988; Ogutu-Ohwayo, 1990b) and of *R. argentea* (J.H.W., pers. obs.). It is not known whether this predation pressure is lower than in the previous system.

In general the disappearance of the haplochromines may well have allowed the increase of their former main food sources. Changes were also observed in the behaviour of birds and mammals. The pied kingfisher, *Ceryle rudis*, changed from a diet of mainly haplochromines to one that consists almost exclusively of *Rastrineobola* (Wanink *et al.*, 1993; Wanink and Goudswaard, 1994). A similar shift in diet was found for the great cormorant, *Phalacrocorax carbo*, and to a lesser extent, for the more inshore-feeding longtailed cormorant, *P. africanus* (Goudswaard and Wanink, 1993). Spotted-necked otters, *Lutra maculicollis*, currently include *Oreochromis niloticus* in their diet while formerly they fed mainly on small haplochromines (Kruuk and Goudswaard, 1990).

6.6 SUMMARY

Obviously, dramatic changes have occurred in the ecosystem of Lake Victoria. Approximately two-thirds of the endemic haplochromine species have disappeared. A unique group of vertebrates has been lost and most of the other indigenous fish species (including the remaining haplochromines) have also shown a strong decline. The formerly diverse ichthyofauna is now dominated by only three fish species (Chapter 5). The slight increase in densities of haplochromines in shallow areas where Nile perch has declined (CIFA, 1990; Chapter 5) concerns only a few species. Preliminary investigations revealed that at least some of these surviving species exhibit morphological and ecological changes (e.g. dwarfism, habitat shift, increased fecundity). Because of the recent observations on deoxygenation of the hypolimnion in the deep-water areas of the lake (Ochumba and Kibaara, 1989; Chapter 3; Hecky, 1993), for the deep-water communities, chances of survival seem less probable.

Although no coherent body of data is yet available, it is plausible that the disappearance of haplochromines which constituted about 80% of the demersal fish biomass and had a biomass estimated at more than 20 t km^{-2} in the sublittoral areas (Kudhongania and Cordone 1974b; pers. obs.), was a major factor in changing the lake ecosystem. The haplochromine cichlids were particularly important as primary and secondary consumers (Witte, 1981; Witte and van Oijen, 1990; Goldschmidt *et al.*, 1993). The currently dominant Nile perch (more than 90% of the demersal fish biomass estimated at <10 t km^{-2} in the sublittoral areas; Ligtvoet and Mkumbo, 1990) and *R. argentea* are secondary and tertiary consumers so that the relative importance of the main food chains has significantly changed (Ligtvoet and Witte, 1991). Compared with the pre-Nile perch situation, we think that the current ecosystem of Lake Victoria is unstable because fast changes in the fauna and flora are still taking place.

The total fish yield in Lake Victoria increased strongly after the Nile perch upsurge. Besides an increased protein production, huge socio-economic benefits have been brought about by the new fisheries regime (Greboval, 1990; Chapters 10, 11). For these reasons, the introduction has been considered a success by many (e.g. CIFA, 1988; Reynolds and Greboval, 1988; Greboval, 1990). After the Nile perch boom, the discussion about the desirability of the introduction (Fryer, 1960b; Anderson, 1961) was raised again. Conservation-minded scientists argued that introductions of this kind are undesirable whatever the material benefit in the short term, because the risks involved with regard to the possible alteration and destabilization of the ecosystem cannot be assessed (Barel *et al.*, 1985, 1991; Balon and Bruton, 1986; Coulter *et al.*, 1986; Ribbink, 1987; Bruton, 1990; Goldschmidt and Witte, 1992; Witte *et al.*, 1992a,b; Goldschmidt *et al.*, 1993). In contrast, the supporters of the introduction emphasized the material benefits, ignoring the time for which these advantages might last (Acere, 1988b; Reynolds and Greboval, 1988; Greboval, 1990). (See Chapters 9, 10, 11 and 23 for other views – *Eds.*)

Whatever the prospects for the fisheries of the lake, we think it important to prevent further damage to the remnant haplochromine fauna. Apart from Nile perch and fishing, haplochromine cichlids and other indigenous fishes in the shallows are currently threatened by a new intruder, the alien water hyacinth, *Eichhornia crassipes*. This species probably entered the lake in the Kagera region (Njuguna, 1991) and subsequently spread along its borders, reaching the Mwanza Gulf in 1990. Hyacinths may drastically affect the coastal habitat by forming dense mats, which cover the surface and cause hypoxic conditions in the underlying water (Njuguna, 1991) (but see another view in Chapter 5 – *Eds*).

The changes that have occurred in Lake Victoria over the past few decades are irreversible. One can only hope that this unfortunate example of an ill-considered introduction will convince even pragmatic people with other priorities than those of conservationists that similar introductions in other lakes are a bad idea.

ACKNOWLEDGEMENTS

We gratefully acknowledge Drs C.D.N. Barel, P.C. Goudswaard, C.E. Hollingworth, O. Seehausen and S.C. Stearns for inspiring discussions and their useful comments on an earlier version of this chapter. We also thank an anonymous referee for his critical comments. We are indebted to Mr M.L. Brittijn for drawing the figures. We thank Aloys, Mhoja, Rueben, Kevin Makonda, Niels Bouton, Yves Fermon, Ole Seehausen and Jan Smits for their assistance in the field. We thank the Director of the TAFIRI Center in Mwanza, Mr E.F.B. Katunzi, and his staff for their enthusiasm, hospitality and cooperation. The Haplochromis Ecology Survey Team (HEST) was financially supported by the Organization for the Advancement of Tropical Research (WOTRO) grants W87-129, W87-161, W87-189, W84-282. The project was also supported by the section for Research and Technology of the Netherlands Ministry of Development Cooperation.

REFERENCES

Acere, T.O. (1988a) Recent trends in the fisheries of Lake Victoria (Uganda Northern part) *F.A.O. Fish. Rep.*, no. 388, pp. 36–45.

Acere, T.O. (1988b) The controversy over Nile perch, *Lates niloticus*, in Lake Victoria, East Africa. *Naga, The ICLARM Quarterly*, **11**, 3–5.

Anderson, A.M. (1961) Further observations concerning the proposed introduction of Nile perch into Lake Victoria. *E. Afr. J. Agric.*, **26**, 195–201.

Arunga, J.O. (1981) A case study of the Lake Victoria Nile perch *Lates niloticus* (Mbuta) fishery, in *Proc. Workshop Kenya Mar. Fish. Res. Inst. Aquatic Resources of Kenya*, July 13–19, 1981, Kenya Natl Acad. Adv. Arts Sci., Nairobi, pp. 165–83.

Balon, E.K. and Bruton, M.N. (1986) Introduction of alien species or why scientific advice is not heeded. *Env. Biol. Fishes*, **16**, 225–30.

Barel, C.D.N., Dorit, R., Greenwood, P.H., Fryer, G., Hughes, N., Jackson, P.B.N., Kanawabe, H., Lowe-McConnell, R.H., Witte, F. and Yamaoka, K. (1985) Destruction of fisheries in Africa's lakes. *Nature, Lond.*, **315**, 19–20.

Barel, C.D.N., Ligtvoet, W., Goldschmidt, T., Witte, F. and Goudswaard, P.C. (1991) The haplochromine cichlids in Lake Victoria: an assessment of biological and fisheries interests, in *Cichlid Fishes: Behaviour, Ecology and Evolution* (ed. M.H.A. Keenleyside), Chapman and Hall, London, pp 258–79.

Benda, R.S. (1981) A comparison of bottom trawl catch rates in the Kenya waters of Lake Victoria. *J. Fish Biol.*, **18**, 609–13.

Bruton, M.N. (1990) The conservation of the fishes of Lake Victoria, Africa: an ecological perspective. *Env. Biol. Fishes*, **27**, 161–90.

Brönmark, C. and Miner, J.G. (1992) Predator-induced phenotypic change in body morphology in Crucian carp. *Science*, **258**, 1348–50.

Bwathondi, P.O.J. (1990) State of Lake Victoria fisheries, Tanzanian sector. *F.A.O. Fish. Rep.*, no. 430, pp. 24–34.

Cadwalladr, D.A. (1965) Notes on the breeding biology and ecology of *Labeo victorianus* Boulenger (Pisces: Cyprinidae) of Lake Victoria. *Revue Zool. Bot. afr.*, **72**, 109–34.

CIFA (1988) Report on the fourth session of the Sub-committee for the Development and Management of the Fisheries of Lake Victoria, Kisumu, Kenya, 6–10 April 1987. *F.A.O. Fish. Rep.*, no. 388, pp. 1–112.

CIFA (1990) Report of the fifth session of the Sub-committee for the Development and Management of the Fisheries of Lake Victoria, Mwanza, Tanzania, 12–14 September 1989. *F.A.O. Fish. Rep.*, no. 430, pp. 1–97.

Corbet, P.S. (1961) The food of non-cichlid fishes in the Lake Victoria basin, with remarks on their evolution and adaptation to lacustrine conditions. *Proc. Zool. Soc. Lond.*, **136**, 1–101.

Coulter, G.W., Allanson, B.R., Bruton, M.N., Greenwood, P.H., Hart, R.C., Jackson, P.B.N. and Ribbink, A.J. (1986) Unique qualities and special problems of the African Great Lakes. *Env. Biol. Fishes*, **17**, 117–18.

Crapon de Caprona, M.D. and Fritzsch, B. (1984) Interspecific fertile hybrids of haplochromine Cichlidae (Teleostei) and their possible importance for speciation. *Neth. J. Zool.*, **34**, 503–38.

Fisheries Department Nairobi, Kenya (1988) State of Lake Victoria Fisheries – Kenya. *F.A.O. Fish. Rep.*, no. 388, pp. 20–28.

Fryer, G. (1960a) The feeding mechanism of some atyid prawns of the genus *Caridina*. *Trans. R. Soc. Edinb.*, **64**, 217–44.

Fryer, G. (1960b) Concerning the proposed introduction of the Nile perch into Lake Victoria. *E. Afr. Agric. J.*, **25**, 267–70.

Fryer, G. (1961) Observations on the biology of the cichlid fish *Tilapia variabilis* in the northern waters of Lake Victoria (East Africa). *Revue Zool. Bot. afr.*, **64**, 1–33.

Fryer, G. and Iles, T.D. (1972) *The Cichlid Fishes of the Great Lakes of Africa: their Biology and Evolution*, Oliver and Boyd, Edinburgh, 641 pp.

Goldschmidt, T. and Witte, F. (1990) Reproductive strategies of zooplanktivorous haplochromine species (Pisces, Cichlidae) from Lake Victoria before the Nile perch boom. *Oikos*, **58**, 356–68.

Goldschmidt, T. and Witte, F. (1992) Explosive speciation and adaptive radiation of haplochromine cichlids from Lake Victoria: an illustration of the scientific value of a lost species flock. *Mitt. int. Verein. Limnol.*, **23**, 101–7.

Goldschmidt, T., Witte, F. and Visser, J. de (1990) Ecological segregation in zooplanktivorous haplochromine species (Pisces, Cichlidae) from Lake Victoria. *Oikos*, **58**, 343–55.

Goldschmidt, T., Witte, F. and Wanink, J.H. (1993) Cascading effects of the introduced Nile perch on the detritivorous/phytoplanktivorous species in the sub-littoral areas of Lake Victoria. *Conserv. Biol.*, **7**, 686–700.

Goudswaard, P.C. (1988) A comparison of trawl surveys in 1969/70 and 1984/85 in the Tanzanian part of Lake Victoria. *F.A.O. Fish. Rep.*, no. 388, pp. 86–100.

Goudswaard, P.C. and Ligtvoet, W. (1988) Recent developments in the fishery for haplochromines (Pisces: Cichlidae) and Nile perch, *Lates niloticus* (L.) (Pisces: Centropomidae) in Lake Victoria. *F.A.O. Fish. Rep.*, no. 388, pp. 101–12.

Goudswaard, P.C. and Wanink, J.H. (1993) Anthropogenic perturbation in Lake Victoria: effects of fish introductions and fisheries on fish eating birds, in *Birds and the African Environment: Proceedings of the Eighth Pan-African Ornithological Congress* (ed. R.T. Wilson), *Annls Mus. r. Afr. cent. (Sci. zool.)*, **268**, 312–18.

Goudswaard, P.C. and Witte, F. (1985) Observations on Nile perch, *Lates niloticus* (L.), 1758, in the Tanzanian waters of Lake Victoria. *F.A.O. Fish. Rep.*, no. 335, pp. 62–67.

Greboval, D. (1990) Socio-economic issues for planning in support of fisheries management. *F.A.O. Fish. Rep.*, no. 430, pp. 75–97.

Greenwood, P.H. (1965) Environmental effects on the pharyngeal mill of a cichlid fish *Astatoreochromis alluaudi* and their taxonomic implications. *Proc. Linn. Soc. Lond.*, **176**, 1–10.

Greenwood, P.H. (1974) The cichlid fishes of Lake Victoria, East Africa: the biology and evolution of a species flock. *Bull. Br. Mus. nat. Hist. (Zool.)*, Suppl. 6, 1–134.

Greenwood, P.H. (1981) *The Haplochromine Fishes of the East African Lakes: Collected Papers on their Taxonomy, Biology and Evolution*, Kraus International Publications, Munich, 839 pp.

Harrison, K. (1991) The taxonomy of East African Nile perch, *Lates* spp. (Perciformes, Centropomidae). *J. Fish. Biol.*, **38**, 175–86.

Harrison, K., Crimmen, O., Travers, R., Maikweki, J. and Mutoro, D. (1989) Balancing the scales in Lake Victoria. *Biologist*, **36**, 189–91.

Hecky, R.E. (1993) The eutrophication of Lake Victoria, Kilham Memorial Lecture, 25th Congress of SIL. *Verh. int. Verein. Limnol.*, **25**, 39–48.

Hoogerhoud, R.J.C. (1986) Taxonomic and ecological aspects of morphological plasticity in molluscivorous haplochromines (Pisces, Cichlidae). *Annls Mus. r. Afr. cent. (Sci. zool.)*, **251**, 131–4.

Hoogerhoud, R.J.C., Witte, F. and Barel, C.D.N. (1983) The ecological differentiation of two closely resembling *Haplochromis* species from Lake Victoria (*H. iris* and *H. hiatus*; Pisces, Cichlidae). *Neth. J. Zool.*, **33**, 283–305.

Hughes, N.F. (1986) Changes in the feeding biology of the Nile perch, *Lates niloticus* (L.) (Pisces: Centropomidae), in Lake Victoria, East Africa since its introduction in 1960, and its impact on native fish community of the Nyanza Gulf. *J. Fish Biol.*, **29**, 541–8.

Kaufman, L. (1992) Catastrophic change in species-rich freshwater ecosystems: the lessons of Lake Victoria. *Bio Science*, **42**, 846–58.

Kruuk, H. and Goudswaard, P.C. (1990) Effects of changes in fish populations in Lake Victoria on the food of otters (*Lutra maculicollis* Schinz and *Aonyx capensis* Lichtenstein). *Afr. J. Ecol.*, **28**, 322–9.

Kudhongania, A.W. and Cordone, A.J. (1974a) Batho-spatial distribution patterns and biomass estimates of the major demersal fishes in Lake Victoria. *Afr. J. trop. Hydrobiol. Fish.*, **3**, 15–31.

Kudhongania, A.W. and Cordone, A.J. (1974b) Past trends, present stocks and possible future state of the fisheries of the Tanzania part of the Lake Victoria. *Afr. J. trop. Hydrobiol. Fish.*, **3**, 167–81.

Kudhongania, A.W., Twongo, T. and Ogutu-Ohwayo, R. (1992) Impact of the Nile perch on the fisheries of Lakes Victoria and Kyoga. *Hydrobiologia*, **232**, 1–10.

Ligtvoet, W. and Mkumbo, O.C. (1990) Stock assessment of Nile perch in Lake Victoria. *F.A.O. Fish. Rep.*, no. 430, pp. 35–74.

Ligtvoet, W. and Witte, F. (1991) Perturbation through predator introduction: effects on the food web and fish yields in Lake Victoria (East Africa), in *Terrestial and Aquatic Ecosystems: Perturbation and Recovery* (ed. O. Ravera), Ellis Horwood, Chichester, pp. 263–8.

Lowe-McConnell, R.H. (1987) *Ecological Studies in Tropical Fish Communities*, Cambridge Univ. Press, Cambridge, 382 pp.

MacDonald, W.W. (1956) Observations on the biology of chaoborids and chironomids in Lake Victoria and on the feeding habits of the elephant-snout fish (*Mormyrus kannume* Forsk). *J. Anim. Ecol.*, **25**, 36–53.

Marten, G.G. (1979) Impact of fishing on the inshore fishery of Lake Victoria (East Africa). *J. Fish. Res. Bd Can.*, **36**, 891–900.

Mbahinzireki, G.B. (1992) Distribution and abundance of two dipteran aquatic larvae in northern Lake Victoria. *F.A.O. Fish. Rep.*, no. 475, p. 39.

Mkumbo, O.C. and Ligtvoet, W. (1992) Changes in the diet of Nile perch, *Lates niloticus* (L) in the Mwanza Gulf, Lake Victoria. *Hydrobiologia*, **232**, 79–83.

Njuguna, S.G. (1991) Water hyacinth: the world's worst aquatic weed infests Lakes Naivasha and Victoria. *Swara*, **14**, 8–10.

Ochumba, P.B.O. and Kibaara, D.I. (1989) Observations on blue-green algal blooms in the open waters of Lake Victoria, Kenya. *Afr. J. Ecol.*, **27**, 23–34.

Ogari, J. and Asila, A. (1990) The state of Lake Victoria fisheries, Kenya waters. *F.A.O. Fish. Rep.*, no. 430, pp. 18–23.

Ogari, J. and Dadzie, S. (1988) The food of the Nile perch, *Lates niloticus* (L.) after the disappearance of the haplochromine cichlids in the Nyanza Gulf of Lake Victoria. *J. Fish. Biol.*, **32**, 571–7.

Ogutu-Ohwayo, R. (1990a) The decline of the native fishes of Lakes Victoria and Kyoga (East Africa) and the impact of introduced species, especially the Nile perch, *Lates niloticus* and the Nile tilapia, *Oreochromis niloticus*. *Env. Biol. Fishes*, **27**, 81–96.

Ogutu-Ohwayo, R. (1990b) Changes in the prey ingested and the variations in the Nile perch and other fish stocks of Lake Kyoga and the northern waters of Lake Victoria (Uganda). *J. Fish Biol.*, **37**, 55–63.

van Oijen, M.J.P. (1982) Ecological differentiation among the piscivorous haplochromine cichlids of Lake Victoria (East Africa). *Neth. J. Zool.*, **32**, 336–63.

van Oijen, M.J.P., Witte, F. and Witte-Maas, E.L.M. (1981) An introduction to ecological and taxonomic investigations on the haplochromine cichlids from the Mwanza Gulf of Lake Victoria. *Neth. J. Zool.*, **31**, 149–74.

Okaronon, J., Acere, T. and Ocenodongo, D. (1985) The current state of the fisheries in the northern portion of Lake Victoria. *F.A.O. Fish. Rep.*, no. 335, pp. 89–98.

Okemwa, E.N. (1981) Changes in the fish species composition of Nyanza Gulf of Lake Victoria, in *Proc. Workshop Kenya Mar. Fish. Res. Inst. Aquatic Resources of Kenya*, July 13–19, 1981, Kenya Nat. Acad. Adv. Arts Sci., Nairobi, pp. 138–56.

Okemwa, E.N. (1984) Potential fishery of Nile perch (*Lates niloticus*) Linne (Pisces: Centropomidae) in Nyanza Gulf of Lake Victoria, East Africa. *Hydrobiologia*, **108**, 121–6.

Reynolds, J.E. and Greboval, D.F. (1988) Socio-economic effects of the evolution of Nile perch fisheries in Lake Victoria: a review. *CIFA tech. Pap.* no. 17. (CIFA = Committee for Inland Fisheries of Africa; Food and Agriculture Organization of the United Nations.)

Ribbink, A.J. (1987) African lakes and their fishes: conservation scenarios and suggestions. *Env. Biol. Fishes*, **19**, 3–26.

Stearns, S.C. (1992) *The Evolution of Life Histories*, Oxford Univ. Press, Oxford, 249 pp.

Stearns, S.C. and Koella, J.C. (1986) The evolution of phenotypic plasticity in life-history traits: predictions of reaction norms for age and size at maturity. *Evolution*, **40**, 893–913.

Turner, J.L. (1977a) Changes in the size structure of cichlid populations of Lake Malawi resulting from bottom trawling. *J. Fish. Res. Bd Can.*, **34**, 232–8.

Turner, J.L. (1977b) Some effects of demersal trawling in Lake Malawi (Lake Nyasa) from 1968 to 1974. *J. Fish Biol.*, **10**, 261–71.

Wanink, J.H. (1991) Survival in a perturbed environment: the effects of Nile perch introduction on the zooplanktivorous fish community of Lake Victoria, in *Terrestrial and Aquatic Ecosystems: Perturbation and Recovery* (ed. O. Ravera), Ellis Horwood, Chichester, pp. 269–75.

Wanink, J.H. and Goudswaard, P.C. (1994) Effects of Nile perch (*Lates niloticus*) introduction into Lake Victoria, East Africa, on the diet of the pied kingfisher (*Ceryle rudis*). *Hydrobiologia*, **279/280**, 367–76.

Wanink, J.H., Berger, M.R. and Witte, F. (1993) Kingfishers at Fishburger Queens: eating Victorian fast food on the wing, in *Birds and the African Environment: Proceedings of the Eight Pan-African Ornithological Congress* (ed. R.T. Wilson), *Annls Mus. r. Afr. centr. (Sci. zool.)*, **268**, 319–26.

Welcomme, R.L. (1967) Observations on the biology of the introduced species of *Tilapia* in Lake Victoria. *Revue Zool. Bot. afr.*, **76**, 249–79.

Welcomme, R.L. (1988) International introductions of inland aquatic species. *F.A.O. Fish. tech. Pap.*, no. 294, 318 pp.

Witte, F. (1981) Initial results of the ecological survey of the haplochromine cichlid fishes from the Mwanza Gulf of Lake Victoria (Tanzania): breeding patterns, trophic and species distribution, with recommendations for commercial trawl-fishery. *Neth. J. Zool.*, **31**, 175–202.

Witte, F. (1984) Ecological differentiation in Lake Victoria haplochromines: comparison of cichlid species flocks in African lakes, in *Evolution of Fish Species Flocks* (eds. A.A. Echelle and I. Kornfield), Univ. Maine at Orono Press, Orono, Maine, pp. 155–67.

Witte, F. and Goudswaard, P.C. (1985) Aspects of haplochromine fishery in southern Lake Victoria. *F.A.O. Fish. Rep.*, no. 335, pp. 81–8.

Witte, F. and Goudswaard, P.C. (1987) Prospects of haplochromine fishery in southern Lake Victoria, in *From Form to Fishery* (F. Witte, 1987), thesis, Rijksuniversiteit Leiden, pp. 166–81.

Witte, F. and van Oijen, M.J.P. (1990) Taxonomy, ecology and fishery of haplochromine trophic groups. *Zool. Verh. Leiden*, **262**, 1–47.

Witte, F. and Witte-Maas, E.L.M. (1987) Implications for taxonomy and morphology of intra-specific variation in haplochromine cichlids of Lake Victoria, in *From Form to Fishery* (F. Witte, 1987), thesis, Rijksuniversiteit Leiden, pp. 33–137.

Witte, F., Barel, C.D.N. and Hoogerhoud, R.J.C. (1990) Phenotypic plasticity of anatomical structures and its ecomorphological significance. *Neth. J. Zool.*, **40**, 278–98.

Witte, F., Goldschmidt, T., Goudswaard, P.C., Ligtvoet, W., van Oijen, M.J.P. and Wanink, J.H. (1992a) Species extinction and concomitant ecological changes in Lake Victoria. *Neth. J. Zool.*, **42**, 214–32.

Witte, F., Goldschmidt, T., Wanink, J., van Oijen, M., Goudswaard, K., Witte-Maas, E. and Bouton, N. (1992b) The destruction of an endemic species flock: quantitative data on the decline of the haplochromine cichlids of Lake Victoria. *Env. Biol. Fishes*, **23**, 1–28.

Zaret, T.M. and Paine, R.T. (1973) Species introduction in a tropical lake. *Science*, **182**, 449–55.

Chapter seven

An analysis of species changes in Lake Victoria: did the Nile perch act alone?

Alida Bundy and Tony J. Pitcher

7.1 INTRODUCTION

The introduction of Nile perch, *Lates niloticus*, to Lake Victoria in the late 1950s has incited much controversy and discussion, primarily because of the subsequent dramatic loss in species diversity and abundance of a unique haplochromine group (Barel *et al.*, 1985, 1991; Coulter *et al.*, 1986; Payne, 1987; Ribbink, 1987; Bruton, 1990; Witte *et al.*, 1992a,b). Four species of tilapias were also introduced, the Nile tilapia, *Oreochromis niloticus*, successfully. For 20 years, while traditional fish catches continued to decline, catches of Nile perch were minimal. Then in the early 1980s the Nile perch population boomed and by 1990 the total annual catch from Lake Victoria had increased fivefold, to over 500 000 t (tonnes) (Fig. 7.1). The lake now supports a thriving fishery and supporting industries (Reynolds and Greboval, 1988; Greboval, 1990; Chapter 10). Catches of traditional species have further decreased and the fishery now largely comprises three species: the total demersal exotics, Nile perch and Nile tilapia and a native zooplanktivorous, pelagic cyprinid, *Rastrineobola argentea*, the dagaa or omena (Chapters 2–11).

The concurrent huge increase in the Nile perch population and decrease in the haplochromines has led many to conclude that the former is the cause of the latter (e.g. Barel *et al.*, 1991; Witte *et al.*, 1992a,b; Goldschmidt *et al.*,

The Impact of Species Changes in African Lakes. Edited by Tony J. Pitcher and Paul J.B. Hart.
Published in 1995 by Chapman & Hall, London. ISBN 0 412 55050 4.

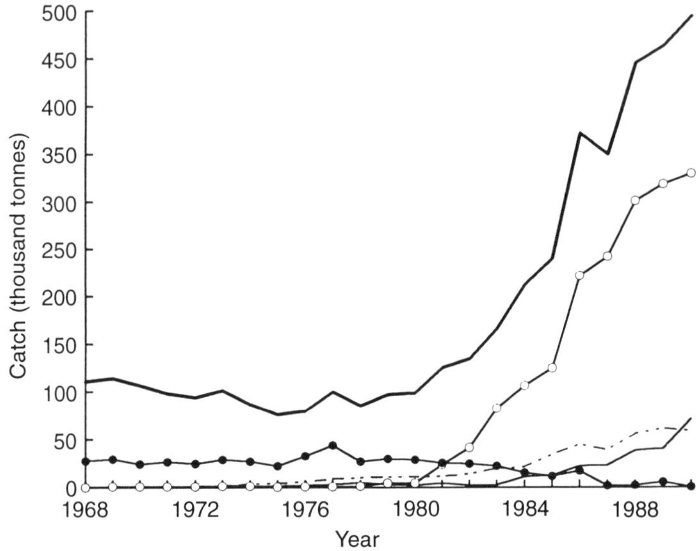

Fig. 7.1 Total annual catch and annual catch of Nile perch, haplochromines, *Oreochromis niloticus* and *Rastrineobola argentea* from Lake Victoria, 1968–1990. Heavy line, total catch; open circles, Nile perch; filled circles, haplochromines; thin line, *Oreochromis niloticus*; broken line, *Rastrineobola argentea*.

1993; Chapters 5, 6). However, continued fishing pressure, the long-term trend of falling catches (Graham, 1929; Garrod, 1960; Mann, 1970; Jackson, 1971; Fryer and Iles, 1972) and environmental changes in Lake Victoria may have had at least equivalent impacts on the aquatic community. In addition, the large time lag of 20+ years between the introduction of Nile perch and their eventual population explosion has yet to be satisfactorily explained. In this chapter we investigate whether we have any clear statistical evidence of the role of the Nile perch as the main agent perturbing the system, and what other factors may be involved in the species changes in Lake Victoria, including the great upsurge of the Nile perch. We have used multivariate analysis (principal components analysis, PCA, and multidimensional scaling MDS) and time series analysis to explore these questions with catch data from Lake Victoria.

7.2 METHODS

We had to assume that Lake Victoria catch data constitute an index of abundance because other data are not available for the time period required.

Clearly CPUE (catch per unit effort) data would be a preferred abundance index but effort data are too diverse, unstandardized and sporadic to use for this analysis. Our use of catch data is supported by experimental CPUE data from trawl surveys which exhibit the same general trends as the catch statistics, suggesting that the commercial catches are tracking abundance (Fig. 7.2).

Virtually all published work has treated the three countries bordering the lake, Kenya, Tanzania and Uganda, separately. Data from tagging studies indicate that Nile perch can travel large distances across Lake Victoria (Ligtvoet and Mkumbo, 1990; Ogutu-Ohwayo, pers. comm.; Chapter 5, Fig. 5.2) which suggests that it should be treated as a single stock. Morphometric evidence suggests that not only may Lake Victoria Nile perch be different from their parent stocks (Harrison, 1991) but also that there may be three morphs of Nile perch in the lake (Bwathondi, 1990; Ogutu-Ohwayo, pers. comm.). However, this is not yet confirmed and for present purposes it is assumed that there is only one species and one stock present. The potamodromous *Barbus* or *Alestes* species may have more than one stock in the lake, but these stocks are also unlikely to be described by national boundaries.

We have analysed the data from each country separately and for the whole of Lake Victoria for the time period 1968–1990. Fisheries data were obtained from the Kenya Fisheries Department *Statistical Bulletins*, the Tanzanian Fisheries Division *Annual Reports* and the Ugandan Fisheries Department. The Kenyan catch data were complete for 1965–1990. The Tanzanian catch data covered 1958–1989, but some of the data for five species were missing during the 1970s. The Ugandan data covered 1965–1990 with seven species having incomplete data. Missing data were interpolated by assuming a linear trend between known data points. There were no Tanzanian 1990 data available so the 1989 figures were used except where there was a clear trend which could be extrapolated to 1990. Tilapia were divided into *O. esculentus*, *O. niloticus* and other tilapia in the Tanzanian and Kenyan data sets, but most tilapia data were combined in Uganda. To make the data compatible, the Ugandan tilapia data were divided into these three groups, using the proportions estimated from the Tanzanian and Kenyan data. These figures were then tuned to available Ugandan data for individual tilapia species (Reynolds and Greboval, 1988). Total Lake Victoria catch data for 1968–1990 were obtained by summing the catch by species from the three riparian states. These total data comprised 15 species groups for which there were 23 years of data.

Principal components analysis

PCA is an ordination technique which can detect a range of different types of underlying data structure (Jongman *et al.*, 1987). It was used here to

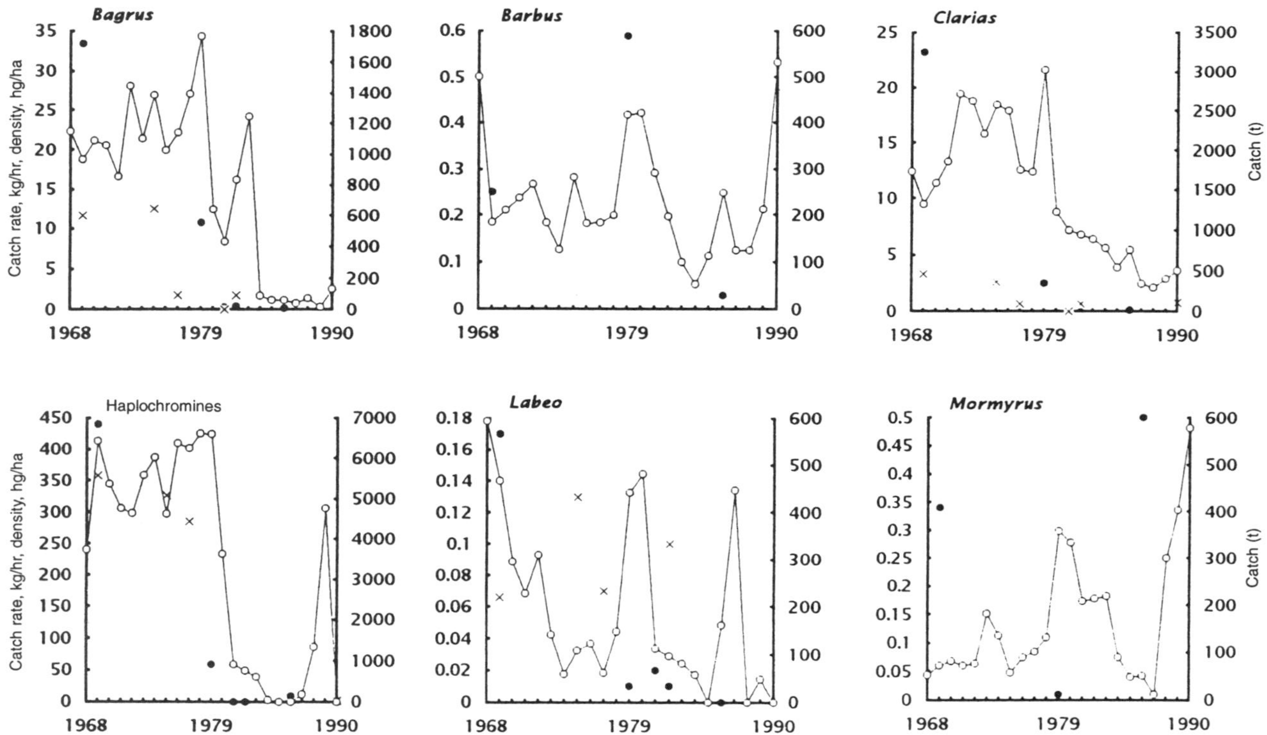

Fig. 7.2 Comparison of annual catch with trawl survey catch rate and stock density in Kenyan and Ugandan waters. Trawl survey and density data come from Kenya (Kudhongania and Cordone, 1974; Kenya Fisheries Department, 1988; Asila, 1991) with the exception of *Oreochromis esculentus* and *Protopterus* which come from Uganda (Kudhongania and Condone, 1974; Okaranon *et al*., 1988; and data obtained from J. Okaranon (UFFRO), pers. comm.). There were no survey data for *Alestes*, other tilapia and *Rastrineobola argentea*. Open circles, annual catch; filled circles, catch rate; crosses, stock density. Note differing axes and starting dates.

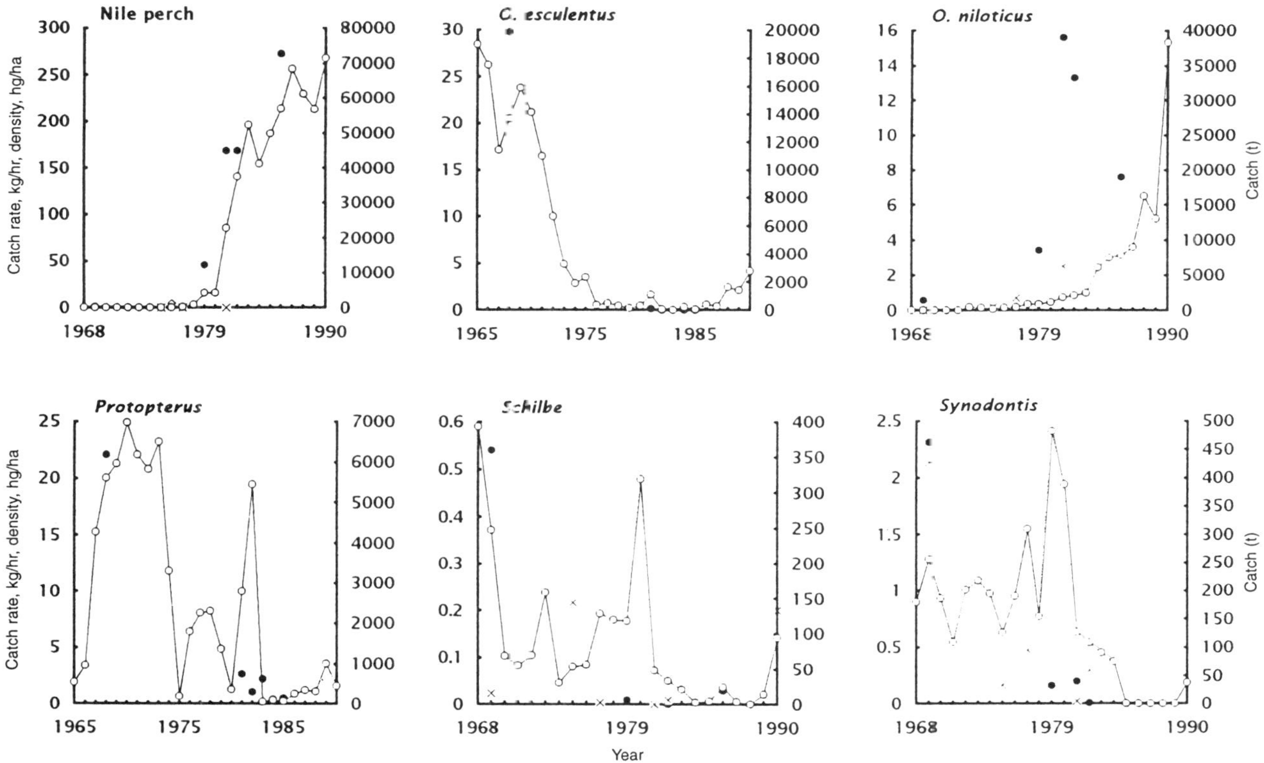

Fig. 7.2 (Continued)

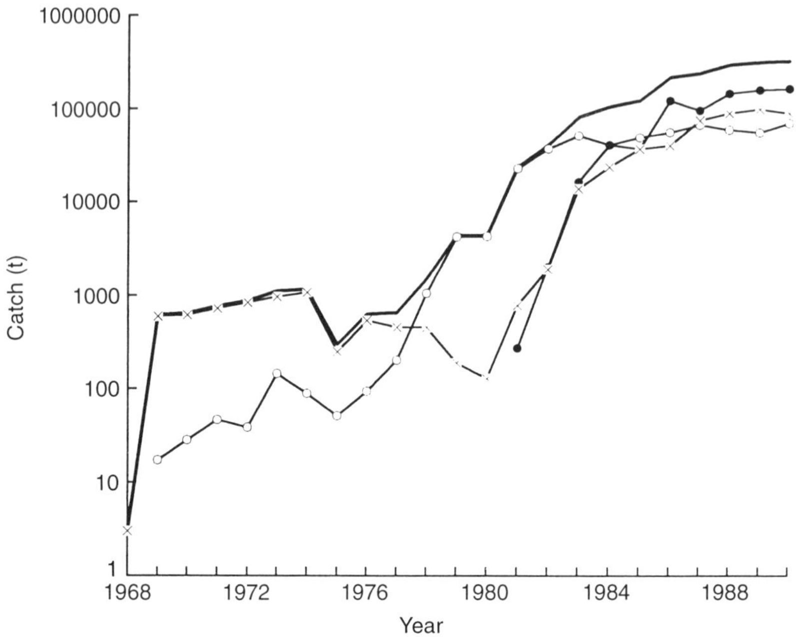

Fig. 7.3 Total Lake Victoria annual catch of Nile perch from Kenya, Tanzania and Uganda. *Y*-axis is log scale to show the detail of catch distribution before the upsurge of Nile perch in the 1980s. Open circles, Kenyan catch; filled circles, Tanzanian catch; crosses, Ugandan catch; heavy solid line, total catch.

examine whether there was an underlying pattern to the change in distribution of species in Lake Victoria by ordinating species composition in time. The PCA would be expected to show Nile perch significant on at least some of the principal components if it has had a major influence on the community structure of the lake since its introduction. PCAs were performed on both the whole-lake catch data and on the catch data from the individual countries. The data were also smoothed with a 3 year moving average.

Multidimensional scaling

MDS is also an ordination technique which explicitly measures dissimilarities between data sets (Jongman *et al.*, 1987) and was used here as an alternative method to PCA for examining the distribution of species through time. The whole-lake and individual-country data sets were used and scaled in two dimensions.

Time series analysis – cross correlation function

The CCF is a method of analysing time series data to examine possible interdependency between pairs of data with time (Diggle, 1990): correlations arc calculated over a range of time lags. We used this method on the Lake Victoria data to examine whether any significant correlations could be determined between Nile perch and the other species in the lake. The whole-lake data were examined for the whole time period and then separately for the pre-upsurge and the upsurge periods. The Uganda data were also used separately because the PCA results suggested that species changes occurred earlier in Uganda than in the two other countries and because we have a longer Nile perch catch data set for Uganda. Nile perch were first introduced into the north of the lake and the pre-1978 catches of Nile perch were higher in Uganda than in Kenya and were not recorded until 1981 in Tanzania (Fig. 7.3).

7.3 RESULTS

Principal components analysis

All the PCAs gave similar results, distinguishing the early years from the 1980s. Figure 7.4 shows the pattern of years ordinated in species space for Lake Victoria on the first two principal components, which explain 34% and 26% of the variation in the data respectively. There appears to be a

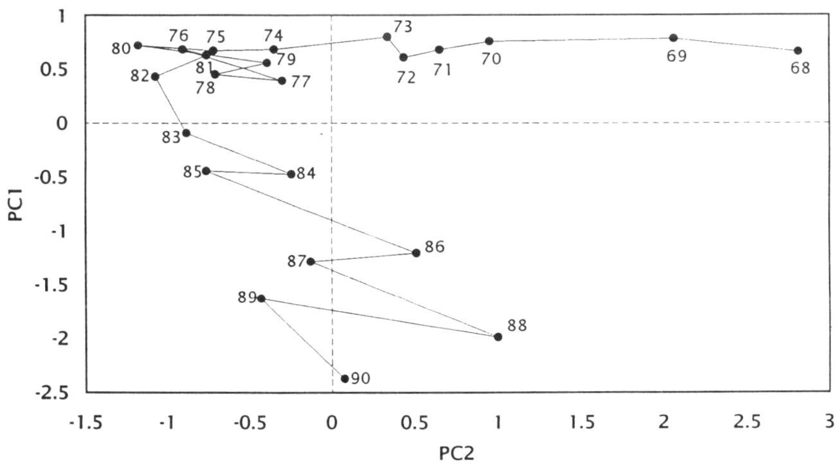

Fig. 7.4 Results of the PCA analysis of the whole lake showing the pattern of years ordinated in species space on the first two principal components.

progressive change in community structure. PC1 and PC3 (which explains 12.48% of the variation in the data) separate the 1980s and PC2 the early years, 1968–1974: the middle years are not well distinguished by any of the first four axes.

The loading of the species on the first three principal components are multivariate and are shown in Fig. 7.5. The most influential species on PC1 are Nile perch, *R. argentea*, *O. niloticus*, haplochromines, *Protopterus* and *Labeo*. *Barbus*, *O. esculentus*, *Alestes*, *Clarias*, *Mormyrus* and *Protopterus* are the most heavily loaded species on PC2, and so the separation of the early years is determined by the traditionally fished species. *Bagrus*, *Schilbe* and *Synodontis* predominate on PC3.

The results of the PCA of the Tanzanian and Kenyan catch data were similar to the whole-lake results, although the separation of the years in species space was not so distinct. The Ugandan results also showed a continuous progression through time, with PC1 separating the late 1960s to the mid 1980s and PC2 separating the 1980s (Fig. 7.6). In this case though, the axis (PC1) which explains most of the variation in the data (46.5%) separates the years up to the mid 1980s whereas the axis (PC2) separating the later years explains 20% of the data. The species loadings on the first two principal components are shown in Fig. 7.7. Both axes are multivariate with all species strongly influencing PC1 except *Bagrus* and *Synodontis*. On the second axis, *O. niloticus*, *R. argentea*, Nile perch, *Barbus*, *Bagrus* and *Synodontis* are influential.

Smoothing the data with a three-year moving average did not alter the general results of the PCA, although it did increase the percentage of variation explained by the principal components. The conclusions were the same.

Multidimensional scaling

The results of the MDS analysis confirmed the results found above. All data sets were scaled in two dimensions and the years were separated into two distinct groups, the 1960s–1970s and the 1980s. In all cases there were one or two years which fell between the two groups, and this varied nationally. For the whole-lake data, the intermediate years were 1981/82, for Kenya 1979/80, Uganda 1982/83 and Tanzania 1983.

Time series analysis – cross correlation function

The results of the time series analysis of the whole-lake data are shown in Fig. 7.8. The CCFs fall into four distinctive patterns of correlation with Nile perch. Group I, *R. argentea* and *O. niloticus*, has a strong positive correlation with Nile perch with no time lag. The second group, the haplochromines, *Protopterus* and *Labeo*, has a strong negative correlation with zero time lag.

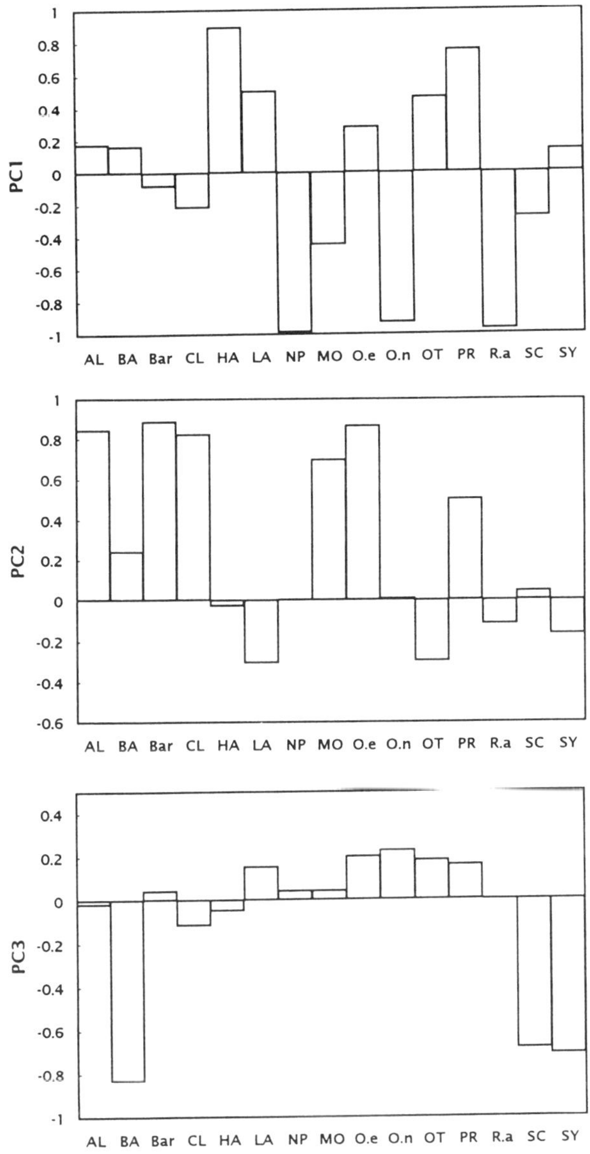

Fig. 7.5 Results of the PCA analysis of the whole lake showing the loadings of the 15 species groups. The percentage of variation explained by the first five principal components is: PC1, 34.08; PC2, 26.25; PC3, 12.48; PC4, 7.82; PC5, 6.73 (PC4 and PC5 not shown). Al, *Alestes*; BA, *Bagrus*; Bar, *Barbus*; CL, *Clarias*; HA, haplochromines; LA, *Labeo*; NP, Nile perch; MO, *Mormyrus*; O.e, *Oreochromis esculentus*; O.n, *Oreochromis niloticus*; OT, other tilapia; PR, *Protopterus*; R.a, *Rastrineobola argentea*; SC, *Schilbe*; SY, *Synodontis*.

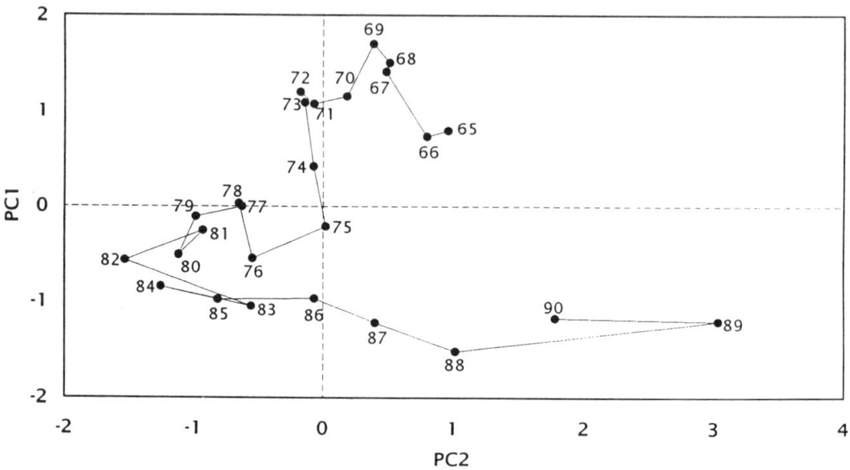

Fig. 7.6 Results of the PCA analysis of the Ugandan data showing the pattern of years ordinated in species space on the first two principal components.

The correlation pattern of group III, *O. esculentus, Barbus, Clarias, Mormyrus* and *Alestes*, changes from a negative correlation with a negative time lag to a positive correlation with a positive time lag. The only significant correlations are the negative correlation between Nile perch and *O. esculentus* at minus 5 years and the positive correlation between Nile perch and *Mormyrus*. Group IV, *Bagrus, Synodontis* and *Schilbe*, shows the opposite pattern, changing from a positive correlation with a negative time lag to a negative correlation with a positive time lag. The positive correlations of Nile perch with *Bagrus* and *Schilbe* are significant at minus 4 years.

The Ugandan results are similar to those above but indicate a greater correlation of Nile perch with the other species (Fig. 7.9). Group I and group IV are the same as above. The species in groups II and III above now have the same correlation pattern with Nile perch, that is, a negative correlation with a negative time lag. With the exception of *Labeo* and *Alestes* all the correlation sets achieve significance.

These species groups produced by the CCF can be traced back to the results of the PCA and are consistent with the species which influenced the loadings on the principal components (Figs 7.5, 7.7). The two analyses are independent because the PCA contains no information about time lags. Thus, for the whole-lake data, Nile perch and group I, *R. argentea* and *O. niloticus*, show a strong negative influence on PC1 while group II, the haplochromines, *Protopterus* and *Labeo*, exert a strong positive influence. Group III can be identified on PC2 as the main species important in explaining the variation

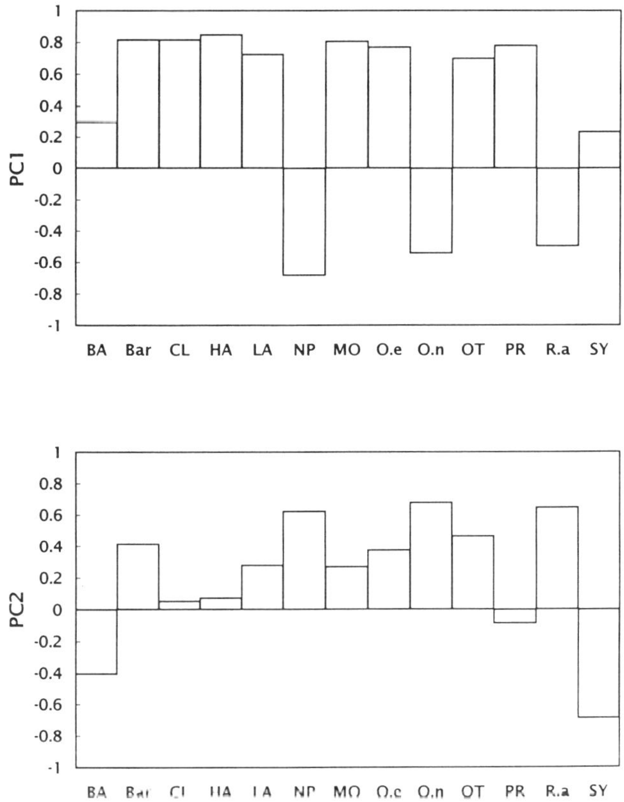

Fig. 7.7 Results of the PCA analysis of the Ugandan data showing the loadings of 13 species groups on the first two principal components. BA, *Bagrus*; Bar, *Barbus*; CL, *Clarias*; HA, haplochromines; LA, *Labeo*; NP, Nile perch; MO, *Mormyrus*; O.e, *Oreochromis esculentus*; O.n, *Oreochromis niloticus*; OT, other tilapia; PR, *Protopterus*; R.a, *Rastrineobola argentea*; SY, *Synodontis*. *Schilbe* is not present in the Ugandan data set and *Alestes* was not included in the PCA analysis because there are no catch data after 1971.

in the data. Group IV consists of the species which influence the loadings on PC3. The Uganda results show a similar connection between the PCA and CCF results. Group I is again identified with the species which have a negative influence on the loadings of PC1, and group II–III the species with a positive influence. Group IV exerts a negative influence on the loadings of PC2.

The data set was split into two and analysed separately to examine the possible influence of Nile perch during the pre (1968–1978) and Nile perch (1979–1990) upsurge periods. The results of the Nile perch upsurge data

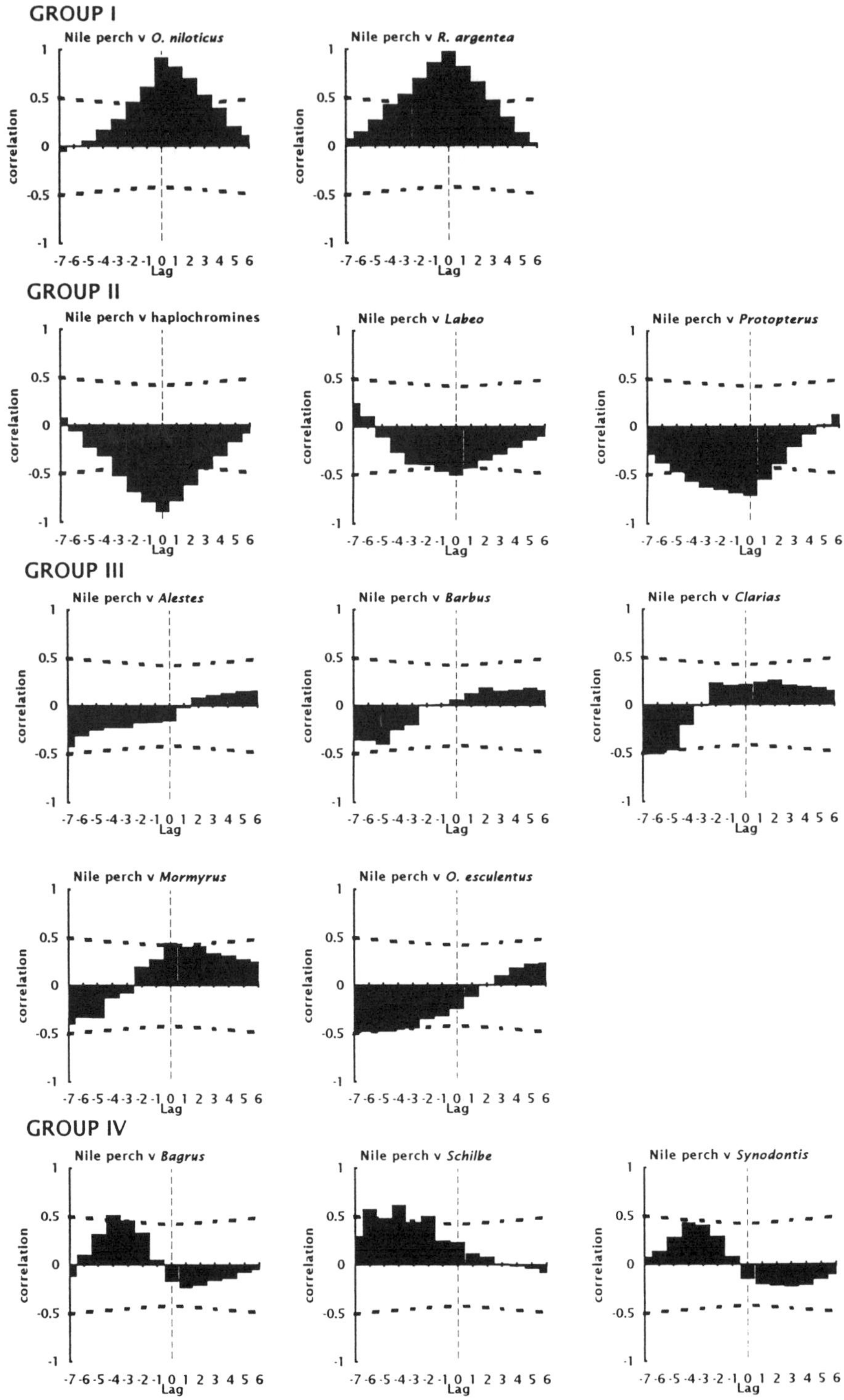

Fig. 7.8 See page 125.

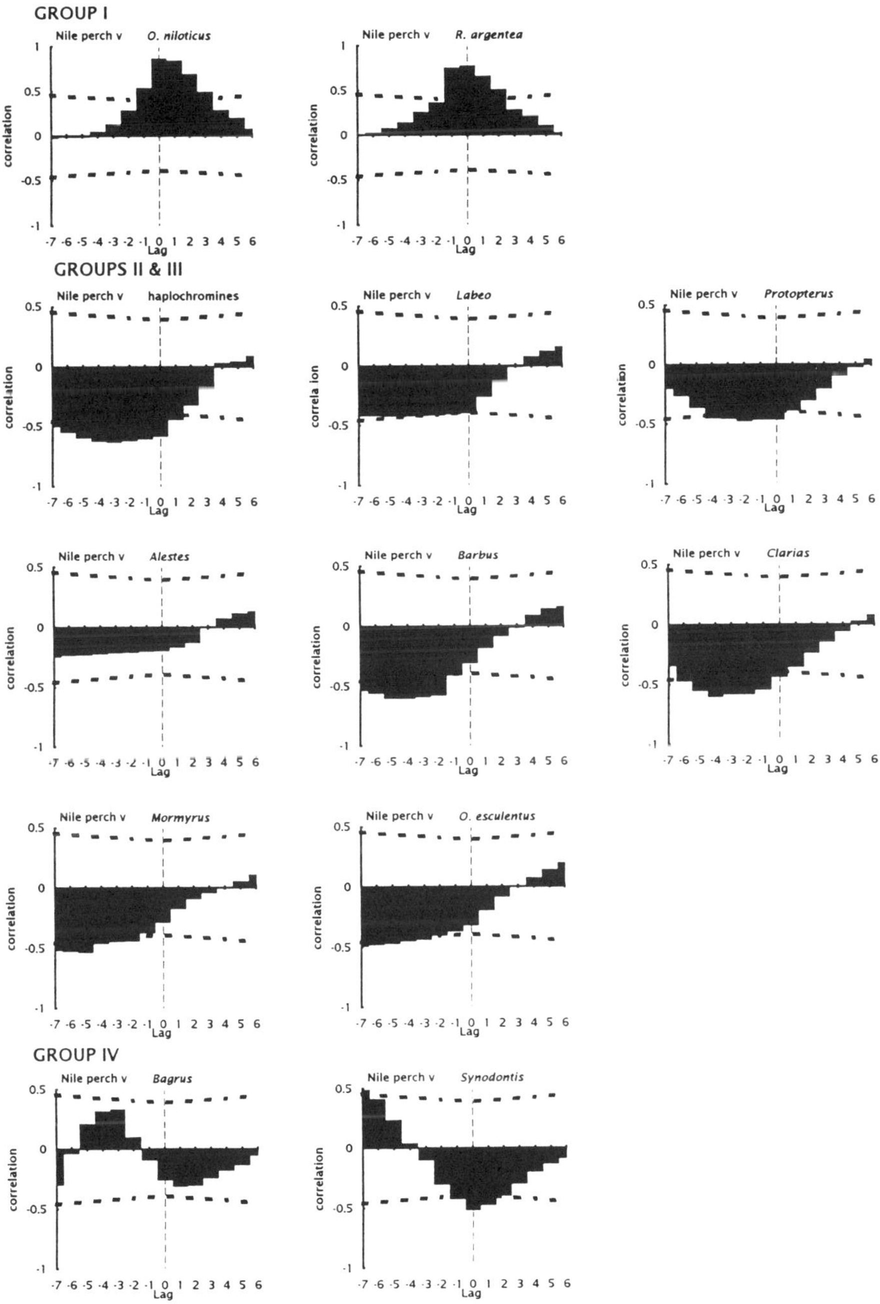

Fig. 7.9 See page 125.

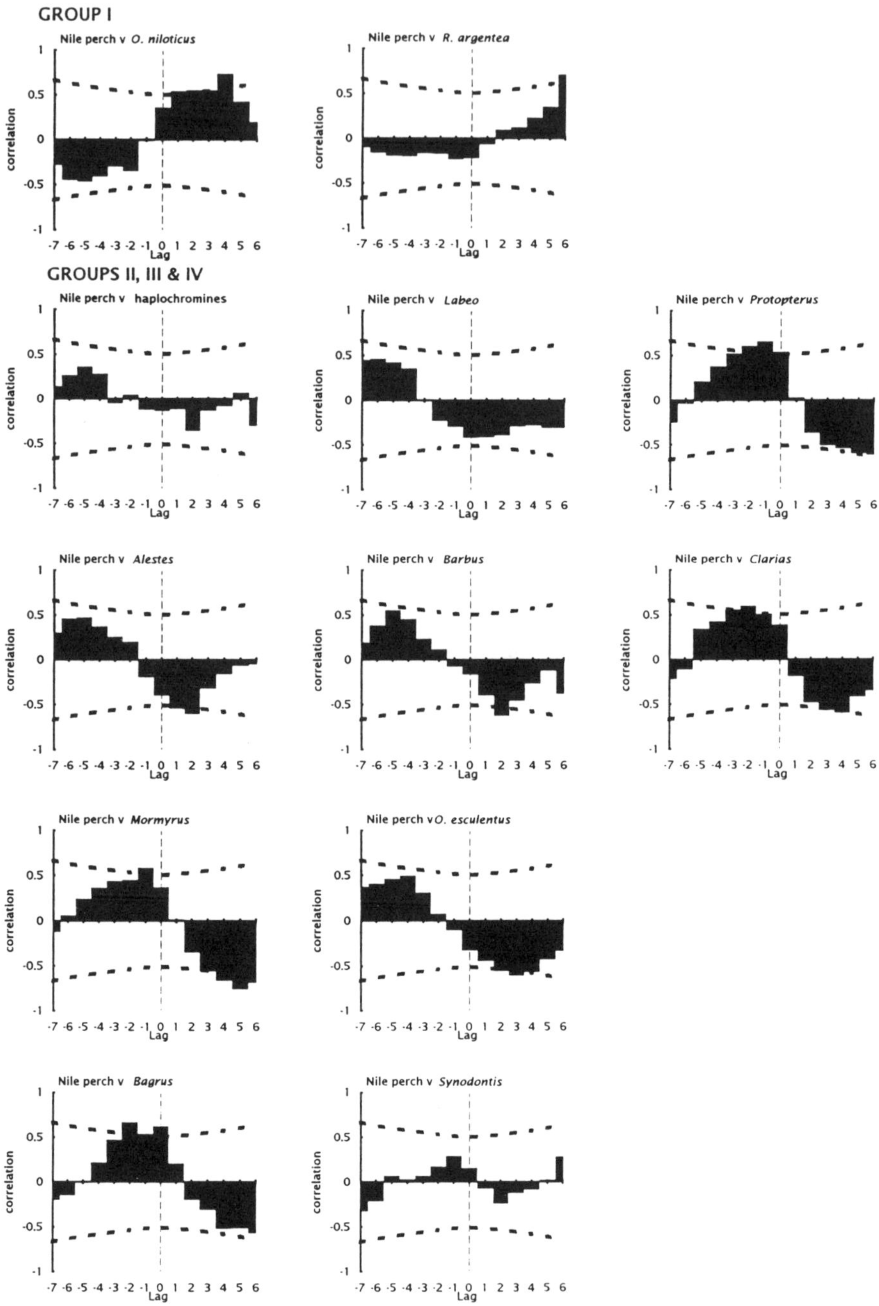

Fig. 7.10 See page 125.

were similar to the whole time series for both data sets, but the pre-Nile perch data sets produced different results. The whole-lake data for 1968–1978 showed no significant correlations but for the Ugandan data (1965–1980), the species fell into two groups (Fig. 7.10). Group I consisted of *R. argentea* and *O. niloticus*, as for the whole time period, this time with a positive correlation at plus 4 years. The second group consisted of all the other species and its general pattern was a swing from a large positive correlation with a negative time lag to a large negative correlation with a positive time lag. Most species' CCFs achieved significance.

7.4 DISCUSSION

While our analysis would certainly have been more rigorous had we been able to use better abundance estimates, we feel that it is nonetheless meaningful and useful in the interpretation of the changes that have occurred in Lake Victoria. The prevailing view that predation by the Nile perch has been the primary cause of the species changes in Lake Victoria, particularly the dramatic loss of many haplochromine species, is, we feel, questioned by the analysis. Instead of a clear-cut picture of the Nile perch as the single most important variable affecting our analysis, what emerges is a less distinct picture with the Nile perch, amongst other factors, as one of the probable causes of change.

The results of principal components analysis and multidimensional scaling confirm the obvious, that is, that there are two distinct time periods, the pre-upsurge and upsurge phases. In addition, however, Figs 7.4 and 7.6 indicate that there has been a progressive change in community structure with no saltatory effects. The direction of this change altered with the Nile perch upsurge in the early 1980s, but the PCA showed that the axis which explained the variation in these years was multivariate, a result that is not so intuitive. Although the Nile perch was an important influence on this axis, so were several other species. Principal components are theoretical

Fig. 7.8 Results of the whole-lake cross correlation analysis showing the correlation pattern of Nile perch with the other species groups, 1968–1990. Heavy dashed lines represent the 95% confidence limits. Any value outside of these lines is statistically significant. The lag is measured in years.

Fig. 7.9 Results of cross correlation analysis of the Ugandan data showing the correlation pattern of Nile perch with the other species groups, 1965–1990.

Fig. 7.10 Results of cross correlation analysis of the Ugandan data showing the correlation pattern of Nile perch with the other species groups, for the pre-Nile perch upsurge period, 1965–1980.

variables explaining pattern and it would be reasonable to expect the Nile perch to appear on at least one axis as the single most important factor affecting the species composition, but this is not the case, even in the 1980s. In all cases all axes were multivariate, for both the whole-lake data and the Ugandan data.

The aim of the time series cross correlation analysis was to determine whether there was any evidence of interactions between Nile perch and the other fish in the lake. The results do suggest that there were significant predatory/prey interactions between many of the species and Nile perch, particularly for the Ugandan data set. In the analysis of the Ugandan data set before the Nile perch upsurge, a cross correlation pattern of a negative time lag switching to a positive correlation with a positive time lag is consistent with classic predator–prey cycles. These were apparent for all species other than group I (Fig. 7.10), supporting the concept of Nile perch as a broad-spectrum predator. However, in the CCF analysis of the data from 1965 to 1990, this cycle is no longer evident for any species other than group IV (Fig. 7.9). All other species (groups II and III) have a negative correlation with Nile perch with a negative time lag, indicating that the decline of these species preceded the increase of Nile perch by 0–7 years. The suggestion is that these species decreased to the extent that they could no longer cycle with Nile perch, which had meanwhile switched to other prey. This implies that the population explosion of Nile perch may not have been fuelled by these species. Part of the decrease of these species will have been due to predation, but these species were also fished during this period and many were in decline since well before the Nile perch introduction.

The CCF analysis of the whole lake gave similar results. Groups III and IV were as described above. Group II (haplochromines, *Labeo* and *Protopterus*), though, had strong negative correlations with a zero time lag. Can this be interpreted as an immediate impact of predation by Nile perch? Neither *Labeo* nor *Protopterus* is recorded as an element in the Nile perch diet (Hughes, 1986; Ogutu-Ohwayo, 1990a), and these species could be declining in parallel for other reasons, such as fishing. Similarly, the decrease in haplochromines may be due to predation or a combination of predation and fishing. The only species that did not seem to suffer from predation by Nile perch were group I (*O. niloticus* and *R. argentea*), which increased at the same time as Nile perch, and were clearly distinguished in all the CCFs.

So the results do confirm the impact of predation by Nile perch on other species in the lake, including haplochromines. However, relating these results back to the PCA results once more, it is apparent that most of the predation effects of Nile perch were evidenced before the boom, not afterwards. For example, group II–III in the Ugandan analysis were associated with PC1 which discriminated the years before the Nile perch upsurge. Group I and Nile perch were also correlated with this axis in the opposite direction (Fig.

7.6). Again, in the whole-lake analysis, group III was associated with the axis (PC2) that explained the variation in the pre-upsurge period, whereas groups I and II were associated with the upsurge period. In both PCAs, though, group IV was also important in discriminating later years.

Our quantitative results suggest that, although Nile perch clearly ate haplochromines and other species in Lake Victoria, additional factors have influenced the pattern of species change. The first two principal components of the PCA differentiated the pre-upsurge period from the upsurge period in the lake, and changes other than the explosion of the Nile perch population occurred during these years. We discuss here changes in fishing pressure and the environment that may have influenced the pattern of species change.

Fishing pressure

There is a long history of fishing in the lake. Graham was sent there in the late 1920s to report on the Lake Victoria fisheries to the British Crown: he recommended increasing mesh sizes (Graham, 1929). Since then, various others have made recommendations (e.g. Beverton, 1959), but little has been done to control fishing pressure. Effectively, all management recommendations have been ignored (Ogutu-Ohwayo, 1990b). The lake is too large and resources of the fishery departments too small (Chapter 10). Even with adequate resources the lake would be difficult to police. So, mesh sizes decreased throughout the years, and fishermen continually switched target to smaller and smaller fish until eventually even the unpalatable bony haplochromines were exploited.

The total annual catch was relatively stable at about 100 000 tonnes (Fig. 7.1). Ligtvoet and Witte (1991) and Ogutu-Ohwayo (1990b) suggest that the pre-Nile perch upsurge fish community had stabilized and could cope with the fishing pressure exerted on it: what it could not cope with was the introduction of a new, efficient piscivore. However, it is questionable whether the pre-upsurge system could really be described as stable, given the changes in fishing effort described above. The composition of the total catch changed continually throughout the years. In the 1970s, before the catch soared due to Nile perch, it consisted largely of the haplochromines, but before this, tilapia. This does not represent a stable system. This instability is reflected in the value of the catch: Greboval (1990) states that the stability in the total catch figure masks a decline in the real value of the fish, "resulting from overfishing".

So we have the situation where traditional fish stocks had long been declining. By the time Nile perch was introduced, we suggest that the stocks of the native fish species were already reduced and vulnerable due to constant overfishing. Can the same reasoning be applied to the dramatic decrease in abundance and species diversity of the haplochromines? That is, were the haplochromines already vulnerable due to fishing effort?

Haplochromines were intensively studied by the Haplochromis Ecology Survey Team (HEST), who conducted various surveys across a transect of the Mwanza Gulf, Tanzania from 1979 to 1990 (for example, see Witte, 1981, Witte *et al.*, 1992a, Witte *et al.*, Chapter 6). There is therefore a reliable detailed assay of the changes in species composition which occurred in this part of the lake over this period.

In 1976, commercial trawling started in the Mwanza Gulf (Witte, 1981) following the recommendations of an FAO/EAFFRO survey (Kudhongania and Cordone, 1974). However, the fishery was not sustainable – catch rates fell (Witte and Goudswaard, 1985), trawling became uneconomical and some boats switched to trawling for Nile perch (Reynolds and Greboval, 1988). Some authors (e.g. Greboval, 1990; Ogutu-Ohwayo, 1990b) have expressed the belief that if trawling for haplochromines had been practised lake-wide as originally intended, then the overall abundance of haplochromines would have declined anyway as a consequence of overfishing. In addition, trophic specialization and life history characteristics of many of the haplochromine species would have rendered them especially vulnerable to exploitation.

Figure 4 in Witte *et al.* (1992a) illustrates the decline in catch rate of haplochromines and the increase of Nile perch from 1972 to 1986 in the Mwanza Gulf. These data come from the Tanzanian Freshwater Fisheries Training Institute research trawl surveys using mesh sizes of 20 mm and 90 mm. Within 4 years of the start of the trawl fishery, the catch rate of larger haplochromines by the 90 mm net fell from $8\,kg\,h^{-1}$ to less than $1\,kg\,h^{-1}$. This means that the abundance of haplochromines larger than 17 cm SL had declined almost tenfold before the explosion in the Nile perch population.

The catch rate of haplochromines by the 20 mm Mwanza trawl net shows a continuous downward linear trend, indicative of a fishery where fishing pressure is too high. By 1984 and 1985, when large amounts of juvenile Nile perch were suddenly recorded in trawl surveys (Goudswaard and Witte, 1985; Goudswaard, 1988) indicating the initial increase of Nile perch due to migration of adults and juveniles from other parts of the lake (Witte *et al.*, 1992b), the catch rate of haplochromines had already fallen by 50–65% from between 1250 and $1550\,kg\,h^{-1}$ to $500–600\,kg\,h^{-1}$. Subsequently, the catch rate fell further to around $300\,kg\,h^{-1}$ in 1986 and to a minimal amount in 1988 (Barel *et al.*, 1991). From the data available, there does not seem to be a sudden impact of the Nile perch on the catch rates of the haplochromines. After the upsurge of Nile perch, the catch rate continued to decline at the same rate as it had done beforehand. Presumably fishing pressure by humans was replaced by predation pressure by Nile perch. But by this stage the abundance of haplochromines had already been much reduced. Similar conclusions can be made from trawl data for the Nyanza Gulf in Kenya (Table 6 in Asila, 1991; Table 2 in Achieng, 1990).

Some authors have categorically stated that the Nile perch is the cause of the haplochromine demise (e.g. Witte *et al.*, 1992a,b and Chapter 6). Witte *et al.* (1992b) identify four observations which "demonstrate" that the impact of Nile perch was much greater than the impact of fishing. We are not trying to suggest in this Chapter that the Nile perch did not predate on the haplochromines (and in fact our CCF results imply that they did), nor that it did not have an impact on them. Our thesis is that Nile perch was not the only cause of the species changes in Lake Victoria and perhaps not the main cause. So with this in mind we address the four points made by Witte *et al.* (1992b).

1. *Haplochromine stocks also vanished from areas without high fishing pressure.* Witte *et al.* (1992a,b) point out that in areas of the lake where there was no intensive haplochromine trawl fishery, e.g. Nyanza Gulf, Speke Gulf, Emin Pasha Gulf and deeper waters, experimental catch rates of haplochromines also declined markedly after the Nile perch upsurge.

The differentiation should first be made between areas that are not fished at all and those which are subject to artisanal fishing. For example, the Nyanza Gulf has been subjected to very intensive beach seine and mosquito seine net fishing since the early 1970s (Marten, 1979; Asila, 1991). Mean lengths of haplochromines had decreased by half in heavily fished areas of the Nyanza Gulf, suggesting that the larger haplochromines were being fished out. Marten in fact suggested, before the explosion of Nile perch, that the haplochromines would not be able to sustain heavy fishing pressure and predation. This pressure has not dropped at any stage since then. Mosquito and seine nets have increased to more than 50% of fishing gear in 1989 (Rabour, 1989) and take more than 50% of the catch by weight (Asila, 1991). Mosquito nets catch both small fish and the juveniles of larger fish, they destroy nests on breeding grounds and disrupt courtship (Ogutu-Ohwayo, 1990b). Both methods of fishing have long been identified as a major problem around the whole lake (CIFA, 1992): it has been recommended that beach seines and mosquito seines with stretched mesh size less than 10 mm should be banned.

So, in Kenyan waters at least, the demise of the haplochromines can also be accounted for by intensive artisanal fishing. Kudhongania and Chitamwebwa (Chapter 2) also complain of a prolonged unrestricted fishery and the use of undesirable fishing methods in Ugandan waters. Haplochromines are mouthbrooders and only produce a few eggs. They are thus very sensitive to perturbation by fishing gear. Anon. (1988) reports a survey of the Kenyan part of the lake, including the Nyanza Gulf, by the National Museums of Kenya and the British Museum, which found that in areas where fishing had been banned, haplochromines were abundant; in areas close by where fishing was allowed, they were scarce. It is thus conceivable that intensive artisanal fishing has contributed to the reduction of haplochromines all around the lake. It is not known how much fishing pressure haplochromines can sustain, but given their reproductive behaviour, it is likely to be low (Chapter 6).

2. *The Nile perch upsurge occurred in different parts of the lake in different years and in each case a severe decline in haplochromines followed shortly afterwards*, and
3. *Before the haplochromines disappeared there was an inverse relationship between Nile perch catches and catches of haplochromines.*

These two observations are connected. The Nile perch appear to have first 'exploded' in the north of the lake in Ugandan and Kenyan waters and then migrated southwards in anticlockwise fashion to Tanzanian waters, presumably migrating as local areas were overpopulated (Witte *et al.*, Chapter 6). After the 'explosion' in the Nyanza Gulf and then in the Mwanza Gulf, catches of haplochromines in both gulfs decreased dramatically. Clearly Nile perch ate some haplochromines during this period. However, we have suggested that in both cases the haplochromines were already depleted due to fishing pressure, and the rate of decline of haplochromines did not differ between the two periods. In little-fished or unfished areas, apart from the 1973 lake-wide survey (Kudhongania and Cordone, 1974), there are no comparative abundance data. For example, the little-fished Emin Pasha Gulf, about 150 km to the west of the Mwanza Gulf (Chapter 6, Fig. 6.1), showed a decrease in experimental haplochromine catch rates from $322 \pm 235\,\text{kg}\,\text{h}^{-1}$ in June 1985 to $64 \pm 115\,\text{kg}\,\text{h}^{-1}$ in August 1986 and an increase in the catch rates of Nile perch from $54 \pm 82\,\text{kg}\,\text{h}^{-1}$ to $107 \pm 89\,\text{kg}\,\text{h}^{-1}$ (Table 2 in Goudswaard and Ligtvoet, 1988). However, because there are no pre-Nile perch upsurge data, there is no evidence that the rate of decline of haplochromines changed during the upsurge period.

4. *Until the haplochromines disappeared they were the main food organisms of the adult Nile perch.*

Diet studies have shown that during the upsurge, Nile perch fed on the haplochromines until they had virtually disappeared, whereupon it switched to smaller food organisms (Hughes, 1986; Ligtvoet and Mkumbo, 1990; Ogutu-Ohwayo, 1990a). Our CCF analyses suggest that before and during the upsurge, Nile perch was a broad-spectrum feeder; it is not in contention that Nile perch ate haplochromines. The question is whether the predation by Nile perch was the only cause of the dramatic loss of haplochromines in Lake Victoria.

Environmental issues

Concurrent with the introduction of the Nile perch to Lake Victoria has been the increasing eutrophication of the lake and the developing presence of a permanent anoxic layer below 30–35 m (Chapter 3). This anoxic layer might explain the disappearance of haplochromines from deeper parts of the lake. Seasonal mixing, and to a lesser extent, daily mixing of the water column would affect the water levels immediately above 30–35 m. Any species that cannot tolerate a low level of dissolved oxygen would be adversely affected by this anoxic layer.

The phytobiomass of the lake has increased (Bugenyi, 1992; Goldschmidt *et al.*, 1993; Hecky, 1993), resulting in an increase in fish kills (Ochumba, 1988). Hecky (1993), on the basis of core sediment samples, has pointed out that fundamental changes to Lake Victoria's limnology were well advanced by the time of the Nile perch population explosion. Lake sediments have undergone dramatic changes that predate the introduction of Nile perch. For example, the diatom *Melosira*'s replacement by *Nitzschia* is symptomatic of increased oxygen in water deeper than 45 m. Core sediments show that in the 1920s nitrogen deposition increased, and in the 1950s phosphate increased. Declining silicon reserves in the lake now favour algal rather than diatom communities.

There have been changes in the zooplankton structure of Lake Victoria, and indeed the whole food web has become simplified: the haplochromine detritivores/phytoplanktivores have been replaced by the freshwater prawn *Caridina nilotica* and the zooplanktivores have been replaced by *Rastrineobola argentea* (Ligtvoet and Witte, 1991; Goldschmidt *et al.*, 1993). The situation is quite complex: trophic cascade theory would predict that the introduction of a top predator would lead to a decrease in zooplanktivores, an increase in zooplankton and therefore a decrease in phytobiomass (Carpenter *et al.*, 1985). However, the phytobiomass has increased.

There are several reasons why predictions from trophic cascade theory do not apply to Lake Victoria. Since *R. argentea* replaced the zooplanktivorous haplochromines and has managed to coexist with Nile perch, likely due to the difference in their reproductive strategies (altricial versus precocial species, Bruton, 1990; *r*-versus *K*-selection, Wanink, 1991), zooplankton may not have increased. In addition, it has been suggested that the large decrease of phytoplanktivorous and detritivorous haplochromines, which previously contributed 50% of the biomass of the haplochromines (Witte *et al.*, 1992b), should be considered as a causal factor in the development of algal blooms (and so, by implication, predation by Nile perch) (Witte *et al.*, 1992b; Goldschmidt *et al.*, 1993). However, Ogutu-Ohwayo and Hecky (1991) and Hecky (1993) consider that the impact of human riparian activities such as clearing and burning of forest and savannah, and agriculture in the watershed are more likely candidates, given the sediment record.

Hecky's hypothesis is that eutrophication favoured the expansion of the freshwater prawn, *C. nilotica*, a major food source for juvenile Nile perch. This increased recruitment success of the Nile perch population, leading to the observed population explosion in the 1980s. Meanwhile, the *K*-selected haplochromines might have a poorer ability to respond to the increased food supply of the eutrophicating lake, unlike the *r*-selected Nile perch and *R. argentea.* This hypothesis concurs with the results of our CCF analysis, which showed that Nile perch increased while lagging some years behind decreases in the original prey species, indicating that another food source had been found, that is, *C. nilotica.*

7.5 CONCLUSIONS

There has been a huge loss in species diversity and abundance of the haplochromines since the 1970s in Lake Victoria, and since the late 1970s the population of the predator Nile perch has increased tremendously. HEST results show that haplochromine species diversity decreased from 30+ to 1–2 species at sublittoral stations along a transect of the Mwanza Gulf, while Nile perch accounted for 90% of the ichthyobiomass in the late 1980s.

The multivariate and time series analyses presented here have been used to question whether the upsurge in Nile perch was the main cause of the decline of haplochromines. The PCA analysis suggested that Nile perch was not the only factor influencing species changes in the lake. While the CCF analysis supported the Nile perch as predator, most of this predation appeared to pertain to years prior to the Nile perch upsurge. We have suggested that fishing pressure and environmental changes may have had as large, if not larger, effects on Lake Victoria as the introduction of Nile perch. Unfortunately all three effects are confounding and it has not been possible to clearly distinguish the effects of predation and fishing. For example, Witte *et al.* (1992b) report that in shallow waters, some recovery of haplochromines has been observed and the authors attribute this to a decrease of Nile perch in these areas. However, an alternative explanation could be that the haplochromines have recovered due to the use of larger meshes in nets directed at the Nile perch and therefore a decrease in haplochromine fishing mortality.

Hecky (1993) suggested that eutrophication led to expansion of *C. nilotica*, which in turn enabled the explansion of Nile perch. Other than hypothetical genetic changes in the Nile perch population, this is the only theory that accounts for the long delay between the Nile perch introduction and explosion. Our analysis, which is largely congruent with Hecky's findings, suggests that introduced Nile perch should be considered as only one of three critical factors that led to the reduction of haplochromine diversity in Lake Victoria. The two other factors are first, an anthropogenic increase in nutrients and production that has favoured *r*-selected species over the *K*-selected haplochromines. Secondly, many of Lake Victoria's haplochromines had low resilience to both commercial and artisanal fishing pressure. Irrespective of Nile perch predation, the latter two factors may have brought about major extinctions of specialized haplochromine fish species in Lake Victoria.

ACKNOWLEDGEMENTS

We thank George Turner and an anonymous referee for earlier comments on an initial draft of this chapter, Eric Reynolds and Dominique Greboval for insightful discussion and John Craig and John Okaranon for making data available. Early work on this project was supported by the Overseas Development Administration of the UK Government.

REFERENCES

Achieng, A.P. (1990) The impact of the introduction of Nile perch, *Lates niloticus* (L.) on the fisheries of Lake Victoria. *J. Fish Biol.*, **37** (Suppl. A), 17–23.

Anon. (1988) Monster fish may be innocent of ecological crimes. *New Scientist*, **1622**, 34.

Asila, A. (1991) Fishery resource base for the Kenyan sector of Lake Victoria, in *Report of a National Seminar on the Development and Management of the Kenyan Fisheries of Lake Victoria, Kisumu, Kenya, 22–24 July 1992.* RAF/87/099-TD/27/91 (EN) (eds G.W. Ssentongo and N.J. Dampha), IFIP, pp. 29–47.

Barel, C.D.N., Dorit, R., Greenwood, P.H., Fryer, G., *et al.* (1985) Destruction of fisheries in Africa's lakes. *Nature, Lond.*, **315**, 19–20.

Barel, C.D.N., Ligtvoet, W., Goldschmidt, T., Witte, F. and Goudswaard, P.C. (1991) The haplochromine cichlids in Lake Victoria: an assessment of biological and fisheries interests, in *Cichlid Fishes: Behaviour, Ecology and Evolution* (ed. M.H.A. Keenleyside), Chapman and Hall, London, pp. 258–79.

Beverton, R.J.H. (1959) Report on the state of the Lake Victoria fisheries. Fisheries Laboratory, Lowestoft. (Mimeo)

Bruton, M.N. (1990) The conservation of the fishes of Lake Victoria, Africa: an ecological perspective. *Env. Biol. Fishes*, **27**, 161–75.

Bugenyi, F.W. (1992) The effect of limnological factors and the changing environment on the fisheries of northern Lake Victoria. In *Report of a National Seminar on the Development and Management of the Ugandan Fisheries of Lake Victoria, Jinja, Uganda, 6–8 Aug. 1991.* RAF/87/099-TD/31/92 (EN) (eds G.W. Ssentongo and F.L. Orach-Meza), IFIP, pp. 36–58.

Bwathondi, P.O.J. (1990) The state of fishing industry in Tanzania with particular reference to inland fishery, in *The Prevailing Activities on the Lake Victoria Basin with Particular Reference to the Lake Fisheries:* Proceedings of the workshop organized by the European Community under the project No. 5100.36.94.401, Mwanza, Tanzania, 8–9 Mar. 1990 (ed. P.O.J. Bwathondi), pp. 5–11.

Carpenter, S.R., Kitchell, J.F. and Hodgson, J.R. (1985) Cascading trophic interactions and lake productivity. *Bioscience*, **35**, 634–9.

CIFA (1992) Report of the 6th session of the Sub-committee for the Development and Management of the Fisheries of Lake Victoria. Jinja, Uganda, 10–14 Feb. 1992. *FAO Fish. Rep.*, No. 475, 48 pp.

Coulter, G.W., Allanson, B.R., Bruton, M.N., *et al.* (1986) Unique qualities and special problems of the Arican Great Lakes. *Env. Biol. Fishes*, **17**, 161–83.

Diggle, P.J. (1990) *Time Series: A Biostatistical Introduction*, Oxford Univ. Press, New York, 257 pp.

Fryer, G. and Iles, T.D. (1972) *The Cichlid Fishes of the Great Lakes of Africa*, Oliver and Boyd, Edinburgh, 641 pp.

Garrod, D.J. (1960) The fisheries of Lake Victoria, 1954–1959. *E. Afr. Agric. For. J.*, **26**, 42–8.

Goldschmidt, T., Witte, F. and Wanink, J. (1993) Cascading effects of the introduced Nile perch on the detritivorous phytoplanktivorous species in the sublittoral areas of Lake Victoria. *Conserv. Biol.*, **7**, 686–700.

Goudswaard, P.C. (1988) A comparison of trawl surveys in 1969/70 and 1984/85 in the Tanzania part of Lake Victoria. *FAO Fish. Rep.*, no. 388, pp. 86–100.

Goudswaard, P.C. and Witte, F. (1985) Observations on Nile perch, *Lates niloticus* (L.), 1758, in the Tanzanian waters of Lake Victoria. *FAO Fish. Rep.*, no. 335, pp. 62–7.

Goudswaard, P.C. and Ligtvoet, W. (1988) Recent developments in the fishery for haplochromines (Pisces: Cichlidae) and Nile perch (Pisces: Centropomidae) in Lake Victoria. *FAO Fish. Rep.*, no. 388, pp. 101–12.

Graham, M. (1929) *The Victoria Nyanza and its Fisheries. A Report on the Fish Survey of Lake Victoria 1927–1928 and Appendices.* Crown Agents for the Colonies, London, 255 pp.

Greboval, D. (1990) Socio-economic issues for planning in support of fisheries management. *FAO Fish. Rep.*, no. 430, pp. 75–97.

Harrison, K. (1991) The taxonomy of East African Nile perch, *Lates* sps. (Perciformes, Centropomidae). *J. Fish Biol.*, **38**, 175–86.

Hecky, R.E. (1993) The eutrophication of Lake Victoria. *Verh. int. Verein. Limnol.*, **25**, 39–48.

Hughes, N.F. (1986) Changes in feeding biology of the Nile perch, *Lates niloticus* (L) (Pisces: Centropomidae), in Lake Victoria, East Africa, since its introduction in 1960, and its impact on the native fish community in the Nyanza Gulf. *J. Fish Biol.*, **29**, 541–8.

Jackson, P.B.N. (1971) The African Great Lakes fisheries: past, present and future. *Afr. J. trop. Hydrobiol. Fish.*, **1**, 35–49.

Jongman, R.H.G., ter Braak, C.J.F. and van Tongeren, O.F.R. (eds) (1987) *Data Analysis in Community and Landscape Ecology*, Pudoc, Wageningen, 299 pp.

Kenya Fisheries Department (1988) State of Lake Victoria Fisheries – Kenya. *FAO Fish. Rep.*, no. 388, pp. 20–28.

Kudhongania, A.W. and Cordone, A.J. (1974) Batho-spatial distribution patterns and biomass estimate of the major demersal fishes in Lake Victoria. *Afr. J. trop. Hydrobiol. Fish.*, **3**, 15–31.

Ligtvoet, W. and Mkumbo, O.C. (1990) Synopsis of ecological and fishery research on Nile perch (*Lates niloticus*) in Lake Victoria, conducted by HEST/TAFIRI. *FAO Fish. Rep.*, no. 430, pp. 35–74.

Ligtvoet, W. and Witte, F. (1991) Perturbation through predator introduction: effects on the food web and fish yields in Lake Victoria (East Africa), in *Perturbation and Recovery of Terrestrial and Aquatic Ecosystems* (ed. O. Ravera), Ellis Horwood, Chichester, pp. 263–8.

Mann, M.J. (1970) A resume of the *Tilapia* fisheries of Lake Victoria up to the year 1960. *EAFFRO Ann. Rep. 1969*, Append. B, pp. 21–7.

Marten, G.G. (1979) The impact of fishing on the inshore fishery of Lake Victoria (East Africa). *J. Fish. Res. Bd Can.*, **36**, 891–900.

Ochumba, P.B.O. (1988) Periodic massiv fish kills in the Kenyan portion of Lake Victoria. *FAO Fish. Rep.*, no. 388, pp. 47–60.

Ogutu-Ohwayo, R. (1990a) Changes in the prey ingested and the variations in the Nile perch and other fish stocks of Lake Kyoga and the northern waters of Lake Victoria (Uganda). *J. Fish. Biol.*, **37**, 55–63.

Ogutu-Ohwayo, R. (1990b) The decline of the native fishes of lakes Victoria and Kyoga (East Africa) and the impact of introduced species, especially the Nile perch, *Lates niloticus*, and the Nile tilapia, *Oreochromis niloticus*. *Env. Biol. Fishes*, **27**, 81–96.

Ogutu-Ohwayo, R. and Hecky, R.E. (1991) Fish introductions in Africa and some of their implications. *Can. J. Fish. aquat. Sci.*, **48** (Suppl. 1), 8–12.

Okaranon, J.O., Acere, T.O. and Ocenodongo, D.L. (1988) The current state of the fisheries in the northern portion of Lake Victoria (Uganda). *FAO Fish. tech. Rep.*, no. 335, pp. 89–98.

Payne, I. (1987) A lake perched on piscine peril. *New Scientist* (27 Aug.), 50–54.

Rabour, C.O. (1989) Annual report on catch and effort assessment for the artisanal fisheries of L. Victoria (Kenyan Waters) – 1988. *KMFRI tech. Pap.*, no. 2. (Mimeo)

Reynolds, J.E. and Greboval, D.F. (1988) Socio-economic effects of the evolution of Nile perch fisheries in Lake Victoria: a review. *CIFA tech. Pap.*, no. 17, 148 pp.

Ribbink, A.J. (1987) African lakes and their fishes: conservation scenarios and suggestions. *Env. Biol. Fishes*, **19**, 3–26.

Wanink, J.H. (1991) Survival in a perturbed environment: the effects of Nile perch introduction on the zooplantivorous fish community of Lake Victoria, in *Perturbation and Recovery of Terrestrial and Aquatic Ecosystems* (ed. O. Ravera), Ellis Horwood, Chichester, pp. 269–75.

Witte, F. (1981) Initial results of the ecological survey of the haplochromine cichlid fishes from the Mwanza Gulf of Lake Victoria (Tanzania): breeding patterns, trophic and species distribution. *Neth. J. Zool.*, **31**, 175–202.

Witte, F. and Goudswaard, P.C. (1985) Aspects of the haplochromine fishery in southern Lake Victoria. *FAO Fish. tech., Rep.*, no. 335, pp. 81–5.

Witte, F., Goldschmidt, T., Wanink, J., van Oijen, M., Goudswaard, K., Witte-Maas, E. and Bouton, N. (1992a) The destruction of an endemic species flock: quantitative data on the decline of the haplochromine cichlids of Lake Victoria. *Env. Biol. Fishes*, **34**, 1–28.

Witte, F., Goldschmidt, T., Goudswaard, P.C., Ligtvoet, W., van Oijen, M.J.P. and Wanink, J.H. (1992b) Species extinction and concomitant ecological changes in Lake Victoria. *Neth. J. Zool.*, **42**, 214–32.

Chapter eight

Analysis of species changes in Lake Victoria using ECOPATH, a multispecies trophic model

Jacques Moreau

8.1 INTRODUCTION

Understanding of the functioning of a diverse ecosystem, and of the possible impacts that species changes have on the system as a whole, calls for quantification of the trophic relationships between its different groups. This is one reason why Polovina (1984) developed ECOPATH, a steady-state model of trophic interactions in ecosystems. Using ECOPATH the system is partitioned into boxes comprising species having a common physical habitat, similar diet and life history characteristics (Polovina, 1984; Christensen and Pauly, 1992). The model produces estimates of mean (annual) biomass (annual) biomass production and (annual) biomass consumption for each of the boxes in the ecosystem. The ECOPATH model assumes that the ecosystem considered is at equilibrium, which means that input to a group should equal output from it for the period considered. This steady-state condition makes it possible to establish a system of biomass budget equations which, for each group, is:

$$\text{Production} - \text{all predation on this species} - \text{non-predatory mortality} - \text{all exports} = 0. \quad (8.1)$$

ECOPATH expresses each term in the budget equation as a linear function of the unknown mean biomass. The resulting budget equations become a

The Impact of Species Changes in African Lakes. Edited by Tony J. Pitcher and Paul J.B. Hart. Published in 1995 by Chapman & Hall, London. ISBN 0 412 55050 4.

system of simultaneous linear equations. The relative simplicity of the ECOPATH model compared with other multispecific models (for instance simulation models such as Andersen and Ursin, 1977; Laevastu and Larkins, 1981) is apparent in its application to several marine and continental ecosystems (Christensen and Pauly, 1993).

The aims of this chapter are:

1. to introduce the reader to the ECOPATH II model, an improved version of the original ECOPATH model of Polovina (1984), as presented by Christensen and Pauly (1992);
2. to describe a severely disturbed ecosystem, Lake Victoria (Beadle, 1981; Payne, 1986; Burgis and Symoens, 1988; Ligtvoet and Witte, 1991; Chapters 2–9);
3. to show how a more generalized use of ECOPATH II can help scientists to interact and to have their research and expertise properly used.

This chapter extends work originally presented at a session of the International Council for the Exploration of the Sea (ICES), held in Copenhagen in October 1990, on ecological modelling (Moreau *et al.*, 1993).

8.2 LAKE VICTORIA: A SEVERELY DAMAGED ECOSYSTEM

In the early 1960s, a large predator, *Lates niloticus*, was introduced into Lake Victoria, the largest lake in Africa (see Chapter 2, Fig. 2.2, for map). The lake hosted a diverse fish community dominated by a species flock of haplochromine cichlids which comprised about 250 species and which formed more than 80% of the demersal fish biomass (Kudhongania and Cordone, 1974). At the end of the 1970s, two decades after its introduction, the *Lates* population expanded rapidly and, simultaneously, a decline of most other fish stocks in the lake was reported (Ssentongo and Welcomme, 1985; Ligtvoet, 1989) (Chapters 2, 6, 7).

The introduction of the Nile perch had an immense impact on the structure of the Lake Victoria ecosystem. To evaluate the strength of this impact, Ssentongo and Welcomme (1985) calculated the trophic exchanges within the lake. They concluded that if the Nile perch biomass amounted to 120 000 t (tonnes) in 1980, they would have consumed about 650 000 t of fish, 85% of which consisted of haplochromines. One disadvantage of this model was the assumption that Nile perch preys exclusively on fish irrespective of its size and age.

Another view, proposed by Ligtvoet (1989) and Ligtvoet and Witte (1991), stresses the trophic importance of haplochromines in the earlier period and the recent increase of *Caridina nilotica*, *Rastrineobola argentea* and juvenile Nile perch in the diet of *Lates niloticus*.

The pre-*Lates* food web (Fig. 8.1(a)) was dominated by the haplochromines.

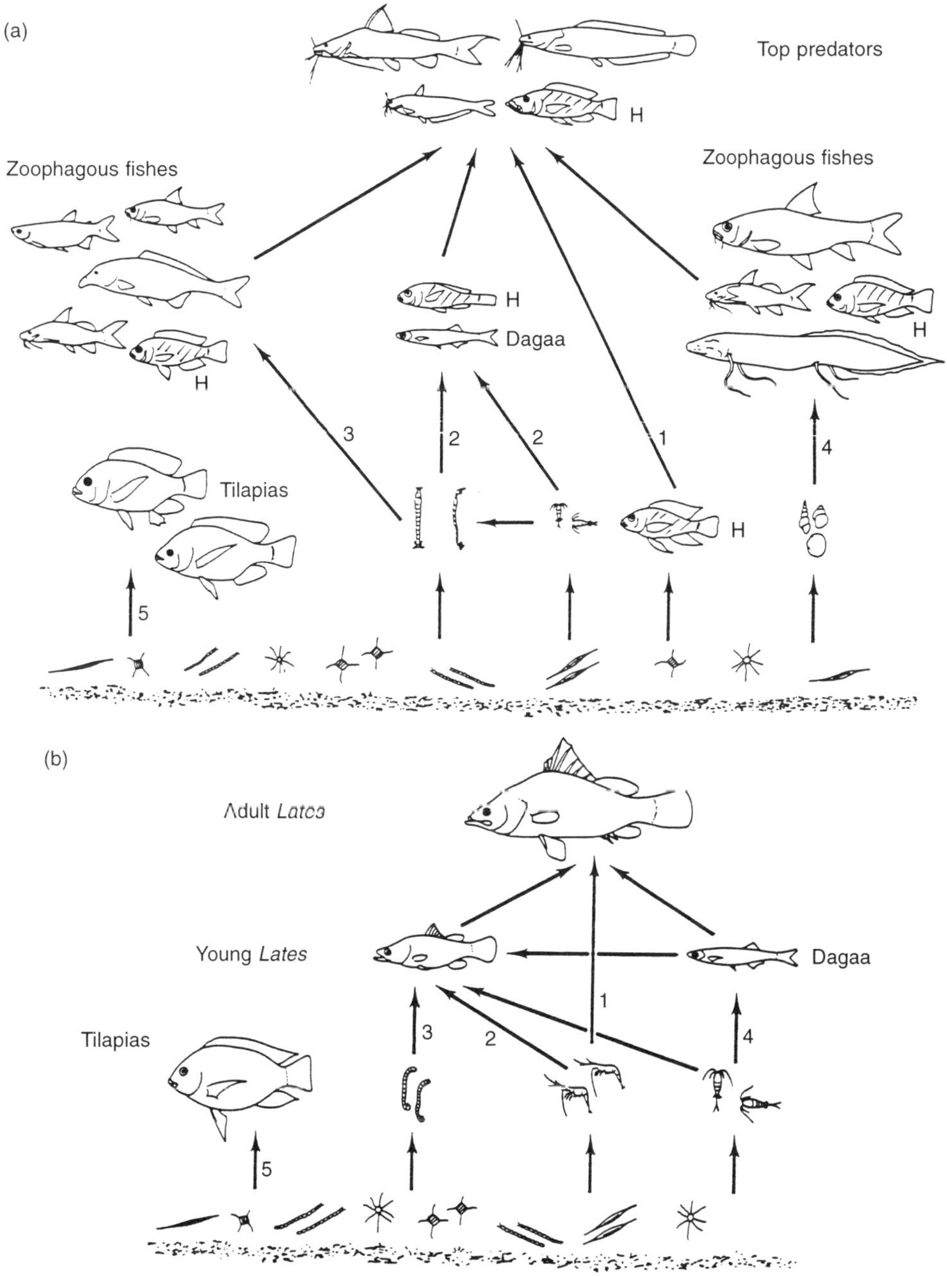

Fig. 8.1 Pre-*Lates (*a) and current (b) food webs of Lake Victoria. Numbers refer to food chains described in the text. H, haplochromines. Redrawn from Ligtvoet and Witte (1991).

In biomass the major trophic groups were the zooplanktivores and the bottom-dwelling detritivores/phytoplanktivores (about 20% and 60% of the total haplochromine biomass, respectively; Witte and van Oijen, 1990). Starting from phytoplankton and bottom deposits, the major food chains through the system were:

1. via haplochromines to the piscivorous catfishes and haplochromines;
2. via zooplankton and insect larvae to haplochromines and *Rastrineobola argentea* and then to piscivores;
3. via insect larvae to various fish taxa (haplochromines, Mormyridae, *Barbus, Alestes, Synodontis*) and then to piscivores;
4. via molluscs to various fish taxa (haplochromines, *Barbus altianalis, Protopterus aethiopicus*) and then to piscivores;
5. a direct flow to several herbivorous tilapiine species.

In the present food web (Fig. 8.1(b)), *Lates* is the top predator, feeding mainly on the atyid prawn *Caridina* (primary consumer), *Rastrineobola* and juvenile *Lates* (both secondary consumers). The main food links are:

1. via *Caridina* to *Lates*;
2. via *Caridina* to juvenile *Lates* and then to adult *Lates*;
3. via insect larvae to juvenile *Lates* and then to adult *Lates*;
4. via zooplankton to *Rastrineobola* and juvenile *Lates* and then to adult *Lates*;
5. directly to *Oreochromis niloticus* (Ligtvoet and Witte, 1991).

Simultaneously, the actual catch and fishing effort significantly increased: 6000 fishers were operating in 1968 and 21 000 in 1986, and the number of fishing nets increased from 90 000 to 250 000. At the same time, the Kenyan catch increased from 16 000 t to 103 000 t (Reynolds and Greboval, 1988).

8.3 MATERIAL AND METHODS

ECOPATH II as used here is a modified version by Christensen and Pauly (1992) of the ECOPATH model proposed by Polovina (1984; Polovina and Ow, 1985). The first step in this modelling exercise is to determine the main components and the feeding network of the groups in the ecosystem. Then the data inputs required by ECOPATH II are assembled and standardized (here to t km^{-2} wet weight) for each component group. It should be noted that the records available for use in this simulation effort describe only the shallow and intensively exploited Kenyan sector of the lake. Thus, these models do not necessarily apply to the lake as a whole. For each group the main assumption of ECOPATH II is: ecological production = actual catch from the group + what is consumed of it predators + the amount of production which goes to the detritus.

The literature cited shows how this leads to the basic equation of ECOPATH II for each group, *i*, included in the model.

$$B_i \times (P/B)_i \times EE_i = Y_i + \sum (B_j \times (Q/B)_j \times DC_{ji}) \qquad (8.2)$$

where B_i is the biomass of *i*, $(P/B)_i$ is its production/biomass ratio, EE_i is its ecotrophic efficiency, Y_i is its yield (= fishery catch), B_j is the biomass of its predators *j*, $(Q/B)_j$ is the food consumption per unit biomass of *j*, and DC_{ji} is the fraction of *i* in the diet of *j*.

Table 8.1 presents the groups we shall use to describe two different periods in the history of Lake Victoria. The first model, describing the early 1970s, emphasizes the role of haplochromine species in the Lake Victoria food web. The second model pertains to the 1980s onwards and shows the ecological importance of Nile perch as it became the dominant predator.

All biomasses were estimated via ECOPATH II, i.e. via a system of linear equations, such as Equation 8.2, for which we estimate most parameters as follows.

Fisheries catches (Y)

Catch estimates pertaining to the periods under consideration were obtained for the fish groups in Table 8.1 from records of the Department of Fisheries and from the literature (Reynolds and Greboval, 1988). They are expressed here, like all other flows, in t wet weight km^{-2} $year^{-1}$.

Production/biomass ratio (P/B)

It was shown by Allen (1971) that at equilibrium, when von Bertalanffy growth applies, P/B is equal to Z, the total mortality rate. Where necessary, we have estimated Z, and hence P/B, from length-frequency data for the fish as outlined in Gayanilo *et al.* (1989). For the other groups, literature values were taken, mainly from Winberg (1971) and Payne (1986). All values of P/B presented here are to an annual base (Table 8.1).

Diet composition (DC)

The average composition of the diet of each consumer organism in Table 8.1 refers to weights of prey eaten and was assembled from published information. Table 8.2 presents the diet matrices used for the pre- and post-*Lates niloticus* periods in Lake Victoria.

Note that haplochromines in the early 1970s constituted the main bulk of the Nile perch diet (Moreau, 1982). In 1979–1985, the diet of the Nile

Table 8.1 Input parameters used in ECOPATH II for periods 1971–1972 and 1985–1986. Note common names as used in Figs 8.4 and 8.5. The source column pertains only to references used to obtain P/B values. From Moreau *et al.* (1993)

Common names	Scientific names	Annual catch[a]		*EE*[b]		P/B		Source
		1971–72	1985–86	1971–72	1985–86	1971–72	1985–86	
Nile perch	*Lates niloticus*	0.001	10.000	0.75	0.95	0.39	0.95	Hughes (1983), Asila and Ogari (1987)
Catfishes	*Bagrus*, *Clarias*	0.770[c]	0.176[c]	0.95	0.95	0.79[d]	0.85[d]	Chilvers (1969), Marten (1978)
Lungfish	*Protopterus aethiopicus*	0.420	0.030	0.95	0.95	0.30	0.30	Merona (1983), Lévêque *et al.* (1977)
Snoutfishes and squeakers	*Synodontis* and mormyrids	0.060[e]	0.030[e]	0.95	0.95	0.90	0.90	Okedi (1969), Lévêque *et al.* (1977)
Predatory dwarf breams	Predatory haplochromines	0.280	0.001[f]	0.95	0.95	1.40	2.50	Maximum observed length from Fryer and Iles (1972), longevity and P/B values from
Planktivorous dwarf breams	Planktivorous haplochromines	0.110[f]	0.001	0.95	0.95	2.50	3.00	
Benthivorous dwarf breams	Benthivorous haplochromines	0.730[f]	0.001[f]	0.95	0.95	2.50	3.00	Lévêque *et al.* (1977) and Witte (1987)
Dagaa	*Rastrineobola argentea*	0.240	4.230	0.95	0.95	1.80	2.20	Wanjala (1978), Lévêque *et al.* (1977)[g]
Nile tilapia	*Oreochromis niloticus*	0.020	1.630	0.90	0.95	0.65	0.90	Getabu (1987)
Other tilapiines	Other *Oreochromis* spp.	0.660[h]	0.380[i]	0.95	0.95	1.00	1.20	Rinne (1975), Lévêque *et al.* (1977)
Zooplankton				0.70	0.80	33.50	33.50	Lévêque (1979)
Lake prawn	*Caridina nilotica*			0.70	0.95	14.00	16.00	Winberg (1971), Hart (1981)

Table 8.1 (Continued)

Common names	Scientific names	Annual catch[a]		EE[b]		P/B		Source
		1971–72	1985–86	1971–72	1985–86	1971–72	1985–86	
Macrozoobenthos				0.95	0.60	5.00	5.00	Symoens *et al.* (1980), Lévêque (1979)[j]
Phytoplankton				0.85	0.95	365.00	365.00	Computed from data by Payne (1986) and Burgis and Symoens (1988)
Benthic producers				0.85	0.85	25.00	25.00	Payne (1986), Burgis and Symoens (1988)

[a] Annual catch of the main species during different periods in Lake Victoria, Kenya waters (6200 km^2) in Reynolds and Greboval (1988). Data are average values for the periods considered, in $t\,km^{-2}\,year^{-1}$.
[b] Assumed values of the ecotrophic efficiency, EE.
[c] 1/4 *Bagrus* and 3/4 *Clarias*.
[d] P/B ($= Z$) ratios given here refer only to *Bagrus docmac* which were obtained with ELEFAN on the data available from the two sources listed. Note that these ratios are assumed to be true for *Clarias* spp. because these two species have similar longevities.
[e] 1/3 *Mormyrus* and 2/3 *Synodontis*.
[f] Partitioned as: 25% predators, 10% phytoplanktivores, 65% benthivores (Greenwood, 1974).
[g] Wanjala (1978) records $t_{max} = 2$ to 2.5 years ($P/B = 1.8$ to 2.2 from method of Lévêque *et al.* (1977)). Slight increase in P/B is due to increased fishing pressure and predation by *Lates niloticus*.
[h] Consisting of 80% *O. esculentus*.
[i] *O. esculentus* replaced by *O. leucostictus* and *T. zillii* recently introduced (Welcomme, 1988).
[j] The longevity was based on the maximum observed length (4 years) and P/B computed using the method of Lévêque *et al.* (1977).

Table 8.2 Diet composition (% per volume) of the groups of species in the Kenyan waters of Lake Victoria. Adapted from Moreau *et al.* (1993)

Predator	Year*	Species group 1	2	3	4	5	6	7	8	9	10	11	12	13	Phyto-plankton	Benthic producer	Detritus	Source
1 Nile perch	1971–72	1.0	8.0	2.0	2.0	35.0	10.0	25.0	2.0	1.0	10.0	1.0	1.0	2.0				Moreau (1982)
	1985–86	5.0	1.0	0.5	0.5	0.3	0.2	0.5	10.0	8.0	0.5	2.5	66.0	5.0				Ogari and Dadzie (1988)
2 Catfishes	1971–72		5.0	2.0	2.0	5.0	5.0	15.0	2.0		5.0	3.0	5.0	50.0			1.0	Greenwood (1966)
	1985–86	4.0	1.0	0.5	0.5	0.1	0.2	0.2	10.0	4.0	0.5	3.0	22.0	50.0			4.0	
3 Lungfish	1971–72		2.0	2.0	2.0	1.0	2.0	2.0	2.0		2.0	5.0	10.0	60.0		5.0	5.0	Ssentongo and
	1985–86	0.5	0.5	0.5	0.5	0.1	0.1	0.3	4.0	2.0	0.5	5.0	15.0	60.0	1.0	5.0	5.0	Welcomme (1985)
4 Snoutfishes and squeakers	1971–72		1.0	1.0	1.0	0.5	0.5	3.0	3.0		4.0	5.0	70.0			1.0	5.0	Ssentongo and
	1985–86		0.5	0.5	0.5	0.1	0.1	0.3	0.5	4.0	0.5	5.0	7.0	70.0		1.5	5.0	Welcomme (1985)
5 Predatory dwarf breams	1971–72					10.0	12.0	18.0	45.0			5.0	5.0	5.0				Fryer and Iles
	1985–86					5.0	10.0	10.0	60.0			8.0	5.0	2.0				(1972)
6 Planktivorous dwarf breams												2.0		2.0	50.0	25.0	21.0	Greenwood (1974)
7 Benthivorous dwarf breams												20.0	10.0	60.0		8.0	2.0	Greenwood (1974)
8 Dagaa												45.0	5.0	5.0	35.0	5.0	5.0	Ssentongo and Welcomme (1985)
9 Nile tilapia												2.0	3.0		85.0	5.0	5.0	Trewavas (1983)
10 Other tilapiines												3.0	2.0		45.0	30.0	20.0	Payne (1986)
11 Zooplankton												5.0			95.0			Lévêque (1979)
12 Lake prawn												40.0		5.0	10.0	25.0	20.0	Hart and Allanson (1981)
13 Macro-zoobenthos												10.0	1.0	10.0	5.0	45.0	29.0	Lévêque *et al.* (1977)

*Years are not stated for groups 6–13 because their feeding habits have remained unchanged for 25 years.

perch shifted to *Rastrineobola argentea*, *Caridina nilotica* and juvenile Nile perch (Ogari, 1988).

Food consumption (Q/B)

This parameter expresses the food consumption (Q) of an age-structured population of fishes relative to its biomass (B), for a conventional period of one year. The estimates of Q/B used here were obtained using the methods and multiple regressions in Pauly and Palomares (1987) and Palomares and Pauly (1989), initially derived from data on marine fishes. Palomares (1991) showed that freshwater and marine fishes have similar Q/B values when their shape, size, food type and environmental temperature are equal. Estimates of Q/B are provided in Table 8.3.

Ecotrophic efficiency (EE)

This is the fraction of the production of any group that is consumed within the system, or caught by the fishery. This parameter is difficult to estimate and is usually assumed to range from low values in apex predators to 0.95 elsewhere (Ricker, 1969). Note that ECOPATH II directs the fraction 1 − EE of production toward the detritus, a feature that is of relevance when attempts are made to equilibrate an ECOPATH II model.

Balancing of model

The equilibrium assumption implicit to Equation 8.2 is very important in that it strongly constrains the possible solutions, i.e. the range of parameters that will satisfy a set of simultaneous equations such as Equation 8.2. We accepted solutions which required the least modification of our initial inputs (including the diet matrix), and yet generated biologically and thermodynamically possible outputs (i.e. all GE and $EE < 1$; GE is gross efficiency).

For further details of the model, the reader is referred to Moreau *et al.* (1993).

Assessment of changes using ECOPATH

Additionally to the original routines of Polovina (1984), Christensen and Pauly (1992) developed several new routines including one dealing with what is called here 'mixed trophic impacts'. It allows assessment of the effect that changes in biomass of a group will have on the biomass of the other groups in the system. The routine was developed by following Hannon and Joiris (1989) and Ulanowicz and Puccia (1990).

Table 8.3 Food consumption per unit biomass estimated from the relationship between asymptotic body weight (W_∞) and gross food conversion efficiency (K_i) presented in Pauly and Palomares (1987) and implemented using the software of Jarre *et al.* (1990). Adapted from Moreau *et al.* (1993)

Species group	Year	W_∞ (g)	K (year)	K_i (year)	Q/B (year)	Source
Lates niloticus	1971–72	200 000	0.08	4.00	2.34	W_∞ and K from Hughes (1983), K_i from Lauzanne (1977)
	1985–86	75 356	0.36	4.50	5.61	W_∞ and K from Asila and Ogari (1987), K_i in agreement with Hughes (1983)
Bagrus docmac	1971–72	3 520	0.25	5.00	5.47	W_∞ and K from Marten (1979) obtained using ELEFAN (Gayanilo *et al.*, 1989), K_i consistent with Chilvers and Gee (1974)
	1985–86	5 785	0.09	5.00	5.56	
Clarias sp.	a	9 500	0.10	5.00	5.33	W_∞ and K from Willoughby and Tweddle (1977), K_i refer to benthic feeding habits, Lauzanne (pers. comm.)
Protopterus aethiopicus	a	13 600	0.10	5.00	4.84	W_∞ and K from Merona (1983), K_i from Lauzanne (pers. comm.)
Mormyrid fishes	a	400	0.50	10.00	11.00	W_∞ and K from maximum size recorded in Merona (1983)
Synodontis sp.	a	650	0.50	10.00	11.62	W_∞ and K suggested in Merona (1983), W_i from Lauzanne (1977)
Haplochromis sp.						
Predators	a	205	0.90	7.00	8.80	W_∞ and K suggested in Merona (1983)
Phytoplanktivores	a	35	1.50	20.00	41.00	K_i from Moriarty and Moriarty (1973)
Benthivores	a	40	1.50	12.00	21.20	$\ln Q/B = -0.1775 - 0.2018 \ln W_\infty + 0.6121 \ln T + 0.5156 \ln A + 1.26 F$ (Palomares and Pauly, 1989) (T = annual temperature in °C, A = aspect ratio, F = food type)

Table 8.3 (Continued)

Species group	Year	W_∞ (g)	K (year)	K_i (year)	Q/B (year)	Source
Oreochromis niloticus	1971–72	2000	0.29	20.00	18.90	W_∞ and K from Rinne (1975), K_i averaged from results in Lauzanne (1977), Moriarty and Moriarty (1973) and Harbott (1975)
	1985–86	13 000	0.25	20.00	24.60	W_∞ and K from Getabu (1987) using ELEFAN, K_i averaged from Moriarty and Moriarty (1973), Harbott (1975) and Lauzanne (1977)
Other *Oreochromis*	1971–72	745	0.37	20.00	32.30	W_∞ and K obtained for *O. esculentus* (dominant species) from Ssentongo (1985), K_i from Moriarty and Moriarty (1973), Harbott (1975) and Lauzanne (1977)
	1985–86	535	0.80	20.00	32.00	W_∞ and K obtained for *O. leucostictus* from Rinne (1975), K_i from Moriarty and Moriarty (1973), Harbott (1975) and Lauzanne (1977)
Rastrineobola argentea	1971–72	6	1.30	11.00	23.50	W_∞ and K obtained from method of Merona (1983), K_i from Lauzanne (1977) and from estimates obtained for zooplanktivores in Lake Tanganyika (Moreau and Nyakageni, 1988)
	1985–86	5	1.50	11.00	20.00	

[a] The data are suitable for both periods: 1971–72 and 1985–86.

8.4 RESULTS

Estimates of biomass and gross conversion efficiency obtained by the model from the input parameters are presented in Table 8.4 (Moreau *et al.*, 1993). Total fish biomass increased from about 27 t km^{-2} in the early 1970s to about 43 t km^{-2} in the mid 1980s. This is mainly due to the increased biomasses of Nile perch and Nile tilapia after their introduction into the lake.

Another important change is the decrease in the biomass of the haplochromine groups. Total haplochromine biomass was estimated by ECOPATH to decline from about 8.5 t km^{-2} in the early 1970s to about 0.5 t km^{-2} in mid 1980s. Reynolds and Greboval (1988) give a similar estimate of 8.8 t km^{-2} haplochromine biomass in the early 1970s. (See Chapter 6 for different figures – *Eds.*) Similar decreases in biomass estimates are recorded for the boxes representing *Bagrus*/*Clarias*, *Protopterus aethiopicus*, mormyrids/synodontids and other *Oreochromis*. Only *Rastrineobola argentea* maintained its biomass over the period.

Estimates for the gross efficiencies for the different fishery groups remain

Table 8.4 Biomass, gross efficiencies (i.e. production/consumption) and food intake for all groups of the Lake Victoria system for two time periods, 1971–1972 and 1985–1986 (from Moreau *et al.*, 1993)

Group	Biomass ($t\,km^{-2}$)		Gross efficiency		Food intake ($t\,km^{-2}\,year^{-1}$)	
	1971	1985	1971	1985	1971	1985
Nile perch	0.0	17.2	0.17	0.17	0.0	96.8
Catfishes	2.8	1.7	0.14	0.16	15.3	9.1
Lungfishes	4.5	2.3	0.06	0.06	21.9	11.2
Snoutfishes and squeakers	1.1	0.8	0.08	0.08	12.5	8.9
Predatory dwarf breams	3.0	0.2	0.16	0.28	26.0	1.5
Planktivorous dwarf breams	1.9	0.1	0.06	0.07	77.6	5.5
Benthivorous dwarf breams	3.6	0.2	0.12	0.14	76.0	5.3
Dagaa	7.6	7.9	0.08	0.11	179.4	158.7
Nile tilapia	0.0	12.1	0.03	0.04	0.17	297.0
Other tilapiines	2.5	0.9	0.03	0.04	80.3	28.3
Zooplankton	10.3	10.5	0.28	0.28	1236.6	1256.0
Lake prawn	2.6	5.9	0.40	0.40	90.9	235.9
Macrozoobenthos	28.0	28.7	0.35	0.34	401.0	402.2
Phytoplankton	4.3	4.5	–	–	–	–
Benthic producers	12.4	12.9	–	–	–	–
Total	84.6	105.9			2218.2	2516.3

more or less similar with the exception of the three haplochromine boxes and that of *Rastrineobola argentea*, which show a general increasing trend.

The introduction of *Lates niloticus* and, to a minor extent, *Oreochromis niloticus* in the early 1970s induced a spectacular change in the structure of the Lake Victoria ecosystem. During the early 1970s, the dominant species in the fishery catch were *Haplochromis* spp., *Bagrus* and *Clarias*. Later in the mid 1980s, the haplochromines had completely disappeared from the fisheries and *Lates niloticus* became the dominant species. The models in Figs 8.2 and 8.3 reflect the changed relationships.

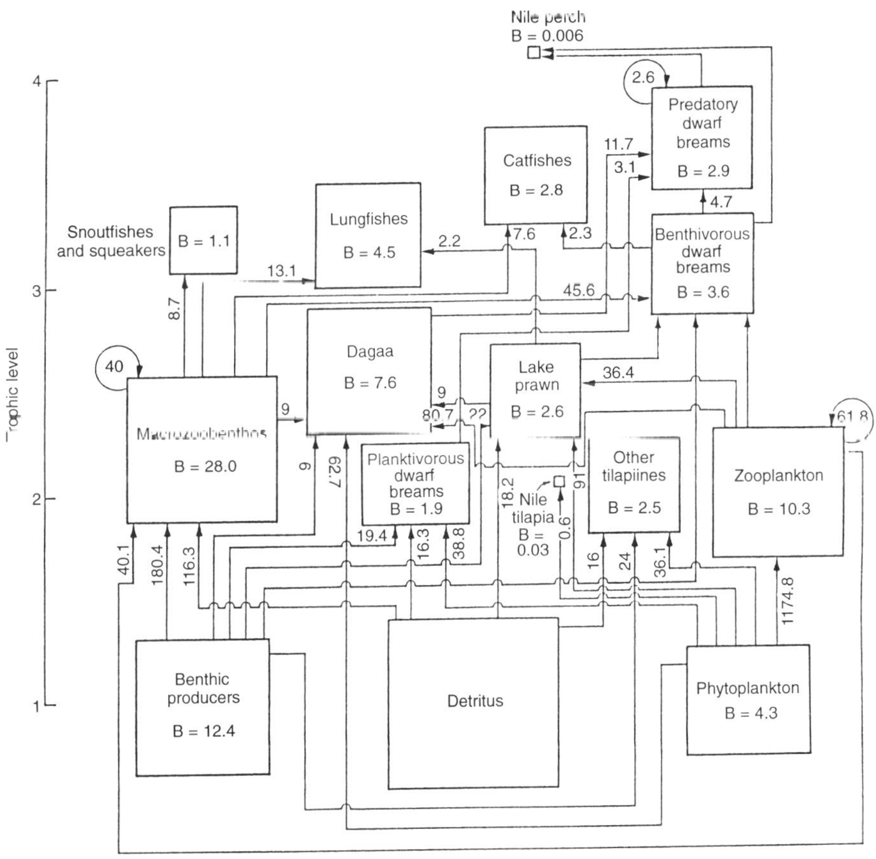

Fig. 8.2 ECOPATH II model obtained for the Lake Victoria system for the period 1971–1972. Flows to the detritus box and fishery catches are not shown. Corresponding Latin names of species groups are included in Table 8.1.

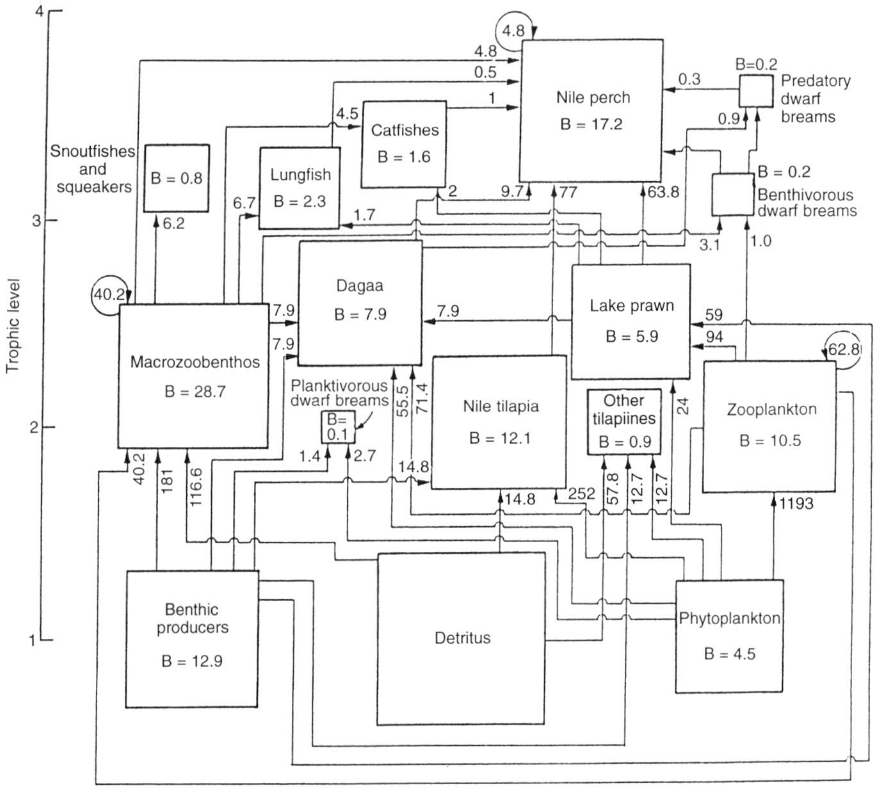

Fig. 8.3 ECOPATH II model obtained for the period 1985–1986 for the Lake Victoria system. Flows to the detritus box and fishery catches are not shown.

It is usually assumed that this decline in the haplochromine biomass can be attributed to the increasing fishing effort in the lake coupled with the big predation losses of haplochromines to Nile perch. Furthermore, the success of the introduced *Oreochromis niloticus* in 1981 (which was restricted to very shallow littoral waters) in the lake offered a strong competition for food, especially for phytoplanktivorous haplochromines.

To try to understand the mechanisms of the changes in the ecosystem of Lake Victoria we used the 'mixed trophic impact routine' of ECOPATH II (Christensen and Pauly, 1992), which shows the effect that an increase in the biomass of a group will have on the biomass of other groups in the ecosystem.

Figures 8.4 and 8.5 and Table 8.5 show the impacts an increase in the biomass of the group named in the rows will have on the groups numbered in the columns. Bars pointing upwards show positive impacts, while the bars

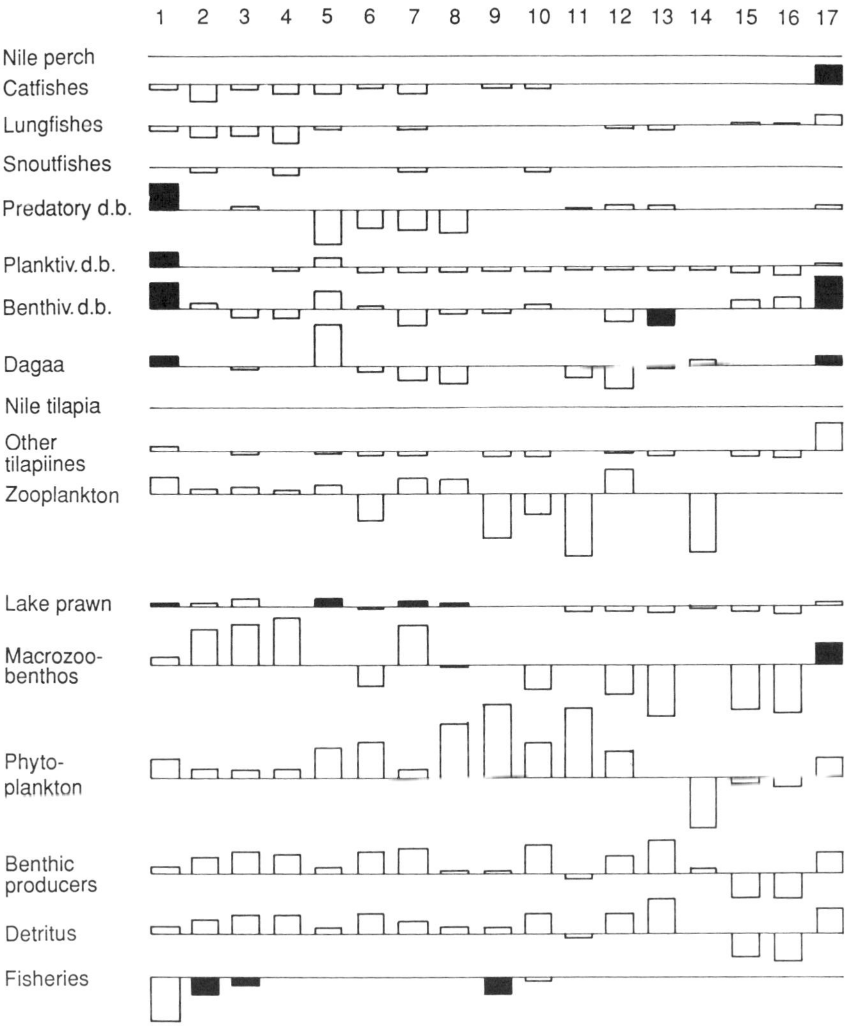

Fig. 8.4 Mixed trophic impacts in Lake Victoria for the period 1971–1972. The figure shows direct and indirect impacts on living groups in the systems that would result from an increase of biomass of the group given on the left. Positive impacts are shown above the baseline, negative below. The impacts are relative but comparable between groups. Numbers at the top refer to groups listed at left, in the same order (see also Table 8.5).

pointing downwards show negative impacts; the method allows impacts to be relative and comparable between groups.

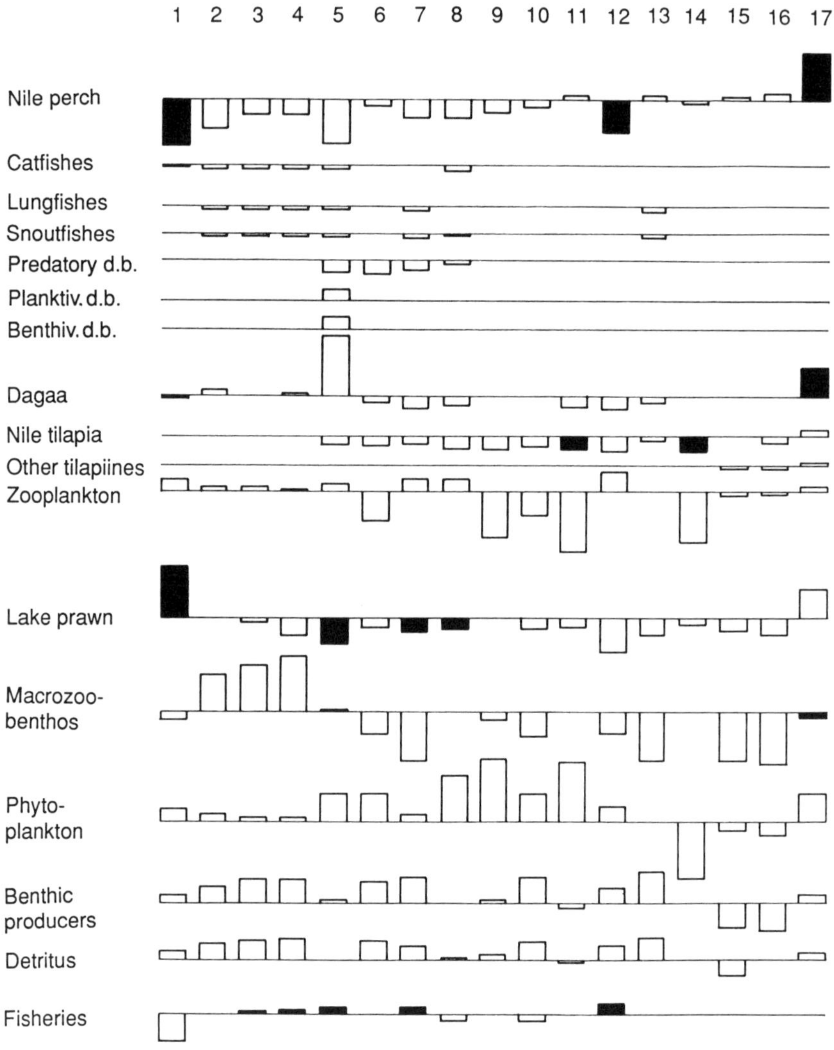

Fig. 8.5 Mixed trophic impacts in Lake Victoria for the period 1985–1986. See Fig. 8.4 for explanations.

Generally, it can be seen that an increase in the abundance of a predator will have a negative impact on the preferred prey, and an indirect, slightly positive impact on their prey in turn. This is an example of a trophic cascade

Table 8.5 Matrices of mixed trophic impacts for two periods, 1971–72 and 1985–86. Figures in bold show strong differences of impacts between the considered periods. Only two significant decimals are shown

Impacting groups	Impacted groups																
	1	2	3	4	5	6	7	8	9	10	11	12	13	14	15	16	17
								1971–1972									
1 Nile perch	**−0.02**	**−0.00**	**−0.00**	**−0.00**	**−0.00**	**0.00**	**−0.00**	**0.00**	**−0.00**	−0.00	−0.00	**−0.00**	0.00	0.00	−0.00	−0.00	**0.00**
2 Catfishes	−0.03	−0.12	−0.04	−0.07	−0.07	−0.02	−0.07	0.01	−0.02	−0.04	0.00	0.00	−0.00	0.00	0.01	0.01	**0.17**
3 Lungfishes	−0.03	−0.08	−0.07	−0.12	−0.02	−0.01	−0.02	−0.00	−0.01	−0.01	0.00	−0.02	−0.03	−0.00	0.02	0.02	0.09
4 Snoutfishes and squeakers	0.01	−0.02	−0.01	−0.04	−0.01	0.00	−0.02	0.00	0.00	−0.02	0.00	0.00	−0.02	−0.00	0.01	0.01	0.00
5 Predatory dwarf breams	**0.21**	0.02	0.02	0.01	−0.25	−0.12	−0.13	−0.17	−0.01	−0.00	0.02	0.03	0.03	−0.00	−0.00	0.00	0.03
6 Planktivorous dwarf breams	**0.10**	0.02	−0.01	−0.02	0.07	−0.04	−0.04	−0.03	−0.02	−0.03	−0.01	−0.02	−0.03	−0.02	−0.04	−0.06	0.02
7 Benthivorous dwarf breams	**0.19**	0.05	−0.07	−0.05	0.11	0.01	−0.11	−0.03	−0.02	0.03	−0.00	−0.09	−0.12	0.00	0.06	0.08	0.21
8 Dagaa	**0.07**	−0.01	−0.02	0.00	0.31	0.04	−0.10	−0.11	0.01	0.01	−0.08	−0.17	−0.02	0.03	0.00	−0.01	**0.06**
9 Nile tilapia	0.01	−0.00	−0.00	−0.00	−0.00	−0.00	−0.00	−0.00	−0.00	−0.00	**−0.00**	−0.00	−0.00	**−0.00**	−0.00	−0.00	0.01
10 Other tilapiines	0.02	−0.00	−0.02	0.01	−0.02	−0.03	−0.03	−0.01	−0.04	−0.04	−0.0[illegible]	−0.02	−0.04	−0.01	−0.05	−0.06	0.18
11 Zooplankton	0.02	0.03	0.04	0.03	0.06	−0.20	0.11	0.10	−0.34	−0.17	−0.44	0.15	0.01	−0.41	0.01	0.01	0.01
12 Lake prawn	**0.03**	0.02	0.05	−0.00	**0.05**	−0.02	**0.04**	**0.02**	0.01	−0.02	−0.04	−0.04	−0.05	0.02	−0.04	−0.06	0.02
13 Macrozoobenthos	0.04	0.25	0.28	0.33	0.00	−0.15	0.27	−0.02	−0.02	−0.18	−0.02	−0.22	−0.39	0.01	−0.32	−0.35	**0.12**
14 Phytoplankton	0.11	0.04	0.04	0.04	0.20	0.25	0.05	0.38	0.50	0.25	0.49	0.15	−0.01	−0.39	−0.06	−0.08	0.13
15 Benthic producers	0.07	0.12	0.17	0.14	0.05	0.16	0.18	0.02	0.02	0.20	−0.03	0.11	0.23	0.01	−0.17	−0.20	0.14
16 Detritus	0.05	0.09	0.13	0.13	0.04	0.14	0.08	0.03	0.03	0.13	−0.02	0.11	0.15	0.00	−0.12	0.00	0.09
17 Fishery	−0.32	**−0.12**	**−0.06**	**0.00**	**−0.01**	**0.00**	**−0.01**	0.00	−0.12	−0.03	0.00	**0.00**	0.01	0.00	−0.00	−0.00	–

Table 8.5 (Continued)

Impacting groups	Impacted groups																
	1	2	3	4	5	6	7	8	9	10	11	12	13	14	15	16	17
	1985–1986																
1 Nile perch	**−0.34**	−0.23	−0.12	−0.12	−0.35	−0.03	−0.13	−0.14	−0.09	−0.05	0.04	**−0.24**	0.04	−0.01	0.03	0.05	**0.35**
2 Catfishes	−0.01	−0.02	−0.02	−0.02	−0.02	−0.01	−0.01	−0.02	−0.00	−0.00	0.00	−0.01	−0.01	−0.00	0.01	0.01	**0.00**
3 Lungfishes	−0.00	−0.02	−0.02	−0.03	−0.02	−0.00	−0.03	−0.01	−0.00	−0.00	0.00	−0.01	−0.02	−0.00	0.01	0.01	−0.00
4 Stoutfishes and squeakers	0.00	−0.02	−0.02	−0.02	−0.02	−0.00	−0.02	−0.01	−0.00	−0.00	0.00	0.00	−0.02	0.00	0.01	0.01	−0.00
5 Predatory dwarf breams	**0.00**	−0.00	0.00	0.00	−0.08	−0.09	−0.07	−0.02	−0.00	0.00	0.00	0.00	0.00	−0.00	−0.00	0.00	−0.00
6 Planktivorous dwarf breams	**0.00**	−0.00	−0.00	−0.00	0.09	−0.01	−0.01	−0.00	−0.00	−0.00	−0.00	−0.00	−0.00	−0.00	−0.00	−0.00	−0.00
7 Benthivorous dwarf breams	**0.00**	−0.00	−0.00	−0.00	0.09	−0.01	−0.01	−0.00	0.00	0.00	−0.00	−0.00	**−0.01**	0.00	0.00	0.01	**0.00**
8 Dagaa	**−0.02**	0.05	0.01	0.02	0.52	0.04	−0.08	−0.06	0.01	−0.00	−0.07	−0.07	−0.03	0.02	0.01	−0.00	**0.23**
9 Nile tilapia	−0.01	−0.00	−0.01	0.01	−0.05	−0.05	−0.03	−0.07	−0.08	−0.05	**−0.07**	−0.09	−0.02	**−0.09**	−0.01	−0.03	0.06
10 Other tilapiines	−0.01	−0.00	−0.01	−0.01	−0.00	−0.01	−0.01	−0.01	−0.01	−0.01	−0.00	−0.01	−0.01	−0.00	−0.02	−0.02	0.02
11 Zooplankton	0.08	0.03	0.04	0.02	0.07	−0.20	0.11	0.11	−0.32	−0.17	−0.42	0.16	0.01	−0.37	−0.02	−0.02	0.05
12 Lake prawn	**0.39**	−0.01	0.03	−0.11	**−0.18**	−0.06	**0.09**	**0.08**	−0.00	−0.07	−0.06	−0.24	−0.10	0.04	−0.08	−0.12	0.21
13 Macrozoobenthos	−0.04	0.28	0.34	0.42	0.02	−0.14	0.34	−0.01	−0.03	−0.16	−0.01	−0.14	−0.34	0.01	−0.34	−0.36	**−0.03**
14 Phytoplankton	0.10	0.05	0.04	0.04	0.23	0.23	0.05	0.36	0.48	0.22	0.46	0.11	−0.00	−0.42	−0.06	−0.09	0.21
15 Benthic producers	0.07	0.13	0.19	0.18	0.02	0.16	0.20	0.01	0.03	0.20	−0.03	0.12	0.26	0.01	−0.18	−0.20	0.06
16 Detritus	0.06	0.12	0.14	0.15	0.01	0.14	0.09	0.02	0.04	0.13	−0.02	0.10	0.16	0.01	−0.12	0.00	0.05
17 Fishery	−0.19	**0.01**	**−0.02**	**0.03**	**−0.06**	**0.01**	**0.05**	−0.04	−0.00	**0.04**	−0.00	**0.08**	−0.01	0.00	−0.01	−0.01	–

effect. Increasing the phytoplankton production has a positive impact on almost all the other groups in the system, with the greatest impact on zooplankton. For each group, an increase of its own biomass has a negative impact on the group itself, reflecting increased within-group competition for resources (the diagonal of the diagrams).

In the 1970s, possible impacts of biomass variations of some groups were very limited: for example, snoutfishes and squeakers, all tilapias and lake prawns. This still held true in the 1980s for snoutfishes and squeakers, catfishes, lungfishes, and for tilapiines other than *O. niloticus*, but not for lake prawns, which became very important (see below).

In the beginning of the 1970s (Fig. 8.4), the development of fishing effort could be expected to have an important negative impact on top predators and on *O. niloticus*, and not so much on haplochromines. Conversely, an increase in biomass of all haplochromines and *Rastrineobola* should have a strong positive impact on *Lates niloticus*. Thus, we can deduce that the availability of a large biomass of haplochromines and *Rastrineobola* has allowed the development of *Lates niloticus*. Simultaneously, the availability of phytoplankton can explain, at least partly, the further development of *O. niloticus* after its introduction. For the early 1970s, the mixed trophic impact routine from ECOPATH II helps to explain how the changes in the ecosystem of Lake Victoria were made possible.

For the mid 1980s (Fig. 8.5), the mixed trophic impact routine shows clearly that *Lates niloticus* had a strong negative effect on its former prey (e.g. haplochromines) and may have been the cause of their decline. Now *Lates niloticus* relies mainly on *Caridina nilotica*, an increase of biomass of which should also increase the biomass of *Lates* (Fig. 8.5). The fishing effort, now directed towards *Lates niloticus*, *O. niloticus* and *Rastrineobola argentea*, is still increasing (E. Okemwa, pers. comm.) and this could reduce the biomass of these groups and contribute to a slight increase of biomass of native zooplanktivorous fishes, including haplochromines. In some restricted parts of the lake, small populations of haplochromines are still present (R.L. Welcomme, pers. comm.; Chapters 5, 6).

The differences of structure and functioning of the ecosystem between the two periods can be analysed and discussed by comparing the two mixed impact matrices and evaluating the large variation in the result of the interactions. The case of *Caridina nilotica* looks particularly illustrative. In the 1970s, it had a small positive impact on native zooplanktivorous fishes and small negative impacts on other groups. But, in the 1980s, the estimated impacts of *C. nilotica* increased enormously. It seems to contribute significantly to *Lates niloticus*, possibly mostly as a food for young fishes (Ogari and Dadzie, 1988). The changes in importance of haplochromines between the two periods can be analysed in a similar way, exemplifying their present limited role in the ecosystem.

8.5 DISCUSSION

The ECOPATH model

Polovina (1984) and Polovina and Ow (1985) considered the sensitivity of the ECOPATH model, in its first version, to be mainly for groups at a high trophic level (i.e. predators). This has been confirmed in a detailed analysis of the sensitivity of ECOPATH in which variations of *P*/*B*, *Q*/*B*, and *EE* for predators have been tested (Moreau and Nyakageni, 1988). In the present analysis, the importance of the feeding matrix for accurate evaluations of biomass should be emphasized. The qualitative information available on the diet composition of each group must be accurately quantified. Here this has been possible by taking into account several sources of information from the literature following previous quantifications (Palomares, 1991; Moreau *et al.*, 1993). Cannibalism, when it occurs, must be very low (most often less than 5% of the food), as suggested by Christensen and Pauly (1992), otherwise the biomass estimates of the group under consideration and its prey become unrealistically high.

We had to assume that equilibrium conditions hold true separately for the two situations analysed here. In reality, we know little about equilibrium states in fish communities and about the sensitivity of the model to perturbations caused by fishing or ecological modifications. This is why Christensen and Pauly (1992) designed the mixed trophic impact routine of ECOPATH II used here. One of the inherent assumptions of this routine is that the diet composition matrix is not altered by changes in the biomass of any groups and this will likely hold true only for small changes. One should therefore consider the impact routine as a tool for indicating the possible impact of direct and indirect interactions (including competition) in a steady-state system. The present Chapter shows that it can also help to explain or to predict short-term variations. However, it cannot be an instrument for making medium- or long-term predictions (Christensen and Pauly, 1992). Future modelling tools may be able to help in simulating medium- or long-term effects of changes in ecosystems.

Future use of ECOPATH II as a tool in fisheries management to predict changes in fish stocks lies in the expansion and modification of the model into a simulation model. The model could then be made to simulate changes to the ecosystem, caused by changes in fishing effort. ECOPATH in its original form is a top-down model, whereas a modified simulation model would be a bottom-up model, with input biomass data being derived from ECOPATH. Such a model must allow for the ability of each trophic group to change its diet composition in response to changing food availability. If the interactions for this can be implemented, then the potential use of ECOPATH would be most beneficial to fisheries management as a tool for predicting changes in the abundance and composition of fish stocks in relation to changing fishing intensity as well as for the introduction of fishes.

Possible interaction among scientists

During the study of Lake Victoria reported in Moreau *et al.* (1993), we used information provided by specialists on different groups who had been working independently. ECOPATH II shows how data on each group can be used to give a quantitative description of the ecosystem. This makes it possible to check the consistency of biomass and production estimates. It shows also the gaps in our present knowledge of this lake. To some extent, these gaps have forced us to use guessed values of some parameters to try to understand the relative importance of transfers of energy and biomass between successive trophic levels. Moreover, correct evaluations of biomass and production are particularly needed for groups which could have a great impact on the ecosystem as a whole, as illustrated in Fig. 8.5.

A short, but not exhaustive list of gaps in our knowledge of Lake Victoria is as follows.

- More information is required on the present demographic structure and ecological production of some groups of fishes including *Rastrineobola* and predatory haplochromines, the zooplankton, the lake prawn and on zoobenthos.
- More data are needed on the diet and food consumption of zoobenthos and benthophagous fishes (catfishes, lungfishes, benthophagous, haplo chromines).
- Quantitative evaluation is required of predation on zooplankton by young fishes (all species considered) in inshore areas.
- An up-to-date evaluation is needed of primary production and its possible seasonal variation for both phytoplankton and benthic producers.

8.6 CONCLUSIONS

ECOPATH II has made possible a proper assessment of the changes in biomass and production of several interactive groups in Lake Victoria. Much of the data required was available from the literature on the lake itself. Other data were adapted from information on other lakes. The accuracy of biomass evaluations, production and actual catch of some major groups has been demonstrated. The possible impacts of some groups on others has been illustrated by the mixed trophic impact routines of ECOPATH II. Gaps in our knowledge about transfers of biomass between groups have been highlighted.

It is the aim of the authors of ECOPATH II (Christensen and Pauly, 1993) to help in identifying such gaps and to encourage a close cooperation among aquatic ecologists.

8.7 SUMMARY

A multispecies trophic model called ECOPATH II can be used to describe the trophic relationships in an aquatic ecosystem on a quantitative basis and to explain the changes in fish stocks after a severe modification of environment such as the introduction of a new species or a significant increase of fishing effort. An example is provided for Lake Victoria. This Chapter also demonstrates the importance of cooperation among specialists.

ACKNOWLEDGEMENTS

Thanks are due to Dr Tony Pitcher for inviting me to prepare this contribution and to the anonymous referees for helpful suggestions and criticism of the first draft of the manuscript.

REFERENCES

Allen, K.F. (1971) Relation between production and biomass. *J. Fish. Res. Bd Can.*, **20**, 1573–81.

Andersen, K.P. and Ursin, E. (1977) A multispecies analysis of the efforts of variations of effort upon stock composition of eleven North Sea Fish Species. ICES Symposium of the changes in the North Sea Fish Stocks and their causes. *Rapp. P.-v. Réun. Cons. int. Explor. Mer*, **112**, 286–91.

Asila, A. and Ogari, J. (1987) Growth parameters and mortality rates of Nile perch (*Lates niloticus*), estimated from length-frequency data in the Nyanza Gulf (Lake Victoria). *FAO Fish. Rep.*, no. 389, pp. 272–87.

Beadle, L.C. (1981) *The Inland Waters of Tropical Africa – An Introduction to Tropical Limnology*, 2nd edn, Longman, London, 475 pp.

Burgis, M.L. and Symoens, J.J. (eds) (1988) *African Wetlands and Shallow Water Bodies: a Directory* (Travaux Doc. ORSTOM), Bondy, France, no. 211, 650 pp.

Chilvers, R.M. (1969) A method for age determination from length frequency data for large fishes with reference to *Bagrus docmac*. *EAFFRO Ann. Rep.*, 32–8.

Chilvers, R.M. and Gee, J.M. (1974) The food of *Bagrus docmac* (Fish) and its relationship with *Haplochromis hilgendorf* in Lake Victoria East Africa. *J. Fish Biol.*, **6**, 483–505.

Christensen, V. and Pauly, D. (1992) ECOPATH II a software for balancing steady state ecosystems models and calculating network characteristics. *Ecol. Modelling*, **61**, 169–95.

Christensen, V. and Pauly, D. (eds) (1993) *Trophic Models of Aquatic Ecosystems* (ICLARM Conf. Proc. 26), Int. Center for Living Aquatic Resources Management, Manila, Philippines, 390 pp.

Fryer, G. and Iles, T.G. (1972) *The Cichlid Fishes of the Great Lakes of Africa. Their Biology and Evolution*, Oliver and Boyd, Edinburgh, 643 pp.

Gayanilo, F.C., Soriano, M. and Pauly, D. (1989) *A Draft Guide to the Compleat ELEFAN* (ICLARM Software 2), Int. Center for Living Aquatic Resources Management, Manila, Philippines, 70 pp.

Getabu, A. (1987) Aspects of Lake Victoria Fisheries with Emphasis on *Oreochromis niloticus* and *Alestes imberi*. *FAO Fish. Rep.*, no. 389, pp. 416–31.

Greenwood, P.H. (1966) *The Fishes of Uganda*, The Uganda Society, Kampala, Uganda, 131 pp.

Greenwood, P.H. (1974) *Cichlid Fishes of Lake Victoria, East Africa: the Biology and Evolution of a Species Flock*. Trustees of the British Museum (Natural History), London, 134 pp.

Hannon, B. and Joiris, C. (1989) A seasonal analysis of the Southern North Sea Ecosystem. *Ecology*, **70**, 1916–34.

Harbott, B.S. (1975) Preliminary observations of the feeding of *Tilapia nilotica* Linnaeus in Lake Rudolf. *Afr. J. trop. Hydrobiol. Fish.*, **9**(1), 27–37.

Hart, R.C. (1981) Population dynamics and production of the littoral freshwater shrimp *Caridina nilotica* (Decapoda) in the littoral of Lake Sibaya. *Freshwat. Biol.*, **11**, 531–47.

Hart, R.C. and Allanson, B.R. (1981) Energy requirements of the tropical freshwater shrimp *Caridina nilotica* (Decapoda, Atydae). *Verh. int. Verein. Limnol.*, **21**, 1597–1602.

Hughes, N.F. (1983) *A Study of the Nile Perch, an Introduced Predator in the Kavirondo Gulf, Lake Victoria*. Rep. Oxford Univ. Nile Perch Project 1983. Oxford, Oxford University Press, 68 pp.

Jarre, A., Palomares, M.L., Soriano, M.L., Sambilay, V.C. and Pauly, D. (1990) A user's manual for MAXIMS. A computer program for estimating the food consumption of fishes from diel stomach contents data and population parameters (ICLARM Software 4), Int. Center for Living Aquatic Resources Management, Manila, Philippines, 27 pp.

Kudhonghania, A.W. and Cordone, A.J. (1974) Bathospatial distribution patterns and biomass estimates of major demersal fishes in Lake Victoria. *Afr. J. trop. Hydrobiol. Fish.*, **3**(1), 15–32.

Laevastu, T. and Larkins, H.A. (1981) *Marine Fisheries Ecosystems. Its Quantitative Evaluation and Management*. Fishing News Books, Farnham, UK.

Lauzanne, L. (1977) Aspects qualitatifs et quantitatifs de l'alimentation des poissons du Lac Tchad, thèse doct. Univ. Paris VI, 284 pp.

Lévêque, C. (1979) Productivité du Lac Tchad. Proc. Int. Workshop on African Limnology. SIL/UNEP, Nairobi (Kenya), December 1979.

Lévêque, C., Durand, J.R. and Ecoutin, J.D. (1977) Relations entre le rapport P/B et la longévité des organismes. *Cah. ORSTOM, Sér. Hydrobiol.*, **11**(1), 17–31.

Ligtvoet, W. (1989) The Nile perch (*Lates niloticus*) in Lake Victoria: a blessing or a disaster? *Annls Mus. Nat. Afric. Cent. Tervuren, Belgique. Sc. Zool.*, **257**, 151–6.

Ligtvoet, W. and Witte, F. (1991) Perturbation through predator introduction effect on the food web and fish yield in Lake Victoria (East Africa), in *Proc. Symp. Perturbation and Recovery of Ecosystems, Sienna, 1990* (eds W. Ligtvoet and F. Witte), Ellis Horwood, Chichester, pp. 263–8.

Marten, G.G. (1978) Calculating mortality rates and optimum yields from length samples. *J. Fish. Res. Bd Can.*, **35**, 197–201.

Marten, G.G. (1979) Impact of fishing on the inshore fishery of Lake Victoria (East Africa). *J. Fish. Res. Bd Can.*, **36**, 891–900.

Merona, B. (1983) Modèle d'estimation rapide de la croissance des poissons d'eau douce d'Afrique. *Revue Hydrobiol. trop.*, **16**, 103–13.

Moreau, J. (1982) Synopsis of biological data on the Nile perch, *Lates niloticus*. (Exposé synoptique des données biologiques sur la perche du Nil *Lates niloticus* (Linnaeus, 1762).) *FAO Fish. Synop.*, no. 132, 44 pp.

Moreau, J., Ligtvoet, W. and Palomares, M.L. (1993) Trophic relationships in fish community of Lake Victoria, with emphasis on the impact of Nile perch (*Lates niloticus*), in *Trophic Models of Aquatic Ecosystems* (Iclarm. Conf. Proc. 26) (eds V. Christensen and D. Pauly), ICLARM, Manila, Philippines, pp. 144–152.

Moreau, J. and Nyakageni, B. (1988) Relations trophiques dans la zone pélagique du lac Tanganyka (secteur Burundi). *Revue Hydrobiol. trop.*, **21**, 357–64.

Moriarty, C.M. and Moriarty, D.J.W. (1973) Quantitative estimation of the daily ingestion of phytoplankton by *Tilapia nilotica* and *Haplochromis nigripinnis* in Lake George, Uganda. *J. Zool., Lond.*, **171**, 15–23.

Ogari, J. (1988) Predator–prey relationship in Lake Victoria with special reference to Nile perch *Lates niloticus* (Linné), in Predator–prey relationships, population dynamics and fisheries productivity of large African lakes (ed. D. Lewis), *CIFA occ. Pap.*, **15**, 53–68.

Ogari, J. and Dadzie, S. (1988) The food of the Nile perch, *Lates niloticus* (L.), after the disappearance of the haplochromine cichlids in the Nyanza Gulf of Lake Victoria (Kenya). *J. Fish Biol.*, **32**, 571–7.

Okedi, J. (1969) Observations on the breeding and growth of certain mormyrid fishes of the Lake Victoria basin. *Revue Zool. Bot. afr.*, **79**(1–2), 55–71.

Palomares, M.L. (1991) La consommation de nourriture chez les poissons: étude comparative, mise au point d'un modèle prédictif et application à l'étude des relations trophiques. PhD thesis, Institut National Polytechnique, Toulouse, 218 pp.

Palomares, M.L. and Pauly, D. (1989) A multiple regression model for predicting the food consumption of marine fish populations. *Aust. J. mar. Freshwat. Res.*, **40**, 259–73.

Pauly, D. and Palomares, M.L. (1987) Shrimp consumption by fish in Kuwait waters: a methodology, preliminary results and their implications for management and research. *Kuwait Bull. mar. Sci.*, **8**, 101–25.

Payne, A.I. (1986) *The Ecology of Tropical Lakes and Rivers*, John Wiley and Sons, New York, 301 pp.

Polovina, J.J. (1984) Model of a coral reef ecosystem. Part I. ECOPATH II and its application to French Frigate Shoals. *Coral Reefs*, **3**, 1–11.

Polovina, J.J. and Ow, M.D. (1985) An approach to estimating an ecosystem box model. *Fish. Bull., U.S.*, **83**, 457–60.

Reynolds, J.E. and Greboval, D.F. (1988) Socio-economic effects of the evolution of Nile perch fisheries in Lake Victoria: a review. *CIFA tech. Pap.*, no. 17, 148 pp.

Ricker, W.E. (1969) Food from the sea, in *Resources and Man* (Rep. of Committee on Resources and Man to US Natn. Acad. Sci.), W.H. Freeman *et al.*, San Francisco, pp. 87–108.

Rinne, J.N. (1975) Age, growth, tagging of *Tilapia* spp. and reproduction biology of Siluroid catfish in Lake Victoria, East Africa. Fresh. Fish. Res. Org., 58 pp. (Mimeo). Publ. East Afric. Fish Res. Org., PO Box 343, Jinja, Uganda.

Ssentongo, G.W. (1985) Recent trends in the fisheries of the Tanzanian sector of Lake Victoria. FAO Fish. Dept, Rome, 13 pp. (Mimeo)

Ssentongo, G.W. and Welcomme, R.L. (1985) Past history and current trends in Lake Victoria. *FAO Fish. Rep.*, no. 335, pp. 123–36.

Symoens, J.J., Burgis, M. and Gaudet, J.J. (1980) *The Ecology and Utilisation of African Inland Waters. UNEP Rep. Ser.*, **1**, 260 pp.

Trewavas, E. (1983) *Tilapiine fishes of the genera* Sarotherodon, Oreochromis *and* Danakila, British Museum (Natural History), London, 583 pp.

Ulanowicz, R.E. and Puccia, C.J. (1990) Mixed trophic impact on ecosystems. *Coenoses*, **5**, 7–16.

Wanjala, B. (1978) The association and distribution of fish species in Nyanza Gulf, Lake Victoria, East Africa, 1975. MSc thesis, Oregon State University, 58 pp.

Welcomme, R.L. (1988) International introduction of inland aquatic species. *FAO Fish. tech. Pap.*, no. 294, 318 pp.

Willoughby, N.G. and Tweddle, D. (1977) The ecology of the commercially important species in the Shire Valley fishery, Malawi. *FAO Fish. Rep.*, **5**, 138–51.

Winberg, G.G. (1971) *Method for the Estimation of Production of Aquatic Animals*, Academic Press, London, 278 pp.

Witte, F. (1987) Ecological differentiation in Lake Victoria haplochromines. A contribution to the comparison of cichlid species flocks, in *Evolution of Fish Species Flocks in Lake Victoria* (ed. F. Witte), University of Leiden, 44 pp.

Witte, F. and Oijen, M.J.P. van (1990) Taxonomy, ecology and fishery of Lake Victoria haplochromine trophic groups. *Med. Leiden*, **262**, 1–47.

Chapter nine

Assessment of the Nile perch fishery in Lake Victoria

Tony J. Pitcher and Alida Bundy

9.1 INTRODUCTION

The aim of this Chapter is to assess, and provide recommendations for, the sustainable exploitation of the Lake Victoria Nile perch fishery, using available data on total commercial catch and fishing effort, incorporating research survey information wherever possible, and acknowledging the considerable uncertainties in the data. Because there are plans for a considerable increase of investment in processing plants and higher-value products for Nile perch fishery exports from the East African region (Bwathondi and Mosille, 1988; Reynolds and Ssali, 1990; Ogunja *et al.*, 1992), it is surprising that a fishery assessment that encompasses the whole lake has not been attempted previously. Moreover, many of those who lay the blame for reduction in Lake Victoria's biodiversity at the door of the Nile perch but have nevertheless acknowledged the fishery benefits (e.g. Witte *et al.*, 1992), have often asserted that the present fishery is not sustainable. In no instance has any evidence for this presumed lack of sustainability been put forward; this Chapter attempts to fill the gap.

Tagging experiments and commercial catches suggest that individual Nile perch make extensive movements over the entire surface and depth range of the whole lake (Ligtvoet and Mkumbo, 1990; Ogutu-Ohwayo, pers. comm.) and so we feel that an assumption that the fish constitute one stock is reasonable. We have used both constrained optimization surplus-production and yield-per-recruit models, together with a range of approximate assessments,

The Impact of Species Changes in African Lakes. Edited by Tony J. Pitcher and Paul J.B. Hart. Published in 1995 by Chapman & Hall, London. ISBN 0 412 55050 4.

to provide as robust an estimate of the relation of the present fisheries catch to maximum sustainable yield (MSY) as possible. Catch and effort data have been derived from the fisheries records of Uganda, Kenya and Tanzania. In addition, research survey data from the Ugandan and Tanzanian Fisheries services has been used to tune the assessment results. Data from 1979 to 1990 were used. This span includes the period of rapid growth of the Nile perch population in the early 1980s and the great expansion of the fishery in the late 1980s: unfortunately, it has proved impossible to obtain data from more recent years.

9.2 ASSEMBLY OF DATA

Commercial catch and effort data

The raw information with which to make an assessment is less than ideal: the fisheries data originate from three different sources with different recording conventions; there are missing blocks of years; the catch data are raised from subsamples; effort data are sparse and often undifferentiated.

We detail here the methods we have used to turn this information into a complete data series for use in the assessment models. In general, we have used linear interpolations and have tuned from research survey data. Where necessary, we have set up alternative scenarios to examine the effects of alternative assumptions. We have used this information to estimate error variance of the catch-and-effort data set so that this may be used in the assessment.

Fisheries data for Nile perch catches were obtained from the Kenya Fisheries Department *Statistical Bulletins*, the Tanzanian Fisheries Division *Annual Reports* and the Ugandan Fisheries Department. There were no Tanzanian 1990 catch data available, so the 1989 figures were used, otherwise the catch data are complete. Total Lake Victoria catch data for 1979–1990 were obtained by summing the catch by species from the three riparian states (Fig. 9.1(b)).

Fishing effort targeted on Nile perch was not differentiated in the catch statistics. Total effort, measured as 'number of canoes', was available from Tanzania until 1989. However, there were only 11 effort data points from Kenya (Kenya Fisheries Department *Statistical Bulletins*, Reynolds *et al.*, Chapter 10) and six effort figures from Uganda in the period 1963–1991 (Kudhongania and Coenen, 1992). The later figures, conducted under an improved frame survey, are considered to be reasonably accurate (Nyeko and Acere, 1990). To estimate the total effort in Lake Victoria, the number of canoes per kilometre of shoreline in Tanzania was used to make a first estimate of the effort for the other two countries (it was assumed that the number of

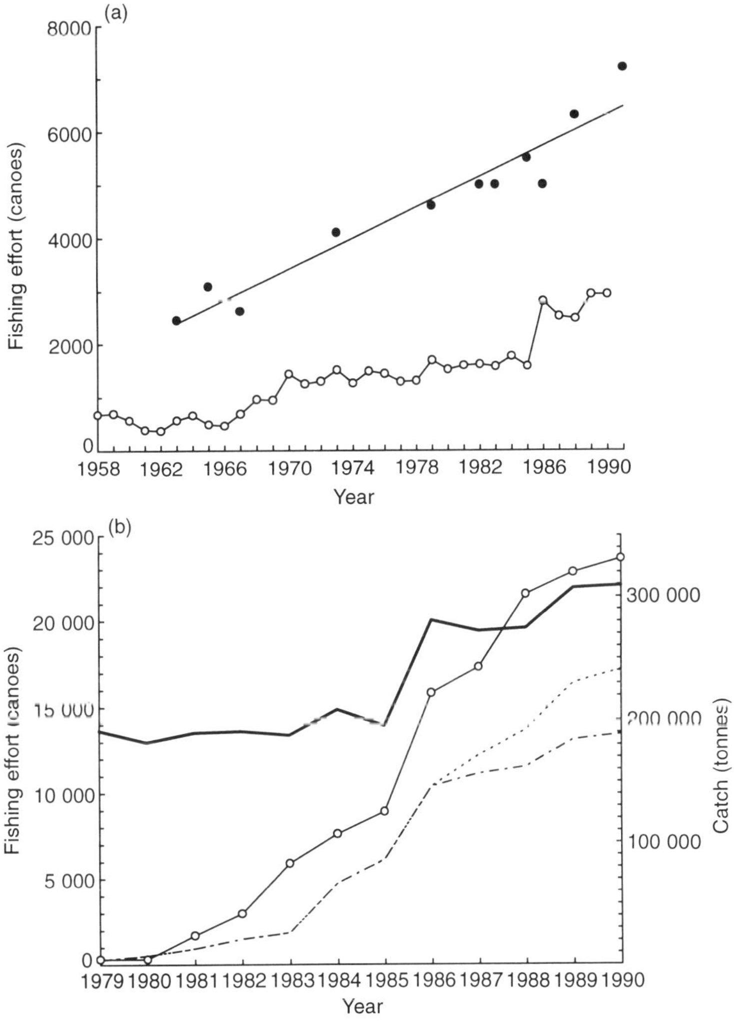

Fig. 9.1 Assembly of Nile perch fishery data for the assessments. (a) Line joining open circles is number of boats in Kenya estimated from the Tanzanian canoe density. Filled circles are surveyed Kenyan boats for all years that were available. Solid line indicates regression used for interpolation in missing years. (b) Solid line with open circles is total Nile perch catch (right axis). Broken lines are estimated numbers of boats targeting Nile perch under two different effort scenarios. Heavy solid line is total number of fishing vessels (see text for further details). Note differing vertical axes and time scales.

canoes in 1990 was the same as in 1989). This estimate was then tuned, as explained below, to real data for Uganda and Kenya to provide interpolations for years with no effort data. Shorelines were measured from a large-scale map.

To illustrate the method, Fig. 9.1(a) shows the estimated number of canoes in Kenya calculated from the Tanzanian canoe density together with the known number of Kenyan canoes for the years available. Both sets of figures follow a similar upward trend but the real data points lie above the estimated points. This may be expected because fishing intensity in Kenya is higher than that in Tanzania. Because the trend in the Kenyan data mirrored that in the estimated data a regression line was fitted, providing an interpolated estimate for years with no data.

As in the Kenyan data, the estimated data for Uganda followed the same trend as the three real data points, but had higher absolute values. The estimated data line was hence lowered by a factor of 0.75, which is the ratio of estimated to actual canoes in 1990, considered the most accurate data point. The numbers of canoes per country were then summed to give the total canoe effort in Lake Victoria (Fig. 9.1(b)).

There were no direct data available on the number of canoes fishing for Nile perch. However, Ligtvoet *et al.* (1988) give the percentage of fishers targeting Nile perch from 1979 to 1986 in Tanzania. In the absence of other information, this figure was used to represent the percentage of total effort directed at Nile perch. The graph was extrapolated from 1986 to 1990 under two scenarios, one with increasing Nile perch effort (standard effort scenario) and one where effort levelled off (lower effort scenario). The proportion of canoes targeting Nile perch was then multiplied by the total number of canoes for the years 1979 to 1990 to give the total effort directed at Nile perch (Fig. 9.1(b)).

There are clearly considerable uncertainties in the catch and effort data thus assembled. Total catch data for the three countries showed fluctuations between 9% and 14% around smooth linear trends. Effort data showed fluctuations of around 5%. We chose an error value of 20% (i.e. COV, coefficient of variation, of ±10%) independently on each catch and effort data point to express likely errors. Uncertainty in the trend of fishing boats targeting Nile perch is reflected in our choice of two alternative effort scenarios from 1986 to 1990.

Research survey data

Research survey data have been used to constrain the fit of the surplus-production assessment in two ways: first, to provide an absolute estimate of total Nile perch stock biomass in one particular year after the main population explosion in the lake (1987), and secondly, to provide an index of relative biomass abundance during the rapid population explosion period from 1979 to 1985.

First, trawl surveys in Tanzania during 1987 estimated Nile perch biomass at 321 540 t (tonnes) in waters shallower than 40 m (Ligtvoet and Mkumbo, 1990), using the swept-area method with an escapement factor of 0.5 (Pauly, 1984; Sparre *et al.*, 1989). About half the lake is deeper than 40 m, but Nile perch rarely inhabits water deeper than 60 m because of deoxygenation (Hecky, 1993). Water shallower than the 60 m depth contour is equivalent to about 80% of the lake's surface area (Chapter 2, Fig. 2.2). This raises the trawl survey biomass estimate to around 578 000 t. Assuming about 20% error in these figures, we estimate that the total Nile perch biomass in 1987 was likely to have been between 500 000 and 700 000 t, and so this range was used to constrain the surplus-production fit.

Secondly, we used research trawl data from each of the three riparian countries. In Uganda, a Fisheries Department research trawl with 20 mm mesh operated out of Jinja from 1980 to 1985. During this period 761 standard trawls were made totalling over 800 h (Acere, 1988). In the Mwanza Gulf of Tanzania, published data cover standard trawls with 20 mm and 70 mm mesh (Witte *et al.*, 1992) from 1979 to 1988. In the Nyanza Gulf, Kenya, 38 mm mesh trawls were made between 1979 and 1986 (Ogari and Asila, 1990). Catch rates of Nile perch in kg h^{-1} from these three data sets were combined to provide an indicator of increase in total Nile perch biomass during its population explosion. Because the absolute catch rates of the Ugandan, Tanzanian and Kenyan trawls were not equivalent owing to differences in mesh, design, operation and vessel, the values for each data set were expressed as relative catch rates. All of the data were then combined and smoothed with a moving average (Fig. 9.2). Consequently, we estimate Nile perch abundance in 1985 to be 8.36 times the 1979 level. This value was used to constrain the surplus-production fit.

9.3 ANALYTICAL METHODS

Yield per recruit

Data on mean length-at-age, growth rate and natural mortality were taken from the literature. The von Bertalanffy growth parameters we used were $L_\infty = 205$ cm; $k = 0.19$ (Asila and Ogari, 1987), taken from a year towards the end of the Nile perch population explosion rather than before or during the increase when growth might arguably have been different. Although there are several other values in the literature, these values were derived from a large set of fish samples, and are concordant with L_{max}, the largest fish landed, of about 200 cm long (Okemwa, 1984). More recently, a two-phase growth model has been fitted to Nile perch growth (Hughes, 1992), but in practice the inflexion in this more accurate growth model makes little

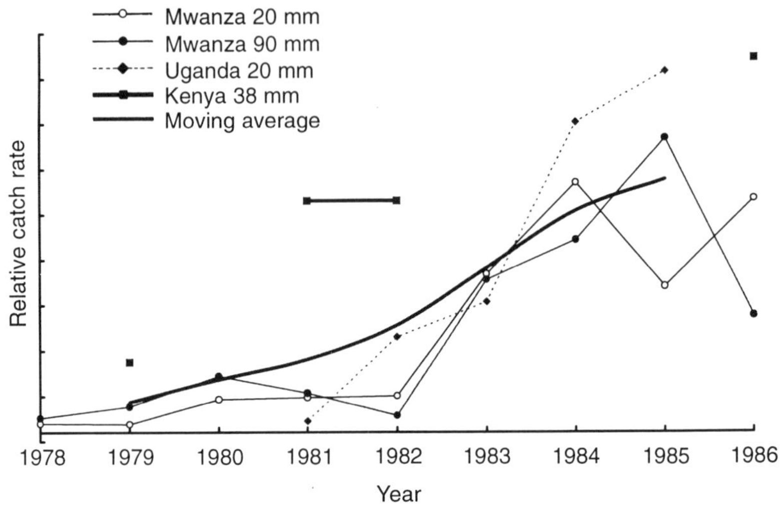

Fig. 9.2 Survey trawl data (relative catch rates) from the three riparian countries used in tuning the fit of the surplus-production model. Solid line joins the only consecutive pair of years in the Kenyan data. Heavy curve is the four sets of trawl catch rates smoothed with a moving average (see text for further details).

difference to a yield-per-recruit analysis. The length–weight relationship was taken as $W = 0.000\,006\,L^{3.17}$ (Ligtvoet *et al.*, 1988), making $W_\infty = 127.8$ kg. The baseline value for natural mortality rate, M, was taken as 0.3 (Ligtvoet and Mkumbo, 1990).

Isopleths of yield-per-recruit and mean-weight-of-fish-in-the-catch were calculated using the growth and mortality rates above (Beverton and Holt, 1957). Fishing mortality of various past years was estimated using the mean weight isopleths by plotting reported mean weights in the catch on the weight isopleth plot. Fishing mortality rate was then read off at the value corresponding to this point on the fishing mortality axis.

Uncertainty in the growth and mortality values used was examined using the bootstrapped variance of the optimal age of entry, t_{opt}, at the peak yield-per-recruit (Y/R) at high fishing mortality, $F = 3$. Because this is a somewhat theoretical figure, we also examined the effect on optimal fishing mortality, F_{max}, at the peak Y/R for the current age of entry, $t_C = 1.5$ year. We chose 2000 bootstrapped sets of parameter values using a Monte Carlo technique and recorded the sensitivity of t_{opt} and F_{max} (Crystal Ball software, Decisioneering Inc., Denver, Colorado, USA). We subjected W_∞ to normal error equivalent to 95% limits of $\pm 20\%$ and the von Bertalanffy growth rate, k, to a log-normal error at 10%: log-normal errors were chosen to avoid negative parameter values. Because we had no clues as to its likely error distribution, the natural mortality rate, M, was set to be equally likely within a uniform distribution range of 0.2 to 0.6.

Surplus production

We fitted a non-equilibrium Schaefer surplus-production model with constrained optimization using the 'Solver' algorithm in the 'Excel' spreadsheet (Microsoft Inc.). This technique supersedes the method we used for previous work on these data (Pitcher and Bundy, 1994). For each year, stock biomass was calculated from the previous year's biomass increased by the population growth rate and depleted by catch. Values optimized in the Solver search were asymptotic population biomass, B_{∞}; the population biomass logistic growth rate, r; and the initial 1979 Nile perch biomass. (Note that the initial biomass is not 'virgin' in sense of B_{∞} here on account of the population explosion.) This enables us to calculate catch per unit effort (CPUE) for each year from the estimated q and the year's biomass. The iterative Solver fit then minimized the sum of squares of the differences between observed CPUE from the catch/effort data and the calculated CPUEs to obtain the best overall catchability, q (method due to C.J. Walters, pers. comm.). Alternative fits were performed with and without biomass constraints from the research surveys, and for the two effort scenarios.

Uncertainties in the catch and effort data set were investigated by examining the sensitivity of maximum sustainable yield (MSY) and the optimum fishing effort (f_{opt}) using Monte Carlo bootstrapping from the catch and effort error distributions. The COV of the catch and effort estimates was assumed to be $\pm 10\%$ as explained above. We ran 1000 Monte Carlo estimates for each of the two effort scenarios.

Approximate estimators

In addition, some approximate methods of maximum sustainable yield estimation were used. These methods do not require a time series of catch and effort data, but do need biomass estimates. Biomass was taken from independent surveys or from the surplus-production fits. In each case we investigated the sensitivity of these methods to the uncertainty in biomass.

Gulland's formula	$0.5\,MB_{\infty}$
Cadima 1	$0.5\,(Y + MB)$
Cadima 2	$0.5\,ZB$
Pauly (1982)	$2.3\,B_{\infty}W'^{-0.26}$
Beddington and Cooke (1983)	graphically determined

where M is natural mortality, B_{∞} is virgin biomass, B is current biomass, Y

is current yield, Z is total mortality, $W' = (W_\infty - W_{mat})/2$ and W_{mat} is weight at maturity. The first three methods are detailed in Sparre *et al.* (1989). Beddington and Cooke estimates were determined using B_∞, L_∞, k and M in the published nomograms. Each of the first four methods was subjected to bootstrapped Monte Carlo sensitivity analysis to errors in the parameters. The error structures were assumed to be similar to those used for the yield-per-recruit and surplus-production analyses as follows.

Parameter	Error distribution	95% CL
B_∞	normal	±10%
B	normal	±10%
Y	normal	±10%
M	uniform	0.2–0.6
Z	log-normal	±10%
W_∞	normal	±10%
W_{mat}	normal	±10%

9.4 RESULTS

Yield-per-recruit model

The yield-per-recruit assessment plot (Fig. 9.3) shows that at $F = 3$ the optimum Nile perch fishery would increase the age of first capture (t_C) to 4.6 years when Nile perch attain about 120 cm length and 10 kg weight. Alternatively, at the current size and age of entry (1.5 years, 50 cm, 1.5 kg), optimal fishing mortality should be around $F_{max} = 0.4$. This is equivalent to the current legal minimum mesh size of 12.5 cm (FAO, 1992), although Nile perch smaller than this are frequently caught inshore. A reasonable target zone for management with an $F_{0.1}$ of 0.3 would be age of entry 2.5 years at 78 cm and 6 kg (equivalent to an 18 cm mesh net).

The mean sizes of Nile perch landed in Uganda (Masese) and Tanzania (Mwanza) are shown in Fig. 9.4(a). The equivalent fishing mortality rates are shown in Fig. 9.4(b). Length-based analysis by Asila and Ogari (1987) suggested that F averaged 1.3 between 1981 and 1985, a value that is compatible with our figures. Figure 9.4 illustrates the great historical increase in fishing mortality from the late 1970s and early 1980s. Because we read the F values from an isopleth plot that assumes an equilibrium fishery, the mortality values must be taken as imprecise, but the clear suggestion is that by the end of the 1980s, fishing mortality was in excess of a desirable management target level. Furthermore, the mesh size for inshore nets is catching male fish at around the age/size at maturity, and female fish before maturity, a practice that can lead to instability in recruitment and possible population collapse (Pitcher and Hart, 1982).

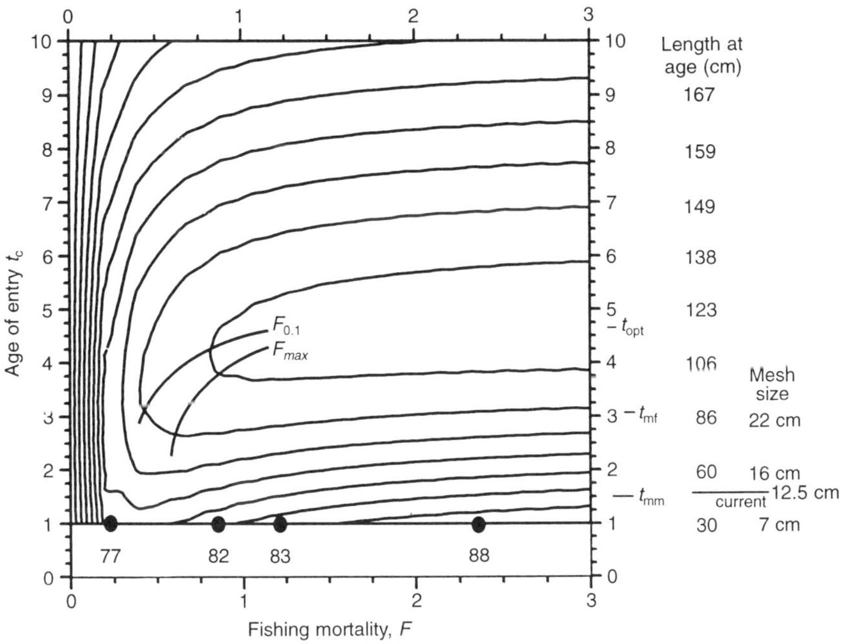

Fig. 9.3 Estimated equilibrium yield-per-recruit isopleths for the Nile perch fishery in Lake Victoria. Parameter values are given in the text. F_{max} and $F_{0.1}$ lines are indicated in the target fishing effort zone. Scales at right indicate Nile perch ages, lengths and approximate equivalent mesh sizes; t_{mm}, age of male maturity; t_{mf}, age of female maturity; t_{opt}, optimal age of entry at $F = 3$. Filled circles along the $t_C = 1$ line show some of the estimated F values for the years indicated.

These general conclusions are not altered significantly by the results of the bootstrapped sensitivity analysis. The 95% confidence limits on t_{opt} of 4.3 years are 3.0 to 5.5 years, while F_{max} at $t_C = 1.5$ years lies between 0.13 and 0.69 and $F_{0.1}$ between 0.1 and 0.65. Over 85% of the variance in optimal age and F_{max} is contributed by uncertainty in the value of M; about 15% is due to uncertainty in k, while W_∞ makes less than 0.5% contribution to the errors.

Surplus-production model

Fitted Schaefer parameters for fits constrained and unconstrained by survey biomass and for the two effort scenarios are listed in Table 9.1. We are inclined to lend more credence to the constrained fits, not just because of the additional non-fishery information that has been used, but also because

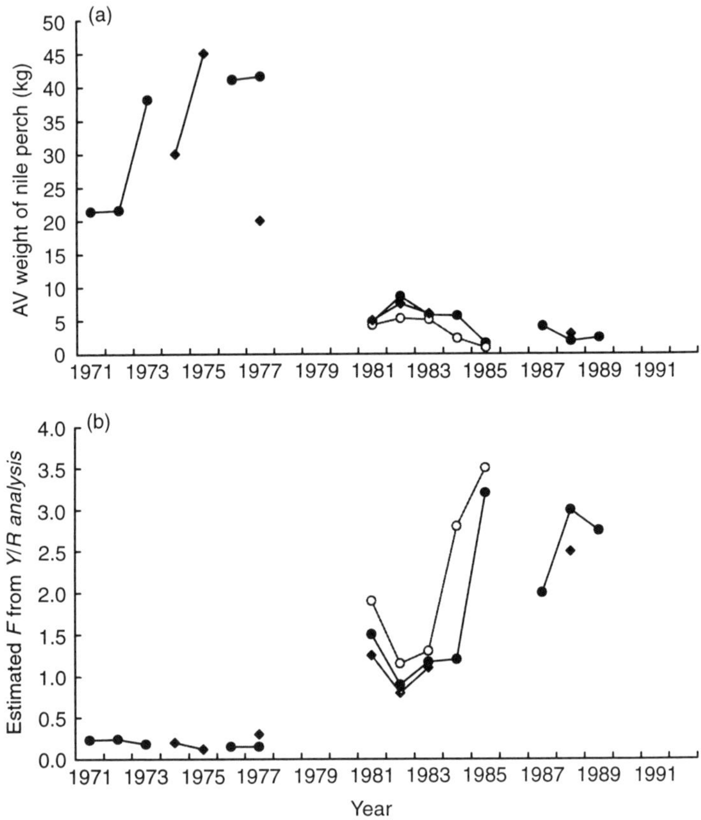

Fig. 9.4 Average weights (a) and estimated fishing mortalities (b) for Lake Victoria Nile perch. Masese, Uganda, weights were obtained from Okaronon (pers. comm.) and the Mwanza, Tanzania, weights from Ligtvoet and Mkumbo (1990). Fishing mortalities were read off an isopleth plot of mean weights from the yield-per-recruit analysis as explained in the text. Filled circles, values in catch, Masese, Uganda; open circles, values in population, Masese; filled diamonds, values in catch, Mwanza, Tanzania.

the trawl survey data suggest that the pre-population-explosion 1979 Nile perch biomass is more likely to have been nearer to 10% rather than 30% of later biomass levels. The population growth parameter is high, but we might expect that for an *r*-selected species. Figure 9.5(a) illustrates the fitted equilibrium yield curve for the normal effort scenario with the data points superimposed, while Fig. 9.5(b) shows the estimated biomass of the Nile perch population in the whole lake together with the catch and total effort data

Table 9.1 Results of fitting the non-equilibrium Schaefer surplus-production model to the catch and effort data for the whole lake. Method of fitting and of incorporating constraints is outlined in the text. Scenarios are alternative effort extrapolations from boats targeting Nile perch. Biomass constraints are derived from research trawl surveys in all three riparian countries, and are explained fully in the text

Scenario and constraints	Fitted values					
	B_∞ (t)	r	MSY (t $year^{-1}$)	f_{opt} (boats)	MSY as % of B_{1979}	Min. sum sqs
Standard scenario						
No constraints	1 049 266	1.06	278 605	19 467	32.4	487
Constraints on biomass*	697 186	1.61	281 409	17 940	10.8	551
Lower scenario						
No constraints	1 870 895	1.05	489 025	34 608	33.2	499
Constraints on biomass*	773 834	1.61	312 381	19 812	11.0	641

* B_{1985} is 8.3 × B_{1979}; B_{1987} is 500 000 to 700 000 t.

for comparison. The lower effort scenario produces, as expected if the same catch results from lower effort, somewhat higher MSY and optimum effort estimates.

Confidence limits from the error bootstrapping analysis are given in Table 9.2: because some of the results produced skewed distributions of MSY estimates, we have taken 95-percentiles for simplicity.

This assessment, which is not altered by the results of the sensitivity analysis, suggests mild overfishing (Fig. 9.5(a)). Walters (pers. comm.) suggests that optimum F should be divided by $\exp(0.5\ CV^2)$ to reflect uncertainty in the data. If we do this the recommended target value for fishing effort is reduced slightly to 16 311–20 545 boats.

Approximate estimators

MSY estimates from the approximate methods are given in Table 9.3. Values vary from 93 128 t $year^{-1}$ (Beddington and Cooke) to 465 567 (Pauly) with a median of 240 874 t $year^{-1}$. Confidence limits from the Monte Carlo error analysis suggest that most of these estimates are not significantly different, with the exception of the Pauly estimator, which is high, and Beddington and Cooke, which is low.

As might be expected, the two Cadima estimators are most sensitive to the mortality rates (89% and 81% of variance in the MSY estimate to M and

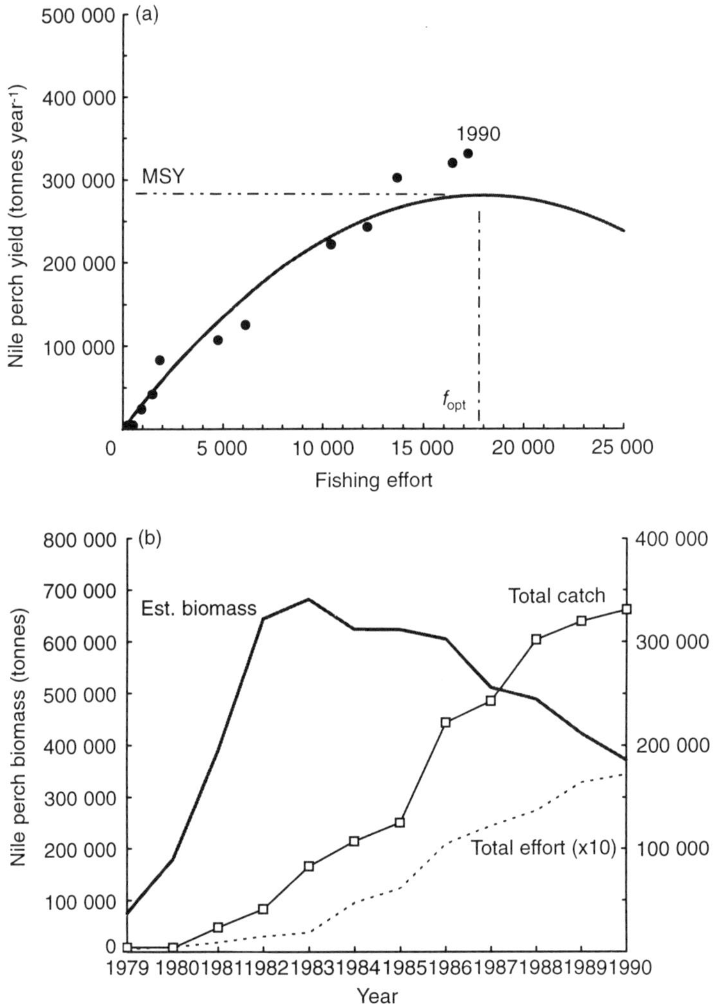

Fig. 9.5 (a) Equilibrium yield on fishing effort curve for the surplus-production model (standard effort scenario) with the fishery data points superimposed. Effort is in fishing vessels. (b) Estimated biomass of the Nile perch population in the whole lake, together with the catch and total effort data for comparison. Effort here is in fishing vessels $\times$ 10, and catch is in tonnes.

Z respectively), but insensitive to error in biomass, as is the Gulland formula (98% to M, 1.4% to B_∞). By contrast, Pauly's estimator is most sensitive to errors in B_∞ (94% contribution to variance).

Table 9.2 Bootstrapped 95% confidence limits for MSY and f_{opt} from the surplus-production assessment

Estimator	Lower CL	Upper CL
Standard effort scenario		
MSY (t year^{-1})	269 600	317 600
f_{opt} (boats)	16 640	20 960
Lower effort scenario		
MSY (t year^{-1})	306 430	318 224
f_{opt} (boats)	18 741	20 794

Table 9.3 Comparison of approximate assessments of MSY (sustainable tonnes per annum), and the increase in yield consequent on the introductions, using a range of different methods, for Lake Victoria Nile perch. Confidence limits (+95%) were obtained by Monte Carlo bootstrapped sensitivity analysis using the method outlined in the text. For B_∞ and current B, estimates from the surplus-production model fit were used

Estimator	MSY	Lower CL	Upper CL
Gulland	143 235	60 789	225 681
Cadima 1	247 305	197 277	297 333
Cadima 2	222 189	172 813	271 565
Pauly (1982)	465 567	419 317	511 817
Beddington and Cooke (1983)	93 128	–	–

9.5 DISCUSSION

Our assessment suggests considerable depletion of the Nile perch biomass by the fishery, the 1990 biomass being about half of B_∞. For MSY, only two of six approximate estimates lie beyond the general range of sustainable levels of exploitation suggested by the surplus-production and yield-per-recruit analyses. Consistency between the yield-per-recruit, surplus-production and most of the approximate assessment methods lends credibility to our results that was not so evident for our previous analyses of these data using older methods (Pitcher and Bundy, 1994). But our overall assessment, that the Nile perch fishery in Lake Victoria is currently overfished, is not altered by the new analysis.

The current (1990) effort is within the confidence interval for F_{opt}, but we are not confident that as it stands this would be a sensible management target. The estimated catchability translates the 95% confidence limits of $F_{0.1}$ from the yield-per-recruit analysis into a target effort zone of 3500 to 15 600 boats, while the equivalent criterion (effort equivalent to an equilibrium yield 90% of MSY) from the surplus-production analysis suggests a target effort range of 13 000 to 15 700 boats. A conservative compromise is to recommend a reduction to an effort of 14 000 boats, representing an approximate 20% reduction on the 1990 fishing effort.

We have carried out projections of catch and stock biomass using the surplus-production assessment for four different effort scenarios from 1990 (Fig. 9.6): scenario a, constant effort at 1990 level; scenario b, effort increases at 5% p.a. from 1990 level; scenario c, effort increases at 7% p.a. (the 1983–1990 rate) from 1990; scenario d, effort is reduced to 14 000 boats by 1993 and held constant. It is evident that scenario c leads rapidly to a dramatic collapse of Nile perch biomass and the probable destruction of the fishery. But it is instructive that the catch is maintained at over 300 000 t almost until the last couple of years (Fig. 9.6). A more moderate increase of effort in scenario b depletes biomass to 10% of B_∞ by the year 2000 and by then reduces catches to about half of MSY, but should not wipe out the fishery within this time frame provided that minimum size limits are implemented. Constant effort in scenario a keeps the Nile perch stock at around half of B_∞ and the yield at around MSY. In practice, the difference in CPUE from scenarios a and b would be indistinguishable from background error for many years. Figure 9.6 shows that scenario d tends to reduce short-term catches by up to 30%, but produces the highest CPUE and therefore profit, and is the safest in maintaining long-term high catches. Scenario d also produces a stock biomass that should be resilient to uncontrollable lake-shore fishing for undersized fish and should provide the best buffer against unforeseen limnological changes.

Our assessment gives a direct and dramatic indication that increasing effort at the rate of the late 1980s could soon lead to a catastrophic stock collapse. But a more modest increase leading to moderate growth overfishing or the maintenance of the 1990 effort should be sustainable. Food and spawning habitat for the juvenile and adult Nile perch seem assured, in contrast to the threat to the Nile tilapia represented by the invading water hyacinth.

A potentially most serious symptom, exemplified in the yield-per-recruit analysis, is the harvesting of immature Nile perch by the inshore netting sector. This could increase the variance of recruitment, leading to instability. Consequently, we recommend that this sector should be restrained as much as possible through the enforcement of minimum Nile perch size limits a little larger than the current 12.7 cm mesh (Ogutu-Ohwayo, pers. comm.). We

---

Fig. 9.6 Projections of Nile perch annual yield (bottom) and total stock biomass (centre graph) in four alternative effort scenarios (top). Solid curves joining open circles are the data for effort and catch; biomass is as estimated from the surplus-production model fit (standard effort scenario). For further details of scenarios a–d, see text.

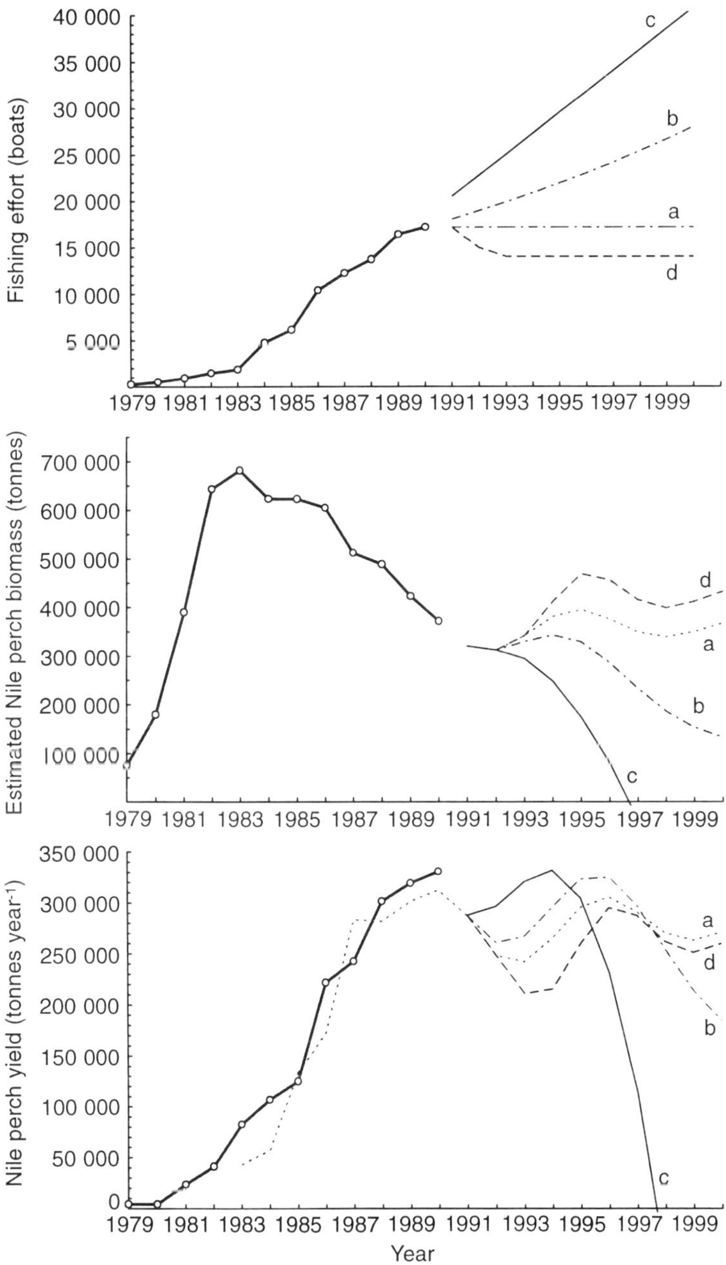

40 000
35 000
30 000
25 000
20 000
15 000
10 000
5 000
Fishing effort (boats)
c
b
a
d
1979 1981 1983 1985 1987 1989 1991 1993 1995 1997 1999
700 000
600 000
500 000
400 000
300 000
200 000
100 000
Estimated Nile perch biomass (tonnes)
d
a
b
c
1979 1981 1983 1985 1987 1989 1991 1993 1995 1997 1999
350 000
300 000
250 000
200 000
150 000
100 000
50 000
0
Nile perch yield (tonnes year-1)
a
d
b
c
1979 1981 1983 1985 1987 1989 1991 1993 1995 1997 1999
Year

recommend a minimum landing size of 3 kg (about 16 cm mesh, Nile perch length about 60 cm, age just under 2 years). Female Nile perch have an average maturation at 3 years old at a size larger than this, about 6 kg, but some trade-off with the inshore catch of Nile tilapia should be allowed. Calculation of a precise optimal minimum size limit requires more precise data from both fisheries.

In the fitted surplus-production model, increase in our fitted biomass is a year too early when compared with data from the fishery in Uganda and Tanzania. But the trawl surveys suggest that peak biomass in 1983 is a reasonable lake-wide average, although the documented expansion was a year later in the far south of the lake at the Mwanza Gulf survey site. Because our assessment represents a lake-wide averaging of stock, catch and effort, we are not too concerned by this. More detailed local information would be needed to partition our lake-wide effort recommendations into their local components.

Our fundamental assumption is that CPUE mirrors stock size, and this may be compromised in an expanding population, although increasing CPUE might justifiably be expected and so the bias may not be too severe. An additional problem is that the almost continuous spawning means that annually recorded catch and effort data may not be fully appropriate partitioning for the time series data.

Reductions in fishing effort for artisanal fisheries are not easy to implement, and in the case of subsistence artisanal fisheries, may be held to be unethical. But the Nile perch fishery in Lake Victoria now has all the trappings of a commercial harvesting operation and the necessary reduction in effort could perhaps be implemented with the cooperation and support of the processing sector.

The critical problem in the management of the Nile perch fishery continues to be lack of resources. Neither finance nor people have been available to effectively implement management policies. The fishery is diverse and extensive, comprising many official and unofficial landing sites, and it requires many field officers to regulate the fishery and to collect fisheries data. Moreover, there is a lack of coherent international organization: at present decisions have to be reached independently by the three riparian countries. There is a recent move toward lake-wide management (FAO, 1992), but the crucial political decisions have yet to be implemented.

Continued, careful monitoring and restriction of fishing effort, and the enforcement of minimum size limits, are essential to the sustainability of the Nile perch resource in Lake Victoria. Conservationists have long anticipated a collapse of the Lake Victoria Nile perch fishery: our assessment suggests that their wish could soon be gratified, not for their stated reason of ecosystem instability, but if lake-wide coordinated management of this valuable fishery resource fails to implement the required controls.

9.6 SUMMARY

Data from the recorded catch in the fishery, effort estimated from census and other data, and published growth and mortality values are used to assess the Nile perch fishery in Lake Victoria. Surplus production, yield-per-recruit and a range of approximate models are employed. Uncertainties are taken into account by bootstrapping using the error distributions of the input parameters. MSY is estimated at around 270 000–320 000 t (around 3.5–4.0 $t\,km^{-2}$) with an optimal effort of 16 000–20 000 boats.

All the assessment efforts indicate that the current Nile perch fishery in Lake Victoria is overexploited. Projections indicate stock collapse within a few years if present expansion of effort continues and immature fish continue to be harvested. It is recommended that effort be reduced to around 14 000 boats and a minimum size limit of at least 50 cm be introduced (pending optimization of mesh sizes with the Nile tilapia fishery) to provide a sustainable Nile perch fishery producing annually around a quarter of a million tonnes of fish suitable for export markets.

ACKNOWLEDGEMENTS

We would like to thank Eric Reynolds and John Craig for making Nile perch data available, and Carl Walters for invaluable advice on assessment. Early work on this project was supported by the Overseas Development Administration of the UK Government.

REFERENCES

Acere, T.O. (1988) Recent trends in the fisheries of Lake Victoria (Uganda Northern part), in Report of the 4th session of the Sub-committee for the Development and Management of the Fisheries of Lake Victoria, Kisumu, Kenya, 6–10 April 1987. *FAO Fish. Rep.*, no. 388, pp. 36–45.

Asila, A.A. and Ogari, J. (1987) Growth parameters and mortality rates of Nile perch (*Lates niloticus*) estimated from length-frequency data on the Nyanza Gulf (Lake Victoria). *FAO Fish. Rep.*, no. 389, pp. 272–87.

Beddington, J.R. and Cooke, J.G. (1983) The potential yield of fish stocks. *FAO Fish. tech. Pap.*, no. 242, pp. 1–47.

Beverton, R.J.H. and Holt, S.J. (1957) *On the Dynamics of Exploited Fish Populations*, Chapman and Hall, London (facsimile reprint 1993).

Bwathondi, P.O.J. and Mosille, O. (1988) The handling, processing and marketing of Nile perch in Tanzania. FAO GCP/KEN/055/NET, 53–6.

FAO (1992) CIFA, Report of the 6th session of the Sub-committee for the Development and Management of the Fisheries of Lake Victoria, Jinja, Uganda, 10–14 February 1992. FAO, Rome, *FAO Fish. Rep.*, no. 475, 18 pp.

Hecky, R.E. (1993) The eutrophication of Lake Victoria. *Verh. int. Verein. Limnol.*, **25**, 39–48.

Hughes, N.F. (1992) Growth and reproduction of Nile perch, *Lates niloticus* in the Nyanza Gulf, Lake Victoria. *Env. Biol. Fishes*, **33**, 299–305.

Kudhongania, A.W. and Coenen, E.J. (1992) Trends in development, prospects and limitations for Lake Victoria (Ugandan sector), *Development and Management of the Ugandan Fisheries of Lake Victoria, Jinja, Uganda, 6–8 August 1992*, (eds G.W. Ssentongo and F.L. Orach-Meza), RAF/87/099 – TD/31/92, pp. 59–76.

Ligtvoet, W. and Mkumbo, O.C. (1990) Synopsis of ecological and fishery research on Nile perch (*Lates niloticus*) in Lake Victoria, conducted by HEST/TAFIRI, in FAO, CIFA, Report of the 5th session of the Sub-committee for the Development and Management of the Fisheries of Lake Victoria, Mwanza, Tanzania, 12–14 Sep. 1989. *FAO Fish. Rep.*, no. 430, pp. 35–74.

Ligtvoet, W., Chande, A.I. and Mosille, O.I.I.W. (1988) A preliminary description of the artisanal Nile perch (*Lates niloticus*) fishery in Southern Lake Victoria, in Report of the 4th session of the Sub-committee for the Development and Management of the Fisheries of Lake Victoria, Kisumu, Kenya, 6–10 April 1987. *FAO Fish. Rep.*, no. 388, pp. 72–85.

Nyeko, D. and Acere, T.O. (1990) Inventory survey of fishing factors in the Ugandan part of Lake Victoria, in Regional Workshop on Fisheries Statistics and Information Systems for Lake Victoria, FAO/UNDP/UGA/007-OP (3):1–8.

Ogari, J. and Asila, A. (1990) The state of Lake Victoria fisheries, Kenya waters, in FAO, CIFA, Report of the 5th session of the Sub-committee for the Development and Management of the Fisheries of Lake Victoria, Mwanza, Tanzania, 12–14 Sep. 1989. *FAO Fish. Rep.*, no. 430, pp. 18–23.

Ogunja, J.A., Werimo, K.O. and Okemwa, E.N. (1992) A case study on high-value Nile perch products, in *Post Harvest Fish Technology* (ed. F. Teutscher). *CIFA tech Pap.*, **19**, 40–50.

Okemwa, E.N. (1984) Potential fishery of Nile perch *L. niloticus* Linne (Pisces: Centropomidae) in Nyanza Gulf of Lake Victoria, East Africa. *Hydrobiologia*, **108**(2), 121–6.

Pauly, D. (1982) Studying single species dynamics in multi-species context, in *Theory and Management of Tropical Fisheries* (eds D. Pauly and G.I. Murphy), ICLARM, Manila, pp. 33–70.

Pauly, D. (1984) *Fish Population Dynamics in Tropical Waters: A Manual for Use with Programmable Calculators*. ICLARM, Philippines, 325 pp.

Pitcher, T.J. and Bundy, A. (1994) Successful species introductions in the African lakes: assessment, uncertainties and strategies for fishery management, in *Management Strategies for Exploited Fish Populations* (eds G. Kruse, D.M. Eggers, R.J. Marasco, C. Pautzke and T.J. Quinn), Alaska Sea Grant College Program 93-02, Fairbanks, AL, pp. 545–70.

Pitcher, T.J. and Hart, P.J.B. (1982) *Fisheries Ecology*, Croom Helm, London, 414 pp.

Reynolds, J.E. and Ssali, W.M. (1990) Lake Victoria fisheries: recent developments in Uganda. FAO/UNDP UGA/87/007-sec-Field-Report 13, 43 pp.

Sparre, P., Ursin, E. and Venema, S.C. (1989) Introduction to tropical fish stock assessment. Part 1 Manual, Part 2 Exercises. FAO Rome, *FAO Fish. tech. Pap.*, no. 306/1 and 2, 429 pp.

Witte, F., Goldschmidt, T., Wanink, J., van Oijen, M., Goudswaard, K., Witte-Maas, E. and Bouton, N. (1992) The destruction of an endemic species flock: quantitative data on the decline of the haplochromine cichlids of Lake Victoria. *Env. Biol. Fishes*, **34**, 1–28.

Chapter ten

Thirty years on: the development of the Nile perch fishery in Lake Victoria

J. Eric Reynolds, Dominique F. Gréboval and Piero Mannini

10.1 THE MAKING OF AN 'ICHTHYO-IMBROGLIO'

Perch plantings

Accounts of the introductions of the large centropomid piscivore, Nile perch, *Lates niloticus*, into the waters of Lake Victoria have been provided by a number of writers (see Reynolds and Gréboval, 1988, for a fuller review). There is a surprising vagueness in many of these accounts when it comes to the actual sequence of events and the main actors (or villains, depending on one's point of view) involved. There is no doubt that the fish was placed into Lake Victoria at some point during the series of Nile perch transplantations to the Upper Nile–Lake Kyoga–Owen's Falls Dam stretch of Nilotic waters, conducted by officials of the Uganda Game and Fisheries Department between 1954 and 1957 (Kinloch, 1972). Stock was captured from Lake Albert and the Nile below Murchison Falls. Unlike the earlier introductions of non-indigenous tilapiines beginning in 1953 (Welcomme, 1966; 1967), the first placement of perch into Lake Victoria seems to have been carried out surreptitiously. It came as a surprise when the fish was discovered in waters close to Jinja (map, Chapter 2, Fig. 2.2) in May 1960, even as debate about whether the introduction ought to be carried out was continuing (EAFFRO, 1960; Fryer, 1960; Hamblyn, 1960; Anderson, 1961; Beadle, 1981). Two official 'deliberate'

The Impact of Species Changes in African Lakes. Edited by Tony J. Pitcher and Paul J.B. Hart. Published in 1995 by Chapman & Hall, London. ISBN 0 412 55050 4.

introductions of Nile perch to Ugandan waters were later conducted in 1962 and 1963 (Gee, 1965).

Thirty years on

Some three decades have now passed since the original introduction of Nile perch into Lake Victoria waters, and the attendant controversy, far from diminishing, seems to have proliferated almost as rapidly as the fish itself. Sometimes assuming quite sensational, strident, and bitter expression, it has spilled over from professional fisheries circles into the popular literature and press.

Debate has expanded beyond initial concerns with energy transfer inefficiencies versus the better utilization of 'trash' haplochromine stocks for the establishment of a new fishery to include broader issues. If one can imagine drawing up a general Nile perch 'charge sheet', it might look something as follows (Reynolds and Gréboval, 1988):

- Nile perch causes ecological disruption by voracious predation of endemic fish species;
- it *ipso facto* causes loss of crucial subsistence and commercial resources for lakeshore communities;
- Nile perch is resented by local people and disliked as food;
- it has adverse social and environmental impacts due to methods used in local processing; and
- its exploitation induces widespread economic displacement of artisanal fishers and small-scale traders.

Without denying the importance of concerns about ecological and other disruptions to the lacustrine system and its dependent human populations, the case has been put forward here that a great deal of value has been realized due to the introduction of Nile perch and associated developments, and that more may be forthcoming.

In the following sections, basic events of change in Lake Victoria are considered in terms of the 'pre-Nile perch' and 'post-Nile perch' fisheries regimes. This comparative perspective on the two regimes is distinct from the approach often taken in other assessments of events in Lake Victoria (but see Chapter 8). These latter tend to dwell on the post-Nile perch period of rapid expansion of effort and particularly on negative consequences, real or alleged, as the fisheries have moved into a phase of overexpansion which has been compounded by a lack of proper management. The comparative perspective adopted here allows an appreciation of the net socio-economic effects of the Nile perch introduction across a fuller historical context.

Trends in catch, effort and prices from the statistical record are reviewed lakewide and for each of the riparian countries. Supplementary evidence from field observations is drawn upon as appropriate. Discussion then focuses on basic economic performance and indicators for both the pre- and post-Nile perch regimes. Estimates are made of the net benefits that have accrued to producers and consumers of the riparian states as a result of the Nile perch introduction. Projections of short- to medium-term accumulations of benefits are also calculated based on differing assumptions for the lake's continued productivity.

10.2 PRE-PERCH, POST-PERCH: THE FISHERIES REGIMES COMPARED

The year 1980 can serve as a convenient dividing line between the 'pre-Nile perch' and 'post-Nile perch' fisheries regimes of Lake Victoria. Although present in the lake by the late 1950s, it was not until around 1980 that its effects began to be felt in the commercial fishery. The figures and tables used here to illustrate changes in catch levels, fishing effort and other aspects of the fisheries generally cover a time series beginning with 1975 and ending in 1989, straddling this 1980 dividing line. The year 1989 is also the latest for which complete official data are available for all three riparian states.

Catch levels

The change in nominal catch tonnages for the whole lake and by each country's contribution is shown in Fig. 10.1. The catches reported for Uganda and Tanzania have been adjusted in the case of *Rastrineobola argentea* to reflect wet weight, as the original figures are believed to be in dry weight (Lymo *et al.*, 1990; Gréboval and Mannini, 1992). A conversion factor of 3 has been used for this purpose. Overall, what is portrayed amounts to quite a dramatic, even astonishing picture. From a 1980 level of some 100 000 t (tonnes), fish catches lakewide rose to over 450 000 t by 1989, well beyond the level of 250 000 t that was considered likely as a peak by regional fisheries authorities in 1984 (CIFA, 1985). On the face of it, then, the lake after the debut of *Lates* has turned into a fish producer that can only be described in gargantuan terms.

'On the face of it' impressions given by available statistics warrant a cautionary note. The difficulties involved with the collection, analysis, and reporting of catch, effort, price and related data are well known to all who have worked in fisheries administration and research around Lake Victoria (e.g. Reynolds *et al.*, 1989; Ssentongo, 1990). They are a subject of some of the earliest official reports, and repeated attempts have been made over the years to resolve them, generally without much long-lasting effect. Monitoring and reporting systems have remained weak, not so much for lack of knowing

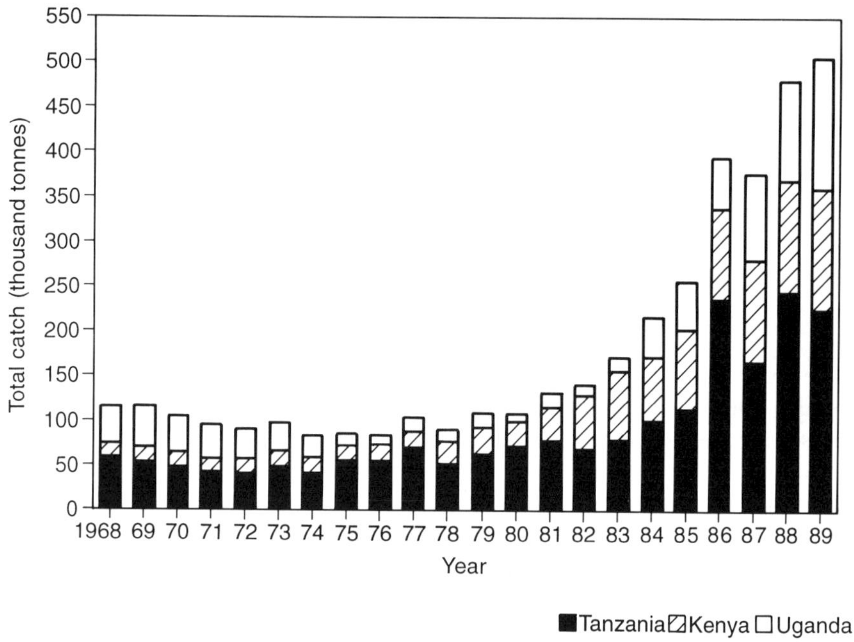

Fig. 10.1 Change in total fish catch, Lake Victoria, 1968–1989. Sources: Annual reports, National Fisheries Departments; CIFA (1982, 1985); Ssentongo (1985).

what needs to be done but for lack of funds, equipment, and trained and committed personnel.

The result is a collection of statistical data from different stations around the lake, the reliability and precision of which is open to question. The record is evidently 'soft' in crucial respects and vast improvement could be made. But on the other hand, there is no case to discard existing data as totally worthless. Broad developments in the fisheries are reflected in the year-to-year catch, effort, and marketing returns even if serious shortcomings exist at the levels of enumeration (e.g. missed counts, improper canoe or market stall sampling, spurious recordings) and compilation-analysis (e.g. lost or misplaced data sheets, use of incorrect raising factors, simple arithmetical mistakes). So the statistical record, with all its deficiencies, can provide indications of trends and orders of magnitude. It is a gauge of coarse rather than fine calibration, and will remain so until the riparian state governments through their respective fisheries authorities seriously dedicate themselves to its improvement.

With these reservations in mind, Fig. 10.1 can again be considered. The 1968–1989 series reflects a relative stagnation in catches until 1980. Catches decreased progressively in the 1960s and 1970s consequent upon the over-exploitation of major commercial species, although increased landings of haplochromines in Tanzania and Kenya in the 1970s tempered this downward trend (Reynolds and Gréboval, 1988). The figure also demonstrates the tremendous increase in fish yield that has taken place since the late 1970s in Lake Victoria, suggesting that it is in the order of a fourfold gain.

Species composition

The most obvious feature in reported species composition is the impressive gain in Nile perch from trace amounts in the late 1970s to levels of over 300 000 t in the late 1980s. A near-reverse pattern is found for the haplochromines: in the late 1970s they were contributing roughly 30 000 t to the overall catch but by the late 1980s this had dwindled to around 5000 t. Substantial rises in the landed tonnages of *R. argentea* and *Oreochromis niloticus* also occurred (catch figures are shown in Chapter 5, Figs 5.2–5.4 – *Eds*).

Trends in species composition represented as percentage of overall catch are shown in Fig. 10.2. This provides a clear picture of changing species balances over the years. The surge in Nile perch is again most obvious: a progression from less than 5% in 1980 to over 60% in 1989. The declining

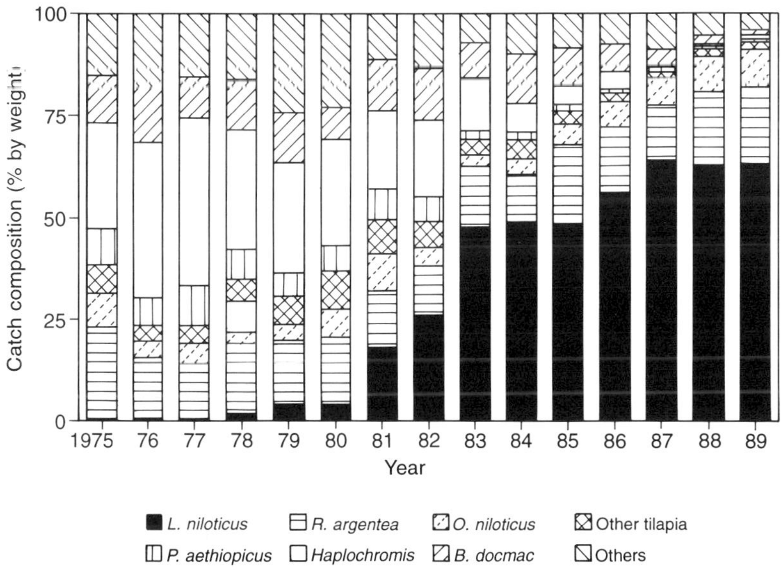

Fig. 10.2 Catch composition by major species, Lake Victoria, 1975–1989.

significance of haplochromines is clear, falling from peak 40% levels in the late 1970s to virtually nothing in 1989. Other points of note are the comparatively steady proportional rates for *R. argentea* (≈10–20%) and for *O. niloticus* (≈5–10%). Also to be noted are the generally declining amounts in the cases of 'other tilapia' (from ≈10% to ≈2% in the last 10 years), the lungfish *Protopterus aethiopicus* (varying from rates of ≈5–10% before 1982 to ≈0.5–2% since), *Bagrus docmac* (≈10% pre-1985, <5% since) and 'others', such as the genera *Clarias, Barbus, Synodontis, Mormyrus* and *Labeo* (generally >10% pre-1982, <10% since).

Figure 10.2 confirms that with the proliferation of *Lates*, the fishery of Lake Victoria has undergone a transformation from a broadly based multispecies fishery to one that is simplified. The commercial fishery is now largely based on the exotic Nile perch and tilapia stocks and the endemic *R. argentea*. It is also producing at levels that dwarf those of the old fishery regime, and producing fish that are more appreciated by riparian populations than the widely disliked haplochromines.

The situation does show some variation from country to country. Figures 10.3 to 10.5 provide breakdowns of catch data for Tanzania, Kenya and Uganda respectively. In Tanzania and Uganda, the Nile perch catch as of

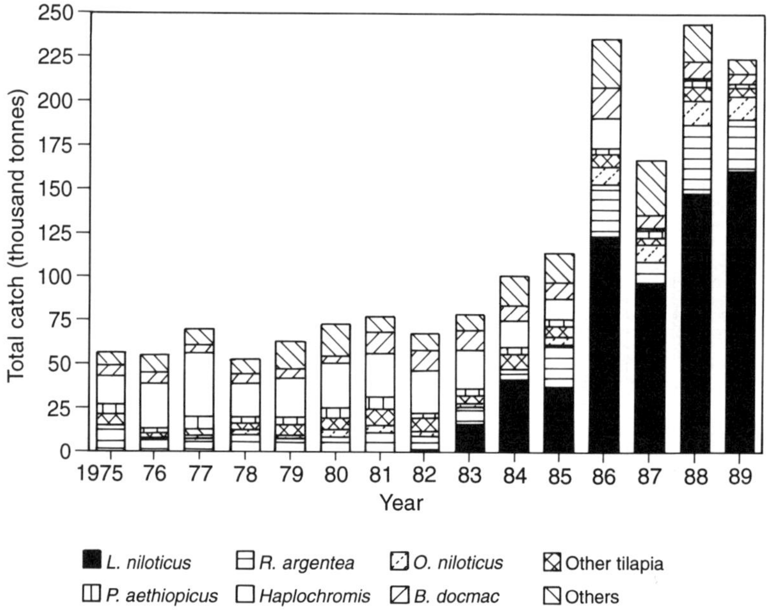

Fig. 10.3 Catch levels and species composition in Tanzanian sector of Lake Victoria, 1975–1989. Source: Annual reports, Fisheries Division, Dar es Salaam.

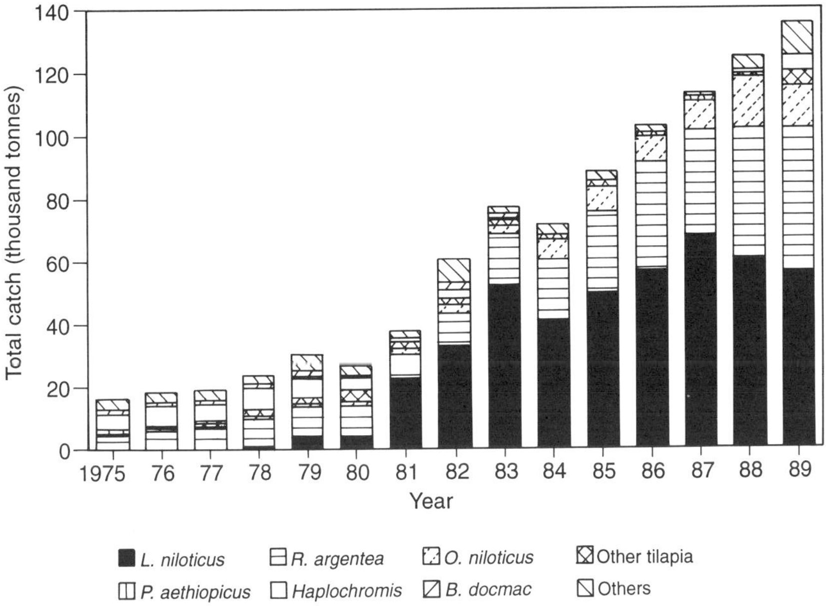

Fig. 10.4 Catch levels and species composition in Kenyan sector of Lake Victoria, 1975–1989. Source: Annual reports, Department of Fisheries, Nairobi.

1989 had reached an historical high whereas Kenya recorded a peak in 1987 and slight drops in both 1988 to 1989. For *R. argentea*, a steady increase in annual tonnage landed is recorded for Kenya whilst in Tanzania there are some marked fluctuations from year to year. In the case of Uganda, catch returns for *R. argentea* are fragmentary. For tilapiines the trend in Tanzania is one of general increase in catch levels. In Kenya there has been an even stronger upward trend. For Uganda, catches have achieved unprecedented levels for the last two years, and have grown consistently on an annual basis from the early 1980s after a period of lowered fishing activity due to civil strife and instability. One additional observation for Kenya is that the last three years have seen successive increases in the catch of haplochromines: some 4700 t are registered for 1989, the highest level since 1979.

Fishing effort

Data on effort in terms of numbers of boats, fishers, nets, or fishing days seem in general to be fragmentary and weak. Tanzania appears to have the most consistent time series for boats, fishers and gear. Kenyan data are intermittent and the Ugandan data show large gaps. Table 10.1 provides a

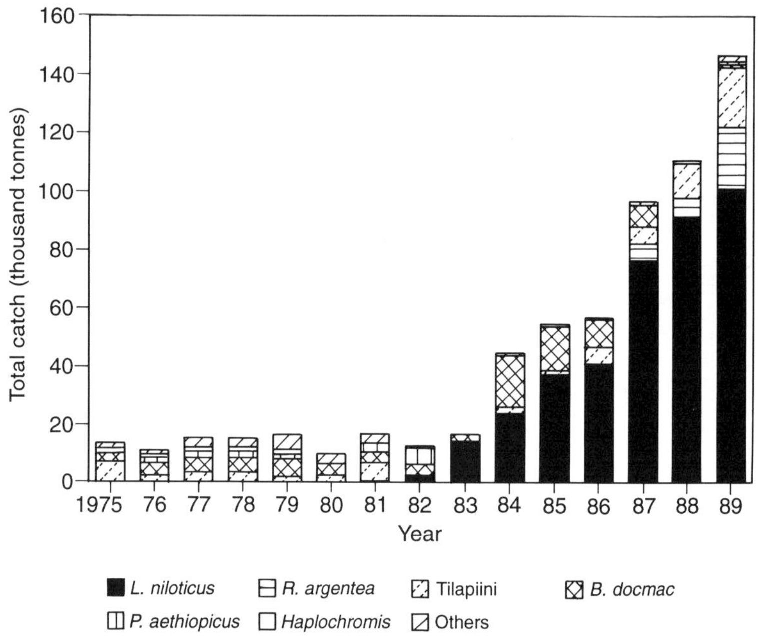

Fig. 10.5 Catch levels and species composition in Ugandan sector of Lake Victoria, 1975–1989. Source: Annual reports and records, Fisheries Department, Entebbe.

breakdown of what figures are available on boat and fisher numbers for each country. Though incomplete, the data are sufficient to indicate a trend towards increasing effort in terms of both kinds of measures (Chapter 9).

For Tanzania the development has been from boat levels of 3000–4000 during the 1970s to levels of 4000–7000 in the 1980s. Fishing vessels in Kenya seemed to have been fewer than 5000 in the 1970s and more than 5000 in the 1980s. The recurrence of the numbers '5000' and '5500' over several years suggests that certain assumptions were used in assembling returns. A recent census conducted with support from the UNDP/FAO Regional Project for Inland Fisheries Planning (IFIP) revealed a total of 6229 boats in the Kenya sector (Hoekstra *et al.*, 1991). For Uganda the data are weakest of all, but readings from the early 1970s and in 1977 and again in 1990 suggest a sharp increase in boat numbers from about 3000 to in the region of 8000.

With regard to numbers of fishers the picture is more or less the same. Official data are completely lacking for Uganda after 1973, when a total of 8100 fishers was reported. A total in the range of 25 1000–30 000 fishers

Table 10.1 Change in fishing effort (number of boats and fishers)*

Year	Tanzania		Kenya		Uganda	
	Boats	Fishers	Boats	Fishers	Boats	Fishers
1970	3 785	12 091	–	–	–	–
1971	3 302	14 131	–	–	3 264	–
1972	3 494	14 690	–	–	2 700	8 100
1973	3 976	15 917	4 100	10 700	–	–
1974	3 332	13 381	–	–	–	–
1975	3 918	15 037	–	–	–	–
1976	3 794	14 733	–	–	–	–
1977	3 398	14 327	–	–	3 000	–
1978	3 418	17 654		–	–	–
1979	4 457	20 937	4 600	18 000	–	–
1980	3 997	20 587	–	–	–	–
1981	4 199	20 787	–	–	–	–
1982	4 245	18 263	5 000	20 000	–	–
1983	4 141	15 194	5 000	20 000	–	–
1984	4 650	17 827	–	–	–	–
1985	4 160	17 556	5 500	21 500	–	–
1986	7 404	24 241	5 000	21 000	–	–
1987	6 667	22 207	–	–	4 000	–
1988	6 546	22 926	5 000	6 487 ?	3 470	–
1989	7 757	29 816	5 500	22 000	8 000	–
1990	–	–	–	–	–	–
1991	–	–	6 229	–	–	30 000

*Sources: Department of Fisheries, Nairobi, Kenya (Fisheries statistical bulletins); Fisheries Department, Entebbe, Uganda (Annual reports); Fisheries Dvision, Dar es Salaam, Tanzania (Annual statistics reports); Kanyike (1972); FAO (1973); Reynolds and Gréboval (1988); Hoekstra *et al.* (1991); Reynolds and Kitakule (1991); Tumwebaze and Coenen (1991). Dashes (–) indicate data not available; question mark (?) indicates improbable data.

can be estimated for recent years using results of a community sample survey conducted in 1991 (Reynolds and Kitakule, 1991). The Kenya time series register a growth in fisher numbers from around 10 000–11 000 in the early 1970s to 21 000–22 000 in the late 1980s. In Tanzania the increase has been from 15 000–20 000 in the latter half of the 1970s to nearly 30 000 in 1989. (A different approach to the estimation of the total fisher numbers lakewide is described later, in connection with employment benefits.)

Despite the overall rise in fisher and boat numbers, the catch per unit effort (CPUE) measured in tonnes per boat per year has shown a remarkable growth (Table 10.2). Tanzanian rates increased from about 15 t in 1975 to over 25 t in the late 1980s. The Kenyan catch rate can be calculated as around 7 t in the late 1970s and around 25 t by the late 1980s. For Uganda the rate has changed from about 5 t in 1977 to roughly 17 t in 1989 (calculated

Table 10.2 Indicative change in CPUE (1970–1989)*. Data are tonnes boat^{-1} year^{-1}

Year	Tanzania	Kenya	Uganda
1975	14.5	–	–
1976	14.6	–	–
1977	20.7	–	5.2
1978	15.4	–	–
1979	14.2	6.7	–
1980	18.3	–	–
1981	18.6	–	–
1982	16.1	12.1	–
1983	19.2	15.5	–
1984	21.9	–	–
1985	27.7	16.1	–
1986	31.9	20.6	–
1987	25.2	–	23.3
1988	37.4	25.0	30.9
1989	29.0	24.6	(16.6)†

*Dashes (–) indicate data not available.

†Figure in parentheses was obtained using 1989 total catch data and results of a fishing boat census carried out in 1990.

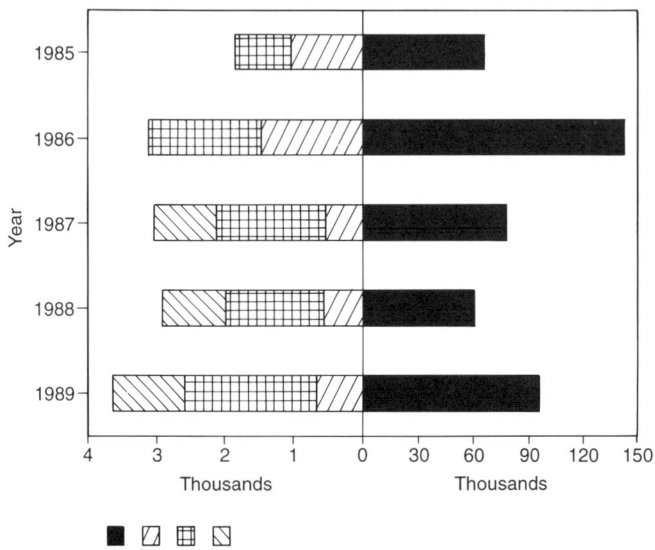

Fig. 10.6 Patterns of major gear composition in Tanzanian waters of Lake Victoria, 1985–1989. Note differing scales. Dark bars, gill nets left-to-right: hatching, beach seines; cross hatching, scoop nets; right-to-left: dagaa seines. Source: Annual reports, Fisheries Division, Dar es Salaam.

on the basis of 1990 boat count figures assembled under the UNDP/FAO FISHIN' Project UGA/87/007, Fisheries Statistics and Information Systems, Tumwebaze and Coenen, 1991).

Data available for the Tanzanian sector of the lake provide some indication of the recent change of effort in terms of major gear composition. In the Mwanza Region, there has been a major development of effort aimed at the 'dagaa' (*R. argentea*) fishery (Fig. 10.6). *R. argentea* gear (scoop nets and dagaa seines) has risen dramatically from fewer than 1000 units in 1985 to nearly 3000 units in 1989. There is also a preponderance of gill nets in Tanzanian gear. About 66 000 gill nets were counted in 1985, which increased to around 143 000 in the following year for reasons that are not clear, decreased sharply to 78 000 in 1987 and 61 000 in 1988, and increased again to about 96 000 in 1989.

Gill net mesh sizes for the entire Tanzanian sector for 1989 (Fig. 10.7) indicate that most units are 100–250 mm mesh, and that a mesh size of 180 mm, suitable for both Nile perch and large tilapia, is the most common. The 'retooling' from smaller nets of the earlier regime to sizes capable of catching larger Nile perch may exercise a positive effect in that it permits the escape of juveniles and smaller traditional food fish and thus helps to

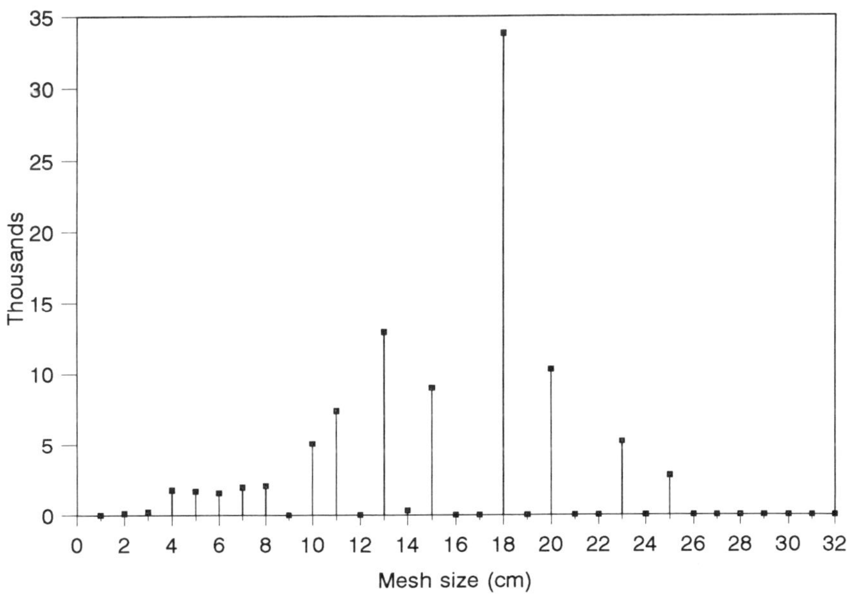

Fig. 10.7 Gill net mesh size distribution in Tanzanian sector of Lake Victoria, 1989. Source: Annual report, Fisheries Division, Dar es Salaam.

sustain or rebuild stocks (Arunga, 1981; Ssentongo and Welcomme, 1985; Chapter 9).

Market expansion

Lake Victoria has always served as an important source of fish for those living within the immediate area but its importance to markets across the wider East African region has until recently been relatively minor. From the 1920s through the 1950s a certain amount of ice-chilled tilapia was sent by rail to Nairobi and other points in Kenya from Kisumu. Improvement of road transport encouraged a gradual expansion of trade in fresh and especially processed fish away from the lakeshore as well. The Lake Victoria trade nevertheless remained fairly localized through the 1970s.

Conditions that developed under the post-Nile perch fishery regime completely transformed this situation. The lake now serves as fish supplier not only to broad sections of the East African region, but also to overseas markets. Interest in possibilities for large-scale commercial processing and overseas export of Nile perch fillets first grew in Kenya from around 1985, and that country still hosts most of the industrial plant activity linked to the lake. Over the last few years, several new processing plants have been established around Entebbe, Kampala and Jinja in Uganda. A few ventures have been launched in Tanzania, but remoteness and lack of good transport facilities have hampered development.

In Kenya and Uganda, processing firms send out insulated vehicles to buy supplies of raw fish from most of the major landing beaches. In some cases vessels are used for collection from island landing sites or catching boats. Most plants are equipped to produce frozen fillets of Nile perch, but there is also a growing trade in fresh chilled whole fish, fresh chilled fillets, and dried Nile perch swim bladders. All fresh fish is exported by air to destinations mostly in Europe; the frozen products are usually shipped by refrigerated container through the port of Mombasa to Europe and the Middle East. Frozen products also reach destinations in the Far East, Australia and North America, as well as other African countries.

The rapid growth in export trade is indicated by Table 10.3. In 1987, Kenyan concerns were already exporting some 4000 t of frozen Nile perch fillets to destinations overseas, with an estimated value of US$ 8 million. By 1991 this traffic had grown to around 10 198 t, worth an estimated US$ 20 million. For Uganda in 1989, official reports on 'documented' overseas exports of Nile perch products indicate that they were still relatively minor, amounting only to some 32.3 t, worth an estimated US$ 70 000. One would expect fairly low volumes for overseas exports at this date because the processing plants were only just beginning operations. It is likely, however, that the actual traffic was greater than official figures suggest. Both the reporting system

Table 10.3 Overseas exports of Nile perch products*

Exporting country	Year	Quantity (tonnes)		Total value (US$ million)
		Fillet	Swim bladders	
Kenya	1987	4082.5	–	8.0
	1989	6479.4	–	11.0
	1991	10198.0	–	20.0
Uganda	1989	18.4	13.9	0.07
	1990	576.5	120.0	1.6

* Sources: National Fisheries Department/Division records. Dashes (–) indicate no data available.

and the processing plants were performing better in 1990, when 696.5 t of fresh and processed fish products from Lake Victoria, mostly frozen Nile perch fillets, were registered as having been sent abroad. This represents an export value of some US$ 1.6 million.

Patterns of intra-regional trade have also undergone fundamental readjustment since the onset of the post-Nile perch regime. One dimension of this relates to overseas trade and the growth of demand for fresh fish by the Kenya industrial processing plants. Because of higher prices and other inducements, large quantities of fresh Nile perch now pass as 'informal' or 'undocumented' exports directly across the water from points in eastern Uganda and north-eastern Tanzania, destined for van collection at points along the Kenyan shoreline. The traffic is particularly heavy from Ugandan waters, with volumes being estimated in tens of thousands of tonnes annually (Ssali *et al.*, 1991).

Also to be taken into account is the considerable volume of sun-dried, smoked and salted fish products that move to remote markets within the three riparian states, or as exports to Zaire, Sudan, Rwanda and Burundi. Maembe (1990) indicated that the quantity of fish exported from the Tanzanian sector of Lake Victoria to Zaire, Rwanda and Burundi rose from a level of 5.2 t in 1986 to 870 t in 1989. In Tanzania there is a particularly heavy traffic in dried dagaa moving to the remote interior. In 1989 the railways alone moved over 3000 t of this product away from the lakeshore. At Kirumba, the largest fish-landing collection market in the Tanzanian sector, it is estimated that some 300 t of dried fish are handled every month.

For Uganda, major outlets for cured fish from Lake Victoria are Kenya to the east and Zaire to the west and north-west. The official data are incomplete and Lake Victoria products are aggregated with those of other national waters, but when considered with information available from direct field observations made during 1989–1991 under the FISHIN' Project (UGA/87/007), it can be estimated that this trade amounts to 5000 to 10 000 t annually.

Change in prices and value of catch

Exchange rate fluctuations, changes in currency, and the difficulty of obtaining reliable information on inflation rates or consumer price indices make the task of charting the prices and value for the Lake Victoria fisheries problematic. Table 10.4 reconstructs the movement of average ex-vessel fish prices from the 1975 Nile perch regime to the post-Nile perch year of 1989. The figures are derived through the use of indices based on the price of tilapia (generally the most valued species), which is assumed to have remained fairly constant over time. The tilapia index (actual price) is set to 100 and index values for the other major species are derived as a percentage of tilapia prices. This procedure is carried out for all major species in the riparian states for the years 1975/76, 1985, and 1989. (Because the Ugandan situation in 1975/76 and 1985 was confused owing to political disruption, rampant inflation and changing currencies, Ugandan prices for those years were reconstructed following Reynolds and Gréboval (1988). For 1989 the prices are averages derived from various surveys conducted under the FISHIN' Project.)

Total index values for each year by country are next computed by multiplying the price indices for each species by the actual landing tonnages for each species (Figs 10.3–10.5). Finally, for any given year, the mean index value is computed by dividing the overall index value (all species combined) by the overall landings (all species combined). The mean index value thus reflects the indexed price of fish in relation to the reference price of tilapia. Changes from year to year are in respect to changes in species composition and relative prices for the major species.

In Kenya, the value of the species mix first decreases over the interval from 1975/76 to 1985 and subsequently rises, so that by 1989 it has almost reached its pre-Nile perch level as the Nile perch becomes more and more sought after by consumers (Table 10.4).

For Tanzania the value of the species mix initially rises (1975/76 to 1985) and then remains stable (1985 to 1989). This is presumably due to greater initial acceptance of Nile perch by riparian and hinterland consumers and also greater resilience of the price of traditional species like *Bagrus*, *Protopterus*, and *Synodontis*. The slight decrease in mean index value that occurs between 1985 and 1989 may be due to the growing importance of the dagaa fishery.

In the case of Uganda a different pattern is apparent. The value of the species mix goes down in the initial period and down again in the subsequent period, despite the relatively high acceptability and value of Nile perch. This downward trend can be attributed to the 1975 species mix consisting mostly of very-high-value fish, and to the appearance of significant tonnages of low-value dagaa in the 1989 catches. Whilst a market developed to some extent in Kenya and Tanzania during the 1970s for haplochromines and related low-value products, this did not occur in Uganda because of higher

Table 10.4 Movement of average ex-vessel prices from 1975 (pre-Nile perch regime) to 1989 (post-Nile perch regime)*

	Price per kg†			Index			Total index value (thousands)		
	1975/76	1985	1989	1975	1985	1989	1975	1985	1989
Kenya									
Tilapia	2.67	6.09	10.39	100	100	100	64.20	944.20	1 780.80
Nile perch	1.59	2.01	4.60	60	33	44	3.04	1 651.20	2 515.17
Dagaa	0.87	1.38	3.09	33	23	30	148.19	586.13	1 352.11
Haplochromis	1.17	3.67	1.35	44	60	13	202.45	0.36	61.84
Bagrus	1.09	3.61	6.22	41	59	60	56.70	3.62	1.08
Protopterus	1.99	4.60	9.88	75	76	95	109.49	11.33	2.28
Others	1.65	3.72	5.00	62	61	48	238.66	185.39	507.60
Totals							822.73	3 382.23	6 220.87
Mean index value							49.62	38.18	45.93
Tanzania									
Tilapia	2.51	19.09	45.67	100	100	100	564.90	1 065.00	1 753.00
Nile perch	–	11.50	22.93	–	60	50	–	2 265.54	8 113.82
Dagaa	0.30	3.25	16.00	12	17	35	181.36	413.49	1 026.64
Haplochromis	0.99	6.90	9.19	39	36	20	636.91	418.27	11.51
Bagrus	1.78	15.15	32.59	71	79	71	413.23	731.31	346.45
Protopterus	1.60	16.40	30.77	64	86	67	393.94	347.93	162.44
Others	1.43	12.35	30.73	57	65	67	440.96	1 146.05	587.75
Totals							2 631.31	6 387.59	12 001.61
Mean index value							46.39	55.50	53.54

Table 10.4 (Continued)

	Price per kg†			Index			Total index value (thousands)		
	1975/76	1985	1989	1975	1985	1989	1975	1985	1989
Uganda									
Tilapia	–	–	100	100	100	100	700.00	126.80	2 021.80
Nile perch	–	–	75	60	70	75	15.00	2 617.20	7 594.28
Dagaa	–	–	17	10	10	17	0.30	–	359.65
Haplochromis	–	–	–	30	30	–	50.70	–	–
Bagrus	–	–	170	120	120	170	351.60	1 802.04	155.21
Protopterus	–	–	65	55	55	65	10.73	3.91	68.46
Others	–	–	95	85	85	95	141.95	71.06	183.92
Totals							1 270.28	4 620.83	10 380.31
Mean index value							92.00	84.66	70.86

* Dashes (–) indicate no relevant figures available.
† Prices shown in the table are expressed in local currency, e.g. KShs kg^{-1}. Currency exchange rates as at December 1989 are as follows: US$ 1 = 21.60 Kenya Shillings (KShs); US$ 1 = 189.73 Tanzanian Shillings (TShs); US$ 1 = 339.90 Ugandan Shillings (UShs).

consumer resistance to such products and the availability of good table fish from Lake Kyoga. Similarly, the recent development of the *Rastrineobola* fishery occurred first in Tanzania, then in Kenya, and only quite recently in Uganda. In the latter two countries *Rastrineobola* is heavily used for fish meal. The lake prawn *Caridina* is now being exploited in the same way in Kenya, a very recent and potentially worrying development.

Overall it can be noted that the increase in average ex-vessel fish prices has led to about the same result in Kenya and Tanzania, with pre- and post-Nile perch regime prices both standing at around half of the price of tilapia. The Ugandan situation is different, probably owing to the effect of consumer preferences.

Turning now to the progress of the ex-vessel value of landings (Table 10.5), average prices in US$ are first derived using the mean index values (Table 10.4) as percentages of the reference price of tilapia in 1989 Shillings, converted using the December 1989 exchange rates of US$ 1 = 21.6 KShs, 189.7 TShs, and 339.9 UShs. In an attempt to relax the assumption of a constant reference fish price, the evolution of the price of tilapia was compared with that of the consumer price index for Kenya. This was only done for Kenya due to the lack of meaningful data for the other two countries. The outcome of the new computations is an adjustment for inflation that shows that the reference prices of fish for 1975 and 1985 are respectively 19.2% higher and 21.5% lower than the 1989 price. These market adjustment factors were applied to Tanzania and Uganda as proxies in the absence of any information that would allow the construction of their country-specific factors.

From 1975 to 1985, fish prices in US$ dropped fairly sharply across the board as a consequence of the sudden increase in supplies on traditional littoral or immediate hinterland markets. But from 1985 to 1989 prices increased as a result of a rapid market expansion, both overseas and especially in urban areas and across the wider East African region (e.g. Nairobi, Dar es Salaam, Dodoma; also Zaire, Rwanda, and Burundi).

Essentially a situation was created in which there was increased competition for fish at landing sites in spite of the continuous increase in landings. Demand remained inelastic: people still bought as prices rose. Prices in 1989 nevertheless remained rather low in comparison with those current elsewhere in the region. The 1989 Lake Victoria price range of US$ 0.10–0.30 kg^{-1} can be set against the 1989 price range of US$ 0.25– 0.45 kg^{-1} for pelagic species in Burundi, for example.

Comparison of price performance between the three Lake Victoria countries also confirms expectations. In Tanzania the consistently lower prices across the 1975–1989 time span reflect geographical remoteness and a more rural population, so that marketing costs are relatively high and there is low opportunity cost of labour. In Uganda the initially higher prices in comparison

Table 10.5 Changes in ex-vessel value of landings from 1975 (pre-Nile perch regime) to 1989 (post-Nile perch regime)

	Mean index value			Average price (US$ t^{-1})			Landings (t)			Value (UD$ millions)		
	1975	1985	1989	1975	1985	1989	1975	1985	1989	1975	1985	1989
Kenya	49.62	38.18	45.93	284.5	144.2	220.9	16 581	88 589	135 431	4.72	12.78	29.92
Tanzania	46.39	55.50	53.34	134.5	106.0	129.9	56 718	115 098	225 011	7.63	12.20	29.23
Uganda	92.00	84.66	70.86	322.6	195.5	208.5	13 765	54 578	146 487	4.44	10.67	30.54
Totals							87 064	258 265	506 929	16.79	35.65	89.69

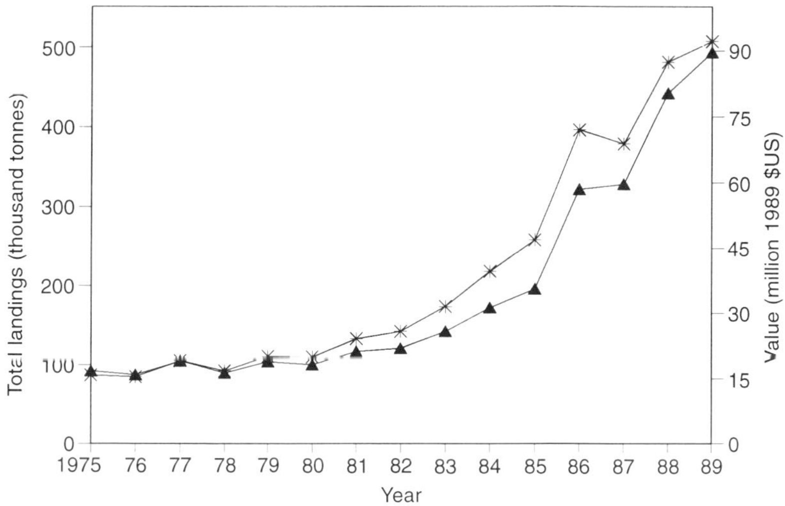

Fig. 10.8 Trends in total landings (×) and landing value (▲), Lake Victoria, 1975–1989.

with those in the other countries reflect the high-value species mix that was found in the 1975 landings. The 1989 price ends up lower than that for Kenya, however, as a result of a change in species composition and also due to very strong export-driven demand in Kenya.

Generally it can be observed that based on the results of the analysis using market adjustment factors, prices in 1989 were still lower than in 1975 but were rising due to an attendant increase in demand from overseas and distant urban and regional markets.

Translated into 1989 US$ value terms, the 1975, 1985 and 1989 landings from the whole of Lake Victoria work out respectively to 16.8, 35.6, and 89.7 million. Clearly these are not trivial sums: the lake is a producer of wealth on a considerable scale. In Fig. 10.8 an attempt is made to trace the change in value from pre- to post-Nile perch days. Ex-vessel value is plotted based on linear extrapolations from the average fish prices between 1975, 1985 and 1989, demonstrating the relative inelasticity of demand through market expansion.

10.3 IMPACT ASSESSMENT

Production and marketing benefits

The pre- and post-Nile perch regimes are contrasted (Fig. 10.9) according to what might have occurred with no change in catch and value levels from

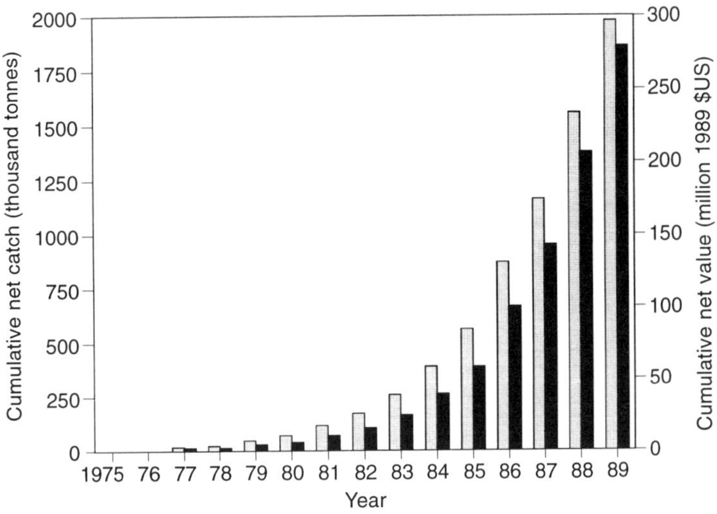

Fig. 10.9 Estimated net cumulative gains in catch (light columns) and value (dark columns) under post-Nile perch regime, Lake Victoria, 1975–1989.

the 1975/76 period (i.e. old regime conditions in force) versus what seems to have occurred with the changes (i.e. new regime conditions in force). The graph plots the net cumulative gains in landings and value/benefits that have attended the new regime. Net gains are computed on the basis of a perpetuation of the pre-Nile perch regime, i.e. of the 1975/76 situation wherein production stood at 87 000 t valued at US$ 16.8 million.

Net gains in value are expressed conservatively using ex-vessel prices. At market prices the value of products is about three times as high considering the typical margins involved (i.e. 100% from ex-vessel to primary wholesaler; 40% for the secondary wholesaler/larger retailer; and about 10% for the small-scale retailer). Value added (the value of output minus the value of externally purchased inputs) reflects the amount of economic value generated by the sector. Essentially, it is the sum of factor incomes, the wages and profits, of the various firms involved in the sector, including taxes and depreciation (gross value added). Due to the high labour-intensity and profitability of the sector, value added can be estimated at about one-third to one-half of market value (Reynolds and Gréboval, 1988). If conservatively set equal to one-third of market value, it is about equal to ex-vessel value. The gain in ex-vessel value can thus serve as a good approximation of net economic benefits, expressed in monetary terms, accruing under the Nile perch regime. Thus, over the period 1975–1989, the net economic benefits amount to about (1989) US$ 280 million.

Employment benefits

A further important dimension of benefit accumulation under the Nile perch regime bears on job creation. At the production level, it is immediately clear that the fishery is supporting more operators and crew, as indicated by the increase in numbers of boats. The effect is multiplied when landing-site activities such as porterage, net making and mending, canoe construction and repair, and fish processing and marketing are taken into account. It should be borne in mind that all of these activities at the harvest and post-harvest levels are very labour intensive in East Africa. For example, mechanization costs tend to be high (favouring the paddle or sail over the outboard engine for canoe propulsion, for example, or the bicycle over the lorry for short-range fish marketing). In addition, the true market value of labour (opportunity cost) is characteristically very low, again favouring manual alternatives to production and follow-on activities. Moreover, technological options are also rather limited so that wood-burning smoking kilns are favoured instead of electrically powered cold-smoking facilities at artisanal landing beaches.

Induced employment in other sectors such as the food and beverage business, input supply, construction and transportation should also be considered, both as service provision at landing settlements and market places, and as a result of many fishers reinvesting some of their earnings in housing and small businesses.

It is rather difficult to derive exact figures on the number of extra jobs and extra dependent-support capability that have been created under the new regime. Though these gains have evidently been quite substantial, the subject is one that warrants further measurement. In an effort to derive a more concrete appreciation of the extent of this gain, an approximate model was devised using numbers of canoes, average crew size, average catch per boat, secondary-to-primary employment ratios, and dependency ratios as elementary variables. These variables were assigned values for 1975 and 1989 respectively, using certain assumptions and more reliable evidence from the literature and field observations. The results are presented in Table 10.6. Beginning with estimated canoe numbers for the pre- and post-Nile perch years, an almost twofold increase from 11 000 (for 1975) to 21 987 (for 1989) is recorded. Working with a crew size (owner included) of 4.8 for both years, it is estimated that the number of fishers in Lake Victoria rose from around 52 800 in 1975 to around 105 000 in 1989, a twofold increase. Values for secondary-to-primary employment ratios were determined on the basis of field observations (Reynolds and Gréboval, 1988) taken in conjunction with a projected influx of processors, fishmongers and business people, to handle, accommodate, or otherwise trade on the almost threefold increase in boat catches per year (from 7.91 t in 1975 to 23.06 t in 1989). In this

Table 10.6 Employment in the fisheries sector, 1975 (pre-Nile perch regime) and 1989 (post-Nile perch regime)

Estimation for	Circa 1978	Circa 1989
Number of canoes*	11 100	21 987
Crew size†	4.4	4.8
Catch per canoe t year^{-1} ‡	7.91	23.06
Secondary-to-primary employment ratio §	2	3
Direct employment	52 800	105 500
Secondary employment	105 600	316 500
Total employment	158 400	422 000
Total fishers + dependants¶	475 200	1 266 000

* Source: Annual reports, National Fisheries Departments/Divisions; Hoekstra *et al.* (1991) for Kenya.

† Source: Tanzania Fisheries Division statistics reports; Kanyike (1972); CIFA (1992). Crew size includes boat owners who manage units even if seldom participating directly in fishing trips.

‡ Actual landings divided by number of canoes.

§ In FAO Fishery Country Profiles this ratio is typically estimated as 2 or 3 for small-scale fisheries in Africa. An estimate of 4 has been made for Uganda (Reynolds and Kirema-Mukasa, 1991). The estimates used here are thus quite conservative, especially considering the likely impact of the dramatic increase in catch per boat per year from 1975 to 1989.

¶ It is assumed that each person employed supports two dependants on average (see text).

way, the secondary-to-primary employment ratio was estimated as 2 for 1975 and 3 for 1989. Total employment in the fisheries sector can be reckoned at around 158 000 in 1975 and 422 000 in 1989 (respectively 3 and 4 times primary employment).

With regard to dependency ratios, it is assumed that two people can be associated with each primary or secondary operator. The figure may seem rather low but it is assumed to allow for the common pattern of joint family involvement in the sector (e.g. a fisher's spouse who is involved in processing or retail trading). This assumption can be taken as a safeguard to avoid any double counting and to keep the estimates on the conservative side. On these bases the number of fishers (primary + secondary) and their dependants are estimated at 475 200 for 1975 and 1 266 000 for 1989.

The model, with conservative values assigned for the variables, yields the information that the net gain in employment under the post-Nile perch regime (264 000 jobs) is over one-and-a-half times as great as the number sustained under the pre-Nile perch regime. In terms of the overall population directly and indirectly benefiting from fishing (fishers + their dependants), the same proportions apply. It is found that the new regime is supporting over a million people (1 266 000), or between two and three times (267%) as many people as the old regime (475 000). This again translates into a net gain of over one-and-a-half times.

Nutritional benefits

Special emphasis should be given to the nutritional welfare benefits that have accrued under the new regime for riparian and regional populations. Far greater numbers of people are able to eat high-quality animal protein in the form of fish than was ever possible before. The relatively sudden and sharp increase in fish from the lake can be regarded as most opportune given the rapid rate of population growth prevailing in the region.

People of the Tanzanian interior early on took to calling the Nile perch 'mkombozi', the Kiswahili for 'saviour' (Reynolds and Gréboval, 1988). The appellation denotes a genuine appreciation for the presence of a palatable new food item in places where fish could never or only rarely be found before, especially at a price most people could afford. In Uganda too, the Nile perch has fulfilled something of a 'saviour' role, becoming available in abundance just at a time when livestock production and agricultural support services in the country had deteriorated markedly owing to prolonged civil strife. In all the major urban centres of the region and across vast stretches of the countryside, Nile perch had become by the end of the 1980s an important and sought-after food.

Export earnings

It is obvious from earlier discussion that the Lake Victoria fisheries are now supporting an export trade of growing significance. This development has multiple techno-environmental and socio-economic implications that are both positive and negative in effect (Ssali *et al.*, 1991; Chapter 11). The processing plants promote the demand for fish and this in turn contributes to the risk of overexploitation of available stocks and spiralling prices for local customers. The plants may also, unless safeguards are taken, become added sources of water pollution. On the positive side, the plants, through their quality control measures, are promoting better handling practices in the artisanal fishery, so reducing post-harvest losses. There may also be gains in terms of forest and fuel-wood conservation arising from the use of product treatments differing from those commonly employed in the traditional processing sector (Ssali *et al.*, 1990).

The leading role of Kenya in the premium export trade was touched upon in earlier discussion. The country's favourable geographical and infrastructural circumstances have been determinative, along with the associated fact that investment was directed towards industrial processing by private entrepreneurs rather earlier in Kenya than was the case elsewhere. But overcapacity now exists in Kenya, and this has ramifications for Ugandan and Tanzanian fish-marketing and distribution patterns that could be interpreted as disruptive to their national economic interests.

A rapid expansion of fish filleting capacity around the lake is occurring as a number of new plants are being built in Uganda and Tanzania. The joint capacity of these two countries will soon be close to that of Kenya, where plants absorb 30 000 t of fresh fish annually. In this context, export earnings can be expected to increase significantly in the short term, but continued unchecked expansion of capacity is bound to exacerbate overexploitation tendencies and other conflicts of fisheries-related interest.

Against the view that growth in the industrial sector undermines and marginalizes the artisanal fishery sector, must be set the observation that the export of premium fish products entails a complex of activities that lead to local fishing operators getting more income, the provision of employment both directly and indirectly to large numbers of people, and the injection of foreign exchange earnings and reinvestment potential into the national economies. In this connection it may be emphasized that the appearance of an overseas-orientated export trade for Lake Victoria fish is a completely new development. It represented around 5% of the lake's production for 1987, with maybe an additional 10% of that year's production being informally exported to neighbouring countries (Gréboval, 1989). This indicates a wide expansion of the market base, with earnings being realized in a way that was not possible before.

Other impacts

We also consider other kinds of socio-economic impacts (alleged or real, likely or remote) that have been associated with developments in the post-Nile perch regime. First on the list is the assertion, quite vehemently made in earlier years and still circulated, that the coming of the Nile perch resulted in a loss of important food sources to riparian folk, who also had or have no liking for the predator as a food in itself. This assertion can be dismissed almost out of hand, having been fabricated on flimsy grounds to begin with and promulgated by those with little or no contact with present lakeshore realities. (The probable origin of this myth is given in Chapter 11 – *Eds.*) Rapid market expansion and the relative stability of prices at very high production levels are clearly demonstrating this assertion as unfounded.

Of a more serious nature are concerns for the impact of the new fishery regime on riparian trees and forests. The increase in landings coupled with the inability of local distribution networks to cope with all or even most of the fish being harvested as fresh product before it spoils means that smoke curing, by far the most popular method of preservation of large table fish around the lake, is now practised on an enormous scale. Excessive demand for wood associated with fish smoking in addition to the domestic requirements of fishing community residents, is leading to localized shortages and deforestation problems as well as the incidence of poorly cured products. An urgent need has therefore been identified to promote the use of more fuel-efficient smoking kilns and alternative preservation techniques (Maembe, 1990; Ssali *et al.*, 1990).

The interrelated questions of overexploitation pressures and socio-economic displacement are also matters warranting serious attention. Observers have frequently noted that increased competition for fresh fish at major landing sites has the effect of driving up ex-vessel prices and that these increases are eventually passed on to the consumer. This tendency is distressing for lower-income people, as fish has in the past been a very attractive buy in comparison with other forms of protein (Reynolds and Gréboval, 1988; Reynolds and Kirema-Mukasa, 1991). One outcome of the situation is a move by consumers to seek lower-grade or cheaper fish, often in the form of undersized or immature Nile perch and tilapia. Such products, easily divided into portions of a few fish each for sale at affordable prices, are increasingly a component of the fresh and cured trade, conducted especially by small-scale fishmongers. These latter are themselves subject to displacement pressure as they are less able to compete in the increasingly more costly bidding process at landing beaches. Higher ex-vessel sales, forced up by the overseas, regional, and domestic market demand, generally lead to stock overexploitation (Gréboval, 1989). In this context, it is particularly worrisome that young and juvenile stock should become a special target for the fishery. Displacement and overexploitation cannot be treated lightly and should become a priority as part of the management strategies for the new fisheries.

10.4 DISCUSSION: THE OUTLOOK

Species composition dynamics

The catch statistics reviewed provide an index of species distribution and abundance between the pre- and post-Nile perch years. The sharp decline in *Haplochromis* spp. populations that greatly concerned and shocked scientists and conservationists around the world (e.g. Barel *et al.*, 1985) can be attributed to two factors, namely fisheries exploitation and Nile perch predation (Chapters 6, 7). Evidence of reduced abundance in catch rates of haplochromines and *O. esculentus* (= *T. esculenta*) dates back at least to the late 1970s (Benda, 1979, 1981; Marten, 1979; Okemwa, 1984). Concern about localized overfishing was even being expressed by some observers much earlier (e.g. Graham, 1929; Beauchamp, 1955), long before Nile perch started to become a serious factor in the lake. Cichlid populations, displaying such life history characteristics as slow growth, low fecundity, and highly developed and complex parental care (Witte and Goudswaard, 1985; Ribbink, 1987), have poor resilience to intensive exploitation. Predation pressure, coupled with the already high fishing mortality in certain locations, quickly led to the reduction of *Haplochromis* stocks to exceedingly low levels. In an inverse fashion, high selective fishing mortality acting on Nile perch apparently can facilitate the recovery of some species of *Haplochromis*, providing that the latter are not further subjected to inshore fishing pressure (CIFA, 1990) (but see Chapter 6).

The upsurge of fish yields in the early 1980s also marked a transformation through which the indigenous multispecies fishery has given way to a simpler three-species regime. Nile perch, *R. argentea*, and the Nile tilapia, *O. niloticus*, are respectively the first-, second- and third-order constituents of this regime. The success of *L. niloticus* is explained by its predator efficiency, flexible and opportunistic diet (Ogari, 1985; Ogari and Dadzie, 1988) and, probably, its high survival rate for offspring during the initial phase of expansion in the new environment (Ogutu-Ohwayo, 1988; Achieng', 1990). Nile perch found a suitable array of prey in Lake Victoria, the haplochromines, which had life history strategies that, having evolved in a relatively, 'enemy-free-space' environment, were ill adapted to the new predator (Fryer, 1986). The rapid depletion of preferred prey did not induce a corresponding decrease in the predator population, as would normally be expected in the case of specialized predators, because Nile perch opportunistically switched to other available food resources, namely the freshwater shrimp *Caridina nilotica* and the small pelagic cyprinid *R. argentea*. Stocks of these two species, the importance of which for the lacustrine ecosystem dynamics has been given relatively scant attention in the past, have greatly contributed to the maintenance of the predator population and its high catch rates.

The Nile tilapia, introduced along with other non-indigenous tilapia species during the 1950s (Welcomme, 1966, 1967), appears to have adapted well to Lake Victoria. It has not been seriously affected by Nile perch predation, owing possibly to different habitat preferences (Ligtvoet, 1989). The Nile tilapia's adaptive success could be related to its relatively high growth rate (Twongo, 1988; Kudhongania, 1990), better efficiency in trophic and nesting competition (Lowe-McConnell, 1987), wide range of habitats (Welcomme, 1967), and flexible food spectrum (Chapter 4).

The small pelagic *R. argentea* is nowadays the second most important commercial species of Lake Victoria and perhaps the only native fish that has not been seriously affected by the presence of *Lates*. It is difficult to establish whether the *Rastrineobola* population has actually expanded during the last decade or so. Its apparently growing abundance could also be attributed to an increase in targeted fishing effort. A likely supposition is that *Rastrineobola* has benefited to some extent from the decline of planktivorous haplochromines due to predation by *Lates* (Wanink, 1989). The population biology of this cyprinid is poorly known. As a small pelagic species it could be presumed to exhibit high growth and natural mortality rates, and high relative fecundity. Such biological characteristics determine elevated production/biomass ratios and fast turnover rates, thus conferring a good tolerance towards exploitation, whether in terms of fishing mortality or of natural predation.

An ecosystem in stress

Diversity and stability as typical features of tropical lake ecosystems (Pitcher and Hart, 1982) are no longer found in Lake Victoria, both having been lost in a very short period. The lake represents a striking case of the transformation of a natural ecosystem by direct human intervention in the form of commercial fishing, species transplantation and nutrient enhancement. Its ecosystem now shows clear signs of stress (*sensu* Rapport *et al.*, 1985), such as decrease in species diversity and retrogression of maturity attendant upon the displacement of specialized native species by more opportunistic species.

According to Margalef, the ratio of productivity to biomass almost always increases in response to stressors (Rapport *et al.*, 1985). For Lake Victoria, such an increase can be postulated on the basis of the community-simplification process marked by the ascendancy of generalists (Nile perch, dagaa, Nile tilapia) and the decline of specialists (the indigenous stocks). A loss of ecosystem productivity in the form of reduced fish yields was supposed to occur in Lake Victoria as a final consequence of decreased energy transfer efficiency (Barel *et al.*, 1985; Ribbink, 1987; Ogari, 1990). The evidence shows that this has not yet happened. At least in the short and medium term, reduction in species diversity has coincided with an enormous increase in useful productivity as expressed in fish yields. Today the central question is whether such yields can be sustained or whether they are merely transitional expressions of imbalance in an ecosystem that is betwixt and between.

Of late, a further perturbation to the lacustrine ecosystem (and to lake observers, planners, and users as well!) has been added to this already confused situation. Lake Victoria is now infested with the notorious water hyacinth, *Eichhornia crassipes*, a development that could conceivably prove to be the most disruptive of all human interventions that have yet been visited upon its waters (Twongo *et al.*, 1991).

Whither the fisheries of Lake Victoria?

The business of making predictions about possible developments for Lake Victoria and its fisheries is clearly something of a venturesome undertaking, to say the least. And now, on top of everything else, comes the vexing presence of the hyacinth.

Whither the lake? To begin with, it seems fairly safe to suppose that the increasing trend in the catches observed over the last decade or so will ease off or even reverse itself. At present the lake is hosting a three-species fishery, over 60% of which amounts to what could be described as a '*Lates* show'. Sustainability would thus seem to depend heavily on the potential of Nile perch stock. *C. nilotica* and *R. argentea* are primary prey for Nile perch and consequently play an important role in maintaining the predator population. Sound prediction for Lake Victoria fisheries development would thus require a comprehensive, lakewide knowledge of population dynamics for these two species, and this is simply not available at present.

Lates and *Rastrineobola* seem strongly bound together in a predator–prey relationship. Together they constituted 80% of the lake's total catch in 1989. It would therefore seem a prudent strategy not to promote a strong development of the *Rastrineobola* fishery until its population dynamics and biomass turnover is well known. In this context, management actions aimed at sustaining the fisheries must encourage selective fishing methods. This means the use of larger-mesh-size gill nets for Nile perch and tilapia (Chapter 9), and the use of boat seines, scoop nets and lift nets in the *Rastrineobola* night (light attraction) fishery in open as opposed to inshore waters.

10.5 SUMMARY AND CONCLUSIONS

An imagined Nile perch 'charge sheet' may list various wrongdoings supposedly perpetrated by the predator since its introduction to Lake Victoria some 30 years ago. Allegations that the fish is responsible for the loss of crucial subsistence and commercial resources for riparian populations, and that it is resented as a food, simply do not stand up in the face of actual performance. In fact, its effect has been exactly the opposite.

In respect to other alleged disruptions to the lake, the situation is more ambiguous. Whilst it can readily be granted that the Nile perch's successful career as a producer of food and wealth came at the cost of a good deal of ecological upheaval, it is difficult to know how to evaluate this cost. Endemic haplochromines have certainly suffered a drastic decline but it is now clear that these stocks suffered setbacks as a result of predation pressure in combination with fishing mortality and environmental change. Nor is it yet conclusively established that the decline in haplochromine stocks is going to result in wholesale extinction. Where Nile perch themselves have been heavily fished there are now indications of some *Haplochromis* resurgence (see Chapters 6 and 7 for further discussion – *Eds*).

With regard to adverse social and environmental impacts of the Nile perch, there would seem to be more reason for concern. Localized problems of deforestation due to heavy fuel-wood demands for fish smoking and domestic needs of fisher communities have been identified, but demand for charcoal as a cooking fuel and the market for sawn timber in large urban

institutes who, through their kind cooperation and interest, made this contribution possible. Views expressed are those of the authors alone.

The designations employed and the presentation of the material in this document do not imply the expression of any opinion whatsoever on the part of the United Nations or the Food and Agriculture Organization of the United Nations concerning the legal or constitutional status of any country, territory or sea area, or concerning the delimitation of frontiers.

REFERENCES

Achieng', A.P. (1990) The impact of the introduction of Nile perch, *Lates niloticus* (L.) on the fisheries of Lake Victoria. *J. Fish Biol.*, **37** (Supp. A), 17–23.

Anderson, A.M. (1961) Further observations concerning the proposed introduction of Nile perch in Lake Victoria. *E. Afr. Agric. For. J.*, **26**, 195–201.

Arunga, J.O. (1981) A case study of the Lake Victoria Nile perch (Mbuta) fishery. *Proceedings of the KMFRI workshop on aquatic resources of Kenya* (Mombasa, 13–19 July 1981), Kenya Marine and Fisheries Research Institute and Kenya National Academy for Advancements of Arts and Sciences, Nairobi, pp. 165–83.

Barel, C.D.N., Dorit, R., Greenwood, P.H., Fryer, G., Hughes, N., Jackson, P.B.N., Kawanabe, H., Lowe-McConnell, R.H., Nagoshi, M., Ribbink, A.J., Trewavas, E., Witte, F. and Yamaoka, K. (1985) Destruction of fisheries in Africa's lakes. *Nature, Lond.*, **315**, 19–20.

Beadle, L.C. (1981) *The Inland Waters of Tropical Africa; an Introduction to Tropical Limnology*, 2nd edn, Longman, London, 475 pp.

Beauchamp, R.S.A. (1955) The danger of overfishing existing stocks of *Tilapia* with particular reference to Lake Victoria and the Kavirondo Gulf. *Rep. E. Afr. Freshwat. Fish. Res. Org.*, Jinja, Uganda, 30–34.

Benda, R.S. (1979) Analysis of catch data from 1968 to 1976 from nine fish landings in the Kenya waters of Lake Victoria. *J. Fish Biol.*, **15**, 385–7.

Benda, R.S. (1981) A comparison of bottom trawl catch rates in the Kenya waters of Lake Victoria. *J. Fish Biol.*, **18**, 609–13.

CIFA (1982) Report of the first session of the sub-committee for the development and management of the fisheries of Lake Victoria, Mwanza, Tanzania, 12–14 October 1981. FAO, Rome, *FAO Fish Rep.*, no. 262, 73 pp.

CIFA (1985) Report of the third session of the sub-committee for the development and management of the fisheries of Lake Victoria, Jinja, Uganda, 4–5 October 1984. FAO, Rome, *FAO Fish Rep.*, no. 335, 145 pp.

CIFA (1990) Report of the fifth session of the sub-committee for the development and management of the fisheries of Lake Victoria, Mwanza, Tanzania, 12–14 September 1989. FAO, Rome, *FAO Fish Rep.*, no. 430, 97 pp.

CIFA (1992) Report of the sixth session of the sub-committee for the development and management of the fisheries of Lake Victoria, Jinja, Uganda, 10–13 February 1992. FAO, Rome, *FAO Fish. Rep.*, no. 475, 48 pp.

EAFFRO (1960) *A. Rep. E. Afr. Freshwat. Fish. Res. Org.*, Jinja, Uganda, 48 pp.

FAO (1973) Lake Victoria fisheries research: Kenya, Tanzania, Uganda. FAO, Rome, *L. Vict. Reg. Fish. Res. Proj. Rep.* FI:DP/RAF/65/049. 44 pp.

Fryer, G. (1960) Concerning the proposed introduction of Nile perch into Lake Victoria. *E. Afr. Agric. J.*, **25**, 267–70.

Fryer, G. (1986) Enemy-free space: a new name for an ancient ecological concept. *Biol. J. Linn. Soc.*, **27**, 287–92.

Gee, J.M. (1965) Nile perch investigation. *A. Rep. E. Afr. Freshwat. Fish. Res. Org.*, Jinja, Uganda, 13–17.

Graham, M. (1929) *The Victoria Nyanza and its Fisheries*. Crown Agents for the Colonies, London, 255 pp.

Gréboval, D.F. (1989) Managing the new fisheries of Lake Victoria. FAO, Rome, *IFIP Tech. Doc.*, UNDP/FAO Regional Project for Inland Fisheries Planning (IFIP), RAF/87/099-TD/04/89 (En), 22 pp.

Gréboval, D. and Mannini, P. (1992) The fisheries of Lake Victoria: review of basic data. UNDP/FAO Regional Project for Inland Fisheries Planning Development and Management in Eastern/Central/Southern Africa (IFIP), Bujumbura. FAO, Rome, RAF/87/099/WP/16/92.

Hamblyn, E.L. (1960) The Nile perch project. *A. Rep. E. Afr. Freshwat. Fish. Res. Org.*, Jinja, Uganda, 26–32.

Hoekstra, T.M., Asila, A., Rabuor, C. and Rambiri, O. (1991) Report on a census of fishing boats and gear in the Kenyan waters of Lake Victoria. FAO, Rome, *IFIP Tech. Doc.*, UNDP/FAO Regional Project for Inland Fisheries Planning (IFIP), RAF/87/099-TD/26/91 (En), 36 pp.

Kanyike, E.S. (1972) The present-day fishery of the Uganda waters of Lake Victoria. Fisheries Dept, Entebbe, Uganda, *Occ. Pap.*, **4**, 2–6.

Kinloch, B. (1972) *The Shamba Raiders; Memories of a Game Warden*. Collins and Harvill Press, London.

Kudhongania, A.W. (1990) The future prospects for the Lake Victoria fishery, in *Fisheries of the African Great Lakes* (Res. paps Int. Symp. Resource Use Conserv. African Great Lakes, Bujumbura, 29 November–2 December 1989). Fisheries and Aquaculture Unit, Int. Agric. Centre, Wageningen, Netherlands, *Occ. Pap.*, **3**, 75–82.

Ligtvoet, W. (1989) Stock assessment of Nile perch in Lake Victoria, in *Fish Stocks and Fisheries in Lake Victoria* (A handbook to the HEST/TAFIRI and FAO/DANIDA regional seminar, Mwanza, January/February 1989). Leiden, The Netherlands, Rep. Haplochromis Ecology Survey Team (HEST) and Tanzanian Fish. Res. Inst. (TAFIRI), no. 53 (App. I).

Lowe-McConnell, R.H. (1987) *Ecological Studies in Tropical Fish Communities*. Cambridge Univ. Press, Cambridge, 382 pp.

Lymo, E., Nhwani, L.B. and Mkisi, M. (1990) The fisheries statistics of Lake Victoria Tanzanian sector, in *Workshop on Fisheries Statistics and Information Systems for Lake Victoria* (ed. G.W. Ssentongo), FAO, Rome, UNDP/FAO Regional Project for Inland Fisheries Planning (IFIP), RAF/87/099-TD/14/90 (En), pp. 23–9.

Maembe, T.W. (1990) A review of the main characteristics of long distance fish trade from Lake Victoria, Tanzania, in *Proc. Symp. Socio-economic Aspects of Lake Victoria Fisheries* (Vol. 5), FAO, Rome, UNDP/FAO Regional Project for Inland Fisheries Planning (IFIP), RAF/87/099-WP/05/89 (En), 68–100.

Marten, G.G. (1979) Impact of fishing on the inshore fishery of Lake Victoria (East Africa). *J. Fish. Res. Bd Can.*, **36**, 891–900.

Ogari, J. (1985) Distribution, food and feeding habits of *Lates niloticus* in Nyanza Gulf of Lake Victoria (Kenya). *FAO Fish. Rep.*, no. 335, pp. 68–80.

Ogari, J. (1990) Introduction and transfer of fish species: a case study of the exotic species found in Lake Victoria (Kenya waters), in *Fisheries of the African Great Lakes* (Res. paps Int. Symp. Resource Use Conserv. African Great Lakes, Bujumbura, 29 November–2 December 1989). Fisheries and Aquaculture Unit, Int. Agric.

Centre, Wageningen, Netherlands, *Occ. Pap.*, **3**, 55–7.

Ogari, J. and Dadzie, S. (1988) The food of the Nile perch, *Lates niloticus* (L.), after the disappearance of the haplochromine cichlids in the Nyanza Gulf of Lake Victoria (Kenya). *J. Fish Biol.*, **32**, 571–7.

Ogutu-Ohwayo, R. (1988) Reproductive potential of the Nile perch, *Lates niloticus* L. and the establishment of the species in Lakes Kyoga and Victoria (East Africa). *Hydrobiologia*, **162**, 193–200.

Okemwa, E.N. (1984) Potential fishery of Nile perch *Lates niloticus* Linne (Pisces: Centropomidae) in Nyanza Gulf of Lake Victoria, East Africa. *Hydrobiologia*, **108**, 121–6.

Pitcher, T.J. and Hart, P.J.B. (1982) *Fisheries Ecology*, Croom Helm, London, 414 pp.

Rapport, D.J., Regier, H.A. and Hutchinson, T.C. (1985) Ecosystem behaviour under stress. *Am. Nat.*, **125**, 618–40.

Reynolds, J.E. and Gréboval, D.F. (1988) Socio-economic effects of the evolution of Nile perch fisheries in Lake Victoria: a review. FAO, Rome, *CIFA tech. Pap.*, no. 17, 148 pp.

Reynolds, J.E. and Kirema-Mukasa, C.T. (1991) Review of fish marketing in Uganda. FAO, Rome, *FISHIN Note. Rec.*, Fisheries Statistics and Information Systems (FISHIN'), UGA/87/007-SEC/WP/1, 54 pp.

Reynolds, J.E. and Kitakule, J.S. (1991) Socio-economic aspects of Lakes Victoria and Albert fisheries: the 1991 FISHIN' community survey. FAO, Rome, *FISHIN Note. Rec.*, Fisheries Statistics and Information Systems (FISHIN'), UGA/87/007-FD/2, 84 pp.

Reynolds, J.E., Wadanya, J. and Nyeko, D. (eds) (1989) Fisheries statistics and information management in Uganda: past approaches, current status, and future prospects. FAO, Rome, *FISHIN' Note. Rec.*, Fisheries Statistics and Information Systems (FISHIN'), UGA/87/007-FD/1, 84 pp.

Ribbink, A.J. (1987) African lakes and their fishes: conservation scenarios and suggestions. *Env. Biol. Fishes*, **19**, 3–26.

Ssali, W.M., Reynolds, J.E. and Ward, A.R. (1990) Fish and fuel, food and forests: perspectives on post-harvest losses in Uganda. FAO, Rome, *FISHIN Note. Rec.*, Fisheries Statistics and Information Systems (FISHIN'), UGA/87/007-SEC/FR/17, 16 pp.

Ssali, W.M., Reynolds, J.E. and Ogutu-Ohwayo, R. (1991) Industrial processing investment and development for the fisheries of Lake Victoria: present and future concerns, in *Resource Stewardship and Multiple Use Interests for a Fisheries in Flux: Observations on Lake Victoria, Uganda* (ed. J.E. Reynolds), FAO, Rome, Fisheries Statistics and Information Systems (FISHIN'), UGA/87/007-FD/4, pp. 1–20.

Ssentongo, G.W. (1985) Recent trends in the fisheries of the Tanzanian sector of Lake Victoria. Fisheries Dept., FAO, Rome. (Mimeo)

Ssentongo, G.W. and Welcomme, R.L. (1985) Past history and current trends in the fisheries of Lake Victoria. *FAO Fish. Rep.*, no. 335, pp. 123–38.

Ssentongo, G.W. (ed.) (1990) *Workshop on Fisheries Statistics and Information Systems for Lake Victoria.* FAO, Rome, UNDP/FAO Regional Project for Inland Fisheries Planning (IFIP), RAF/87/099-TD/14/90 (En), 36 pp.

Tumwebaze, R. and Coenen, E.J. (1991) Report on the frame survey conducted in the Ugandan part of Lake Victoria (1990). FAO, Rome, *FISHIN Note. Rec.*, Fisheries Statistics and Information Systems (FISHIN'), UGA/87/007-BIOST/FR/22, 13 pp.

Twongo, T. (1988) Recent trends in the fisheries of Lake Kyoga, Uganda, in *Predator–prey Relationships, Population Dynamics and Fisheries Productivities of Large African Lakes* (ed. D. Lewis), FAO, Rome, *CIFA Occ. Pap.*, **15**, 140–151.

Twongo, T., Reynolds, J.E. and Mwene-Beyanga, P. (1991) Management issues, options, and strategies for Lake Victoria fisheries, in *Resource Stewardship and Multiple*

Use Interests for a Fisheries in Flux: Observations on Lake Victoria, Uganda (ed. J.E. Reynolds), FAO, Rome, Fisheries Statistics and Information Systems (FISHIN'), UGA/87/007-FD/4, pp. 103–132.

Wanink, J.H. (1989) The ecology and fishery of dagaa, *Rastrineobola argentea* (Pellegrin) 1904, in *Fish Stocks and Fisheries in Lake Victoria* (A handbook to the HEST/TAFIRI and FAO/DANIDA regional seminar, Mwanza, January/February 1989). Leiden, The Netherlands, Rep. Haplochromis Ecology Survey Team (HEST) and Tanzanian Fish. Res. Inst. (TAFIRI), no. 53 (App. II).

Welcomme, R.L. (1966) Recent changes in the stocks of *Tilapia* in Lake Victoria. *Nature, Lond.*, **212**, 52–4.

Welcomme, R.L. (1967) Observations on the biology of the introduced species of *Tilapia* in Lake Victoria. *Revue Zool. Bot. afr.*, **76**, 249–79.

Witte, F. and Goudswaard, P.C. (1985) Aspects of the haplochromine fishery in southern Lake Victoria. *FAO Fish. Rep.*, no. 335, pp. 81–8.

Chapter eleven

Socio-economic impacts of introduced species in Lake Victoria fisheries

Craig K. Harris, David S. Wiley and Douglas C. Wilson

11.1 INTRODUCTION

One of the basic principles of introductory ecology is the YCCJOT Principle: you cannot change just one thing. The fundamental notions of ecosystem analysis imply that all elements are connected, directly or indirectly, so that a change in one element will eventually have some impact on every other element (Mwebaza-Mdawula, 1990).

Ecosystems are not isolated physical systems, and the biological and limnological changes in them do not occur without important human dimensions. Almost any aquatic ecosystem includes significant human influence (Balirwa, 1990). Usually humans have developed ways of obtaining sustenance and/or income from the natural resource base, and ways of adapting to and managing the functioning of the ecosystem (McEvoy, 1986; McCay and Acheson, 1987). If we accept that human actors are equally significant as elements of the fisheries ecosystem (Harris, 1992), it is not surprising that changes in the biotic and aquatic resource base ramify throughout riparian societies, with impacts on households, communities, national governments and international agencies (Vanderpool, 1987).

The Impact of Species Changes in African Lakes. Edited by Tony J. Pitcher and Paul J.B. Hart. Published in 1995 by Chapman & Hall, London. ISBN 0 412 55050 4.

The social and economic impacts which might follow from the introduction of new species of fish into a body of water are many and varied. These impacts may occur at any level of social organization, from the household to the global level. Their effects may be manifold, including technology requirements for the fishers, quality of life, levels of income, forms of organization of production and distribution, relations of employment, modes of regulation, and even the diets of the fishers and the populations they supply. They may be viewed as beneficial or detrimental, permanent or temporary, unavoidable or mitigable.

Changes in the non-human components of the ecosystem generate two types of impacts for the human elements in the system (Schnaiberg, 1980). First, the altered ecosystem includes new stocks of potential resources, which can generate streams of benefits for potential consumers. We can analyse this type of impact by asking how these streams of benefits are distributed among the potential users; thus we refer to these as **distributional** impacts. Second, alterations in the ecosystem also remove or seriously alter some of the former resources, or change them in ways which make them less accessible to former users. We can analyse these latter types of impacts by asking how these changes result in the redistribution of the benefits among the human actors in the ecosystem; for this reason we refer to these as **redistributional** impacts.

In this chapter we can only sketch an outline of the full set of impacts which we hypothesize will follow from the introduction of new species into Lake Victoria. We are currently investigating the impacts; the material in this chapter is based upon preliminary observations from our research in midstream. We view the impacts as occurring at four levels: the individual and the household, the community, the nation, and the international level. At each of these levels, we examine impacts in the areas of economics, social relationships and political structures.

Recent socio-economic impacts result primarily from the soaring catches of Nile perch in the 1980s, and the less spectacular increases in the catches of tilapiines (Greboval and Mannini, 1992). However, it is important to remember that for almost a century, portions of Lake Victoria have been the subject of intense fishing pressures, rapid socio-economic and political change, and rising inputs of soil, nutrients and chemicals from the turbulent changes in the activities of the lake basin. Frequently, it is difficult to disaggregate the changes specifically resulting from the introductions and alterations in species composition from other sources of change. This is especially true for the Nile perch, the great production of which in the 1980s arrived simultaneously with soaring world market demand for good-quality fish. The perch, and the foreign exchange paid for it, combine to create a striking series of impacts on all the humans linked to the lake.

11.2 INDIVIDUAL AND HOUSEHOLD IMPACTS

Categories of roles in the Lake Victoria fishery

First, we examine the impacts at the level of individuals and the households in which they live. For purposes of this discussion we focus on several distinct categories of human roles which have direct relationships with the Lake Victoria fishery; a more complete analysis would include others (Harris, 1991). For this analysis, we distinguish between the roles, and therefore the distinct skill requirements, and needs of: (1) boat owners, (2) boat managers, (3) boat operators, (4) boat crew members, (5) processors, local and industrial, and (6) collectors and marketers. In addition, we discuss briefly the members of the families and households of the occupants of each of these roles.

The first category is the owners of the fishing boats. Even before the fish introductions of the 1950s and 1960s, a major shift to more capitalist economic relations in the artisanal fishery began in the first years of this century. Changes in capture technology occurred with the arrival in the Kenya fishery of the 'Irish linen nets' and the entry into the lake fishery of European and Asian fishers (Wiley and Yongo, 1992). The transition from the earlier subsistence and barter economy of lakeside peoples probably began even before this, although we have no oral or documentary evidence to provide approximate dates.

More recently, the elaboration of the perch–tilapia fishery has intensified this transition. (We shall use the term 'perch–tilapia fishery' to refer to the state of the Lake Victoria fishery in the late 1980s and early 1990s. Although Nile perch is the predominant species, catches of tilapia are still considerable, and the *Rastrineobola* fishery is increasing in importance. Other species are of marginal significance, although a *Caridina* fishery seems to be emerging; Chapters 6, 11). As a result, we suggest that, for most of the lake, the commercialization of the fishery has proceeded so far that the term 'artisanal fishery' serves only to camouflage the extensive cash-economy relations which have developed in both the capture and marketing sectors (Ferguson *et al.*, 1993). The separation of the function of boat ownership from the other functions in harvesting is one consequence, and seems to be less prevalent in Kenya than in Uganda. Whereas Butcher and Colaris (1975) reported that 55% to 60% of the boats operating on Lake Victoria were operated by their owners, Jansen (1979) reported that almost all full-time fishers in Kenyan waters owned their own boats. In contrast, a 1991 survey in Uganda found that only 60% of fishing boats are operated by their owners (Table 49 in Reynolds and Kitakule, 1991).

The perch–tilapia fishery has the potential to produce such a large marketable surplus that, in the context of the high prices for fish in the local and global markets, boats and nets become attractive as investments, both for persons who have gained investable savings from their fishery activities and for investors from outside the fishery (Asowa-Okwe, 1989). So far on Lake Victoria this type of differentiation is still emerging. However, there is growing differentiation among owners within the small-scale sector, contrary to the expectations of Platteau (1989). This differentiation probably has occurred primarily because of the rising prices for gear (Asowa-Okwe, 1989). In the 1980s, Nile perch was harvested by four or five commercial trawlers in the Kenyan waters and 13 trawlers in Tanzanian waters (Siwo, 1988; Greboval, 1989), but they are no longer operating. Since 1986 two pair-trawler units have operated out of Entebbe as a joint venture between the governments of Uganda and the People's Republic of China.

Wherever it occurs, this separation of boat ownership from boat operation has two consequences. First, the relationship between the boat owner's household and the fishery is changed. In many parts of the world, kinship has traditionally played a crucial role in the social organization of fishing (Acheson, 1981). In the Lake Victoria fishery, owners have always felt pressure to find crew members who are 'trustworthy', usually meaning a member of their immediate or extended family (Luhmann, 1982). The families of owner/operators have often been crew members, and female kin often process and market the fish which are caught and landed by the men. Jansen (1979) reported that owners who did not work in their own canoes would always have a crew member to whom they were closely related, but kinship-based relations are becoming less important. In Tanzania, 63% of crew members are not related to the owner or renter of the fishing boat. Yongo (1991) also reports that an increasing number of the women processors in the Kenyan fishery are single parents rather than wives of the fishers.

Second, the boat owner arranges for the management and operation of the boat. Although in theory the roles of manager and operator could be combined, in practice they tend not to be. Thus the perch–tilapia fishery has given rise to the role of boat manager, a person who lives at or near the beach or landing and is responsible for the maintenance and oversight of the owner's investment. Kinship may still attach to the role of manager, but some community leaders in Uganda have suggested that managers are unlikely to be close relatives of owners, and unlikely to involve other members of their households in fishery activities.

When he does not operate the boat himself, probably the most important decision the manager makes is the selection of an operator for the fishing boat. This third type of actor on the lake, the operator, decides where and how to set the nets and takes responsibility for the security of the catch and gear while on the lake. So a manager seeks a person who is knowledgeable, skilful and trustworthy (Nemec, 1972). Although in the past, the roles of boat owner, manager and operator frequently were combined in one individual, we find that this is decreasing in both Kenya and Uganda.

Table 11.1 Crew size, from a Tanzanian sample. Source: Wilson and Ntara (1995)

Crew Size*	Frequency
2	1
3	34
4	25
5	8
>6	3

* Includes operator.

The fourth significant category of actors in the Lake Victoria fishery is the crew members. In the past, members of the crew might have been siblings or children of the owner-operator, but that is less true now, for similar reasons to that of the choice of managers. Most fishing boats have two to four crew members in addition to the operator (Table 11.1), but now these often are temporary rather than long-term relationships. Further, being a member of a crew member's household does not necessarily generate direct involvement in the fishery. Increasingly, crew members may be seasonal labourers from outside the fishing community, migrant labour from distant communities, and casual or contractual labour (Asowa-Okwe, 1989; Babikwa, 1992).

The increasingly narrow restriction of operator and crew roles to transactions mediated by the labour market has led to pressure for an increased share of the value of the catch. The operator and crew members formerly expected that they would share in future disbursement of family wealth, which was being generated in part by fishing, and so they may have been content to receive a smaller share of current earnings. Without that expected future gain, they are more insistent upon a larger percentage of the daily income. This has led to strikes by crew lasting as long as a week at some landings in Uganda.

The fifth category of roles in the Lake Victoria fishery is the fish processors. Traditionally, processing was limited to cleaning and preserving, usually by smoking or sun-drying. To some extent, these have continued; Bwathondi (1990a) and others have noted that dried Nile perch are in demand in both local and in Central and Western African markets. The high fat content of the Nile perch means that large fillets or whole fish can turn rancid before sun-drying is completed and makes it difficult to smoke in traditional, semi-ventilated artisanal smoking kilns. (This is the origin of the story that Nile perch is not acceptable to lakeshore communities; see Chapters 2 and 10 – *Eds.*) The advent of the larger Nile perch has therefore necessitated other forms of processing. At least three new forms of fish processing have emerged on Lake Victoria as a result of the expansion of the perch–tilapia fishery. The two emergent small-scale forms of processing include fish frying (Akersie, 1991; Kirema-Mukasa and Dhatemwa, 1992) and, subsidiary to the industrial processors removing the fillets, perch fish frame roasting, frying, and oil extraction.

Industrial processing is the emergent large-scale form; indeed, a commodity chain analysis of the Lake Victoria fishery suggests that the industrial processors are the dominant force in the restructuring of the fishery (Harris, 1991; Odongkara, 1993). The principal forms of industrial processing for Nile perch and tilapia are chilled fresh fish and fish fillets, frozen fillets, frozen whole fish, cold-smoked fillets, fish meal, fish maws and swim bladders (Kirema-Mukasa and Dhatemwa, 1992). For *Rastrineobola*, factory pelleting of the dried catch for animal and aquaculture feeds is predominant. The final marketing of fresh fish locally remains an article of small-scale commerce (Kirema-Mukasa and Dhatemwa, 1992), with the possible exception of a few urban centres.

This is not to imply that industrial processing has totally replaced artisanal processing, which traditionally had been a major provider of work for women in the fishing villages and in the markets. In addition to the frying of fish, sun-drying of *Rastrineobola* is still widespread, and large quantities of smoked tilapia are produced by artisanal processors in Uganda for distribution in the areas distant from the lake and for export to Kenya. Nevertheless, in 1991 only 30% of Ugandan fishing communities reported that fish processing provided an income-generating activity for women of the community (Reynolds and Kitakule, 1991). In Tanzania, none of the dryers, half the smokers, but most of the fryers are women (Akersie, 1991). The volume and value of fish passing through the industrial processing conduit probably accounts for well over half of the total annual catch from the lake (Okaronon, 1990). At the moment these products are largely intended for export (Twongo *et al.*, 1991; Kirema-Mukasa and Dhatemwa, 1992).

Detailed discussion about the social and economic characteristics of the owners, managers, and workers engaged in both industrial processing and the secondary and tertiary employments deriving from the processing of by-products and commercial marketing is beyond the scope of this chapter. In general, the owners of industrial processing plants came to the fishery from the food industry rather than from the fishery. Their capital is a mixture of foreign aid, foreign investment and domestic savings from corporations and individuals. Their labour force is largely indigenous, but again frequently not with fishery backgrounds. Thus, at least in the processing sector, the benefits of the extended Nile perch fishery apparently have been distributed to those who were not formally involved in the fishery (Goodman and Redclift, 1991). Indeed, the large-scale commercial sector, built on the rise of the perch in the 1980s, probably constitutes a new form of export of protein from the lake to foreign consumers (KCBS, 1991; UNICEF, 1991), and of profit from the lake to distant investors, both within and outside the riparian countries. However, the benefits of processing in the tilapia fishery have not been redistributed, but have remained largely with the traditional small-scale processors in or near the fishing villages.

The sixth set of roles in the Lake Victoria fisheries includes the very diverse collectors and marketers of fish, including the fish dealers or fishmongers. In the past, most local fish dealers operated in the vicinity of the fish landings and only occasionally sent fish inland to distant markets. Over time, a second set of marketers developed in both the nearby and distant urban areas, where demand for fish was high. The expansion of the perch–tilapia fishery has led to two significant changes. First, in part in response to the expanded fishery (CICS, 1989), feeder roads to fish landings are generally in better condition than they were 25 years ago. As a result, at most fish landings the opportunities for transport include small pick-up trucks, lorries, and taxis as well as bicycles. Second, the demand for fish for industrial processing has led to the development of the role of fish collector. These persons may be commissioned agents or they may operate on speculation, using a boat or a truck to pick up a large quantity of fish and transport it to a processing plant.

For present purposes, we identify three types of dealers in the Lake Victoria fishery: local mongers, long-distance dealers, and industrial collectors. (It is certainly possible to elaborate these further; Kirema-Mukasa and Reynolds, 1991, following Balarin, 1985, distinguish ten types of dealers in the Ugandan–Lake Victoria fishery, and these more detailed classifications are very important for a more refined analysis of the economics and sociology of trade relations and networks; Reeves, 1989.) These three types are not necessarily distributed within the fishery. While Bwathondi, (1990a) suggests that almost all dealers in Tanzania are still of the first type, data from our surveys indicate more variation (Table 11.2). In Uganda, less than 60% of the fresh fish moves from landing to market by large-scale means such as lorry or collecting boat (Kirema-Mukasa and Dhatemwa, 1992).

As with the various harvesting roles, research to date has revealed little

Table 11.2 Destination of fish bought from fishing boats, broken down by type of beach, from a Tanzanian sample. Source: Wilson and Ntara (1995)

Destination of fish	Type of fishing beach	
	% Isolated ($n = 34$)	% Not isolated ($n = 48$)
Retail	44	27
Local district	35	17
Large in-country town	12	10
International	9	46

about the social characteristics or the economic returns of marketers and collectors. In Uganda in 1991, only 28% of communities reported that fish trading offered an income-generating activity for women (Reynolds and Kitakule, 1991). Again, it would be interesting to know whether collectors are former owner-operators, or if some local mongers amassed the capital and became long-distance traders. Anecdotal evidence from Tanzania suggests that some entrepreneurs operating local retail shops and transport companies near the lake have used their capital to enter the fish trade after the advent of the Nile perch. However, because a collecting boat, engine, and working capital require the equivalent of approximately US$ 500, we hypothesize that not many local mongers would be able to make that transition.

Specific impacts of the species changes in the fishery

In discussing the various roles of the Lake Victoria fishery, we have shown that the expansion of the perch–tilapia fishery has resulted in gross alterations to the capture, processing and marketing operations. These alterations have had differing beneficial and detrimental impacts on the various social strata and roles (Freeman *et al.*, 1982). Much more research must be completed to yield precise details about these impacts. Nevertheless, we suggest that at least seven areas of significant impacts at the household and individual levels can be discerned: (1) the scale and patterns of income, (2) ease of entry into the fishery, (3) the quality of work life of participants in the fishery, (4) patterns of geographic mobility, (5) opportunities for employment in the fishery, (6) the availability of fish for household consumption, and (7) the health of members of fishing households.

Scale and patterns of income

Almost all researchers reviewing socio-economic impacts, including Reynolds *et al.* in this volume (Chapter 10), have noted the increased aggregate income to the fishery (Reynolds and Greboval, 1988; Greboval and Mannini, 1992). Less consensus exists on how this aggregate income is distributed among the harvesting, marketing, transporting and processing sectors; on the equity of that division; and on whether the income is benefiting general well-being in the riparian societies.

Reynolds and Kitakule (1991) conclude that, even in comparison with other rural Ugandans, members of Ugandan fishing communities generate modest levels of income from their fishery activities. Although crew members in Uganda receive a relatively high share of the catch in comparison with non-African fisheries, Ugandan crew members are particularly disadvantaged in comparison with other Ugandan fisher's roles and other Ugandan rural residents. With Peyton (1988), we suggest that the new fishery regime has had the effect of concentrating income among a smaller proportion of participants in the fishery, in spite of increases in aggregate cash income from the fishery. If catches in the years ahead are unstable (Ogutu-Ohwayo and Hecky, 1991; Chapter 9), even these aggregate gains in wealth could be short-lived.

We also believe that personal and household incomes in the fishery are being grossly and differentially distorted by the effects of inflation, devaluation, general economic decline, and the early stages of structural adjustment programmes (Cornia *et al.*, 1987; UNECA, 1989). Thus, for many fishery workers, any increases of cash incomes are being more than offset by increasing prices for food, commodities, transport, and all public services, especially health and education. For example, in western Kenya there is much anecdotal evidence of increases of income for some workers accompanied by a declining quality of life for most.

Successful exploitation of the perch has required new fishing gear, particularly synthetic multi-filament large mesh gill nets. These nets are produced largely in foreign factories, and secondarily in urban industrial plants seeking to substitute locally manufactured gear for the imported nets. Both foreign and local factories require capitalization adequate to purchase foreign machinery to produce nets, hooks and lines. This in turn requires an export of profits from the fishery and creates employment in the core industrial nations for the production of the manufacturing machinery (Hancock, 1989). The gross terms of trade for the 'exchange' of local products for foreign manufactured goods also favours the economies of the core nations. For instance, using foreign boat motors, spares and petroleum products also exports capital and increases employment for foreign suppliers (Bailey, 1988). In the competition for fish, the advantages of using imported and industrially manufactured gear have diminished the ability of poorer villagers to enter the fishery utilizing low-cost traditional gear. In Tanzania, 55% of gill nets in the Nile perch fishery are owned by boat owners.

The industrial processors apparently are making significant profits, but specific details are not available from these private companies. Nor do we have significant information about the repartition of sectoral incomes between capital, labour and government within each sector. Akersie (1991) indicates that owners of gill net outfits are receiving good net incomes, but crew members are not. The strikes by fishing crews in Uganda suggest that they perceive their current share to be unfair. Traditional fish processors also complain about their incomes, viewing the processing plants as encroachers on their profits and believing they are disadvantaged in competing with the greater market power of the processing plants (Kirema-Mukasa and Dhatemwa, 1992). Dressers and filleters in the processing plants receive wages that are locally acceptable, but considerably lower in absolute terms than those which would be paid for comparable work in European and North American plants.

Some marginally important new opportunities are afforded by the extracted perch oils, which have provided some poorer households in Tanzania with lamp and cooking oil to substitute for expensive imported paraffin oils (kerosene) and oilseed cooking oils.

Ease of entry to the fishery

While the separation of ownership, management and operation has made it easier for those having non-fishery capital but lacking fishery expertise to own boats and gear, the increased cost of boats and nets has made it more difficult for operators or crew members to save the amount required to obtain the benefits of ownership in that sector (Akersie, 1991). If the average crew member in Tanzania were able to save one-quarter of his fishery wages, he would have to work for 2 years to purchase the average boat and gear.

We have already noted the relatively high cost of boats, even though transport by water probably is less expensive per kilometre than road transport, which requires trucks of entirely foreign origin. While some bicycle transporters are still employed in the fishery, they do not earn sufficient income to expand their operations. Akersie (1991) notes that most fish dealers in Tanzania have invested capital less than US$ 100.

In the Ugandan processing sector, even small-scale modern smoking technology is beyond the reach of most processors, and has been possible only with the capital supplied by foreign aid programmes (CICS, 1989). In contrast, Akersie (1991) indicates that the invested capital for smoking fish in Tanzania is about US$ 100, which approximates Yongo's (1991) data on Kenya.

While fish processing plants are not very expensive relative to the costs of facilities in industrialized countries, only a few in the riparian countries are capable of assembling sufficient capital from domestic sources to finance a plant and its operations. These operations also require equipment of foreign manufacture and, therefore, require an allocation of scarce international exchange capital for export to foreign industries. We hypothesize that the industrial processors can more easily absorb and amortize those costs than can small-scale processors orientated to local markets, because of the relatively higher earnings from world export markets.

Quality of fisher's working life

Operators and crew members feel strongly that their work is now more burdensome than formerly. While theft of gear has always been a problem (EACSO, 1961; Jansen, 1979; Kongere, 1979), the Nile perch has made it worse, because better-quality gear is more valuable. To guard against theft, fishers must remain out on the lake with their nets all night; furthermore, boats are now more likely to fish 7 days a week instead of the former 5 or

6 days. The impacts of changes in the quality of working life in the marketing and transporting sector are more difficult both to assess and to evaluate, but in the processing sector the contrast is much greater. Whereas small-scale traditional processing was intermixed with other productive and social activities in public areas of the fishing community, industrial processing occurs on assembly lines in mechanized factories. In 1991 in Uganda, approximately 60% of the members of each of the major groups of fishery actors living in fishery communities reported that conditions of work were becoming worse over time (Reynolds and Kitakule, 1991).

Geographic mobility

A fourth type of impact, which accompanies the quality of working life, is geographic mobility. Traditionally, most of the Lake Victoria fishing peoples mixed fishing with a blend of agricultural and pastoral economies, and some even engaged in medium- and long-distance trade. Before the recent market and species changes in the fishery, fishers were able to shift their harvesting activities among several species as their local availability waxed and waned (Bwathondi, 1990a; Mwebaza-Mdawula, 1990; Ogari, 1990). While human migration is hardly a new phenomenon in East Africa, in the past it was usually related to avoiding overcrowded fisheries or to differences in regulation (EAHC, 1958). With the shift to a perch–tilapia fishery, some fishing groups have developed patterns of migration which entail extended periods away from their home landings (Reynolds and Kitakule, 1991). This implies not only that operators and crew will be away from their households for periods of 2 or 3 months, but also that marketing and distributing relationships must be reconstituted each time an operator relocates to a new landing. Selling the catch to long-distance transporters fits well with this pattern of geographic mobility (Maembe, 1990); otherwise the relationships established are presumably impersonal market changes, rather than the more multiplex social relationships characteristic of residential communities. Migration has also been linked to difficulties in capital formation (Yongo, 1991).

These first four impacts focus on the persons involved directly in fishery activities; the other three impacts are effects on the other members of fishing households.

Employment opportunities

The fifth impact is on the employment opportunities for members of families in which the head is a fisher. In the past, fishing was as much a household enterprise as it was an individual occupation. If a male head of household was a fisherman, his spouse, mother, and/or his sister was a fish processor and/or fish dealer, and his son was a crew member. Now, however, it is common for fathers to express a desire that their sons do not become fishers

(Jansen, 1979). The increase in geographic mobility of fishing concerns has diminished the opportunities for male relatives to work as crew members. This is less true in the Tanzanian fishery than in the other two countries (Akersie, 1991). At the same time, the decline of local marketing and small-scale traditional processing implies that the opportunities for female kinfolk in ancillary activities have diminished. Away from the beaches, some new employment opportunities are found in the industrial processing operations and in the re-processing of de-filleted Nile perch frames.

Availability of fish for household consumption

For households engaged in fishing, the fishery was not only a source of income from wages or sales, but also a source of sustenance in the form of fish for household consumption. Several interviewees have suggested that this is less so now. The relatively high price for perch and tilapia diminish the willingness of a manager to see the fish taken home by crew members. When fish are kept, it is for special occasions. Another common source of fish for home consumption in the past was beach seining; but with the perception that this is diminishing the reproduction of the perch and tilapia, Department of Fisheries officials have put more effort into stopping beach seining (Siwo, 1988). In many areas, the amount of table-sized fish available for catching is variable or small, resulting in consumption of many immature fish prior to spawning, including even catching and eating Nile perch fingerlings as a substitute for *Rastrineobola.*

In urban areas of the three riparian nations, many complaints are heard from the lower and middle classes about the exorbitant prices charged in the markets for even small fish. Indeed, in Kenya in 1992, the price per kilogram of de-filleted Nile perch frames (with only head, bones, some residual meat, digestive system, and tail remaining) was as high as that for a kilogram of whole Nile perch in 1989 (pers. obs.). In Kenya, this occurs against a backdrop of increasing child malnutrition in the late 1970s and early 1980s (UNICEF–Kenya, 1984) which may be continuing.

Health of members of fishing households

The decline of employment in the absence of other opportunities, and the decreased availability of fish for household consumption in the absence of other sources of protein, in turn probably have impacts on the health of members of fishing households. We suggest that the high price and decreased availability of fish results in a decrease of protein and an increase in starches in the diets of many poorer families throughout the riparian nations, especially in regions adjoining the lake. Indeed, as the market for perch and tilapia has drawn fish out of local markets, preliminary results from a study in Kisumu District in Kenya suggest a higher level of child stunting and wasting than found in most other regions of the nation (KCBS, 1991; UNICEF–Kenya, 1991). This occurs against a background of much higher rates of child mortality in the lake regions of Kenya than is found in the highland areas because of the greater prevalence of malaria, schistosomiasis, other water-borne disease, and even poor water sources (pers. comm., Chief Medical Officer of Health of Kenya, 1990). Likewise, in Uganda, the lakeside is where tsetse-borne trypanosomiasis is prevalent. In Uganda and Tanzania, protein is derived almost exclusively from fish in lakeside communities; Bwathondi (1990a) notes that the per capita consumption of fish is at least 50% greater in the areas bordering Lake Victoria than for Tanzania as a whole.

11.3 COMMUNITY AND ORGANIZATIONAL IMPACTS

Sociologists define a community as a group of people who perceive themselves as living together with some interactive relationships. In contrast to the broad scope of community, an organization is a group of people who interact for the accomplishment of some narrowly defined, mutually agreed goal. Of the many types of communities in Uganda, the ones which are most affected by the changes in Lake Victoria are fishing communities.

The expansion of the perch–tilapia fishery has had at least two significant positive impacts on fishing communities on Lake Victoria. First, in Uganda these communities have prospered relative to the changes which have occurred in other rural communities (Kitakule and Reynolds, 1991). Similar explosive growth and new wealth is found on a number of beaches and communities in Kenya. In some of these communities, the new wealth creates a boomtown atmosphere with the usual attendant raucous culture (Gramling and Brabant, 1986). The income earned by boat owners, managers, operators and crew members who live in these communities has supported local shopkeepers and sellers of services. At least in the Ssese Islands (map, Chapter 2, Fig. 2.2), the increased catches of perch and tilapia have provided increased employment for fish smokers. The expansion in the perch–tilapia fishery has enabled some landings to grow from fishing communities into trading communities. These are usually landings that are readily accessible to motorized vehicles during all seasons. In these communities, as many as half the boats on the beach may be transport boats rather than fishing boats.

The second significant positive impact of the expanded perch–tilapia fishery is on the formation of cooperative societies in the fishing communities. Fishing communities generally did not provide fertile grounds for the establishment of cooperatives during the colonial and post-colonial periods (TDRI, 1983). In 1976 in Kenya, it was estimated that only 4.1% of the earnings in the fishery were channelled through cooperatives, primarily because the cooperatives were too small to be viable (Kongere, 1979). There also are many reports of corruption in the Kenyan cooperatives, poor leadership and confusion because these enterprises are administered by the Ministry of Cooperatives rather than the Department of Fisheries, even though many Kenyan officials hope for a larger role for the cooperatives within the fishery.

The perch–tilapia fishery expanded at a time when few funds were available to expand the harvesting capacity of the fishery (CICS, 1989). In this situation, cooperatives were generally welcomed by people in the fishery, both for the acquisition of boats, engines and gear, and for the construction of processing kilns. These cooperatives contributed to the integration and solidarity of the fishing communities. In surveys in Uganda in 1991, many residents of fishing communities mentioned cooperatives as an important vehicle for the solution of their problems (Reynolds and Kitakule, 1991).

On the other hand, the expansion of the perch–tilapia fishery contributes significantly to at least five negative impacts on fishing communities on Lake Victoria. First, the increase in the amount of more valuable fish has led to increased incidence of theft of fish on the Lake. Although overt piracy has not yet been reported in Kenya and Uganda, it is common for gill nets to be lifted and the fish removed; sometimes the entire net is stolen with its catch. In Uganda, the existence of "good discipline" was one of the main factors differentiating growing fishing communities from ones that were not growing (Reynolds and Kitakule, 1991); preliminary surveys in Tanzania even suggest that some fishers choose well-organized beaches because they provide security of gear (pers. obs.). Interviewees in Uganda indicated that the thieves could even be fellow fishers in their own communities.

Second, the increased incidence of theft has led not only to greater distrust among local fishers, but also to greater distrust between owners and operators. When an operator reports that a net has been stolen, some owners may think that the operator sold it to a relative or a friend elsewhere. When an operator tells the owner or manager that only a few fish were caught during the previous night, the owner is likely to think that the operator and the crew sold most of the catch to a collecting boat on the Lake. Babikwa (1992) suggests that operators and crew also may engage in sabotage; laying the nets badly or destroying them. Thus some of the bonds of community solidarity which may have linked owners and operators in the past are now being attenuated by suspicion and perceptions of inequity (Gambetta, 1988).

The third negative impact of the expanded perch–tilapia fishery is the increase in temporary migration which has been noted above. Not only does the absence of the boat operator affect the fisher's family (Dixon *et al.*, 1984), it also affects the fishing community by depriving it of the leadership of some of its most economically active members. The increased number of distant owners ensures that the profits of many boats will be 'exported' to distant communities and entrepreneurs rather than being retained in the local community. It is these owners and operators who might be expected to be most motivated to help generate long-term stability of a community; however, if they are absent, internal dissension can overwhelm the bonds of community solidarity.

Fourth, at the same time that changes in the fishery have taken potential leaders out of their communities, outsiders have moved into the lakeside communities. Jansen (1979) reported that at one time almost all the fishers came from the vicinity of the lake. Now, many of them come from further inland, and foreigners from outside the riparian nations have begun owning and operating fishing outfits. The absence of local owners and operators and the presence of outside operators and crew members reduce solidarity within the fishing communities. As one specific example of this decline of community solidarity, the high levels of suspicion and distrust in the fishing communities, coupled with the increased frequency and duration of migration, has led to the dissolution of most of the fishers' cooperative societies in Uganda (Odongkara, 1990; Reynolds and Kitakule, 1991). In 1991, fewer than 5% of fishing communities had functioning cooperative societies.

A fifth negative impact of the expanded fishery is located more in the offshore islands than in the mainland fishing communities. Especially in the Ugandan Ssese Islands, the landings are not served by daily transport boats, and the long distance to mainland markets makes transporting fresh fish a risky proposition. For these reasons, most fish landed on the islands is smoked for preservation. This has resulted in extensive deforestation of the islands (*New Vision*, 1991), intensifying a trend begun in pre-Nile perch days (Temple, 1965). The deforestation of accessible lands has increased the pressure for harvesting firewood in forest reserves, which in turn has resulted in conflicts between wardens and woodcutters. We also suggest that, in some areas, deforesting of wood and charcoal production may have been given further impetus by the collapse of the tilapia fishery and the search for alternative sources of income by those who could not afford the new and more expensive gear required of the perch fishery. At the same time, responses to the economic and political crises of the region, including the recent impacts on living standards of the World Bank and International Monetary Fund plans for 'structural adjustment', may also be significant sources of deforestation as the needy exploit these resources (UNECA, 1989; Mugyenyi, 1991).

One additional impact of the expanded fishery on fishing communities is that the increased value of the harvested fish has increased the importance of regulating conflict among the local fishers. Communities now are more likely to regulate local fishing activity (e.g. how early in the evening boats can leave the landing, how close to nets of other fishers one can set one's nets in the water, how early in the morning boats can return to the landing). The increased value of fish means that the motivation of boat operators to violate these local fishing regulations also is high. This, in turn, has increased the importance of the beach leader or head fisherman and the fishermen's committee in the enforcement of local regulations; 70% of the landings in Uganda have a designated head fisherman (*gabunga*), and 14% have a fishing committee (Reynolds and Kitakule, 1991). Although our research to date suggests that these efforts to regulate competition are successful, we suspect that, especially in Uganda, an increasing proportion of the fishing is being done by outlaw operators who do not land at formal landings, do not register their boats and gear, and do not sell their fish through regulated channels.

This examination of the impacts of the expanded fishery has focused largely on fishing communities, landings where the marketing and processing of fish constitute a significant fraction of the community's economic activity, and/or where a significant fraction of the labour force participates in fishery activities. But these are not the only types of communities to have experienced impacts from the changes in the fishery. We have noted above that some lakeshore communities might be better termed trading communities rather than fishing communities. Other communities are affected by industrial processing plants. Some of these historically were fishing communities. In other cases the plants are located inland to take advantage of fish supplied from many landings. To the extent that the plant is a major source of employment for the community, the community will be affected by changes in the fishery.

Communities with the potential to attract large numbers of tourists also could be influenced by an expanded perch fishery. Advocates for introducing the Nile perch into Lake Victoria argued that it would stimulate the development of a sport fishery (Ian Dunn, pers. comm.); the ancient Egyptians called the Nile perch 'the fighter fish' (Acere, 1988). But to date we have identified no significant recreational fishing communities along the lakeshore. In interviews in Nairobi, Kenyan tourist agents reported that neither tourists nor they themselves had much interest in fishing safaris to Lake Victoria. Such recreational communities could develop, however, if land transport to lakeshore sites improves, if the problems of bilharzia and chloroquine-resistant malaria are reduced, if tourist facilities can be improved and if the fishery is managed to allow giant sport fish to occur in sufficient numbers. Indeed the Kenyan Lake Basin Development Authority has already planned for such improvements.

There have been distributional impacts at the community level. Some fishing communities have become trading communities, whereas most have not. Some companies have been chosen as sites for processing plants, but most have not. Some communities were selected for aid projects in support of cooperatives or processing experiments, and some were not. It is less clear if there have been redistributional effects at the community level. One could argue that the absolute decline in the catch of traditional species (*Clarias, Bagrus, Protopterus*) deprived artisanal processors of the fish which they could process. To the extent that these species survive in the lake, they may function largely as prey for the Nile perch: *Clarias* are sought as bait for the perch by fishers who use longlines. In this sense, a resource has been taken away from one group of users and redistributed to two other groups of users; the processors with chorkor kilns and the industrial processing plants. (Chorkor kilns were developed by an Italian fisheries development project. Their construction permits less air exchange so there is less problem with flaming and fuel is used more efficiently.)

The expanded fishery has led to the establishment of at least one new non-governmental organization (NGO) in Uganda. In 1991, the industrial fish processors began establishing a formal organization to act on their behalf. In the meantime, they have continued to discuss informally matters of mutual interest, and to meet with relevant officials to seek favourable treatment under government policies and programmes.

The changes in the fishery may have also stimulated the development of a similar organization of small and medium-size fish dealers in Uganda. As the demand for fresh fish in Kenya expanded, fish were transhipped initially through the Sigulu Islands (map, Chapter 2, Fig. 2.2). Subsequently, collecting boats were sent to fishing sites off the shoreline of Iganga District. Eventually boats and trucks went as far west as Mukono District. But when the fish dealers in Mukono District realized that they were having greater difficulty obtaining supplies of fish at the prices they were willing to pay, they complained to the District Administrator. In response to their complaints, the Mukono District Administrator banned the trucks of Kenyan processors from the landings, and required that fish be sold to local dealers. Although this could be seen as effectively reestablishing the monopoly position of the local dealers, its probable effect will be to increase sales to collecting boats on the lake.

11.4 IMPACTS AT THE NATIONAL LEVEL

From the point of view of the Kenyan, Tanzanian and Ugandan governments, the most important impact of the rise of the perch catch probably has been the generation of new revenues and products for the nations, especially in increases of foreign exchange from fish sales on the world market (MAIF, 1983, 1989). Some biologists and ecologists predict that this will be a short-lived gain from a brief period of great, but one-time, fecundity before the fishery collapses into greatly reduced productivity (W.E. Cooper, pers. comm.; Kaufman, 1992; Chapter 9). Nevertheless, the benefits have been large in the 1980s and 1990s (Chapter 10).

One of the expected results from an expanded fishery is the elaboration of physical or organizational infrastructure or the development of capacity in certain subsectors of the economy, with benefits then spilling over into other sectors. In the literature, this is termed export-led development (Roemer, 1970). For example, the main purpose for upgrading roads from the main highway to the landings must be to facilitate the transport of the increased supply of fish. But once the road has been improved, it can be used for transporting other crops to market, or for transporting tourists to the lakeshore; thus it can stimulate or support the development of a fresh fruit industry or tourism. Critics of this approach, however, argue that such infrastructures are not often easily transferable to other sectors (Delacroix and Ragin, 1981) and that, in many cases, development based on the export of resources results in an economy with both exhausted resources and little resultant useful investment (Bunker, 1985).

A full investigation of the role of the Lake Victoria fishery in export-led development is beyond the scope of this chapter. At least at first glance, it would seem that feeder roads are the main infrastructural consequence of the expanded perch–tilapia fishery that has the potential to be a motor for further development. The exporting of chilled, fresh fish has provided some stimulus to the air cargo subsector, but the service is not yet dependable in Uganda (EPADU, 1991). If it becomes dependable, it might provide a channel for, and thereby stimulate the production of, vegetable and floral crops for export (Gruhn, 1991). Air export of quality vegetables and cut flowers is already well developed in Kenya. Although the Nile perch is also a source of oil, skin for tanning, and glue from scales (Kirema-Mukasa and Dhatemwa, 1992), all of these products pass through existing industries and have not stimulated the development of new industries. Indeed one of the weaknesses of the Lake Victoria fishery is that it relies so heavily on imports such as synthetic twine for nets, net machines, fuel and engines for boats, and machines for fish processing, cold storage and making ice. Even wax-coated boxes to meet international standards for fish shipments, which formerly were made in Uganda, are now imported. The expansion of the fishery has increased the demand for boats, but almost all are wood-plank boats, so this is not likely to stimulate any further economic development.

There are, however, three areas at the national level where the effects of fishery expansion can be seen clearly. The most significant is in fishery management where there has been an explosion of demand for management activity. When the riparian nations achieved their independence in the 1960s, the colonial fisheries departments were transferred almost unchanged to the national governments (Gruhn, 1971, 1984) and the three fishery departments continued their activities into the 1980s with little alteration. Although a few projects for fisheries development were proposed (e.g. trawling for haplochromines in Tanzanian waters for reduction to fish meal), the organization of the departments, the training of the officers, and the content of the regulations remained basically unchanged.

The past several years have seen a dramatic change. Concerns for the sustainability of the perch and tilapia fisheries have led to revisions in the fishing regulations. The fear of exceeding sustainable harvest levels has led to demands for expanded management efforts and for research to provide the information for sound management decisions (Greboval, 1989). The need for more data from the fish landings and the need to increase fishermen's understanding of the regulations have required changes in the training of fisheries officers. The heads of the fisheries departments in the three nations increasingly find themselves forced to make decisions which entail complex socio-economic issues, including allocating the potential harvest among competing uses (Siwo, 1988). They must make these decisions with almost no indication of what level of harvest is sustainable, and with no information on the ecological, economic or social impacts of different allocations of that harvest (Ogutu-Ohwayo, 1990b).

In Uganda, this has led to the formation of ad hoc government committees that make difficult and complex decisions with few data. For example, faced with approximately 20 applications for licences to construct processing plants and additional applications for the use of non-traditional gear such as purse seines, the Commissioner of Fisheries appointed the Lake Victoria Exploitation Committee to provide advice on appropriate responses. Similarly, the perception of problems with the fishery led to the formation of a regional Fishery Committee which includes the fisheries officers, district administrators, and industrial processors in the Iganga, Jinja and Mukono districts (Greboval, 1989). One area in Uganda where no action has been taken is the conflict between the pair-trawling activities of the Sino–Ugandan Joint Venture and the deep-set gill nets of the small-scale fishing operations (Odongkara, 1990).

The second area of national impact is in food security (Trainer, 1989). In some situations, the expansion of the perch–tilapia fishery may have led to an increase in the national consumption of fish. As one example, in Uganda fish currently provide 20% of the animal protein consumed by humans (MAIF, 1989). As a source of protein, fish have the advantage of being less affected by drought than either livestock or legumes. In the late 1980s, a kilogram of Nile perch cost less than one-third the cost of a kilogram of beef, mutton, pork or poultry (Acere, 1988). Nevertheless, some Ugandans have voiced concerns that the expanding perch–tilapia fishery on which Ugandans depend for protein may be either reallocated or shifted to processing for export (Ogutu-Ohwayo, 1990a; Twongo *et al.*, 1991). Indeed the Fisheries Department acknowledges this potential danger, and seeks to avert its consequences in the long run by expanding agriculture in Uganda. Even in the short term, a segmented market appears to be emerging. In this dual market, the freshest and most valuable fish go for export. For example, fish for export are examined by Government inspectors while fish for local sale are not (Kirema-Mukasa and Dhatemwa, 1992). At the same time, the lower-quality perch and tilapia, the *Rastrineobola*, and the defilleted fish frames remain for the domestic market

(Twongo *et al.*, 1991). As Kirema-Mukasa and Dhatemwa (1992) note about Uganda:

> When the public realized that the 'good' fish meat is exported to overseas consumers, and knowing very well that these factories are the cause of fish shortages within the markets, they started to reject the fish frames on the basis that they have been relegated to eating 'scraps'.

Serious concerns are being raised about the long-term sustainability of the current levels of exploitation (Kirema-Mukasa and Dhatemwa, 1992; Kaufman, 1992; Chapter 9).

A third area of national impact results from the dependence on fish exports as a source of foreign exchange. Given the extensive imports which are required to support the fishery, it is not yet clear whether or not the Lake Victoria fishery is a net earner of foreign exchange. Even if there is a net benefit to the nation, it is not clear which local populations benefit, directly or indirectly, from the foreign sales. Regardless, the Government of Uganda views the expansion of fishery exports as an important element in developing its foreign trade. From 1990 to 1991, exports of fish and fish products quadrupled in dollar value, and increased from 0.77% of all exports to 2.88% (MFEP, 1992: Table 14). If this export policy is successful, it creates two potential dangers. First, it leaves the national economy susceptible to fish shortages produced by normal fluctuations in the stock of fish in the lake (McEvoy, 1986; Siwo, 1988). Second, it leaves the national economy at the mercy of shifts in the international demand for, and supply of, fish which can be substituted for Ugandan perch and tilapia in world markets. As exporters try to expand their markets by producing specific commodities that compete in existing markets, they enhance the danger that competitors may increase supplies or decrease prices to the detriment of the Ugandan economy. This risk may not be as threatening as the danger of supply fluctuations, because most forecasters see little danger of long-term oversupply of fish in the world markets.

One potential national impact which has not occurred is the expansion of the recreational sector in response to the availability of Nile perch in the lake. Nevertheless, the potential for such a development continues to exist; were it to be realized, it would greatly complicate the problems already facing fisheries managers. Problems between artisanal fishers and tourist fishing have been documented in places as diverse as Bermuda (Barret, 1991) and Malaŵi (Ferguson *et al.*, 1990).

11.5 INTERNATIONAL AND TRANSNATIONAL IMPACTS

The expansion of the perch–tilapia fishery has clearly increased the competition between the riparian nations, both for the raw resource and for the economic benefits to be derived from its processing and marketing. At the moment, Kenya and Kenyans are obtaining the bulk of the benefits from processing and exporting fishery products. Large quantities of fish that have been harvested in Ugandan waters move, legally or illegally, from Uganda to Kenya (Kirema-Mukasa and Dhatemwa, 1992), and smaller quantities move from Tanzania to Kenya (Maembe, 1990). While Kenya has by far the largest processing capacity of the three nations, a significant portion of this capacity is used for ocean fish. Uganda is rapidly expanding its processing capacity. Indeed, some of the Ugandan plants aim at processing and/or handling more than 50 tonnes of fish per day (Okaronon, 1990). Although foreign trade for Nile perch was small in Tanzania as recently as 1990 (Bwathondi, 1990a), an industrial processing capability is starting to develop using both foreign and local capital.

To date this competition for fishery benefits has provoked no more than minor incidents of conflict. Both Tanzania and Uganda have attempted to respond by requiring Kenyan buyers to purchase fish from local middlepersons, rather than directly from fishers. Indeed, the heads of all three countries continue to talk about reducing trade and transportation barriers between the countries, rather than maintaining national privileges. However, one can only expect that governments that support the construction of processing plants will seek to ensure that the plants receive a sufficient supply of fish. To the extent that there is a limit to the sustainable harvest, the fish for new plants must either be diverted from foreign markets or be reallocated from small-scale dealers and processors. In this way, competition between the countries may eventually have negative impacts on the domestic dealers and processors.

Furthermore, in the midst of 'structural adjustment', devaluation of local currencies and the wider African economic crisis, there is great emphasis in each nation to exploit to the maximum its 'comparative advantage' (Cornia *et al.*, 1987; Ochieng, 1991). In this context, the resources of the fishery could be readily subjected to overexploitation (UNECA, 1989). This seems especially dangerous and unsustainable if the processing capacity of the three nations is pegged to the currently high levels of perch catch and those levels are not sustained (Ogutu-Ohwayo, 1990a). If yields are not sustainable at current levels, then, under the pressure to amortize the costs of the expanded industrial processing capacity, catch either will go increasingly to the plants, and not for local consumption, or it will be maximized unsustainably, or both.

Nevertheless, it would be misleading to suggest that the needs for fish in the riparian nations have generated only hostile competition. The acknowledgement that the fishery is a shared resource, composed of somewhat mobile stocks, has led to cooperative activities in the area of fisheries management, regulation and research. Since 1987, under the stimulus of a grant from the (then) EEC, researchers from the three nations have collaborated on research.

This collaboration has also been supported by the Inland Fisheries Planning Project of the Food and Agriculture Organization of the United Nations, and by the International Development Research Council of Canada. Since 1979 representatives of the three nations have met together in a forum provided by the Lake Victoria Subcommittee of FAO's Committee on Inland Fisheries of Africa. As a result of these interactions, the three nations are now considering a proposal for harmonizing their fisheries regulations (Greboval, 1992). At the same time because none of the three countries is enforcing existing regulations in more than a minimal way, harmonization may be more symbolic than real.

A third area of impact at the international level concerns the types of foreign aid which have been stimulated by the expansion of the Lake Victoria fishery (Bailey, 1988). Since 1980, Kenya has received aid for fisheries development and assessment from Japan, the United States and the World Bank; Tanzania has received aid from the United Kingdom, the Netherlands, the United Nations Development Programme, Finland and Japan; and Uganda has received aid for fisheries development from Italy, the European Economic Community, the World Bank and the Agency for International Development of the United States. While the orientation of these projects has generally been congruent with the expressed desires of the three governments (e.g. MAIF, 1983), the specific content of the projects has often been determined by the potential for benefits to firms in the donor countries (Hancock, 1989). Thus chilling and freezing plants have been favoured by countries that would benefit from a supply of inexpensive fish, smoking plants by countries that manufacture smoking equipment, and motorization projects by countries that manufacture engines,

These international aid efforts have two consequences. On the one hand, the increased export of chilled and frozen fish to Europe indicates the possibility of a shift in global fishery production from European marine waters, where labour and operating costs are high, to African inland waters where these costs are low (EPADU, 1991). On the other hand, the importance of foreign aid for African fisheries development and the importance of foreign markets for sales of African fish mean that control over African fisheries is shifting from local communities and national agencies to transnational firms and global organizations (Bailey *et al.*, 1986; Platteau, 1989).

Indeed, most international attention to the expanding perch–tilapia fishery has been focused on the impacts of Lake Victoria's endemic population of some 300 species of haplochromine fishes (University of Leiden, 1986). Some authors suggest that as many as 200 of the species may have been rendered extinct, and it is generally agreed that the stocks of these small colourful fish, which formerly constituted 80% of the ichthyomass in the lake, have been seriously diminished, at least in part as a result of predation by the Nile perch. Understandably, this has generated intense controversy between the conservers, who seek to reverse the decline of the haplochromines, and the users, who advocate continuing to expand the perch–tilapia fishery. The conservers wave the banners of biodiversity and conservation while the users wave the flags of food for the people and foreign exchange for the developing countries. Both sides of the issue can be found in East Africa, in the wealthier industrial nations, and in international organizations. From the perspective of a social impact analysis, competing goods must be balanced; however, and more importantly, the management of Lake Victoria has attracted the heavy involvement of transnational and global organizations that are beyond the control of the riparian states.

11.6 SUMMARY

This chapter has only just begun the exploration of the social and economic impacts of the changing Lake Victoria fishery. We have not discussed many new developments, such as the growing *Rastrineobola* fishery and the nascent *Caridina* fishery. Until recently the *Rastrineobola* fishery was relatively small in scale at all its stages; small boats used cast nets or set beach seines to harvest a catch which was sun-dried at the landing, often by the wife of the operator. After drying it was packed in a bag or a metal tin and taken to market, where it was sold for human consumption in sauce.

Recently, this fishery has greatly increased in scale at each stage. Harvesting is now carried out with larger boats at night using pressure lamps to attract the fish. On a good night a boat may need to return to the landing two or three times to unload the catch. The sun-dried fish are packed in large gunnysacks for transport to the mainland. In Kenya and Uganda, an increasing portion of the *Rastrineobola* catch is being used to make fish meal for animal feed, even including feed for farmed tilapia.

The same questions addressed to the perch–tilapia fishery could be explored in the *Rastrineobola* fishery (Okedi, 1982). Who is involved in the fishery? Do *Rastrineobola* fishers also fish for perch or tilapia at other times of the month or the year? Who owns the boats, engines, nets, lamps and other gear of this fishery? To what extent has this fishery attracted investment from outside the fishery? As perch and tilapia are being shifted from local consumption, is *Rastrineobola* filling the demand for fish protein for human consumption, especially for households with lower incomes, as we understand anecdotally is true? Are the economics of fish meal sufficiently profitable that even *Rastrineobola* may be priced out of human consumption and devoted to animal feed? It is not the case that all *Rastrineobola* goes for domestic consumption; Bwathondi (1990a) reports that dagaa from Tanzania is finding markets in southern and western Africa, and proposals have been put forward to harvest these fish in Lake Victoria for animal feed in Zimbabwe's game parks, presumably because prices of local Zimbabwean sun-dried sardines have greatly increased.

Fishery ecosystems are dynamic; the presence of human elements and the perturbations caused by climatic influences imply that the attainment of a stable climax is rare, especially in tropical regions (McEvoy, 1986). The impacts of the changes resulting from the Nile perch and tilapia introductions will continue to ramify through the lake ecosystem for many years to come and thereby affect both the fishery and the human societies which have depended on the lake for many centuries. Socio-economic research is needed to track those changes, to provide information to policy makers on the distributional and redistributional impacts. Who is benefiting from the changes and who is being hurt by them in both the long term and the short term? These are the questions that require our attention.

ACKNOWLEDGEMENTS

The work on which this chapter is based was supported in part by a grant from the McArthur Foundation to the African Studies Center at Michigan State University, and in part by the resources of the MSU African Studies Center and the Department of Sociology. We would like to express our appreciation to our colleagues in the Program On the Lakes of East Africa, William Derman and Anne Ferguson, for the thoughts and insights they have shared with us throughout the Project; to Stephen Bunker and Chris Vanderpool for their comments and suggestions on this chapter; and to our colleagues at KMFRI, TAFIRI and UFIRI for their collaboration on the research reported here.

REFERENCES

Acere, T.O. (1988) The controversy over Nile perch, *Lates niloticus*, in Lake Victoria, East Africa. *Naga* (October), 3–5.

Acheson, J.A. (1981) Anthropology of fishing. *A. Rev. Anthropol.*, **10**, 275–316.

Akersie, W.O. (1991) Preliminary Analysis of Socio-Economic Pilot Survey Tanzania Fisheries. Project URT/87/016 Report. Dar Es Salaam.

Asowa-Okwe, C. (1989) Capital and conditions of fisher-labourers of Lakes Kyoga and Victoria canoe fisheries. Labour Studies Project – Paper No. 2. Center for Basic Research Working Paper No. 3. Kampala: Center for Basic Research.

Babikwa, J. (1992) The changing social and economic relations in Uganda's private (informal) fishing economy: the case of the fisher communities of Kalangala 1971–1992. Research proposal for MA degree in Development Studies. Inst. Development Studies, Univ. Dar es Salaam.

Bailey, C. (1988) The political economy of fisheries development in the Third World. *Agric. Human Values*, **5**(1/2), 35–48.

Bailey, C., Cycon, D. and Morris, M. (1986) Fisheries development in the Third World: the role of international agencies. *World Development*, **14**(10/11), 1269–75.

Balarin, J.D. (1985) National reviews for aquaculture development in Africa. No. 10: Uganda. Rome, FAO, *FAO Fish. Circ.*, (770.10).

Balirwa, J.S. (1990) The present status of the fishery for tilapia in Lake Victoria (Uganda), pp. 61–4 in Bwathondi (1990b), q.v.

Barret, G. (1991) The fish pot ban: artisanal overfishing and state mismanagement in Bermuda. Paper presented at Conf. Int. Ass. Study of Common Property, Winnipeg, 26–29 September 1991.

Bunker, S. (1985) *Underdeveloping the Amazon*, Chicago, Univ. Chicago Press.

Butcher, D.A.P. and Colaris, J.C.G. (1975) A sociological survey of the fishermen population around Lake Victoria. Report prepared for the Lake Victoria Regional Fisheries Research Project. Rome, FAO, FI:DP RAF/71/242/4.

Bwathondi, P.O.J. (1990a) The state of fishing industry in Tanzania with particular reference to inland fishery, pp. 5–11 in Bwathondi (1990b), q.v.

Bwathondi, P.O.J. (ed.) (1990b) *The Prevailing Activities on the Lake Victoria Basin with Particular Reference to the Lake Fisheries*. Proceedings of a Workshop On That Topic Held at Mwanza, 8–9 March 1990, Organized as Part of the Lake Victoria Research Project of the EEC. Tanzania Fisheries Research Institute, Dar es Salaam.

CICS (Centro Internazionale di Cooperazione allo Sviluppo) (1989) *Program for the Support of Production, Technical Assistance to the Cooperatives, and Training In the Fishery Sector in Uganda*.

Cornia, G.A., Jolly, R. and Stewart, F. (eds) (1987) *Adjustment with a Human Face, Vol. I, Protecting the Vulnerable and Promoting Growth*, Oxford, Clarendon Press.

Delacroix, J. and Ragin, C. (1981) Structural blockage: a cross-national study of economic dependency, state efficacy, and underdevelopment. *Am. J. Sociol.*, **86**, 1311–47.

Dixon, R.D., Lowery, R.C. and Sabella, J.C. (1984) Fishermen's wives: a case study of a middle Atlantic coastal community. *Sex Roles*, **10**(1/2), 33–52.

EASCO (East African Common Services Organization) (1961/62) *Annual Report*.

EAHC (East African High Commission) (1958) *Annual Report*, London Colonial Office.

EPADU (Export Promotion and Development Unit) (1991) *Seminar on Fish Exports and Quality Assurance*. Ministry of Planning and Economic Development, Kampala, Uganda.

Ferguson, A., Derman, B. and Mkandawire, R. (1990) Whose commons? Fishermen, developmentalists and conservationists on Lake Malawi. Paper presented at Int. Ass. Study of Common Property, Duke University, Durham, NC, Sept. 28–30.

Ferguson, A.E., Derman, W. and Mkandawire, R.M. (1993) The new development rhetoric and Lake Malawi. *Africa*, **63**(1), 1–18.

Freeman, D.M., Frey, R.S. and Quint, J.M. (1982) Assessing resource management policies: a social well-being framework with a national level application. *Environmental Impact Assessment*, **3**, 59–73.

Gambetta, D. (ed.) (1988) *Trust: Making and Breaking Cooperative Relations*, Oxford, Basil Blackwell.

Goodman, D. and Redclift, M. (1991) Food into freezers: women into factories, in *Refashioning Nature* (by . Goodman and M. Redclift), London, Routledge, pp. 1–46.

Gramling, R. and Brabant, S. (1986) Boom towns and offshore energy impact assessment: the development of a comprehensive model. *Sociological Perspectives*, **29**, 177–201.

Greboval, D. (1989) Managing the new fisheries of Lake Victoria: major socio-economic issues. UNDP/FAO Regional Project for Inland Fisheries Planning Development and Management in Eastern/Central/Southern Africa (IFIP), RAF/87/099/TD/94/89 (En). Bujumbura.

Greboval, D. (1992) Report of a regional meeting for the management of Lake Victoria and the creation of a Lake Victoria Fisheries Commission. UNDP/FAO Regional Project for Inland Fisheries Planning Development and Management in Eastern/Central/Southern Africa (IFIP), RAF/87/099/TD/45/92(En). Bujumbura.

Greboval, D. and Mannini, P. (1992) The fisheries of Lake Victoria:: review of basic data. UNDP/FAO Regional Project for Inland Fisheries Planning Development and Management in Eastern/Central/Southern Africa (IFIP), RAF/97/099/WP/16/92(En). Bujumbura.

Gruhn, I. (1971) The transition years: the commission for Technical Cooperation In Africa. *J. Mod. Afr. Stud.*, **9**, 17–36.

Gruhn, I. (1984) Towards scientific and technological independence. *J. Mod. Afr. Stud.*, **22**, 1–12.

Gruhn, I. (1991) The development of vegetable and floral markets in East Africa. Paper presented at the Workshop on the Globalization of the Fresh Fruit and Vegetable Industry, Santa Cruz, 5–8 December 1991. Santa Cruz, University of Santa Cruz College Eight.

Hancock, G. (1989) *Lords of Poverty: The Power, Prestige, and Corruption of the International Aid Business*, New York, Atlantic Monthly Press.

Harris, C.K. (1991) *Science and Technology in the Fishery System of Uganda*, Jinja, Uganda Freshwater Fisheries Research Organization.

Harris, C.K. (1992) *Species and Niche In the Analysis of Lake Victoria Fisheries*, East Lansing, Michigan State University.

Jansen, E.G. (1979) Report to EAFRO on The Fishing Population in the Kenyan Part of Lake Victoria. Jinja, EAFRO.

Kaufman, L. (1992) Catastrophic change in species-rich freshwater ecosystems. *BioScience*, **42**, 846–58.

KCBS (Kenya, Republic of, Central Bureau of Statistics) (1991) Household Welfare Monitoring and Evaluation Survey of South Nyanza District.

Kirema-Mukasa, C.T. and Reynolds, J.E. (1991) Marketing and consumption of fish in Uganda. Fisheries Statistics and Information Systems (FISHIN) – Uganda Occ. Pap., no. 4. Entebbe.

Kirema-Mukasa, C.T. and Dhatemwa, C.M. (1992) Present aspects of marketing fish from Lake Victoria, Uganda. Paper presented at the Sixth Session of the CIFA Subcommittee for the Development and Management of Lake Victoria, 10–14 February 1992, Jinja.

Kitakule, J.S. and Reynolds, J.E. (1991) Fishing communities of Lake Victoria-Uganda: a first assessment. Fisheries Statistics and Information Systems (FISHIN) – Uganda Socio-Economic Working Paper, no. 2. PO Box 201, Entebbe, Uganda.

Kongere, P. (1979) Production and socio-economic aspects of fisheries in the Lake Victoria Basin (Kenya). Seminar Series on Lake Victoria Basin Development (no. 6). Nairobi, University of Nairobi, IDS.

Luhmann, N. (1982) *The Differentiation of Society*, New York, Columbia Univ. Press.

McCay, B. and Acheson, J.M. (eds) (1987) *The Question of the Commons: The Culture and Ecology of Communal Resources*, Tucson, Univ. Arizona Press.

McEvoy, A.F. (1986) *The Fisherman's Problem*, Cambridge, Cambridge Univ. Press.

Maembe, T. (1990) Social dimensions of fisheries management in Tanzania. Dar es Salaam, *Dept Fish. Res. Rep.*, 90–140.

MAIF (Ministry of Animal Industry and Fisheries) (1983) Blueprint for fisheries development. PO Box 201, Entebbe, Uganda.

MAIF (Ministry of Animal Industry and Fisheries) (1989) Livestock and fisheries policy and development program: toward the year 2000. PO Box 201, Entebbe, Uganda.

MFEP (Ministry of Finance and Economic Planning) (1992) *Key Economic Indicators.* Ninth Issue, April 1992. Statistics Department, PO Box 362, Entebbe, Uganda.

Mugyenyi, J.B. (1991) IMF conditionality and structural adjustment under the national resistance movement, in *Changing Uganda: The Dilemmas of Structural Adjustment and Revolutionary Change,* (eds H.B. Hansen and M. Twaddle) London, James Currey, pp. 61–76.

Mwebaza-Mdawula, L. (1990) The role of invertebrate organisms in the fishery potential of Lake Victoria, pp. 56–60 in Bwathondi (1990b), q.v.

Nemec, T.F. (1972) I fish with my brother: the structure and behavior of agnatic-based fishing crews in a Newfoundland Irish outport, in *North Atlantic Fishermen* (eds R. Andersen and C. Wadel), St. Johns, Memorial University of Newfoundland, pp. 9–34.

(The) New Vision (1991) Fisheries boom in Ssese Islands. 23 May 1991, p. 4.

Ochieng, E.O. (1991) Economic adjustment programmes in Uganda 1985–9, *Changing Uganda: The Dilemmas of Structural Adjustment and Revolutionary Change* (eds H.B. Hansen and M. Twaddle), London, James Currey, pp. 43–60.

Odongkara, O.K. (1990) Socio-economic aspects of the Kome Island fisheries: report of a trial frame survey. Fisheries Statistics and Information Systems (FISHIN) FAO/UNDP Project UGA/87/007 Socio-Economic Field Report no. 12, February 1990.

Odongkara, O.K. (1993) Report of the First National Fisheries Production to Consumption System Research Workshop. Jinja, Uganda Freshwater Fishery Research Organization.

Ogari, J. (1990) The co-existence between *Oreochromis niloticus* and *Lates niloticus* in Lake Victoria (Kenya Sector), pp. 36–42 in Bwathondi (1990b), q.v.

Ogutu-Ohwayo, R. (1990a) Changes in the prey ingested and the variations in the Nile perch and other fish stocks of Lake Kyoga and the northern waters of Lake Victoria (Uganda). *J. Fish Biol.*, 37, 55–63.

Ogutu-Ohwayo, R. (1990b) The decline of the native fishes of Lakes Victoria and Kyoga (East Africa) and the impact of introduced species, especially the Nile perch, *Lates niloticus*, and the Nile tilapia, *Oreochromis niloticus. Env. Biol. Fishes*, **27**, 81–96.

Ogutu-Ohwayo, R. and Hecky, R.E. (1991) Fish introductions in Africa and some of their implications. *Can. J. Fish. Aquat. Sci.*, **48**, 8–12.

Okaronon, O. (1990) Recent observations on the exploitation of the fish stocks of Lake Victoria, Uganda, pp. 65–71 in Bwathondi (1990b), q.v.

Okedi, J. (1982) Integrated management strategy for the 'dagaa' fishery of Lake Victoria, in *Aquatic Resources of Kenya* (Proc. Workshop Kenya Mar. Fish. Res. Inst., 13–19 July 1981), Nairobi, Kenya Mar. Fish. Res. Inst. and Kenya National Acad. Adv. Arts and Sciences, pp. 440–44.

Peyton, J. (1988) Technical and social aspects of traditional preservation of Nile perch, in *Technical Papers Presented at the Project Seminar on Improved Utilization of the Nile Perch*, Kisumu, 28–31 March 1988, Rome, FAO, pp. 31–6.

Platteau, J.-P. (1989) The dynamics of fisheries development in developing countries. *Development and Change*, **20**, 565–97.

Reeves, E. (1989) Market places, market channels, market strategies: levels for analysis of a regional system, *Human Systems Ecology* (eds S. Smith and E. Reeves), Boulder, Westview Press, pp. 58–80.

Reynolds, J.E. and Greboval, D.F. (1988) Socio-economic effects of the evolution of

Nile perch fisheries in Lake Victoria: a review. Rome, FAO, CIFA Tech. Pap., no. 17, 188 pp.

Reynolds, J.E. and Kitakule, J.S. (1991) Socio-economic aspects of Lakes Victoria and Albert fisheries: the 1991 FISHIN community survey. FAO/UNDP Project UGA/87/007 FISHIN Notes and Records Field Document No. 2. PO Box 201, Entebbe, Uganda.

Roemer, M. (1970) *Fishing for Growth: Export Led Development in Peru, 1950–1967*, Cambridge, MA, Harvard Univ. Press.

Schnaiberg, A. (1980) *The Environment*, New York, Oxford Univ. Press.

Siwo, J. (1988) Nile perch fishery in Lake Victoria, in *Technical Reports Presented at the Project Seminar On Improved Utilization of Nile Perch*, Kisumu, Kenya, 28–31 March 1988. Rome, FAO, GCP/KEN/055/NET, pp. 93–103.

TDRI (Tropical Development and Research Institute), (1983) *Fisheries Rehabilitation Study, Uganda.*

Temple, P.H. (1965) Lolui fishermen: a study of migratory groups on Lake Victoria. *Proc. E. Afr. Acad.*, **3**, 119–27.

Trainer, T. (1989) *Developed to Death: Rethinking Third World Development*, London, Green Print.

Twongo, T., Reynolds, J.E. and Mwene-Beyanga, P. (1991) Management issues, options, and strategies for Lake Victoria fisheries, in *Resource Stewardship and Multiple Use Interests for a Fisheries In Flux: Observations On Lake Victoria-Uganda* (ed. J.E. Reynolds). FAO/UNDP Project UGA/87/007 FISHIN Notes and Records Field Document no. 4.

UNECA (United Nations Economic Commission for Africa) (1989) African alternative framework to structural adjustment programmes for socio-economic recovery and transformation. E/ECA/CM/.15/6/Rev.3. Addis Ababa: UNECA.

UNICEF–Kenya and the Central Bureau of Statistics, Republic of Kenya (1991) *Kisumu District Monitoring Study*. Nairobi, Republic of Kenya. (Mimeo)

UNICEF–Kenya (1984) Situation analysis of children and women in Kenya: section 4, The well-being of children. Nairobi, Republic of Kenya. (Mimeo)

University of Leiden (1986) *The Decline of Lake Victoria's Cichlid Species Flocks: Survey of the Press Publications*, Leiden, Zoologische Laboratorium.

Vanderpool, C.K. (1987) Social impact assessment and fisheries. *Trans. Am. Fish. Soc.*, **116**, 479–85.

Wiley, D.S. and Yongo, E. (1992) Trends in social and economic responses of Kenya fishing peoples to changes in the Lake Victoria fishery. Paper for National Science Foundation Workshop Conference, Jinja (Uganda), August 1992.

Wilson, D.C. and Ntara, M. (1995) The implications for fisheries management of the changing situation in Lake Victoria fishing communities. *J. Afr. Rural Urban Stud.* (in press)

Yongo, E.O. (1991) Socio-economic aspects of fish utilization and marketing. Kisumu, KMFRI. (Mimeo)

Part Two

Evaluation of Species Changes in Other African Lakes with Introduced Fish Species

Chapter twelve

Inshore fish population and species changes in Lake Kariba, Zimbabwe

Lawrence Karenge and Jeppe Kolding

12.1 INTRODUCTION

Lake Kariba (Fig. 12.1) is artificial and probably one of the most studied freshwater impoundments in Africa, as judged from the bibliographical listings (Marshall, 1979; Sanyanga, 1988; Hutton, 1991). However, relatively few analyses or reviews have been published on the fish population changes (Jackson, 1961; Donnelly, 1970, 1971; Balon, 1974b; Kenmuir, 1984; Marshall, 1984; Moyo, 1986), and these describe mainly the early transformations. The last detailed analysis ends in 1975 (Kenmuir, 1984) and since 1982 no data have been published (Marshall, 1984). Moreover, as with all artificial dams and reservoirs, changes in the fish fauna are to be expected during the initial phases (Petr, 1967; Lelek, 1973; Lowe-McConnell, 1973; Balon, 1974b; Lewis, 1974) and so the critical question is when stabilization will occur and the rate of biological changes will decrease in Lake Kariba.

Balon (1974b), using Lake Kariba as an example, described typical changes in biological development which new impoundments seem to undergo. These are (1) the filling phase, characterized by eutrophy and high productivity but low diversity, (2) the decreased production but increased diversity phase, and (3) the stabilization phase. After these phases 'maturity' was reached and

The Impact of Species Changes in African Lakes. Edited by Tony J. Pitcher and Paul J.B. Hart. Published in 1995 by Chapman & Hall, London. ISBN 0 412 55050 4.

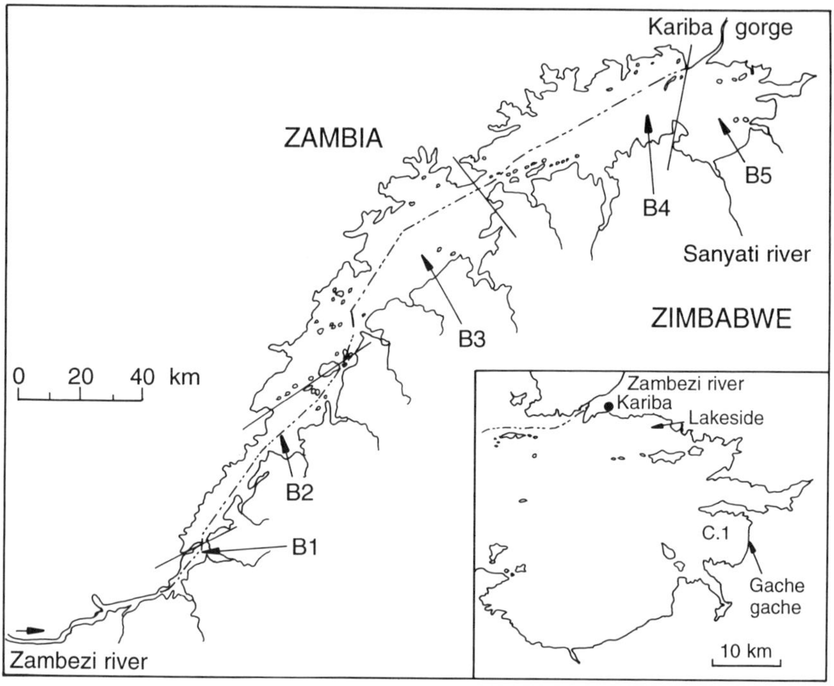

Fig. 12.1 Lake Kariba on the Zambezi river between Zambia and Zimbabwe. Arrow at lower left shows direction of flow. Inset, position of Lakeside sampling station in basin 5. From Karenge and Kolding (1994).

new unbalanced eutrophy would be from anthropogenic disturbances. According to Balon, the stabilization of Lake Kariba was reached around 1972. Kenmuir (1984, p. 195) suggested that for Lake Kariba: "The year 1975 can be regarded as the end of the period when observed fish populations reflected the outcome of mainly natural evolutionary processes ... and 1976 as the beginning of a period of intensive mono-species commercial fishing, when further evolutionary and largely man-induced trends were probably put into motion". These views were based partly on growing concerns about increased loss of nutrients from the inshore community into the rapidly expanding population of the introduced pelagic sardine (Balon, 1974a), and partly because of increased fishing pressure on this stock (Kenmuir, 1984).

This review aims at testing these statements by describing and attempting to explain the observed trends in species composition and population changes between 1969 and 1992 in a non-exploited area of Lake Kariba in the Sanyati basin. It builds largely on the study of Karenge (1992) and uses a continuation of the data series analysed by Kenmuir (1984) which covered the period from 1960 to 1975.

12.2 BACKGROUND INFORMATION

Lake Kariba (277 km long; 5364 km^2; 160 km^3; 29 m mean depth and 120 m max. depth) is located on the Zambezi river at latitudes 16°28′ to 18°04′S and longitudes 26°42′ to 29°03′E. The catchment area covers 663 817 km^2 extending over parts of Angola, Zambia, Namibia, Botswana and Zimbabwe. The dam wall (128 × 580 m) was completed in 1960 and the filling phase lasted from December 1958 to September 1963 when the water reached the mean operation level around 485 m a.m.s.l. The lake levels fluctuate annually from 1 to 5 m as a function of inflowing floods between December and June and continuous drawdowns through the turbines and, before 1981, spillage through the sluice gates. Since 1982 lake levels have declined due to a series of droughts and the lowest level recorded was in December 1992 at 476 m (Fig. 12.2). The lake is divided naturally into five basins (Fig. 12.1) and is almost equally shared by two riparian countries, Zambia and Zimbabwe.

The physico-chemical and various biological characteristics of Lake Kariba are well described elsewhere (reviews: Balon and Coche, 1974; Marshall *et al.*, 1982; Marshall, 1984; Machena, 1988). The most salient features are: a warm (mean surface temperature of 26 °C) oligotrophic monomictic lake with overturn occurring in June–July. Stratification begins around September with a thermocline around 15 m which gradually moves deeper to around 35 m at the time of the turnover. The oxygen concentration in the hypolimnion declines steadily during the stratification period. The volume-to-inflow ratio is low, giving a mean water retention time of only 2.6 years. The Zambezi provides about 80% of the inflow but its water is very clear and low in plant nutrients. Phytoplankton blooms occur just after turnover and locally at the onset of floods from tributaries, but during most of the year primary production may be very low. Zooplankton abundance shows variation in response to phytoplankton biomass and zooplankton undergo diurnal migrations (Begg, 1976). During the early eutrophic years the floating fern, *Salvinia molesta*, colonized large areas of the lake and retained large amounts of plant nutrients (Mitchell, 1973). From the 1970s it has gradually decreased and been replaced by increasing amounts of rooted macrophytes down to a depth of around 10 m (Fig. 12.2). Also the benthic fauna increased rapidly in number of species and biomass during the early years with the establishment of rooted flora, so the present biomass may be high among macrophytes and submerged trees (Machena, 1989) (Fig. 12.2).

Jackson (1961) listed 28 fish species in the middle Zambezi from a study before inundation. Balon (1974b) recorded 40 species in Lake Kariba but

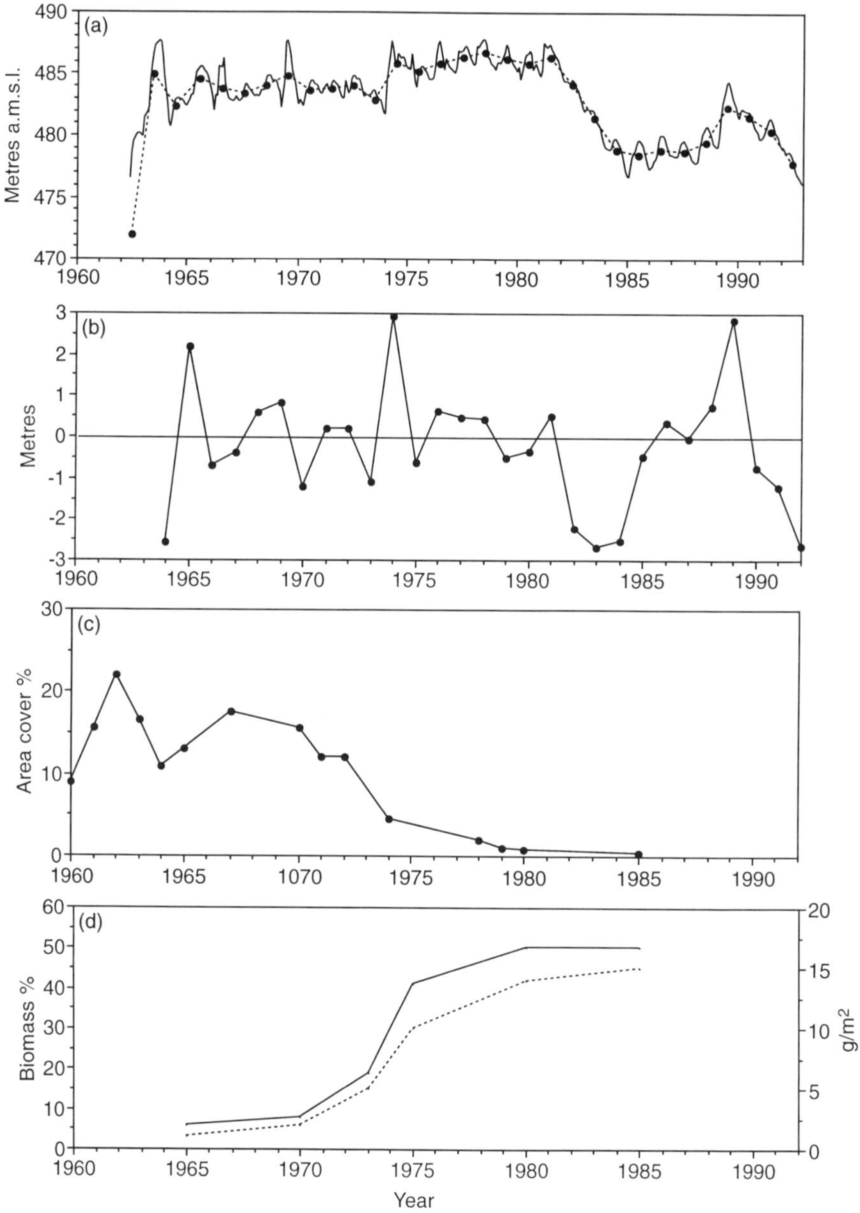

Fig. 12.2 (a) Mean monthly lake level (solid curve) and mean annual lake levels (filled circles and broken curve), (b) mean annual lake level fluctuations, (c) coverage of floating fern *Salvinia molesta* and (d) increase in benthos (Lamellibranchiata, dotted curve, right axis) and macrophyte biomass (*Lagarosiphon ilicifolius*, solid curve, left axis). Lake level data from Karenge and Kolding (1994). Macrophyte and benthos figures modified from Machena (1989).

missed *Oreochromis macrochir* which was introduced in the 1960s and is now fairly common in the lake. Balon's record was updated to 45 species by Marshall (1984). Balon (1974b) explained the natural expansion of some species by suggesting that they had invaded the lake from the upper Zambezi, above Victoria Falls, which had been considered a barrier to downstream migrations. There are, however, indications that the so-called 'upper Zambezi invaders' may also have been present in the middle Zambezi before inundation but were missed by the sampling. They may also have been accidentally introduced or could have reached the lake through the Victoria Falls power station overflow (Jubb, 1976; Kenmuir, 1984; Bell-Cross and Minshull, 1988).

12.3 BRIEF HISTORY OF THE COMMERCIAL FISHERIES

The lake was primarly constructed for hydroelectricity but additional expectations were also derived from various estimates of the potential fish yields. Before inundation, an area of some 950 km^2 (about 18% of the lake area) was cleared of bush at various places within the 20 m contour to establish inshore fishing grounds. Artisanal inshore gill netting began in 1962 on the Zimbabwean side and catches rose to a peak of some 2500 t in 1964. From then catches declined almost linearly to around 1000 t in 1970 and have remained fairly stable around this level (Marshall *et al.*, 1982; Sanyanga *et al.*, 1993). The species composition in the initial high landings resembled very much the composition of dominant species in the pre-impoundment survey, consisting mainly of *Labeo* spp., *Distichodus* spp., *Hydrocynus*, *Mormyrus* and *Clarias* (Kenmuir, 1984). The cichlids, particularly *Oreochromis mortimeri* and *Serranochromis codringtonii*, have now replaced *Labeo* and *Distichodus* from being the most important species in the inshore fishery (Sanyanga *et al.*, 1992), although there is a west–east gradient in species composition from the 'riverine' basin 1 at the Zambezi inflow to the 'lacustrine' basin 5 near the dam wall (Begg, 1974).

Jackson (1961) predicted that the pelagic habitat of Lake Kariba would remain empty because the species present in the Zambezi evolved in a riverine habitat and would therefore only inhabit the shallow littoral zones. This was later confirmed by Coke (1968) and Mitchell (1978). Following Jackson's recommendation kapenta, *Limnothrissa miodon* (a small pelagic clupeid) was introduced into Lake Kariba between 1967 and 1969 from Lake Tanganyika (Bell-Cross and Bell-Cross, 1971; Junor and Begg, 1971). The introduction was a success and kapenta fishing, or the offshore fishery, began in July 1973 with a single purse-seiner. But effort grew rapidly and the fishery has now developed into a million-dollar industry, with more than 30 000 t landed annually using light-attraction to lift nets, and with some potential for further expansion (Anon., 1992; Marshall, 1992, 1993; Machena *et al.*, 1993;

Pitcher and Bundy, 1994). Thus, with the introduction of the pelagic fishery, Lake Kariba has lived up to even the most optimistic pre-impoundment predictions, that the lake might produce as much as 30 000 t per year (Maar, 1959, quoted by Marshall, 1985) (Chapter 24).

12.4 THE EXPERIMENTAL FISHERY

The Lake Kariba Fisheries Research Institute (LKFRI) has since 1960 conducted an experimental gill net fishery at Lakeside station close to Kariba town in the Sanyati basin (Fig. 12.1). Commercial fishing was only permitted here between 1962 and 1972 (Kenmuir, 1984) and limited fishing took place from 1977 to 1980, but this probably had little effect on the results (Marshall, 1984). Experimental fishing during the early years was sporadic (Kenmuir, 1984), but from 1970 to 1975 settings were twice a month and since 1976 sampling has been done on a weekly basis, except during 1981 when no fishing took place (Karenge, 1992). The sampling method has remained constant over the whole period and consists of multifilament nets in multi-mesh fleets with 12 panels ranging from 38 to 178 mm stretched mesh sizes with 12.5 mm increments. The fleet, 550 m in total length, was set overnight perpendicular to the shore with the smaller meshes inshore. Each individual fish caught was identified to species, and the mesh size, weight, length, sex and maturity stage were recorded. All experimental gill net data from 1969 to 1992 have been computerized into a database which now contains more than 130 000 individual fish records and represents more than 10 000 panel settings (Table 12.1).

Catch per unit effort (CPUE) in the experimental fishery is expressed as catch (kg wet weight or number of fish) per standard unit per setting. The standard unit is one mesh panel of 45.7 m (50 yards) long. CPUE can be used as an index of abundance or biomass of a fish stock if the catchability coefficient (q) is constant (Gulland, 1983). In this review q is considered constant because the calculated CPUE values are annual means and the nets and sampling programme have remained constant over the whole period.

As a measure of relative abundance or commonness of each species (j) in the gill nets, an index of relative importance (*IRI*, Kolding, 1989) is used:

$$IRI_j = \frac{(W_j + N_j)F_j}{\sum_{i=1}^{m} [(W_i + N_i)F_i]} \times 100 \qquad (12.1)$$

where W_j and N_j are percentage weight and number of each species of total catch, F_j is percentage frequency of occurrence of each species in total number of settings and m is total number of species. This index combines the numeric

Table 12.1 Overall catch composition in experimental gill nets at Lakeside station, Kariba, between 1969 and 1992. Total number of settings (individual mesh panels) was 10 367. The percentage index of relative importance (*IRI*, see text) gives a rough indication of the abundance and commonness of a species (from Karenge and Kolding, 1995)

Family	Species	Number*	*N*†	*W*‡	*F*§	*IRI*
Characidae	*Hydrocynus vittatus*	39 520	29.4	21.9	44.1	34.8
	Brycinus imberi	18 381	13.7	1.5	13.1	3.1
Cichlidae	*Sarranochromis codringtonii*	21 678	16.1	17.1	49.2	25.1
	Oreochromis mortimeri	14 060	10.4	16.5	37.0	15.3
	Tilapia rendalli	2 650	2.0	2.7	16.3	1.2
	Serranochromis macrocephalus	2 099	1.6	0.7	8.9	0.3
	Oreochromis macrochir	135	0.1	0.1	1.1	–
	Pharyngochromis darlingi	4	–	–	–	–
	Serranochromis robustus	1	–	–	–	–
Mormyridae	*Mormyrus longirostris*	5 671	4.2	19.0	28.7	10.3
	Hippopotamyrus discorhynchus	6 610	4.9	1.0	15.9	1.4
	Marcusenius macrolepidotus	2 413	1.8	0.4	6.8	0.2
	Mormyrops deliciosus	560	0.4	2.2	4.7	0.2
Clariidae	*Clarias gariepinus*	3 620	2.7	8.3	19.4	3.3
	Heterobranchus longifilis	98	0.1	0.6	0.9	–
Schilbeidae	*Schilbe mystus*	8 315	6.2	1.4	17.0	2.0
Mochokidae	*Synodontis zambezensis*	6 040	4.5	3.2	18.3	2.2
	Synodontis nebulosus	122	0.1	–	0.8	–
Cyprinidae	*Labeo altivelis*	2 400	1.8	3.1	7.7	0.6
	Labeo cylindricus	89	0.1	0.1	0.7	–
	Labeo congoro	19	–	0.1	0.2	–
	Barbus marequensis	10	–	–	–	–
	Varicorhinus nasutus	1	–	–	–	–
Distichodontidae	*Distichodus schenga*	65	–	0.2	0.6	–
	Distichodus mossambicus	1	–	–	–	–
Malapteruridae	*Malapterurus electricus*	2	–	–	–	–
Clupeidae	*Limnothrissa miodon*	1	–	–	–	–

*Number of records (total 134 565).
† Percentage number of each species in total catch.
‡ Percentage weight of each species in total catch.
§ Percentage frequency of occurrence of each species in total number of settings.

abundance (*N*), the average size (*W*) and the commonness (*F*) of a species (Pinkas *et al.*, 1971; see also Caddy and Sharp, 1986), and gives this new value in percentage to the combined value of all the species present.

To analyse the demographic structure of selected species an attempt was made to split them into age or size groups. For the three species, *Hydrocynus vittatus*, *Oreochromis mortimeri* and *Serranochromis codringtonii*, a knife-edge length-to-age conversion was done using the von Bertalanffy growth parameters given in Karenge (1992). For other species, where reliable growth parameters were not available, splitting was simply done by grouping the catches by mesh size into three groups: fine (38–76 mm), medium (89–127 mm) and coarse (140–178 mm). Mean weights and mean lengths were obtained from Karenge (1992).

12.5 CHANGES IN SPECIES COMPOSITION

Altogether, 27 fish species belonging to 10 families have been caught in the experimental gill nets at Lakeside during the study period (Table 12.1). Of the 14 or so fish families present in the lake, only six are commonly caught in experimental gill nets and these are, in order of overall relative importance (*IRI*), Cichlidae (42%), Characidae (38%), Mormyridae (12%), Clariidae (3%), Schilbeidae (2%) and Mochokidae (2%) (Table 12.1). The two families Cyprinidae and Distichodontidae, which were the most important in the filling phase, are now of minor importance. The remaining two families in the gill net catches, Malapteruridae and Clupeidae, are represented only by one species each and few specimens, *Malapterurus electricus* and the introduced, pelagic *Limnothrissa miodon*. The latter, however, is abundant offshore and is the most important commercial species, targeted by its own fishery.

During the early period in this study, the fish community at Lakeside had a relatively low species diversity. From 1970 to 1974 characids and cichlids together made up about 90% of the total *IRI* (Fig. 12.3) constituting largely of three species: tigerfish, *Hydrocynus vittatus*, Kariba tilapia, *Oreochromis mortimeri*, and green tilapia, *Serranochromis codringtonii*. *H. vittatus* was by far the most abundant species, alone constituting between 50% and 80% of the *IRI*. The mormyrids during this period were dominated by *Marcusenius macrolepidotus* and *Hippopotamyrus discorhynchus* (together 5.1% *IRI* in 1971 whereas *Mormyrus longirostris* had only 0.2% *IRI* the same year). *Schilbe mystus* had an *IRI* of 6% and 7% in 1970 and 1971, respectively, and these values are relatively high for this species when the whole period 1969 to 1992 is taken into consideration.

During the period 1970 to 1978, there was a gradual decline in the relative importance of *H. vittatus*. The cichlids increased in importance, contributing 50%, whereas by 1978 *H. vittatus* contributed only 25% of the total *IRI*.

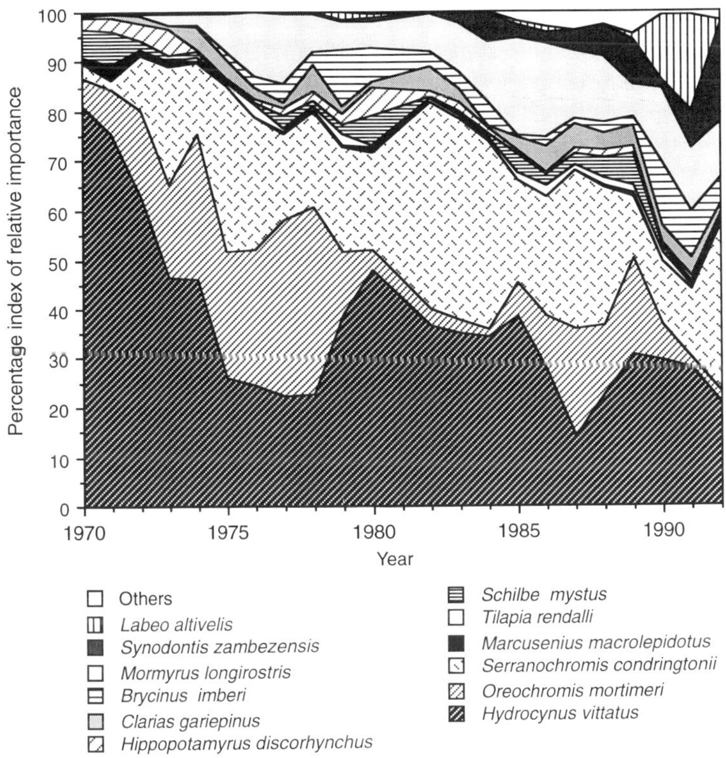

Fig. 12.3 Changes in index of relative importance (*IRI*) of the 12 most dominant fish species caught at Lakeside between 1970 and 1992. Note the trend from a few dominant species to more species with a more even composition.

Among the cichlids, *S. codringtonii*, which was uncommon throughout the lake in the early years (Kenmuir, 1984), steadily became more important than *O. mortimeri* (except for the period 1977 to 1978 when the reverse was true). This phase was also marked by the gradual decline of *M. macrolepidotus* and *H. discorhynchus* but with a substantial increase of the larger mormyrid species *M. longirostris*.

Between 1980 and 1992, two other species, *Synodontis zambezensis* and *Labeo altivelis*, were gradually becoming important in the inshore system and the so-called 'upper Zambezi invader' (Balon, 1974b), *Serranochromis macrocephalus*, appeared in the lake and increased in abundance. The appearance of new species and the increasing abundance of other species (like *L. altivelis* and *S. zambezensis* which were common in the Zambezi before inundation) resulted in increasing species diversity. Species richness increased to 14 species that were commonly caught in gill nets during the 1980s, compared with nine species towards the end of the 1960s. Another feature during the 1980s was the reduction in importance of the cichlid *O. mortimeri*, with *H. vittatus* and *S. codringtonii* annually exchanging the role as most abundant species. The mormyrid *Mormyrus longirostris* was also more abundant than in previous years (with an *IRI* of around 19% of the total in 1985–1986). The small characid *Brycinus imberi* has lately shown a strong presence with an *IRI* of around 10% of the total in 1990s. Only two species, *Schilbe mystus* and *Tilapia rendalli*, did not show any notable changes during the period 1970 to 1992. All in all there has been a continuous change in species composition and relative abundance in the inshore community, as shown by these experimental gill net catches, and there is every indication that future changes are to be expected.

12.6 TRENDS IN CATCH PER UNIT EFFORT

Combined catch per unit effort (CPUE), by weight for all mesh sizes and species, is shown in Fig. 12.4. The overall CPUE has fluctuated over the investigated period, with two peaks in 1974 and 1986, but the overall mean of around 4.5 kg per standard net setting (SD 1.8 kg) has not changed significantly. By weight, four species, *H. vittatus, O. mortimeri, S. codringtonii* and *M. longirostris*, contributed most to these fluctuations, but by numbers smaller species also fluctuated substantially (Fig. 12.5). The main groups are treated separately below.

Characidae

The two species, *H. vittatus* and *Brycinus imberi*, were the most commonly caught in the small mesh sizes (38–76 mm). *H. vittatus* was also the most abundant fish during the early years and the CPUE (Fig. 12.6), for all mesh sizes, resembles closely that of the cichlids in having a sharp peak in 1974 (about 10 fish or 3 kg per setting). However, the overall time series of the CPUE of *H. vittatus* since 1969 (Fig. 12.6) exhibits a significant decline of 0.23 per year ($r^2 = 0.47, P < 0.0001$), although the trend is most pronounced in the late 1970s and is not significant from 1979 onwards. Looking at the population split into age groups (Fig. 12.7), the situation looks rather different. Age group 1, taken as a recruitment index, shows again that 1974 apparently was an exceptional year, but then there is a clear decline with time ($b = -0.11, r^2 = 0.64, P < 0.0001$) and the recruitment index even reaches zero in 1986–87 and 1992. Adult fish (age group 3+), however, show a remarkably steady occurrence ($b = -0.01,\ r^2 = 0.02,\ P = 0.49$) in the catches over the whole time period of between 0.5 and 1 fish per standard

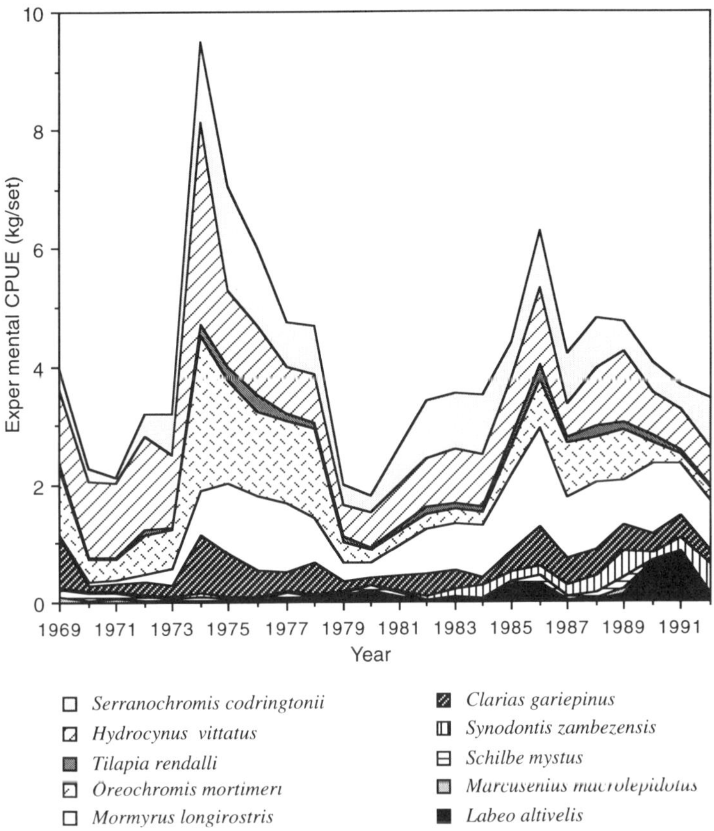

Fig. 12.4 Experimental CPUE (kg per standard setting) of the ten most abundant fish species at Lakeside station, 1969–1992. From Karenge and Kolding (1994).

setting. The decline in the relative abundance of small *H. vittatus* was associated with an increase in the abundance of *B. imberi* (Fig. 12.5), and the two species have a significant negative correlation (Table 12.2). It might be that these two species are mutually competing in their young stages.

Cichlidae

Apart from the gradual shift in the relative abundance between the two major cichlids, *O. mortimeri* and *S. codringtonii*, over the years, and interannual fluctuations, the cichlids as a group have remained stable overall from 1970 to 1992 ($r^2 = 0.001$, $P = 0.916$, Fig. 12.6). Both species showed a sharp increase reaching a peak, like tigerfish, in 1974. The abundance of *T. rendalli*

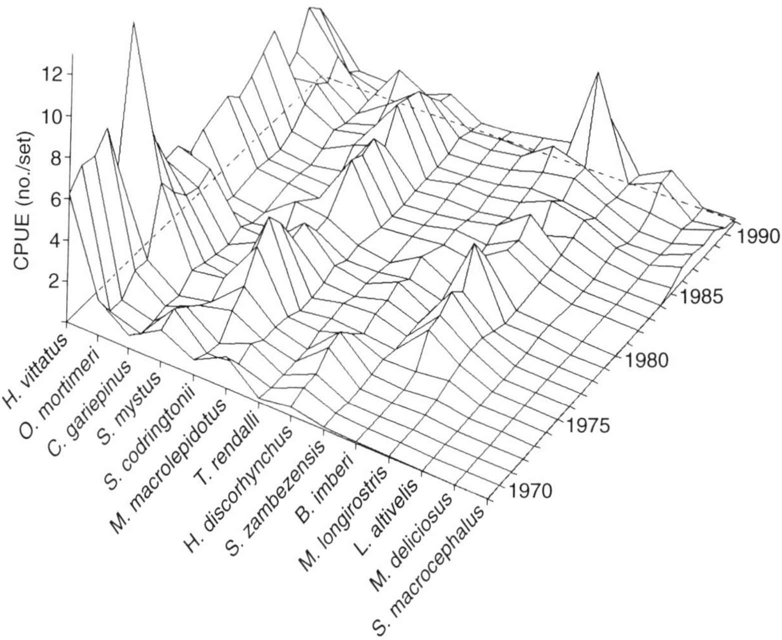

Fig. 12.5 Trends in abundance (no. per standard setting) for 14 species commonly caught in experimental gill nets at Lakeside between 1969 and 1991. From Karenge (1992).

followed a similar pattern but CPUE was much lower compared with the former two cichlids (Fig. 12.5). However, *T. rendalli* is suspected to evade gill nets (Kenmuir, 1984), so the catches might not reflect true abundance. *S. macrocephalus* showed a peak similar to that of *S. codringtonii* during the late 1980s (Fig. 12.5) but after the prominent 1974 peak the cichlids declined in abundance to reach lower values between 1979 and 1980. The decline was more noticeable for *O. mortimeri* than *S. codringtonii*, and the latter showed a much quicker recovery (Fig. 12.7). Splitting into age groups for the two major cichlids (Fig. 12.7) shows more or less the same picture with no significant trends. However, recruitment of *O. mortimeri* has two distinct humps around 1974–1976 and 1987–1989 with very low values in between, whereas that of *S. codringtonii* is more even. All in all there are no significant changes in the cichlid populations, although *S. codringtonii* seems to be slowly replacing *O. mortimeri* as the most important cichlid at Lakeside. Since 1975 *O. macrochir* has been caught in very low, although regular, numbers at Lakeside.

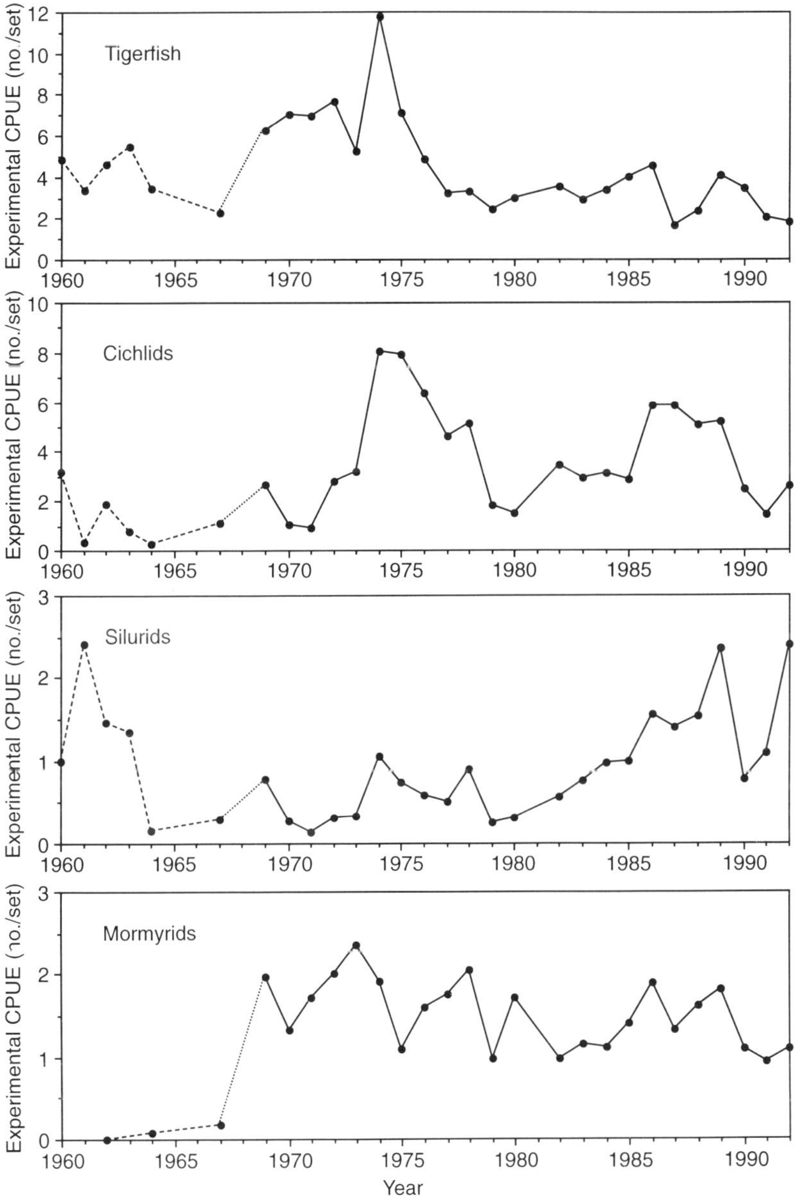

Fig. 12.6 Trends in experimental catch per unit effort at Lakeside between 1960 and 1992 for the most important fish species or families (tigerfish, cichlids, silurids and mormyrids). Data for the period 1960–1967 calculated from Kenmuir (1984), dotted line between 1967 and 1969 interpolated. Note differing scales.

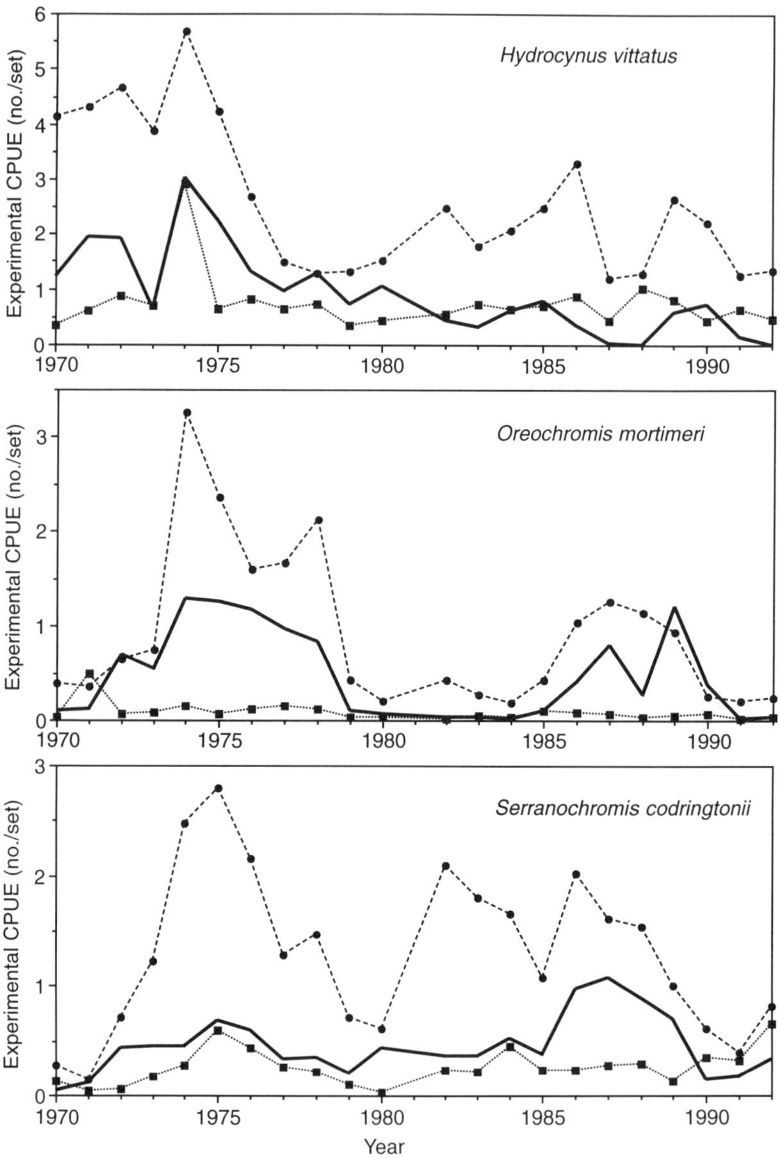

Fig. 12.7 Experimental CPUE (no. per standard setting) between 1970 and 1992 split into three age groups for the species *Hydrocynus vittatus*, *Oreochromis mortimeri* and *Serranochromis codringtonii*. Note differing scales. Solid curves, age group ⩽1 year; ●, 2 years; ■, ⩾3 years.

Table 12.2 Results of Spearman rank-correlation analysis (rho) among various species and species groups and overall experimental CPUE (exp). All values in catch in numbers per standard setting. Number of years = 22. Significant values in bold. From Karenge (1992)

Species		Exp.	Cichlids	Silurids	*H. vit.*	*M. lon.*	*S. cod.*	*O. mor.*	*H. dis.*	*M. mac.*	*B. imb.*
Cichlids	rho	**0.66**									
	$P <$	0.002									
Silurids	rho	0.36	**0.53**								
	$P <$	0.10	0.015								
H. vittatus	rho	**0.53**	0.12	−0.27							
	$P <$	0.015	0.59	0.22							
M. longirostris	rho	0.40	**0.67**	**0.67**	−0.22						
	$P <$	0.06	0.002	0.002	0.31						
S. codringtonii	rho	**0.47**	**0.92**	**0.48**	0.016	**0.74**					
	$P <$	0.03	0.000	0.029	0.94	0.001					
O. mortimeri	rho	**0.87**	**0.77**	0.28	0.37	0.39	0.53				
	$P <$	0.000	0.000	0.20	0.09	0.08	0.015				
H. discorhynchus	rho	0.09	−0.04	−0.34	0.19	−0.40	−0.16	0.12			
	$P <$	0.68	0.85	0.12	0.38	0.07	0.47	0.57			
M. macrolepidotus	rho	0.28	−0.22	−0.40	**0.49**	**−0.58**	**−0.47**	0.35	**0.60**		
	$P <$	0.20	0.33	0.07	0.03	0.008	0.031	0.11	0.006		
B. imberi	rho	−0.08	0.023	0.20	**−0.58**	0.38	0.11	−0.12	−0.35	**−0.48**	
	$P <$	0.71	0.91	0.91	0.008	0.08	0.61	0.57	0.11	0.028	
L. altivelis	rho	−0.14	−0.07	0.41	**−0.60**	**0.42**	−0.02	−0.26	−0.34	**−0.53**	**0.70**
	$P <$	0.54	0.75	0.06	0.006	0.054	0.95	0.24	0.11	0.015	0.001

Mormyridae

This family shows a remarkably steady, although slightly negative ($b = -0.025$, $r^2 = 0.0181$, $P = 0.048$), trend in CPUE by numbers since 1969 (Fig. 12.6). The low numbers in the early 1960s might be an artefact as most of the nets were top-set up to around 1965, since when nearly all netting was bottom-set (Kenmuir, 1984). The most important mormyrid at Lakeside is now *M. longirostris*, the second largest member of the family in Lake Kariba after the uncommon *Mormyrops deliciosus*. As the abundance of the two small species, *M. macrolepidotus* and *H. discorhynchus*, gradually declined over the study period ($b = -0.05$, $r^2 = 0.422$, $P < 0.0001$, Fig 12.5), that of *M. longirostris* increased slightly ($b = 0.01$, $r^2 = 0.025$, $P = 0.03$). The abundance of *M. longirostris* has been setting a pattern closely related to that of the cichlids. *M. longirostris* is a benthic feeder like *S. codringtonii* and the two species' abundance is significantly positively correlated (Table 12.2). The predatory *M. deliciosus* is one of the largest fish in Lake Kariba but occurs in very small numbers at Lakeside. *M. deliciosus* increased slightly during the later half of the 1980s.

Cyprinidae

After prominent catches in 1960–61 (Kenmuir, 1984; Marshall, 1984), the larger cyprinids *Labeo altivelis* and *Labeo congoro* had disappeared from the Lakeside station by around 1967–68. By 1974 they started reappearing in small numbers together with a third smaller species, *Labeo cylindricus*. The greatest change came from *L. altivelis* with a remarkable increase during the late 1980s (Fig. 12.8). This increase is difficult to explain as the species is anadromous and essentially fluvial (Marshall, 1984), and the increase was during a period with droughts and reduced river flows.

Clariidae

The catfish, *Clarias gariepinus*, is the most abundant member of the family and its fluctuations in abundance are basically similar to that of the cichlids (especially *O. mortimeri*) with no significant trends (Fig. 12.8), although Marshall (1977) suggests that this species is not well adapted to lacustrine conditions. In fact, all the silurids are significantly positively correlated with the cichlids (Table 12.2). *Heterobranchus longifilis* is the largest fish species in Lake Kariba, but appears in such small numbers that its overall contribution to the biomass of fish in the lake is small.

Mochokidae

Two species belonging to the family Mochokidae, *Synodontis zambezensis* and *Synodontis nebulosus*, were caught at Lakeside. Of these two, the brown

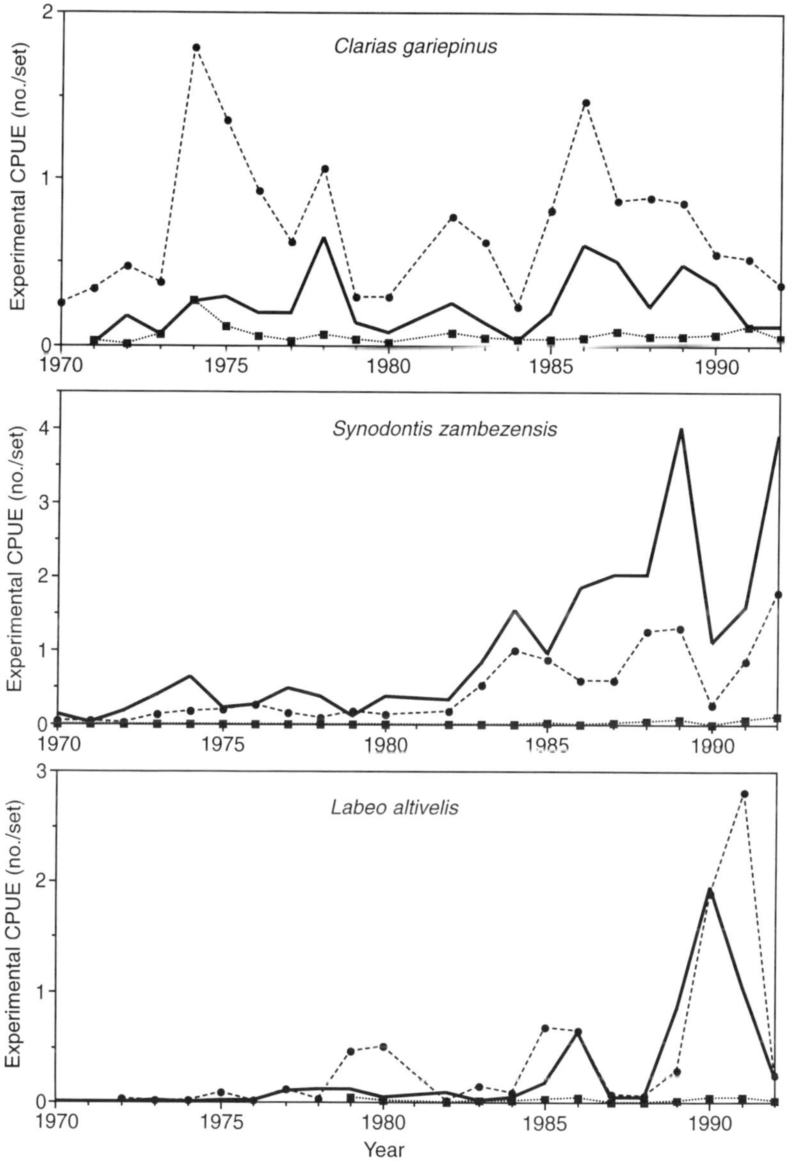

Fig. 12.8 Experimental CPUE (no. per standard setting) between 1970 and 1992 split into three mesh size classes for the species *Clarias gariepinus*, *Synodontis zambezensis* and *Labeo altivelis*. Note differing scales. Solid curves, mesh size 38–76 mm; ●, 89–127 mm; ■, 140–178 mm.

squeaker, *S. zambezensis*, is the most common (Table 12.1). After being absent or uncommon in catches between 1960 and 1965 (Kenmuir, 1984; Marshall, 1984) this species has gradually increased at Lakeside ($b = 0.06$, $r^2 = 0.63$, $P < 0.0001$), especially during the last decade (Fig. 12.8). In 1992 it was the second most important species after *Brycinus imberi*, by numeric abundance. *S. zambezensis*, like *L. altivelis* and *S. macrocephalus*, have contributed significantly to the increased species diversity at Lakeside. Kenmuir (1984) reported *S. nebulosus* as largely disappeared, but this species has been caught in small, but significant, numbers at Lakeside every year from around 1980 (Karenge, 1992).

Schilbeidae

The one member of this family, *Schilbe mystus*, was uncommon in early catches at Lakeside (Kenmuir, 1984). Since 1965 it increased, and has always been present in the lake, but in small numbers only (the mean CPUE rarely exceeded 2 fish per setting). However, there is a strong seasonal variation in CPUE during one year for *S. mystus*, which must be attributed to seasonal changes in catchability and perhaps migrational behaviour (Karenge, 1992).

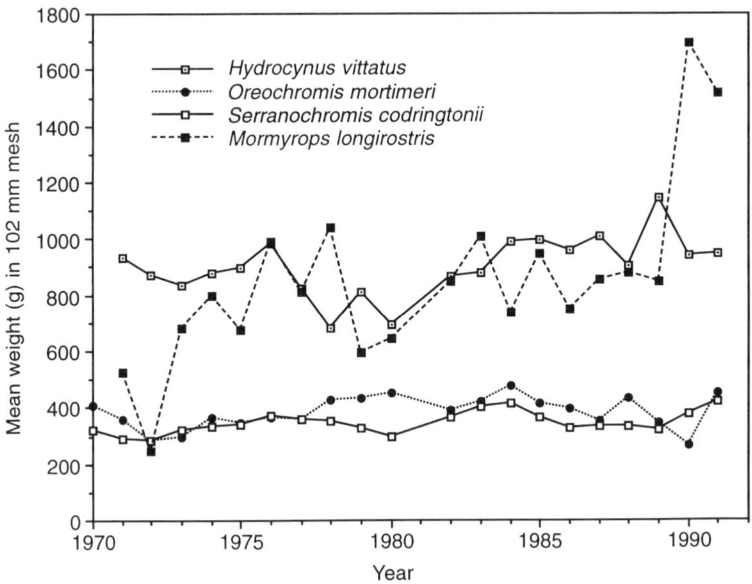

Fig. 12.9 Trends in mean annual weight (grams) in the 102 mm mesh size between 1970 and 1991 for the four commercial species *Hydrocynus vittatus*, *Oreochromis mortimeri*, *Serranochromis codringtonii* and *Mormyrops longirostris*.

Changes in mean length and weight with time

Changes in mean length or weight in the 102 mm mesh size (the smallest mesh size permitted in the inshore artisanal fishery) were analysed separately for some species (Fig. 12.9) to see if there was any impact on the population structure from fishing. The mean length of the two benthic-feeding species of commercial importance, *S. codringtonii* and *M. longirostris*, showed a significant increase with time (Table 12.3). For *O. mortimeri* and *H. vittattus*, mean length was more or less constant with time. The mean weight of *H. vittatus, O. mortimeri, S. codringtonii, M. longirostris, S. mystus, S. zambezensis, C. gariepinus* and *L. altivelis* in 102 mm mesh nets either increased or remained constant with time (Karenge, 1992). Against this background there are no indications that selective mortality from fishing is having any visible impact on the fish populations around Lakeside.

12.7 DISCUSSION

The gill net data from the Lakeside sampling programme form the longest and probably one of the most reliable data series from Lake Kariba. The weekly routine is well incorporated in the activities of LKFRI and the sampling design has remained constant over the whole period. This means that the catchability coefficient can be assumed constant and that the CPUE is likely to be proportional to the stock abundance of individual species.

A drawback, however, is that catchability by gill nets is species specific and this might affect samples of species suspected of evading gill nets. The relative abundance of *S. zambezensis, Malapterurus electricus, T. rendalli* and *C. gariepinus* are therefore probably underestimated in gill net catches. This seems confirmed by comparisons with other sampling surveys using poison or explosives (Balon, 1974b; Mitchell, 1976, 1978; Langerman, 1984) which recorded higher proportions of these species. *T. rendalli* and *C. gariepinus* prefer shallow vegetated areas (Jackson, 1961; Donnelly, 1971; Marshall, 1977)

Table 12.3 Results of linear regression analysis of mean length versus time for the most common and commercially important fish species caught at Lakeside in 102 mm mesh experimental nets (from Karenge, 1992)

Species	*n*	a	b	r^2	*P*-level
Serranochromis codringtonii	20	−230	0.13	65.2	<0.001
Mormyrus longirostris	19	−1038	0.54	54.3	<0.001
Hydrocynus vittatus	20	−261	0.15	16.3	0.07
Oreochromis mortimeri	20	−80	0.05	5.5	0.31

where it is difficult to set nets. *S. zambezensis* has been reported to be most abundant in deep water (Machena *et al.*, 1993) and *M. electricus* has a very low catchability by gill nets due to its body form. Also small inshore species are not caught in the experimental gill nets, for example small cichlids, cyprinids and the characid *Brycinus lateralis*.

The Lakeside station in basin 5 represents a lacustrine habitat close to the Kariba Dam wall. Hence the results cannot easily be extrapolated to the whole lake, especially to the riverine habitats of basins 1 to 2 (Begg, 1974; Marshall, 1984). Similarly, the Lakeside station is not in a commercially fished zone, therefore the impact of fishing cannot be assessed directly. The Lakeside data are, however, extremely important in showing the relative changes in species composition and population structures over time and could prove valuable in population dynamics and community ecology studies. One strong advantage of a long time series is that the possibility of making erroneous deductions or spurious correlations decreases with the length of the series. For example, a study from 1975 to 1980 would give the impression of a consistent decline in biomass (Fig. 12.4), whereas one from 1980 to 1986 would produce the opposite conclusion.

Species diversity and ecological succession

The successional changes that artificial impoundments seem to undergo (Petr, 1978) have been repeated in Lake Kariba (Balon, 1974b). After a first phase of high productivity, as abundant submerged terrestrial vegetation decomposed, there was a second phase characterized by low productivity and species diversity as the riverine species adjusted to the lacustrine environment. The Lakeside data used in this study indeed showed relatively low productivity and species diversity from 1964 to 1972 when three species, particularly the predator, *H. vittatus* and the two cichlids, *O. mortimeri* and *S. codringtonii*, made up some 90% of the *IRI*, with only six other species. Such an increase in predators during and soon after the filling phase was also observed in Lake Kainji (Lewis, 1974; Balogun, 1986) and Lake Volta (Lelek, 1973). According to Balon (1974b), a third phase, characterized by increased species diversity and stabilized production, began in Lake Kariba around 1968 and lasted to 1972. The Lakeside data, however, show that diversity continued to increase from 1972 to 1992 but overall production fluctuated as a result of environmental conditions. By 1992, 14 species were commonly caught in gill nets and relative abundance was more evenly distributed among the species than during the late 1960s (Fig 12.3).

After the decline of the large shading mats of *Salvinia molesta* in the 1970s the macrophytes and benthic fauna increased (Fig. 12.2). So did benthic-feeding fish species like *M. longirostris*, *S. zambezensis*, *S. codringtonii*, *S. macrocephalus* and *L. altivelis*. Such a trend towards increased benthic-feeding fish species was also observed in Lake Volta (Lowe-McConnell, 1973), in Lake Kainji during the early, post-impoundment phase (Lewis, 1974) and in Lake Kyle (Zimbabwe), where mormyrids increased gradually in the lake (Karenge, pers. obs.). Increased diversity within the inshore fish community in Lake Kariba could be attributed to the development of a mosaic of new niches among the aquatic macrophytes, periphyton and submerged trees where a variety of benthic gastropods, molluscs, insects and other invertebrates provided a new food resource to fish.

However, stable conditions and the existence of a rich aquatic flora along the lake margin is not a permanent habitat for Lake Kariba's fauna. Every other year disturbance from lake level change occurs and engenders changes in species composition. Aquatic macrophytes may then colonize a new marginal surface in shallow areas at a different level. The speed of colonization and the resultant biomass are governed by prevailing conditions. Hence, the belief that Lake Kariba will stabilize after the two initial phases (Balon, 1974b) may not be correct. As in many impoundments, in Lake Kariba the influence of fluctuations in lake level is still important 30 years after its construction (Karenge and Kolding, 1995). It is possible that events that took place during the filling may be repeated, on a smaller scale. For example, a drop and a rise in lake level separated by some years may allow terrestrial vegetation to grow between the two water levels. In addition, Begg (1970) noted that Lake Kariba will never become a water body of uniform characteristics because the mean water retention time is less than 3 years, and there are strong river influences at the river mouths, especially in basins 1 to 2. Thus there are several possible interactive factors with different effects shaping the community structure and productivity of the inshore system.

Effects of fluctuations in lake level

Karenge and Kolding (1995) analysed the relationship between the fisheries and hydrology in Lake Kariba and found that experimental, artisanal and offshore kapenta catch per unit of effort were all significantly correlated with mean annual fluctuations in lake level, but not with the absolute lake levels (Table 12.4). This strongly suggests that nutrient inputs from rivers and the flooded marginal areas are important when stimulating fish production. When the lake level rises, terrestrial vegetation is submerged and nutrients leaching from decomposing organic matter (dung, terrestrial grass, shrubs and trees) and directly from rivers, result in increased plankton and fish production (McLachlan, 1970, 1974; Marshall, 1982; Chapter 14). Flooding of marginal areas of the lake and increased river flow result in better spawning conditions, especially for potamodromous species, and excellent conditions for growth and survival of juveniles of inshore fish species (Jackson, 1961). The breeding

Table 12.4 Pearson product-moment correlation coefficient (r) and Spearman rank-correlation coefficient (rho) and their significance level for association between various variables describing the fisheries and hydrology of Lake Kariba between 1982 and 1992. 'y' in subscript parentheses gives the time lag between variables. Modified from Karenge and Kolding (1995)

Variable 1	Variable 2		Pearson		Spearman	
		n	r	$P <$	rho	$P <$
Exp CPUE (no.)*	Δ level†	11	0.829	0.002	0.845	0.007
Exp CPUE (no.)	Lake level	11	0.088	0.798	0.000	1.000
Kapenta CPUE	Δ level	11	0.808	0.003	0.761	0.016
Kapenta CPUE	Lake level	11	−0.176	0.604	0.137	0.965
Artisanal CPUE	Δ $\text{level}_{(y-1)}$	11	0.756	0.007	0.755	0.017
Artisanal CPUE	Lake $\text{level}_{(y-1)}$	11	−0.121	0.722	−0.064	0.840
Exp CPUE (no.)	Kapenta CPUE	11	0.764	0.006	0.811	0.013

*CPUE (no.), catch per unit effort, expressed in numbers per set.
† Δ level, mean annual change in water level.

activity of most important inshore species (*H. vittatus, M. longirostris, S. codringtonii, O. mortimeri* and *C. gariepinus*) is at its peak between November and March, coinciding with the local rains (Jackson, 1961; Karenge, 1992). Changes in recruitment with time also show that juvenile fish abundance coincides with years of increasing lake level and inflow of local rivers (Karenge and Kolding, 1995). Improved conditions for fish growth after a flood may not be common to artificial lakes alone, for example in Lake Victoria (Pitcher and Hart, 1982) and Lake Turkana (Kolding, 1992, 1993 and Chapter 16), flooding of shallow marginal areas results in input of detritus and nutrients and an increase in fish production. The fact that artificial lakes become less fertile as the initial supply of nutrients is washed out or fixed in sediments is a major piece of evidence that lakes do not maintain their fertility unless an external loading of nutrients is continually applied (Moss, 1988). As an example, the exceptionally high production of most species around the mid 1970s in Lake Kariba (Fig. 12.4) might have been enhanced by nutrient release from the rapid decline and decomposition of *Salvinia* mats in this period (Marshall and Junor, 1981) (See Chapter 14 for more on the results of changes in lake level.)

Effects of macrophytes

Various authors (Donnelly, 1971; Mitchell, 1976; Marshall, 1984, 1985; Moyo, 1986; Machena, 1989) have stressed the importance of cover for juvenile fish and there is a marked correspondence between submerged macrophytes and fish abundance in Lake Kariba (Mitchell, 1976; Marshall, 1984). Aquatic macrophytes have a limited capacity to adjust to changes in lake level (Bowmaker, 1973b; Machena, 1989). Bowmaker recommended a maximum rate of 0.6 m per month water level change and suggested that to promote full development of macrophytes, the range of water levels should not exceed 3.5 m in a single year. Juvenile fish are presumed to be exposed to increased predation when rapid drawdown occurs (as happened during the early 1980s and 1990s), exposing the submerged macrophytes, which form a narrow bed along the lake's margin limited to about 6–12 m deep (Bowmaker, 1973b; Machena, 1989). Similarly, after a sudden rise in lake level when submerged terrestrial vegetation decomposes before aquatic macrophytes get established, cover from predators would not be available.

During the 1960s, full development of macrophytes was probably inhibited by the presence of the floating fern *Salvinia molesta*, which competed with rooted macrophytes for nutrients and also had a strong shading effect where it accumulated (Begg, 1970). But from the early 1970s macrophytes and benthic fauna increased in abundance (Fig. 12.2) along the margins of the lake, although conditions were not suitable during some years (the excessive lake level rises of 1974 and 1989 and frequent droughts during the 1980s).

With the development of aquatic macrophytes, cover from juveniles and small fish species now exists in the lake and smaller species such as *S. macrocephalus*, *B. imberi* and *S. zambezensis* have increased during the 1980s. However, the increased presence of small species like *B. imberi* was also significantly correlated with the decline of small *H. vittatus*. In addition, the small species *H. discorhynchus* and *M. macrolepidotus* have declined in abundance since the early 1970s. Hence the contribution of macrophytes to changes in abundance of small species is not unequivocal.

Effects of predation and competition

In the Zambezi, before inundation, Jackson (1961) described the tigerfish, *H. vittatus*, as the dominant species in the area, having a commanding influence on the other fishes by predation. Thus the negative relationship (Table 12.2) between *B. imberi* and *H. vittatus* has been explained as a result of a decreased predation pressure on *B. imberi* as the tigerfish declined (Marshall *et al.*, 1982). But, although tigerfish is still the foremost top-predator in Lake Kariba, its role and diet have changed. The introduced kapenta ('sardine') is now the predominant prey, constituting at least 70% of the food intake (Kenmuir, 1973; Cochrane, 1976; Mitchell, 1976; Langerman, 1984). However, simulation studies (Anon., 1992; Machena *et al.*, 1993) showed that, with this diet, predation mortalities on kapenta from tigerfish accounted for less than 15% of the total sardine mortality. Interestingly, even a low proportion of cannibalism (2% of food intake) in the sardines incurred a much higher mortality than that from tigerfish predation. With regard to the cichlids, Donnelly (1969), Kenmuir (1973) and Moyo (1986) found that predation by *H. vittatus* had no significant effect on the populations because the cichlid juveniles did not leave cover until they reached a length of around 17 cm. Thus the tigerfish seems to have lost much of its governing role in the Lake Kariba ecosystem.

The observed significant decline in small tigerfish, despite the enormous amount of food (kapenta) available for the adult population, suggests that density-dependent processes are mainly operating on the young and juvenile part of the population and might be a result of competition rather than predation. Most inshore fish species, when very small, feed on zooplankton and so does the introduced sardine: competition for food at the juvenile stages should therefore exist. Marshall *et al.* (1982) suggest that the reason why kapenta CPUE is low between September and March is that this is its main breeding season, and that kapenta then move to shallow water thus contributing to reduced catches in the pelagic zone. Hence the breeding season of sardines coincides with that of most inshore fish (Karenge, 1992). However, experimental CPUE and sardine CPUE from the fishery were positively correlated (Table 12.4) and this suggests that kapenta does not necessarily gain at the expense of inshore fish, because both pelagic and inshore fish wax and wane synchronously with the limiting food conditions.

The gradual decline in CPUE of *M. macrolepidotus* was significantly negatively correlated (Table 12.2) to CPUE of *M. longirostris*, and competition between the mormyrid species may occur because their diets tend to overlap (Bowmaker, 1973a; Joubert, 1975; Marshall *et al.*, 1982). The decline of *M. macrolepidotus* and increase of *M. longirostris* among the mormyrids is, however, of great benefit to the artisanal fishery because the latter species is bigger and is caught efficiently by large mesh sizes. Also the diets of *L. altivelis* and *O. mortimeri* tend to overlap, and as *L. altivelis* was increasing in the 1980s the abundance of *O. mortimeri* was declining although not significantly correlated (Table 12.2). Burne (1971) indicated that, although diatoms appear to form their basic diet, competition between these species should not exist because they tend to feed on slightly different algal species.

To sum up, it is difficult to explain the various positive or negative correlations between the species abundances (Table 12.2), as what is cause and what is effect is always a problem. However, compared with Jackson's (1961) description of the 'cut-throat' conditions that existed in the Zambezi before inundation, it seems that predation pressure is now less important than interspecific competition, perhaps particularly on the lower trophic levels, for the fluctuating resources in Lake Kariba.

Effect of fishing

Commercial gill net fishing at Lakeside started late in 1962 and was stopped in 1972, and hence commercial netting could not have had much influence on the declining abundance of nettable fish species until at least later than 1963 (Kenmuir, 1984). According to Kenmuir, the decline in abundance of various species was already well advanced by 1963 and he suggested natural causes as being responsible for the decline in CPUE between 1961 and 1972 at Lakeside. Also, because riverine cyprinids continue to constitute a large proportion of the commercial catch in basins 1 and 2, these species may be capable of sustaining fishing pressure if conditions are right (Kenmuir, 1984). According to Marshall *et al.* (1982), limited commercial fishing also took place at Lakeside between 1977 and 1980, but the strong decline in CPUE in this period had started in 1975, again before fishing recommenced. The mean length of four important species in the inshore fishery, *H. vittatus, S. codringtonii, O. mortimeri* and *M. longirostris*, at harvestable sizes either increased or remained the same over the whole period investigated. This indicates that the overall population structure of these species did not change in the large mesh sizes (Fig. 12.9). If selective fishing mortality is having an impact, the mean length or weight of the above species would be expected to decrease with time.

Marshall (1984) attributed the rapid decline in tigerfish biomass after 1975 to commercial purse seining for sardines where they occurred as bycatch. Similarly, Marshall (1987) observed a positive correlation between the relative abundance of 0+ tigerfish juveniles and the sardine CPUE and concluded that, because juvenile tigerfish depend on small sardines in shallow marginal zones, the decline of sardines, as a result of commercial fishing, was responsible for the decline of tigerfish. However, CPUE declined in that period (1975–1980) for many other species as well, such as *O. mortimeri, S. codringtonii* and *M. longirostris* (Fig. 12.7), and those have no obvious direct link with *Limnothrissa*. We believe that the decline in sardine CPUE during the same period can hardly be attributed to fishing. This deduction is based on two reasons: (1) the fishing effort in that period was less than a third of the present effort (Sanyanga *et al.*, 1993), and (2) the present CPUE shows no declining trends and fishing mortality is estimated as only about 20–25% of total mortality (Anon., 1992; Machena *et al.*, 1993). Thus the explanation of the decline (Marshall, 1987) in tigerfish could be a result of a spurious correlation.

Tigerfish are very vulnerable to exploitation during their spawning runs upriver. For example, experimental gill netting in Naodza river during the 1976 spawning run yielded 22 kg fish per hour per 100 m of net (Marshall *et al.*, 1982). Illegal fishing in the rivers is a persistent and increasing problem in Lake Kariba (Marshall *et al.*, 1982; Moyo, 1986; Sanyanga *et al.*, 1993) and recruitment overfishing on the spawning stock has been suggested as another cause of the decline in the tigerfish population, particularly from the recreational fishermen. The picture is not clear, however. Although there has been a decreasing trend in recruitment (age group 1, Fig. 12.7) since the early period, this appears to be independent of the size of the mature part of the population (age group 3+, Fig. 12.7), which has remained remarkably stable over the last two decades. The stock–recruitment relationship is thus not clear from these data, which might suggest that density-dependent processes are taking place mainly on the young and juveniles, as Marshall (1987) observed, but which are not subject to fishing, and this suggests a more general cause.

In summary, the results of this study show no evidence of overfishing in the Lakeside area, but it is not possible to extrapolate this finding directly to other parts of the lake. Moreover, Machena *et al.* (1993), using the ECOPATH II model to assess the interrelationships and community structure of Lake Kariba, conclude that, although perhaps locally heavily cropped, the inshore commercial species seemed not to be overfished. This was also substantiated by a working group (Anon., 1993).

12.8 CONCLUSION

Of the various agents having possible impact in the Lake Kariba ecosystem, it appears that the hydrological regime, with its connected nutrient input, is having a major governing role at present (Table 12.4). Both the inshore and offshore production chains seem highly geared to fluctuations in lake level, which might also have an impact on the community structures in two other ways.

- When water level in Lake Kariba remains more or less constant as a result of drawdown regulations or below-average river inputs, macrophytes and periphyton develop along the lake's shallow margin and a mosaic of new niches develops. Fish species diversity increases but production remains low due to nutrient limitation. The increase in diversity is mainly a result of an increase of benthic-feeding fish or species that depend on macrophytes and periphyton, either as food or as cover. Potamodromous mud-sifting and predatory fish species decrease in abundance.
- When a significant rise in lake level occurs, the nutrient status of the lake improves, both from inflow and from flooding of submerged land. Fish production, in both inshore and pelagic areas, increases. From a fishery point of view, conditions following a water level rise are the most desirable. A rise in lake level initiates the increase of sardines, cichlids, silurids and characids, which leads to increased fishery yields. Despite this, species diversity might suffer due to stronger predation pressure on small species

and because benthic feeders might decrease. This is because fluctuations in water level have a negative effect on marginal aquatic flora and associated benthic fauna.

It is questionable whether the observed changes over the last decade are primarily a consequence of human intervention through the introduction of sardines, as foreseen by Balon (1974b) and Kenmuir (1984). Looking at the history of the inshore species community, as represented by the Lakeside data series, the introduction of the pelagic *Limnothrissa miodon* and the subsequent large increase in landings from this stock do not seem to affect the inshore production to any visible degree. This observation might be important because the sardine introduction to Lake Kariba has been an undeniable success, and this species will undoubtedly be considered for introduction in other artificial reservoirs in the future (Chapter 24). Overall, total fish production is fluctuating with hydrological changes (Karenge and Kolding, 1995), but there are no significant changes between 1969 and 1992 in the inshore fish community biomass. The continued changes in species composition seem more a result of natural ecological succession and development than a result of human interventions.

The hypothesis that a stable phase (Balon, 1974b) will eventually occur in Lake Kariba and other artificial lakes could also be strongly challenged. Lake level fluctuations still seem as important in regulating fish abundance as was the case during the filling phase. Weather conditions, interspecific competition and general impoundment regulations governing lake level are important in determining which species increase. Because relatively stable conditions are temporary and can be disrupted any time due to weather or a change in drawdown conditions, Lake Kariba, 30 years after filling, can hardly be described as being stable. Changes are still taking place.

12.9 SUMMARY

An experimental fishery data series covering the period 1969 to 1992 from Lakeside station in the Sanyati basin of Lake Kariba was used to assess trends in species changes and catch per unit effort. Since the creation of the lake there has been a continuous change in species composition and increase in species diversity, whereas the production is fluctuating with nutrient input into the system. Impacts on the inshore fish community are discussed from the effects of fluctuations in lake level, macrophyte and benthos succession, predation and competition, species introduction and fisheries. Of these various factors, it is concluded that lake level fluctuations and associated nutrient inputs probably have the strongest driving role with simultaneous effect on several of the other factors. The introduction of the pelagic kapenta does not seem to have affected the inshore production and, from the present data, there are no immediate signs of overfishing.

ACKNOWLEDGEMENTS

The fishery data used in this study were collected by Lake Kariba Fisheries Research Institute's staff who deserve to be sincerely thanked, particularly Dr Cecil Machena for letting us have access to the database and for the use of the Institute's facilities. The collaboration between the two authors was a result of the Zambia/Zimbabwe SADC Fisheries Project sponsored by NORAD and DANIDA, and we are indebted to these two institutions. This chapter is submitted with the approval of the Department of National Parks and Wildlife Management, Zimbabwe, but the views expressed are the responsibility of the authors.

REFERENCES

Anon. (1992) Working group on assessment of Kapenta (*Limnothrissa miodon*) in Lake Kariba (Zambia and Zimbabwe). Zambia/Zimbabwe SADC Fisheries Project Report, no. 11. Lake Kariba Fisheries Research Institute, Kariba, 76 pp. (Mimeo)

Anon. (1993) Working group on assessment of the inshore fisheries stocks of Lake Kariba. Zambia/Zimbabwe SADC Fisheries Project Report, no. 26. Kariba 13–22/10 1993. Lake Kariba Fisheries Research Institute, Kariba, 110 pp. (Mimeo)

Balogun, J.K. (1986) Fish distribution in Kainji Lake, Nigeria. *J. Fish Biol.*, **29**, 489–98.

Balon, E.K. (1974a) *Fishes of Lake Kariba, Africa*, T.F.H. Publications, Neptune City, NJ, 144 pp.

Balon, E.K. (1974b) Fish production of a tropical ecosystem, in *Lake Kariba: A Man-made Tropical Ecosystem in Central Africa* (eds E.K. Balon and A.G. Coche) (*Monographiae Biologicae* **24**), Dr. W. Junk, The Hague, pp. 249–573.

Balon, E.K. and Coche, A.G. (eds) (1974) *Lake Kariba: A Man-made Tropical Ecosystem in Central Africa (Monographiae Biologicae* **24**), Dr. W. Junk, The Hague, 766 pp.

Begg, G.W. (1970) Limnological observations on Lake Kariba during 1967 with emphasis on some special features. *Limnol. Oceanogr.*, **15**, 776–88.

Begg, G.W. (1974) The distribution of fish of riverine origin in relation to the limnological characteristics of the five basins of Lake Kariba. *Hydrobiologia*, **44**, 277–86.

Begg, G.W. (1976) The relationship between the diurnal movements of some of the zooplankton and the sardine *Limnothrissa miodon* in Lake Kariba. *Limnol. Oceanogr.*, **21**, 529–39.

Bell-Cross, G. and Bell-Cross, B. (1971) Introduction of *Limnothrissa miodon* and *Limnocaridina tanganicae* from Lake Tanganyika into Lake Kariba. *Fish. Res. Bull. Zambia*, **5**, 207–14.

Bell-Cross, G. and Minshull, J.L. (1988) *The Fishes of Zimbabwe*, National Museums and Monuments of Zimbabwe, Harare, Zimbabwe, 291 pp.

Bowmaker, A.P. (1973a) An hydrobiological study of the Mwenda river and its mouth. PhD thesis, University of Witwatersrand, Johannesburg, 576 pp.

Bowmaker, A.P. (1973b) Hydrophyte dynamics in Mwenda Bay Lake Kariba. National Museums and Monuments of Zimbabwe, *Kariba Stud.*, **3**, 43–58.

Burne, R.H. (1971) A comparative study of the diets of *Labeo congoro*, *Labeo altivelis* and *Tilapia mossambica* in the Sanyati Basin of Lake Kariba. Lake Kariba Fish. Res. Inst. Proj. Rep., no. 12. (Mimeo)

Caddy, J.F. and Sharp, G.D. (1986) An ecological framework for marine fishery investigations. *FAO Fish. tech. Pap.*, no. 283, 151 pp.

Cochrane, K.L. (1976) Catches of *Hydrocynus vittatus* Castelnau during sardine fishing operations in Lake Kariba. National Museums and Monuments of Zimbabwe, *Kariba Stud.*, **7**, 98–108.

Coke, M. (1968) Depth distribution of fish on a bush cleared area of Lake Kariba, Central Africa. *Trans. Am. Fish. Soc.*, **97**, 460–5.

Donnelly, B.G. (1969) A preliminary survey of Tilapia nurseries on Lake Kariba during 1967–68. *Hydrobiologia*, **34**, 195–206.

Donnelly, B.G. (1970) The fish population changes between 1960 and 1968: Part I, Cichlidae. Lake Kariba Fish. Res. Inst. Proj. Rep. (Mimeo)

Donnelly, B.G. (1971) The fish population changes on Lake Kariba between 1960 and 1968: Part II, Characidae and Citharinidae. Lake Kariba Fish. Res. Inst. Proj. Rep. (Mimeo)

Gulland, J.A. (1983) *Fish Stock Assessment – A Manual of Basic Methods* (FAO/Wiley Series on Food and Agriculture, Vol. 1), John Wiley & Sons, Chichester, 223 pp.

Hutton, J.M. (1991) A bibliography of Lake Kariba and related topics, Paper VI *in* Lake Kariba, Zimbabwe, A background to planning. Zambia/Zimbabwe SADC Fisheries Project Rep., no. 10. Lake Kariba Fisheries Research Institute, Kariba, 36 pp. (Mimeo)

Jackson, P.B.N. (1961) Ichthyology. The fish of the Middle Zambezi. National Museums and Monuments of Zimbabwe, *Kariba Stud.*, **1**, 1–36.

Joubert, C.S.W. (1975) The food and feeding habits of *Mormyrops deliciosus* Leach 1818 and *Mormyrus longirostris* Peters 1852 Pisces: Mormyridae in Lake Kariba, Rhodesia, National Museums and Monuments of Zimbabwe, *Kariba Stud.*, **5**, 68–85.

Jubb, R.A. (1976) Unintentional introductions of fishes via hydro-electrical power-stations and centrifugal pumps. *J. Limnol. Soc. sthn Afr.*, **2**, 29–30.

Junor, F.J.R. and Begg, G.W. (1971) A note on the successful introduction of *Limnothrissa miodon* the Lake Tanganyika sardine to Lake Kariba. *Newsl. Limnol. Soc. Sth Afr.*, **16**, 8–14.

Karenge, L.P. (1992) Inshore fish population changes at Lakeside, Kariba, between 1969–1991. MPhil thesis, University of Bergen, 54 pp.

Karenge, L.P. and Kolding, J. (1995) On the relationship between hydrology and fisheries in Lake Kariba, Central Africa. *Fish Res.*, **22**, 205–26.

Kenmuir, D.H.S. (1973) The ecology of the tigerfish, *Hydrocynus vittatus* Castelnau, in Lake Kariba. *Occ. Pap. natn. Mus. Rhod.*, **B5**(3), 115–70.

Kenmuir, D.H.S. (1984) Fish population changes in the Sanyati basin, Lake Kariba, Zimbabwe. *S. Afr. J. Zool.*, **19**, 194–209.

Kolding, J. (1989) The fish resources of Lake Turkana and their environment – thesis for the Cand. Scient degree in fisheries biology and final report of KEN 043 trial fishery 1986–1987, University of Bergen, 262 pp.

Kolding, J. (1992) A summary of Lake Turkana: an ever-changing mixed environment. *Mitt. Int. Verein. Limnol.*, **23**, 25–35.

Kolding, J. (1993) Population dynamics and life history styles of Nile tilapia (*Oreochromis niloticus*) in Ferguson's Gulf, Lake Turkana, Kenya. *Env. Biol. Fishes*, **37**, 25–46.

Langerman, J.D. (1984) Optimum harvest strategies for tigerfish in Lake Kariba. MSc thesis, University of Witwatersrand, Johannesburg, 186 pp.

Lelek, A. (1973) Sequence of changes in fish populations of the new tropical man-made lake, Kainji, Nigeria, W. Africa. *Arch. Hydrobiol.*, **71**, 381–420.

Lewis, D.S.C. (1974) The effects of the formation of Lake Kainji (Nigeria) upon the indigenous fish population. *Hydrobiologia*, **45**, 281–301.

Lowe-McConnell, R.H. (1973) Summary: reservoirs in relation to man – fisheries, in *Man-made Lakes: Their Problems and Environmental Effects* (eds W.C. Ackermann, G.F. White and E.B. Worthington) (Geophysical Monographs **17**), Am. Geophys. Union, Washington, DC, pp. 641–54.

Machena, C. (1988) Predator–prey relationships, fisheries productivity and fish population dynamcs in Lake Kariba – a review, in Predator–prey relationships, population dynamics and fisheries productivities of large African lakes (ed. D. Lewis). *CIFA Occ. Pap.*, **15**, 26–44.

Machena, C. (1989) Ecology of the hydrolittoral macrophyte communities in Lake Kariba, Zimbabwe. PhD thesis, University of Uppsala.

Machena, C., Kolding, J. and Sanyanga, R.A. (1993) A preliminary assessment of the trophic structure of Lake Kariba – a man-made Lake, Zimbabwe, *Trophic Models of Aquatic Ecosystems (ICLARM Conf. Proc.*, no. 26) (eds V. Christensen and D. Pauly), ICLARM, Manila, pp. 130–37.

McLachlan, A.J. (1970) Some effects of annual fluctuations in water level on the larval chironomid communities of Lake Kariba. *J. Anim. Ecol.*, **39**, 19–90.

McLachlan, A.J. (1974) Development of some lake ecosystems in Tropical Africa with special reference to the invertebrates. *Biol. Rev.*, **49**, 365–97.

Marshall, B.E. (1977) On the status of *Clarias gariepinus* (Burchell) in large man-made lakes in Rhodesia. *J. Limnol. Soc. Sth Afr.*, **3**, 67–8.

Marshall, B.E. (1979) A bibliography of Lake Kariba 1969–1978. Lake Kariba Fish. Res. Proj. Rep. (Mimeo)

Marshall, B.E. (1982) The influence of river flow on pelagic sardine catches in Lake Kariba. *J. Fish Biol.*, **20**, 465–70.

Marshall, B.E. (1984) Kariba (Zimbabwe/Zambia) in Status of African reservoir fisheries (eds J.M. Kapetsky and T. Petr) *CIFA tech. Pap.*, **10**, 105–53.

Marshall, B.E. (1985) A study of population dynamics, production and potential yield of the sardine, *Limnothrissa miodon* Boulenger in Lake Kariba. PhD thesis. Rhodes University, Grahamstown, 219 pp.

Marshall, B.E. (1987) Research note on the relationship between tigerfish and sardines in Lake Kariba. *J. Limnol. Soc. Sth Afr.*, **132**, 78–9.

Marshall, B.E. (1992) The relationship between catch and effort and its effect on predicted yields of sardine fisheries in Lake Kariba. *NAGA, The ICLARM Quarterly*, Oct. 1992, 36–9.

Marshall, B.E. (1993) Biology of the African clupeid *Limnothrissa miodon* with reference to its small size in artificial lakes. *Rev. Fish Biol. Fisheries*, **3**, 17–38.

Marshall, B.E. and Junor, F.J.R. (1981) The decline of *Salvinia molesta* on Lake Kariba. *Hydrobiologia*, **83**, 477–84.

Marshall, B.E., Junor, F.J.R. and Langerman, J.D. (1982) Fisheries and fish production on the Zimbabwean side of Lake Kariba. National Museums and Monuments of Zimbabwe, *Kariba Stud.*, **10**, 175–271.

Mitchell, D.S. (1973) Supply of plant nutrient chemicals in Lake Kariba, in *Man-Made Lakes: Their Problems and Environmental Effects* (Geophysical Monographs **17**), (eds W.C. Ackermann, G.F. White and E.B. Worthington), Am. Geophys. Union, Washington, DC, pp. 165–9.

Mitchell, S.A. (1976) The marginal fish fauna of Lake Kariba. National Museums and Monuments of Zimbabwe, *Kariba Stud.*, **8**, 109–62.

Mitchell, S.A. (1978) The diurnal activity patterns and depth zonation of marginal fish as shown by gill-netting in Lake Kariba. National Museums and Monuments of Zimbabwe, *Kariba Stud.*, **9**, 163–73.

Moss, B. (1988) *Ecology of Fresh Waters, Man and Medium*, 2nd edn, Blackwell Scientific Publications, Oxford, 417 pp.

Moyo, N.A.G. (1986) Changes in the cichlid population abundance in the C2 area of Lake Kariba. *Zimbabwe agric. J.*, **86**, 171–4.

Petr, T. (1967) Fish population changes in the Volta Lake in Ghana during its first sixteen months. *Hydrobiologia*, **30**, 193–200.

Petr, T. (1978) Tropical man-made lakes – their ecological impact. *Arch. Hydrobiol.*, **81**, 368–85.

Pinkas, L., Oliphant, M.S. and Iverson, I.L.K. (1971) Food habits of albacore, bluefin tuna and bonito in Californian waters. *Fish Bull. Calif. Dep. Fish Game*, **152**, 1–105.

Pitcher, T.J. and Hart, P.J.B. (1982) *Fisheries Ecology*, Croom Helm, London, 414 pp.

Pitcher, T.J. and Bundy, A. (1994) Successful species introductions in the African Lakes: assessment, uncertainties and strategies for fisheries management, in *Management Strategies for Exploited Fish Populations* (eds G. Kruse, D.M. Eggers, R.J. Marasco, C. Pautzke and T.J. Quinn), Alaska Sea Grant College Program 93-02, pp. 545–70.

Sanyanga, R.A. (1988) Bibliography of Lake Kariba (1979–1987), published works and project reports. Lake Kariba Fish. Res. Inst. Proj. Rep., no. 57, 7 pp. (Mimeo)

Sanyanga, R.A., Jokonya, P. and Chimanga, R. (1992) 1991 fisheries statistics Lake Kariba – Zimbabwe shore. Lake Kariba Fish. Res. Inst. Proj. Rep., no. 70. 21 pp. (Mimeo)

Sanyanga, R.A., Jokonya, P. and Chimanga, R. (1993) 1992 fisheries statistics Lake Kariba – Zimbabwe shore. Lake Kariba Fish. Res. Inst. Proj. Rep., no. 73. 21 pp. (Mimeo)

Chapter thirteen

The impact of an introduction of sardine into Lake Kivu

Hans H. de Iongh, Petronella C. Spliethoff and Fritz Roest

13.1 INTRODUCTION

To a biologist the interesting aspects of Lake Kivu are its ecological immaturity and the relative poverty of its fauna. The fish fauna is practically unrelated to that of Lake Tanganyika and there are but few similarities to Lake Edward. There are only 16 species in the lake itself, belonging to the families Cyprinidae, Clariidae, and as a result of recent introduction, Clupeidae. As an ecosystem the lake has low species diversity and production so that traditional fishing activities were at a low level. The pelagic niche, occupied in Lake Tanganyika by the sardines *Stolothrissa tanganicae* and *Limnothrissa miodon*, was not occupied in Lake Kivu and so Verbeke (1957b) suggested the introduction of a suitable species like the Tanganyika sardines to fill this niche in Lake Kivu, in the hope that more organic matter would thereby be converted into edible fish and less lost in the mud.

In the period 1958–1960, thousands of sardines were transported from Lake Tanganyika to Lake Kivu (Collart, 1960). In July 1974 the presence of *Limnothrissa* in the southern part of the lake was demonstrated by Collart and in 1976 an expedition recorded it over the whole lake (Frank *et al.*, 1977). *Stolothrissa* was never encountered in experimental catches, so it was assumed that this species had not adapted to the specific conditions of the lake. However, it remains uncertain whether *Stolothrissa* was effectively introduced, because the fry transported from Lake Tanganyika were not identified to species level.

The Impact of Species Changes in African Lakes. Edited by Tony J. Pitcher and Paul J.B. Hart. Published in 1995 by Chapman & Hall, London. ISBN 0 412 55050 4.

Based on the promising results of the 1976 survey, an artisanal fisheries development programme was initiated in 1980 by FAO/UNDP (United Nations Development Program). Food habits, growth, reproduction and migration patterns of the newly introduced clupeid were studied in relation to limnological conditions. At a later stage, acoustic surveys were carried out to assess the biomass distribution and the maximum sustainable yield of *Limnothrissa*. This chapter summarizes the results of the biological studies and evaluates the socio-economic impact of the artisanal fisheries for *L. miodon* which have developed around the lake. Particular attention is given to food preferences and feeding habits of different length classes of *L. miodon* in the lake and the possible impact of clupeid introductions on the lake ecosystem.

13.2 CHARACTERISTICS OF LAKE KIVU

Geographical characteristics

Situated in the western Rift Valley, between latitudes 1°40′S (North Basin) and 2°30′S (Bukavu Bay), Lake Kivu forms a natural border of more than 100 km between Zaire to the west and Rwanda to the east (Fig. 13.1).

Since the Pleistocene (20 000 years ago), during which volcanic eruptions gave rise to the formation of a new volcanic range, the waters of Lake Kivu that used to flow north towards Lake Albert have been blocked. After a considerable rise in the water level, the water flow was reversed towards the south into northern Lake Tanganyika through the Ruzizi River. The water surface of Lake Kivu is at an altitude of 1463 m and the lake covers a surface of 2699 km^2 (Welcomme, 1972) of which 42% is Rwandan territorial waters. The lake has a mean depth of 285 m (Verbeke, 1957a) with a maximum of 485 m in the north. The littoral zone is very narrow because the shores are steep. The littoral zone, however, has a length of 1200 km, as a consequence of the sinuous shoreline. The pelagic zone is the most important and almost touches the lake sides (Beadle, 1974).

Limnological characteristics

The temperature of surface waters shows small seasonal fluctuations between 23.1 °C and 24.1 °C. The temperature gradually decreases with water depth (23 °C at 40 m, 22.1 °C at 70 m). Below 100 m the water is rich in salts and because of its density it maintains a stable and permanent thermal stratification (Verbeke, 1957a). Regular winds blow south-east to east, crossing the lake in an oblique or lateral way. During the night, winds coming down from the volcanoes modify this pattern until about 07.00 or 08.00. The upward

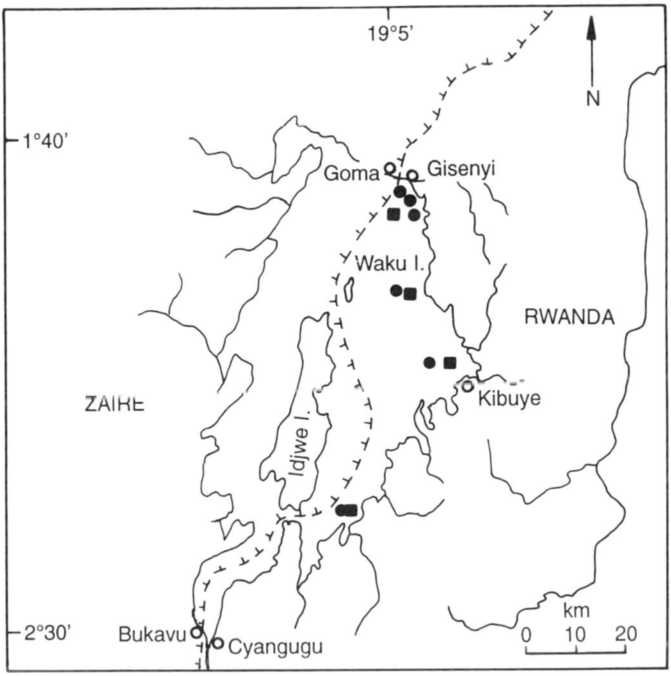

Fig. 13.1 Map of Lake Kivu, with locations of biological sampling stations (●) and physical/chemical sampling stations, including echosounding transects (■).

recycling of nutrients from the lower water into the euphotic zone only occurs to a certain degree at the beginning of the wet season. The transparency of the water varies from 2.5 m to 9.35 m. These figures are lower than those of Lake Tanganyika (maximum of 22 m; Chapman, 1975) and may affect the efficiency of night fishing with lamps (less penetration of the light in the water and less attraction of the fishes).

The salinity of the water is high (1‰) and has a correspondingly high concentration of bicarbonate, which is continuously precipitated as carbonate. As a result, a calcareous deposit is formed all along the shoreline, covered with the calcareous precipitate and dense mats of filamentous *Cladophora* (Beadle, 1974). The pH of the inflowing water varies around 7.5. The pH of the surface waters and the epilimnion ranges from 8.1 to 9.2. Below 50 m the water is slightly acid, probably due to high concentrations of carbon dioxide (Van Meel, 1954). Most members of the invertebrate fauna, such as Nematoda, Turbellaria, Cladocera, Ostracoda, Insecta and Hydracarina, are limited to this zone (Verbeke, 1957a). The quantities of dissolved oxygen in the upper layers (20 m) of the lake water vary little (from 6 to 7.5 $mg\,l^{-1}$ or 85% to 110% saturation) throughout the year (Verbeke, 1957a). The oxygen content declines after 20 m depth to 3 $mg\,l^{-1}$ at 50 m. Below 70 m (thermocline), the water is anaerobic.

According to the studies of Degens *et al.* (ex Beadle, 1974) and Jannash (quoted by Frank *et al.*, 1977), it appears that the primary production (240 to 540 $gC\,m^{-2}\,year^{-1}$ of gross photosynthesis) is relatively low but similar in magnitude to that of other African lakes of the rift. Large seasonal fluctuations in primary production occur as a result of the recirculation of nutrients from the mixolimnion. The phytoplankton biomass in Lake Kivu is the same range as that in Lake Tanganyika (Hecky *et al.*, 1977) or slightly lower in deeper parts because of the more severe nutrient limitation in Lake Kivu.

Biological characteristics

Until 1980 fewer than 20 species of fishes had been recorded in Lake Kivu. This poverty can be explained by the relatively recent formation of the lake in its present form, and as a result of the violent volcanic activities which occurred in the area. Another source of faunal poverty is the waterfalls on the outflowing Ruzizi River, which prevent most fishes from migrating up from Lake Tanganyika. Only one fish species from this lake (*Barilius moorii*) has succeeded in colonizing Lake Kivu.

According to Verbeke (1957a), several niches in the ecosystem of Lake Kivu are not occupied. For example, there were no plankton-feeding fish in the lake. He added that when the large biomass of zooplankton living in the pelagic waters die, they fall to the sediments, where their decomposition enriches the waters in mineral salts and gas. For zooplankton, the volume collected per unit of surface was only slightly less than that in Lake Edward and fluctuated with variation in phytoplankton production (Verbeke, 1957a). This niche is occupied by the small sardines *Limnothrissa miodon* and *Stolothrissa tanganicae*, and the cyprinid *Engraulicypris minutus* in Lake Tanganyika, by *Aplocheilichthys* spp. and *Engraulicypris* spp. in Lake Mobutu Sese Seko (Albert) *Rastrineobola* in Lake Victoria, and finally by *Alestes baremose* in most Sudanese waters (Beadle, 1974). In Lake Kivu none of these species was present. Echosoundings and fishing by means of explosives during the exploratory survey of 1952–1953 did not reveal any fish in pelagic waters.

Because of the absence of limnetic predators it is assumed that the fluctuations in the biomass of the clupeids after successful introduction into Lake Kivu would be largely determined by variations in the production of plankton (Verbeke, 1957a).

13.3 BIOLOGY OF *LIMNOTHRISSA*

Growth

Because *Limnothrissa* has no fixed spawning time, length-frequency distributions are complex. The clupeid is characterized by relatively fast growth and a maximum life span of 2 years. Growth is fastest during the first 6 months (20 mm month^{-1}), followed by an asymptotic growth averaging 10 mm month^{-1}.

The rate of growth during early life is unknown because juveniles only appeared in mosquito net samples as fingerlings from 11 mm upwards and eggs and yolk-sac fry were never sampled. Sardines larger than 105 mm were only occasionally caught by the fishermen, probably because they are too fast to be caught by the lift nets, but they often appeared in gill nets near the shores. The growth curve of the Kivu sardine is similar to the growth curve in Lake Tanganyika, but quite different from the ones obtained for Lake Kariba and Cahora Bassa (Manini, 1990; Tables 13.1, 13.2; but see Chapter 24). The asymptotic maximum length (L_∞) for the Kivu clupeid is on average 20 mm less than that for Lake Tanganyika, while the k value shows strong fluctuations in both lakes. It is generally assumed that differences in growth rate, L_∞ and reproductive activity can be attributed to differences in ecological conditions. Limnological studies show that ecological conditions in Lakes Kivu and Tanganyika are similar (Hecky *et al.*, 1977), but ecological conditions in the artificial lakes Cahora Bassa and Kariba are more unstable due to large fluctuations in water inflow and water level (Chapter 12). On the other hand, due to a greater nutrient influx, production levels of sardines are expected to be higher in these artificial reservoirs (Chapter 23).

Maturity

In Lake Tanganyika, Ellis (1971) found an average length at first maturity (50% of the gonads in stage II) of 64 mm fork length (FL) for the male fish and 75 mm FL for females. In Lake Kivu, Spliethoff *et al.* (1983) concluded

Table 13.1 Parameters of the von Bertalanffy growth equation for *L. miodon* in Lake Kivu

Year	L_∞ (mm)	k	t_0
1980	149	1.30	−0.142
1983	154	1.31	−0.144
1986	138	1.32	−0.147
1989	144	1.28	−0.149

Table 13.2 Growth parameters of *L. miodon* from different lakes. From Manini (1990)

Lake	L_∞ (mm)	k	Source
Kivu	144	1.28	Manini (1990)
Kivu	138	1.32	Manini (1990)
Kivu	154	1.31	Manini (1990)
Kivu	149	1.30	Manini (1990)
Kivu	145	1.20	Spliethoff *et al.* (1983)
Kivu	105	2.33	Kaningini (1990)
Kivu	106	2.26	Kaningini (1990)
Tanganyika	115	2.16	Matthes (1965–66)
Tanganyika	164	0.96	Pearce (1988)
Tanganyika	160	1.15	Munyandorebo (1989)
Tanganyika	172	1.31	Munyandorebo (1989)
Tanganyika	172	1.15	Munyandorebo (1989)
Kariba	74	3.05	Marshall (1987)
Cahora Bassa	70	5.40	Gliwicz (1984)

that the length at first maturity for male fish is 60 mm and for females 62 mm FL. The sex ratio (males:females) is constant from 75 mm onwards (sex ratio = 0.35, SD = 0.33). No relationship could be found between sex ratio and catchment area.

Although Matthes (1968) stated that spawning takes place in the inshore zone, this could not be confirmed in Lake Kivu by Spliethoff *et al.* (1983). The percentage of stage III fish inshore was no higher than in the offshore catches. Figure 13.2 shows the monthly percentages of clupeids in maturity stage III during the year. Although spawning is not restricted to any one

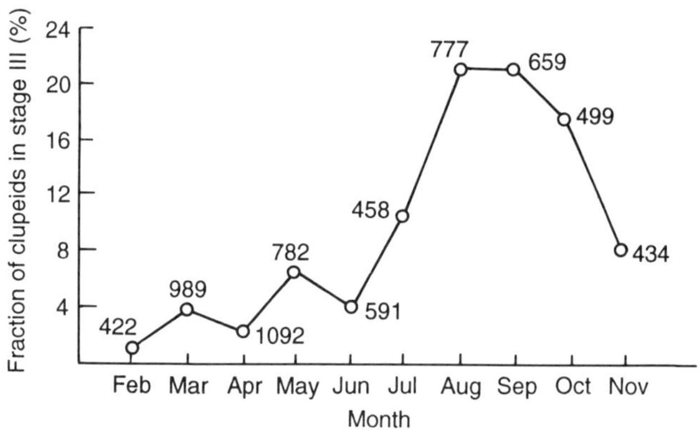

Fig. 13.2 Monthly percentages of clupeids in maturity stage III during 1980; numbers denote sample size. From Spliethoff *et al.* (1983).

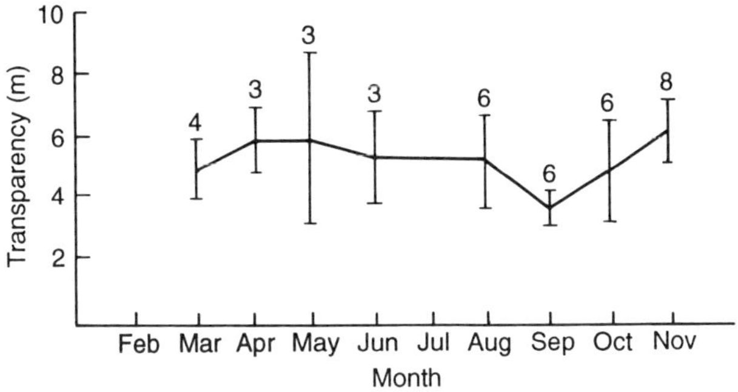

Fig. 13.3 Average monthly transparency figures during 1980. Vertical bars indicate standard deviation; numbers above bars indicate sample size. From de Iongh *et al.* (1983).

period, peaks in sexual activity can be seen during August–October during seasonal plankton blooms and less strong activity can be seen during February–May. Matthes (1968) concluded that *Limnothrissa* in Lake Tanganyika had an extended breeding season, with breeding peaks during May–June and September–December. From the data of Verbeke (1957c) for Lake Kivu, a peak in zooplankton production occurred in August/September. Secchi disc readings during 1980 also showed a non-significant decrease in visibility in September (Fig. 13.3). This plankton bloom occurs at the end of the dry season (Fig. 13.4) when winds caused a partial recirculation of the deeper, nutrient-rich layers. This plankton bloom corresponds with the sardine's breeding peak. Planktonic food availability thus seems to be a predominant stimulus for reproduction. This is confirmed by observations on fat content during the year, as the percentage of sardines with fat in their ventral cavity

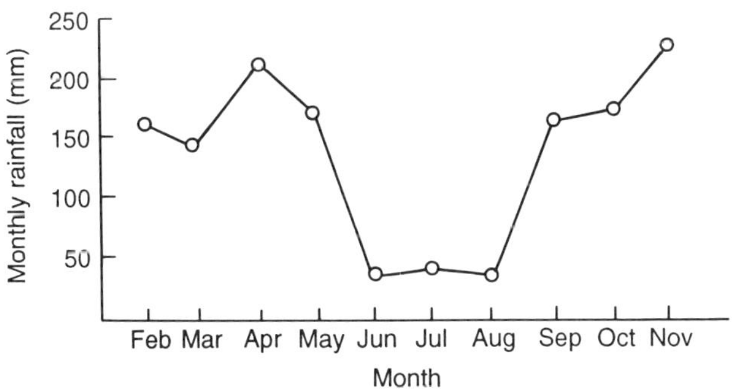

Fig. 13.4 Monthly rainfall at Gisenyi airport during 1980. Source: meteorological station, Gisenyi airport (from de Iongh *et al.*, 1983).

falls to a minimum during June–September. Because of the availability of enormous amounts of plankton in September, their condition is soon restored and reaches a high level by the end of November (Spliethoff *et al.*, 1983).

Condition factor, *K*

The condition factor *K* was calculated by Spliethoff *et al.* (1983) for clupeids longer than 50 mm. Length groups were discerned with clearly different condition factors (Fig. 13.5). The equation used for calculating the condition factor *K* is: $K = P \times 10^5/L^3$ where *P* is wet weight in g and *L* is fork length in mm.

Fish smaller than 40 mm could not be weighed but fingerlings of about 40 mm showed a low condition, possibly caused by feeding difficulties. When the fingerlings grow, *K* shows an upward trend (Fig. 13.5).

When the fish reach the length of maturity their condition factor is about 0.95. At that period their fat content is high and the circumstances are favourable for gonad development. When they start spawning at 80 mm, energy reserves are used so that body mass is reduced. The condition factor then drops to an average of 0.77 and rises again after they attain 107 mm in length. This might be attributed to a change in diet; larger specimens tend to become piscivorous, predating their own offspring in the littoral zone.

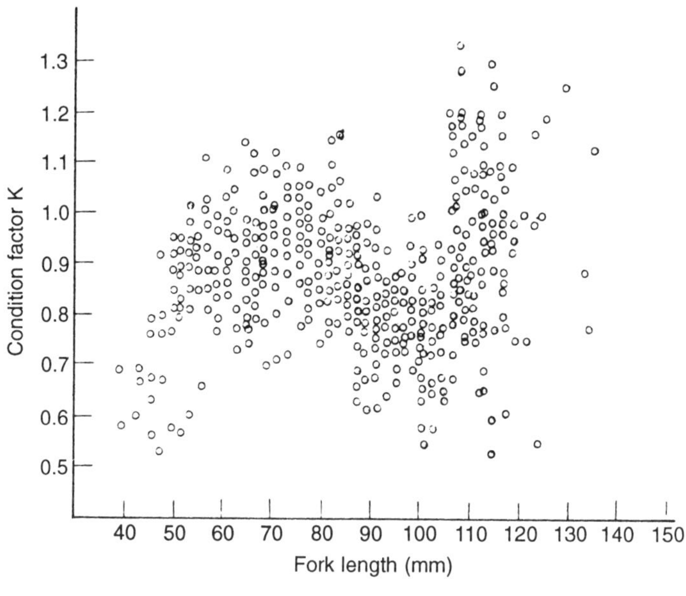

Fig. 13.5 Relationship between condition factor *K* and fork length of clupeids sampled from artisanal catches during 1980. From Spliethoff *et al.*, 1983).

Larger clupeids can be found at maturity stage III all year round and it is reasonable to assume that spawning is spread over long periods in sardines over 107 mm. Because of this, variation in the condition factor is rather large in comparison with that in Lake Tanganyika.

Length–weight relation

Lays (1979) found in his study of Lake Kivu during the first half of 1979 a length–weight relation for *Limnothrissa* of

$$W = 0.000\,002\,9L^{2.7} \qquad (n = 805) \qquad (13.1)$$

where W is weight, L is fork length and n is the number of samples.

A comparison of the data from Lays (1979) with those of Spliethoff *et al.* (1983) shows that fish condition in Lake Kivu, as revealed by the length–weight exponent b = 2.71, is slightly lower than in Lake Tanganyika, where b = 2.98 for fish 65–145 mm (Mann, 1975), but the same as fish from Lake Kariba, where b = 2.37 (Cochrane, 1978). Based on the relation between condition factor K and fork length L as shown in Fig. 13.5, the division into three length classes for the calculation of different length–weight relationships seemed appropriate: I, 40–80 mm (b = 3.5); II, 80–105 mm (b = 2.7); III, >105 mm (b = 3.2). The condition factor seems to correspond with the data obtained on growth rate and the length–weight relation. According to these calculations, the division into three length classes corresponds with different stages in the life cycle of the sardine. Until the clupeids reach the reproductive stage at a length of 80 mm, their growth is fast (20 mm month^{-1}) and their condition factor is high as shown by b = 3.5. The low value of b = 2.7 in the second length group (80–105 mm) corresponds with the decline in condition and a growth rate that has been reduced. Once they reach a length of 107 mm the sardines tend to become more piscivorous and the value of b increases to 3.2 from 80 mm onwards.

Migration and feeding

Spliethoff *et al.* (1983) also provided a general picture of sardine offshore migration in Lake Kivu (Fig. 13.6). Fingerlings of 11–40 mm are found along the coast at 1–1.5 m depth. From a study on the feeding habits of *Limnothrissa* (de Iongh *et al.*, 1983) it was observed that the stomachs of these inshore fingerlings were filled with Copepoda, Cyanophyta and Rotatoria, which corresponded to the species composition of inshore plankton samples. From 28 mm onwards the clupeids start to migrate to deeper waters and from a length of 65 mm they can be caught 100 m or more offshore. The stomachs of these offshore fish were filled with Copepoda and Cyanophyta in similar

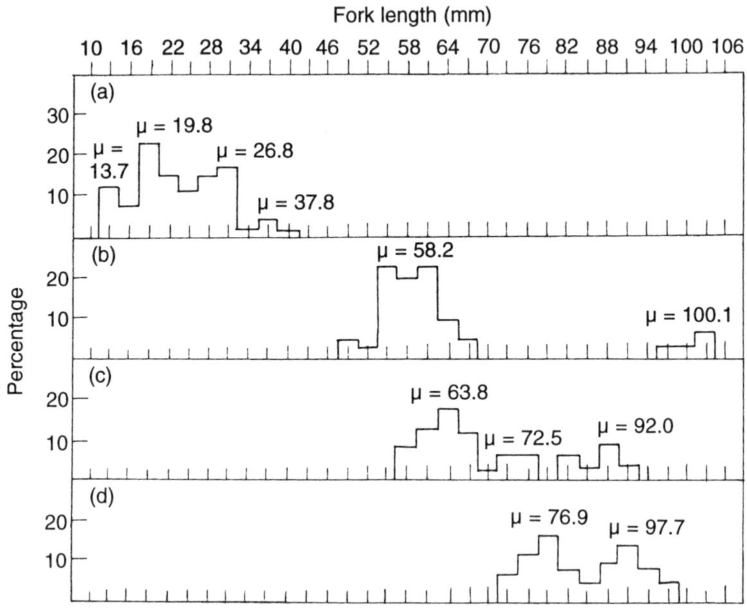

Fig. 13.6 Length distributions of clupeids sampled at different distances from the shoreline: (a) 5 m ($n = 511$); (b) 50 m ($n = 456$); (c) 100 m ($n = 371$); (d) 2560 m ($n = 323$). Samples taken on 13/14 August 1980. From Spliethoff *et al.* (1983).

percentages to those in pelagic plankton samples. Fish with a length of 80–110 mm can be found in all zones, except the lake margins (Fig. 13.6). Stomach examinations of this length class revealed, besides copepods and *Microcystis*, a relatively large fraction of chironomid pupae and juvenile sardines. Analysis of the stomach contents of sardines larger than 110 mm showed cannibalistic feeding behaviour and as the distribution of these fish does not show a preference for a special zone, it is difficult to say what happens after the spawning inshore. The hypothesis is that migration patterns for sardines longer than 110 mm depend on the availability of food. Fishing methods other than lift netting might reveal more about the life cycle of *Limnothrissa*.

Food availability and periodicity

Several studies were conducted to examine the relation between plankton production, migration pattern and reproductive activities of the clupeids. Data obtained so far support the hypothesis that the biology of the clupeids is largely determined by fluctuations in plankton production. Secchi disc readings were used as a measure of plankton density in the pelagic sampling stations. It is appreciated that some bias could occur, especially in the littoral stations, due to the presence of organic particles. Data on transparency during 1980 at pelagic sampling stations are presented in Fig. 13.3. A minimum was reached during September, which coincided with a plankton bloom during this period. This plankton bloom occurs at the end of the dry season, when strong winds stir up the waters of Lake Kivu and cause a turnover. Verbeke (1957a) also reported a similar increase in plankton biomass during the windy period of August–September, presumably for similar reasons.

During October and November, de Iongh *et al.* (1983) described plankton samples taken from littoral and pelagic sample stations in the Bay of Gisenyi (Fig. 13.1). In both littoral and pelagic samples, Copepoda and Cyanophyta together accounted for more than 50% of the abundance (Table 13.3). In the littoral samples, Copepoda were relatively less abundant whilst Rotatoria were more abundant than in pelagic samples.

The predominant copepod species was *Tropocyclops confinis* (Kiefer). Less abundant were *Thermocyclops consimilis* (Kiefer) and *Mesocyclops* (non *leuckarti* (Claus) (identified by Dr I. van de Velde, Laboratory for Morphology and Taxonomy, University of Ghent, Belgium). Cyanophyta were almost exclusively represented by *Microcystis viridis* but in November a bloom of *Anabaena flos aquae* was observed in a littoral sample. Cladocera, such as *Ciaphanosoma excisum* and *Moina micrura*, were encountered in low numbers, mainly in pelagic samples. The species composition of the Copepoda is similar to that observed by a Belgian survey in 1952–1954 (Verbeke, 1957a). During this period *Tropocyclops confinis* and *Mesocyclops consimilis* (non *leuckarti*) were less abundant.

De Iongh *et al.* (1983) also reported that analyses of the results of chemical and physical sampling supported findings concerning the seasonality of plankton abundance in the lake. NO_3–N and PO_4–P especially showed peaks

Table 13.3 Relative abundance (%) of different groups of organisms in littoral and pelagic samples taken during 1980. From de Iongh *et al.* (1983)

	Littoral samples			Pelagic samples			
	23 Oct.	4 Nov.	25 Nov.	9 Oct.	23 Oct.	5 Nov.	25 Nov.
Cladocera	0	1	0.5	8	8	4	0
Copepoda	15	32	10	64	64	74	17
Rotatoria	50	16	12	9	9	4	10
Ciliata	0	0	0	0	0	0	5
Cyanophyta	35	41	68	44	15	17	67
Chlorophyta	0	10	5	1	4	1	1
Chrysophyta	0	0	4.5	0	0	0	0

Table 13.4 Percentages of *Limnothrissa* with full stomachs, encountered during different hours of the night in pelagic samples of 13–14 August. From de Iongh *et al.* (1983)

Sample no.	Time	Sample size	% with full stomach
1	22.10	50	64
2	22.30	35	63
3	22.40	36	35
4	24.30	38	13
5	03.00	41	12
6	06.00	54	4

of 3.5 and 3.0 ppm respectively at the pelagic sampling stations during September/October. pH varied between 8.2 and 8.7, which corresponds with the high concentrations of calcium bicarbonate observed by several authors (quoted in Beadle, 1974). Oxygen values ranged between 6.5 and 7.7 $mg\,l^{-1}$ (92–109% saturation when corrected for altitude); no peak values were observed. A maximum temperature of 25.5 °C was registered during the dry season, in July, while a minimum of 24.0 °C occurred during the monsoon period in November.

The results of an experiment to gather information about the periodicity of feeding of pelagic *Limnothrissa* and the digestion rate are summarized in Table 13.4. *Limnothrissa* was sampled at intervals throughout the night and the stomachs examined. Those stomachs which still had a determinable amount of food were classified as 'full', while others were 'not full'. The percentage of 'full' stomachs was highest after sunset and decreased gradually during the night. Obviously no feeding took place during the night, and shortly before sunrise all food was digested.

13.4 ECOLOGICAL IMPACT OF THE SARDINES

Collections of zoo- and phytoplankton were first made in the early 1950s (Kiss, 1959), and the first year-round survey was made by Verbeke (1957c). The most recent collections were taken by Reyntjens (1982). Dumont (1986) presented an overview of the historical data on limnology available for Lake Kivu, and suggested a structural change in species composition and individual size of the zooplankton community.

In 1953, the zooplankton community was composed of large, distinctly pigmented (easily visible to predators) Copepoda and Cladocera (average length 0.6 mm, maximum 1.5 mm) (Dumont, 1986). Using length–weight regression equations, Reyntjens (1982) obtained a mean yearly plankton biomass of 7.5 g dry weight per m^2 for Lake Kivu before the introducton of the clupeids.

By July September 1981, Reyntjens (1982) found that this zooplankton community had changed completely: at least eight rotifer species now occurred, in addition to four small species of Cladocera previously not noted by Verbeke (1957c). The cyclopoid species found earlier were still present, but their average sizes had been nearly halved (from 1.03 to 0.55 mm). The *Daphnia* had entirely disappeared, but there was a great abundance of a ciliate, *Stentor coeruleus*, which is normally a sessile, littoral species. One of the Cladocera, *Alona* cf. *cambouei*, previously known only from the littoral weed-beds of the lake, had now invaded the pelagic zone. This increase in diversity was deceptive when expressed in terms of biomass: the standing crop had dropped to an average of 0.15 g dry weight per m^2, while the average length of plankton organisms had decreased to 0.2 mm, with maxima rarely above 0.5 mm (Reyntjens, 1982). A further sample taken by Dumont from Lake Kivu in 1986 contained no animals longer than 0.2 mm, and those in it had an average length of 0.1 mm and a total biomass of 0.05 g dry weight, per m^2.

The changes can be interpreted as the result of heavy size-selective predation on the plankton by *Limnothrissa*.

Echosounding surveys

Frank (1977) reported the first echosounding trials in Lake Kivu in 1976. By utilizing a Furuno echosounder, he concluded that sardines had colonized the entire lake, both inshore and offshore. During July, August and September 1980, additional echosoundings were performed with a Furuno FE 400 instrument on different transects, opposite Kibuye, Wahu Island and Gisenyi (Fig. 13.1). The transects covered both inshore and offshore waters (de Iongh *et al.*, 1983).

In 1980 clupeids were again present both inshore and in pelagic waters. But between the inshore zone and the offshore part of the lake there was always a 'gap' with low clupeid densities. The echosoundings did not show any significant differences among either vertical distribution or sardine densities in different parts of the lake. This confirms the results of the echosoundings done by Frank (1977).

Continuous diurnal and nocturnal echosoundings, following a transect in the Bay of Gisenyi, contributed to a better picture of the feeding habits of *Limnothrissa* in Lake Kivu. Vertical distribution follows a fixed pattern. In the early morning the clupeids rise and concentrate near the surface in a layer between 0 and 10 m. During daytime the clupeids form dense shoals, which are spread over a wide layer between 10 and 50 m in depth. In the late afternoon the shoals disperse and the clupeids rise to the surface 20 m, where they feed on abundant plankton such as the copepod *Tropocyclops confinis* and the bluegreen alga *Microcystis viridis*. During the night sardines go down to 10–35 m, disperse and digest their food. In the early morning they rise again to the upper water layer for feeding. This pattern is disrupted during full moon, when they are at 0–10 m all night.

Vertical migration late in the afternoon and early in the morning was obvious but horizontal inshore/offshore migration was not observed by echosounding, presumably because this occurred on a longer time scale. Biological analyses of the inshore and offshore catches, however, demonstrated inshore/offshore migration as an integral part of the life cycle of *Limnothrissa*.

These findings are similar to events in Lake Kariba, where sardines were spread over a continuous layer between 7 and 20 m during the night but 10–18 m during the day (Balon and Coche, 1974). Conversely, Mann *et al.* (1973) found an average nocturnal depth of 26 m in Lake Tanganyika and compact shoals of average depth 17.6 m during the day. (Recent echosurveys in Lake Kariba show dense sardine shoals at about 30 m depth during the day, dispersing and rising to 5–10 m at night – *Eds.*)

13.5 THE FISHERY

In October and November 1976 the first trials were made to catch *Limnothrissa* in Lake Kivu. An artisanal light-fishing technique, already commonly in use by fishermen on Lake Tanganyika (Collart, 1954; Frank, 1977) was employed by this nocturnal fishery; the clupeids are attracted by the light from 200 W kerosene lamps which are attached to a catamaran or trimaran fishing unit (Fig. 13.7).

A lift net with a depth of 15 m hangs beneath the boat. The net is lowered to 40 m and left for 2 h. During this period the clupeids are concentrated beneath the boat by the kerosene lamps. The net is then lifted with ropes and beams and the catch is then sorted. By using this technique, eight to ten fishermen can be employed on each unit. Each fishing campaign lasts 20 days, because it is impossible to concentrate the clupeids in the period around full moon.

The first experimental catches in Lake Kivu during October and November 1976 by an artisanal catamaran yielded 833 kg of *Limnothrissa* for one campaign. The average catch per night per unit was 43.8 kg, which was comparable to similar catamaran catches in Lake Tanganyika. Based on this survey it was recommended that the exploitation of the clupeids by artisanal light-fishing should commence.

By December 1980, after some further experiments, six trimarans were still in operation and by 1987 the total number of fishing units had increased to 97 in the Rwandese part of the lake. Sardine catches are recorded

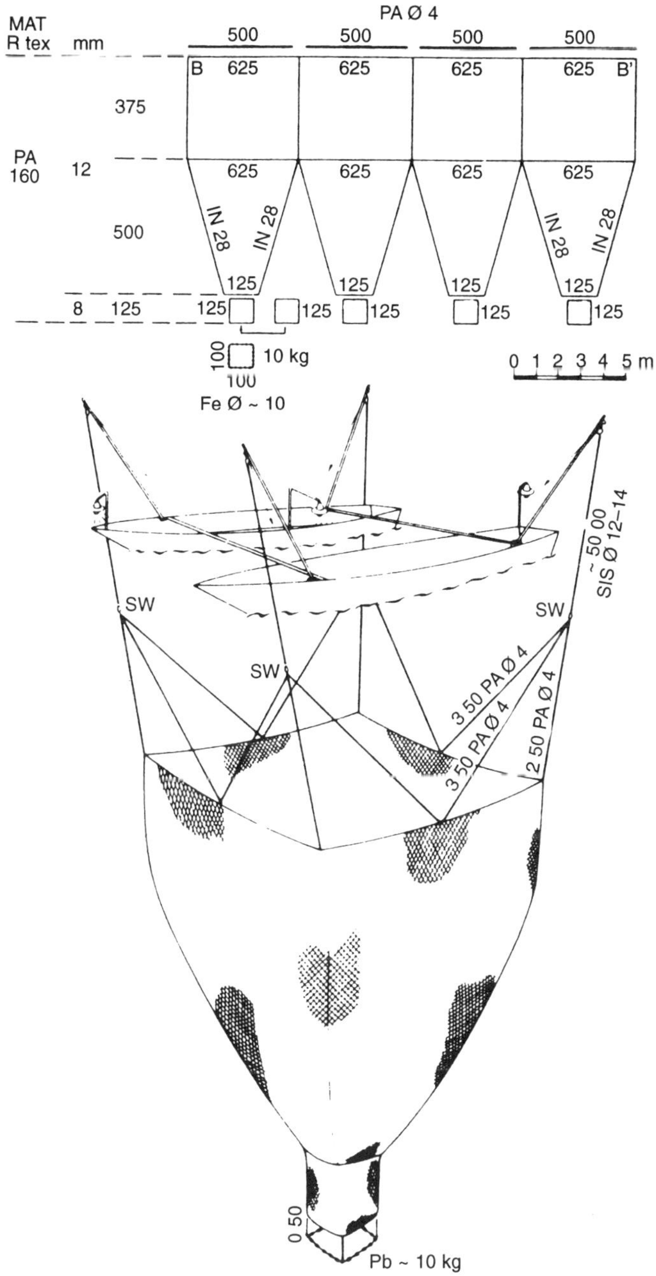

Fig. 13.7 Design drawing of an artisanal catamaran with kerosene lamps and lift net. Reproduced with permission from Nedélec (1975).

throughout the year. The average catches per unit showed a peak value every year during September; minima occurred during July and August, partly caused by bad weather conditions at the end of the dry season.

The average catch per unit per night in Lake Kivu was 43.6 kg but declined from 59 kg in 1980 to 26 kg in 1988, and then increased to 48 kg in 1989. Production increased from 66.5 t in 1981 to 370.1 t in 1987.

Acoustic resource surveys in the lake, carried out by FAO in 1980, 1984–1987 and in 1989, resulted in biomass estimates between 6000 and 10 000 t (tonnes), with an average of 7500 (Johannesson and Lamboeuf, 1987). Further investigations by Johannesson and Lamboeuf (1987) indicated that the clupeid biomass was probably below previous estimates at 3039 t. Based on this new biomass estimate, Johannesson and Lamboeuf (1987) calculated the maximum sustainable yield (MSY) of clupeids in the length class 4–12 cm to be 74% of the total biomass at 2250 t. It has to be stated, however, that this is well below the estimate of 30 000 t by Welcomme (1972) and the 11 300 t estimated by Spliethoff *et al.* (1983) (and see Chapter 23). We think that, as the annual production of clupeids far exceeds their average biomass, catches could probably be much higher, possibly as high as indicated by Spliethoff *et al.* (1983).

In view of the short life span of the clupeids and their high production over short periods, we think it safe to continue and expand the present fishing effort. The collection and analysis of biological data remain, however, important to monitor the impact of fishing on clupeid stocks.

13.6 SOCIO-ECONOMIC BENEFITS OF THE SARDINE FISHERY

Rwanda is one of the most densely populated areas in Africa, with 283 persons per square kilometre. The economy is exclusively based on agriculture or agro-industries.

The socio-economic benefits of the sardine exploitation are twofold: (1) because the percentage of unemployed people is very high, labour-intensive-exploitation and marketing is beneficial; and (2) the fish catch helps to alleviate the lack of protein in Rwanda.

In December 1980, 1 year after the introduction of artisanal light fishing on Lake Kivu, six fishing units were in operation employing 48 fishermen; by 1987 these numbers had increased to 97 fishing units (42 catamarans and 55 trimarans) and well over 800 fishermen. Each unit of eight fishermen elects a group leader, who becomes fishing master and is responsible for boats, lamps, nets and outboard engine given to him on loan by the FAO project.

The human settlement pattern is dispersed and within the extended family there is an explicit sentiment of solidarity based on mutual assistance and communal organization (compare Chapter 11 on Lake Victoria fishers).

Despite their large contribution to agriculture and family affairs, women do not have any decision-making power and are not entitled to inherit land, but they do have an important role in the fishery.

The aim of the FAO project dictated that catches be landed at the fishing centre in Gisenyi, where they were sold by project employees to 'middle women' for a fixed price. These 'middle women' formed a link with local markets and consumers, where the fish is sold for 100–150 Rw. Fr. kg^{-1} ($1–1.5, 1980 rate) if still fresh. Most fish is smoked or dried, to be sold later at 210–250 Rw Fr. ($2.2–2.6, 1980 rate). The income generated by the women is above the average wages for fishermen.

Soon after the start of the FAO project, fish began to be landed and marketed outside the project fishing centre at Gisenyi because prices offered at unofficial landings were higher than the fixed prices offered by the project. Moreover, fish that could not be sold elsewhere could always be disposed of at the official centres. The consequence is that, nowadays, it is assumed that 75% of the catch is marketed away from the official fishing centres.

In the second phase of the project, the responsibility for the credit schemes and the recovery of the loans was handed over to the Rwandese banking system. The conditions set by the bank, however, proved to be too rigid, and were not adapted to fluctuations in catch and income of the fishermen.

Privatization of the supply services such as for boats, lamps, nets and maintenance has not developed satisfactorily. As a result, the official centres are still needed for those supplies. Marketing of fish is restricted to the zone bordering the lake, mainly because the 'middle women' do not have the time or facilities to sell fish far from their homes. Intervention by the project is still needed to market fish in remote areas because people in Rwanda are not used to eating fish and traditional marketing networks do not exist.

Surveys by nutritionists revealed a high percentage of malnutrition amongst the Rwandese population, due to the consumption of nutrient-poor staple foods. Extension and awareness campaigns, to increase knowledge about the nutritional value and composition of the daily diet as well as the contribution of fish (in different forms) to improve protein intake, have been undertaken on a small scale, but still have to be continued on a wider scale. (This need will be even more acute after the disruption caused by the recent war — *Eds.*)

The daily revenues of each fishing unit were recorded. Each month the operating costs (mainly kerosene and petrol) were subtracted from income and the remaining gross profit was divided so that one-quarter went to the fishing master, while one-quarter was retained for the repayment of loans for boats, nets, lamps and engines. The remaining half represented the 'wages' of the seven fishermen. Repayment of credit would take 3–4 years under the prevailing conditions. This period is probably too long when the high rate of depreciation of the outboard engines is taken into consideration. The average yearly income for crew members during 1980 was Rw. Fr. 19 622 ($207, July 1980 rate), while the fishing masters had an average yearly income of Rw. Fr. 68 817 ($724). The crew members' wages are low by Rwandese standards. That they accept this relatively low income can perhaps be explained by the fact that most fishing teams were formed by younger kin (compare traditional Lake Victoria fishers, Chapter 11). The high percentage of unemployment in Rwanda is also an important factor. The motivation of crew members to go fishing seems to grow during the year because of the increasing number of effective fishing nights.

Several experiments were conducted to test more advanced fishing techniques such as purse seining. The introduction of semi-industrial fisheries such as that on Lake Kariba would benefit only a small number of people. For this reason purse seining should not be allowed to compete with the established artisanal lift net fishery. Pelagic purse seining may not influence the artisanal fisheries directly as they are restricted to the coastal zones. But the offshore migration of the sardines means that the two kinds of fisheries will influence each other. Close monitoring of the interaction of purse seining and the artisanal lift net fishing will be necessary before any new method of fishing is introduced.

13.7 SUMMARY

The freshwater clupeid *Limnothrissa miodon* (in Kiswahili and Shona: 'Kapenta'; in Kinjiruwanda: 'ndagala' or 'lumpu') was transplanted successfully from Lake Tanganyika to Lake Kivu during 1958–1960. Surveys conducted in 1974 showed successful colonization and adaptation of *Limnothrissa* to the ecological conditions in Lake Kivu. An FAO project to exploit the clupeid stock started in 1980. Biological research revealed information about growth rate, condition factor, fat content, migration, distribution and reproduction of the sardine, and the plankton and limnology of the lake. Growth rates of *Limnothrissa* in Lake Kivu are similar to those recorded for Lake Tanganyika. The asymptotic maximum length (L_∞) of the Kivu sardines is about 20 mm less than the value for Lake Tanganyika. The length–weight relationship in Lake Kivu indicates that food availability in Lake Kivu seems to extend over a longer period than in Lake Tanganyika. However, a collapse of the clupeid biomass, as predicted by some authors, has not occurred to date. Echosoundings and periodic observations of stomach contents support the hypothesis that *Limnothrissa* feeds in the late afternoon and digests its food during the night, while a second feeding/digestion cycle starts in the early morning at sunrise. Horizontal migration and distribution seem determined by feeding habits.

In the absence of natural predators it seems most likely that sardine biomass is largely determined by fluctuation in plankton production. The main spawning peak coincides with a plankton bloom caused by the turnover of the lake at the end of the dry season in September. At that time a small percentage of clupeids had fat in their body cavity.

Comparison of plankton samples in 1953, before the introduction of *Limnothrissa*, with those taken in 1981 after the introduction, indicated a strong decrease in size of cyclopoid species and a significant drop in standing crop. These changes are indicative of size-selective predation on the plankton by the sardine.

Consecutive echosounding surveys in the lake provided estimates for the biomass and maximum sustained yield of the sardine stock. In 1989, clupeid biomass in Lake Kivu was estimated between 6000 and 10 000 t, well below earlier estimates of 11 000–30 000 t (Spliethoff *et al.*, 1983). Since the early 1980s, a viable artisanal fishery has developed on the lake, using catamarans and trimarans with lift nets. The number of fishing units in Rwandese waters increased from six in 1980 to 144 in 1991. Recorded Rwandese catches increased from 65 t in 1981 to 370 t in 1987. Socio-economic and nutritional benefits of the artisanal fishery are described and the effect of *Limnothrissa* on the lake ecosystem is analysed.

ACKNOWLEDGEMENTS

The authors gratefully acknowledge the help of L.J.J. Haling, W. Scheffers and P. Hanek, and J.B. Mushiymimana, Director of the Fishery Project at Lake Kivu. Special gratitude is due to Dr J.M. Kapetsky, Fishery Resource Officer at the FAO, Rome, Dr D.F.E. Thys van den Audenaerde, Director of the Royal Museum for Central Africa in Tervuren, Belgium and M.J. Mann, Senior Project Operations Officer at the FAO, Rome. The fishermen's community of Gisenyi and the project staff gave us, often under difficult conditions, their full assistance during nocturnal sampling and echosoundings. For them, we hope *Limnothrissa* will provide a means of livelihood now and in the future.

REFERENCES

Balon, E.K. and Coche, A. (1974) *Lake Kariba, a Man-made Tropical Ecosystem in Africa.* Dr W. Junk, The Hague, 767 pp.

Beadle, L.C. (1974) *The Inland Waters of Tropical Africa*, Longman, London.

Chapman, D.W. (1975) Characteristics of layers and schools of pelagic fish in Lake Tanganyika. *Lake Tanganyika Fish. Res. Dev. Project Work Pap.*, no. 16, 10 pp.

Cochrane, K.L. (1978) Seasonal fluctuations in the catches of *Limnothrissa miodon* (Boulenger) in Lake Kariba. *Lake Kariba Fish. Res. Inst.* Project Rep. no. 29.

Collart, A. (1954) La Pêche au Ndagala au Lac Tanganyika. *Bull. Agric. Congo Belge*, **45**, 671–720.

Collart, A. (1960) L'Introduction du *Stolothrissa tanganicae* (Ndagala) au Lac Kivu. *Bull. Agric. Congo Belge*, **61**, 975–85.

Dumont, H.J. (1986) The Tanganyika sardine in Lake Kivu; another ecodisaster for Africa? *Env. Conserv.*, **13**, 143–8.

Ellis, C.M.A. (1971) The size at maturity and breeding seasons of sardines in southern Lake Tanganyika. *Afr. J. trop. Hydrobiol.*, **1**, 59–66.

Frank, V.G. (1977) Transplant sardines thrive in Lake Kivu. *Fish Farming Int.*, **4**, no. 3.

Frank, V.G., Diedrichsen, H., Herman, C. and Roest, F. (1977) Evaluation de la réussite de l'introduction d'un clupeide (*Limnothrissa miodon*) dans de lac Kivu, Rwanda. *FULREAC, Université de Liège, Belgique*, unpub. rep.

Gliwicz, Z.M. (1984) Limnological study of Cahora Bassa reservoir with special regard to sardine fishery expansion. FAO/GCP/MOZ/012/DOC/TR/06.

Hecky, R.E., Fee, E.J., Kling, H. and Rudd, J.W.M. (1977) *Studies on the Planktonic Ecology of Lake Tanganyika*. Western Region Fisheries and Marine Service, Department of Fisheries and Environment, Winnipeg, Manitoba, Canada, 88 pp.

de Iongh, H.H., Spliethoff, P.C. and Frank, V.G. (1983) Feeding habits of the clupeid *Limnothrissa miodon* (Boulenger) in Lake Kivu. *Hydrobiologia*, **102**, 113–22.

Johannesson, K. and Lamboeuf, M. (1987) Estimation acoustique de l'abondance de stock d'Isambaza (*Limnothrissa miodon*) du lac Kivu, par méthode basée sur le comptage des echos de poissons. Gisenyi project RWA/87/012, 64 pp., RWA/87/012/DOC/TR/13.

Kaningini, M. (1990) Etude de la biologie et de l'exploitation de *Limnothrissa miodon* (Boulenger) au lac Kivu (Zaîre). (Rapport de Doctorat) 1988–1989. UNFCED, FUNDP, Namur, 25 pp.

Kiss, R. (1959) Analyse quantitative du zooplankton du lac Kivu. *Folia Sci. Afr. Centralis*, **4**, 78–80.

Lays, J.F. (1979) Contribution a l'étude biologique d'un poisson clupeide *Limnothrissa miodon* introduit dans de lac Kivu, Ruanda. Thèse pour obtenir le grade d'ingenieur agronome à l'Université Catholique de Louvain, Belgique, 54 pp.

Manini, P. (1990) Parametres de la population de *Limnothrissa miodon* du Lac Kivu (1980–1989). *Rapport Projet de Developpement de la Pêche au Lac Kivu*. RWA/87/012, 46 pp., RWA/87/012/DOC/TR/32.

Mann, M.J. (1975) Rapport préliminaire sur la biologie des poissons et l'estimation des stocks dans le lac Tanganyika. Projet de Recherches et Développement BP 1940. Bujumbura, Burundi, 46 pp.

Mann, M.J., Bashirwa, F., Ellis, C.M.A. and Wahabakomeye, J.B. (1973) Rapport préliminaire sur la biologie des poissons et l'estimation des stocks dans le Lac Tanganyika (Burundi). FAO, Rome, Proj. mech. Dev. Pêche, Fi: DP BDI/73/020/5, 24 pp.

Matthes, H. (1968) Preliminary investigation into the biology of the Lake Tanganyika Clupeidae. *Fisheries Research Bulletin Zambia*, **4**, 39–46.

Munyandorebo, J. (1989) Estimation des paramètres démographiques des clupeidae exploités dans les eaux burundaises du lac Tanganyika. *Univers. Bretagne Occidentale, Fax. Sc. Techn. de Brest*, 30 pp.

Nedélec, C. (1975) *FAO Catalogue of Small-scale Fishing Gear/Catalogue des Engins de Pêche nationale*, Fishing News (Books) Ltd, West Byfleet, Surrey, 191 pp.

Reyntjens, D. (1982) Bijdrage tot de Limnologie van het Kivu meer. MSc thesis, University of Gant, Belgium, 95 pp.

Spliethoff, P.C., de Iongh, H.H. and Frank, V. (1983) Success of the introduction of the freshwater clupeid *Limnothrissa miodon* in Lac Kivu. *Fish Manage.*, **14**, 17–31.

Van Meel, L. (1954) La phytoplancton. Résultats scientifiques de l'exploration hydrobiologique du Lac Tanganika (1946–1947). Institut Royal des sciences naturelles de Belgique, Text A and Atlas B, **4**(1), 1–681.

Verbeke, J. (1957a) Recherche écologique sur la faune des grands lacs de l'Est du Congo Belge, in *Exploration, Hydrobiologiques des Lacs Kivu, Edouard et Albert*, Inst. r. Sci. nat. Belg., Tervuren, Belgium.

Verbeke, J. (1957b) Chaoboridae (Diptera, Nemotocera) Stades immatures et adultes, in *Exploration Hydrobiologiques des Lacs Kivu, Edouard et Albert*, Inst. r. Sci. nat. Belge, Tervuren, Belgium, pp. 185–202.

Verbeke, J. (1957c) Recherches écologiques sur la faune des grands lacs de l'Est du Congo, in *Exploration Hydrobiologiques des Lacs Kivu, Edouard et Albert* (1952–1954), Vol. III, Fasc. 1, Royal Museum of Central Africa, Tervuren, pp. 1–176 + pl.

Welcomme, R.L. (1972) The inland waters of Africa. FAO, *Committee for Inland Fisheries of Africa tech. Pap.*, no. 1, 117 pp.

Chapter fourteen

The persistence of two introduced tilapia species in Lake Naivasha, Kenya, in the face of environmental variability and fishing pressure

S. Mucai Muchiri, Paul J.B. Hart and David M. Harper

14.1 INTRODUCTION

Unlike most of the small lakes in the Rift Valley, Lake Naivasha is fresh despite the absence of an exit river. The most likely reason is that the dissolved substances in the input are dilute, biochemical and chemical processes are in place to remove ions such as sulphates and carbonates and there is seepage from the lake bottom (Harper *et al.*, 1990). Because the lake is fresh, there might be an expectation of a rich fish fauna, in line with the diversity found in other lakes such as Victoria and Malaŵi. It comes as a surprise to many that the fish fauna is sparse and contains mostly introduced species. In the past, these have died out after only a few years. A clue to the low fish diversity, and to the frequent extinctions, may lie in the variability that is characteristic of the water level, the abundance of the aquatic macrophytes and the limnological conditions. In the face of this variability, three fish species and a crayfish form the basis of a commercial fishery that has lasted for 30 years.

The Impact of Species Changes in African Lakes. Edited by Tony J. Pitcher and Paul J.B. Hart.
Published in 1995 by Chapman & Hall, London. ISBN 0 412 55050 4.

Why did the three fish species, largemouth bass, *Micropterus salmoides*, and two tilapias, *Tilapia zillii* and *Oreochromis leucostictus*, establish themselves in the lake and then persist in the face of exploitation? Not much will be said in this chapter about the bass, but data will be presented that suggest a reason why the tilapia species have survived. Initially, we will outline recent changes in the lake and describe the history of the fishery. Some comment will be made at the end on Lake Naivasha as a place in which to study the role of environmental variability in determining extinction rates and persistence of species.

14.2 VARIABILITY OF THE LAKE NAIVASHA ECOSYSTEM

The recent history of the Lake Naivasha ecosystem has been described by Harper *et al.* (1990), Muchiri and Hickley (1991) and Harper (1992). Variability in water level has been a strong influence on the littoral habitat for as long as the lake has been studied scientifically. Oral histories by the Masai tell of the lake drying out shortly before the European immigration into the area. The lake was also much bigger about 10 000 to 12 000 years ago. In the last 100 years the lake level has had a high of 1896 m a.s.l. in 1895 and a low of 1882 m a.s.l. in the 1945–1955 period (Harper *et al.*, 1990). Changes like this alter the greatest depth of water in the lake, but the most important ecological effect is on the littoral. The lake is shallow around its edges (Chapter 1). A drawdown of 14 m would change drastically the area of shallow water available for macrophytes. If water levels fluctuate rapidly, there could be an important influence on the abundance and diversity of organisms living in the littoral, including fish which may use the area as a spawning and feeding ground. The reduction of diversity is true for storage reservoirs that have heavy drawdown.

Since 1970, there have been marked changes in the abundance and distribution of the floating and submerged macrophytes. The reasons for the changes may be several (Harper *et al.*, 1990) and are not well understood but may include changes in water levels, predation by coypu, *Myocaster coypus*, and by red swamp crayfish, *Procambarus clarkii* Girard. The changes in macrophytes have been grouped into three periods by Harper (1992). From 1930 up to the late 1970s the shallow-water macrophyte community had 25 species, 10 submerged, 8 floating and 7 associated with floating vegetation (Gaudet, 1977). This period is regarded by Harper (1992) as a 'normal' condition.

From 1975 to 1983 no submerged or floating-leaved aquatic vegetation was found in the main lake. Floating vegetation was dominated by the introduced *Salvinia molesta*, which was often associated with large floating rafts of papyrus, *Cyperus papyrus*. These drifted around the lake, driven by the wind, eventually becoming beached when the water levels fell to their lowest for the period in 1987 and early 1988. From about 1986, submerged macrophytes began to return to the littoral. Recolonization began on the eastern shores of the lake, around Crescent Island and near the Safariland Lodge (map, Fig. 14.7). This recolonization was accompanied in the latter part of 1988 by increasing water levels.

During the period of rapid change, the clarity of the water decreased and phytoplankton abundance increased. This was accompanied by an increase in gross primary production in the upper layers of the lake (Harper, 1992). The benthic fauna of the lake is discussed by Clark *et al.* (1989).

14.3 THE FISH COMMUNITY: SPECIES AND THEIR CHANGES

Some of the introduced fish species have been mentioned already. A list of fish species that have been recorded from the lake is given in Table 14.1. Only *Aplocheilichthys antinorii* and *Barbus amphigramma* have arrived in the lake without human intervention. Lakes Nakuru and Elmenteita, once part of the same lake as Naivasha, have no endemic species (Richardson and Richardson, 1972). The remaining nine species or species groups have been introduced and several of them, such as largemouth bass and the rainbow trout, *Oncorhynchus mykiss*, are originally from outside Africa. The average persistence time is 36.9 $\pm$ 29.7 years for the ten species in Table 14.1 for

Table 14.1 A history of the fish community in Lake Naivasha since the beginning of the 20th century (from Muchiri and Hickley, 1991)

Species	Date of introduction/disappearance
Endemic	
Aplocheilichthys antinorii	Last reported 1962
Introduced	
Failed	
Oreochromis spirulus niger	1925
O. leucostictus × *O. s. niger* hybrid	Gone by 1972
O. niloticus	1967
Still present	
Micropterus salmoides	1929, 1940s and 1951
Tilapia zillii	1956
Oreochromis leucostictus	1956
Gambusia sp. and *Poecilia* sp.	Date unknown
Lebistes reticulata	Date unknown
Oncorhynchus mykiss	R. Malewa, date unknown
Barbus amphigramma	1982 (rivers)

which the start and end of their presence can be deduced with any certainty. In the case of the only known endemic, it is assumed that the fish had been present for 100 years before extinction in 1962. This would take account of the waterless lake just prior to the arrival of Europeans, when the fish was presumably absent. The reciprocal of the persistence time, 0.027, can be regarded as an extinction rate (Schoener and Spiller, 1992). Over recorded history, five out of eleven species have become extinct which is 45% of the fish that have been known to live in the lake. This appears to be a poor record.

The present fish fauna is composed of five species: largemouth bass, the two cichlids, *Barbus* and the guppy. The first three form the basis of the commercial fishery for which records exist since 1963. The crayfish, introduced in 1970, is also fished for commercially, but we will not discuss its fishery further. The commercial catch of tilapias declined through the 1970s. At the same time crayfish populations were increasing, eventually reaching about $4\,m^{-2}$ in the littoral. This increase has been suggested as a possible contributor to the decline in macrophyte populations (Harper *et al.*, 1990), a decline which possibly contributed to the decrease of the cichlid stocks. The crayfish increase may have led to a gradual recovery of the bass populations towards the end of the 1970s because crayfish form an important part of the bass diet. Possibly as a result of a 10 cm minimum mesh size rule in the fishery and a return of the macrophytes, the cichlid catches began to increase towards the end of the 1970s. By the end of the 1980s the crayfish population had fallen back to about $0.25\,m^{-2}$, a change possibly coupled with the fall in the bass catch at the same time. Tilapia catches had stabilized at around 200 t (tonnes) a year.

14.4 THE FISHERY FOR TILAPIA

Commercial fishing is mostly by gill nets set from glass-fibre or wooden canoes (Muchiri and Hickley, 1991). These are powered by sail, oars or outboard motor. Each fisherman is permitted to use 10 multifilament gill nets, each of 100 m. The mesh must be no less than 100 mm. The catch must be landed at a point on the eastern shore near the town of Naivasha where inspectors from the Fisheries Department of the Kenya Government record the weight of the catch from each vessel. Records began in 1963 and the catch was separated into bass and 'tilapia' from 1974 on. Landings were not separated into *T. zillii* and *O. leucostictus* until 1987. Bass are caught by both commercial and sport fishermen, the latter using rod and line. The only other species caught commercially is *B. amphigramma*, which is caught using scoop nets as it attempts to ascend the River Malewa to spawn.

The total catch in tonnes from 1963 to 1989 is shown in Fig. 14.1 (from Muchiri, 1990). This also shows the number of canoes in the fishing fleet

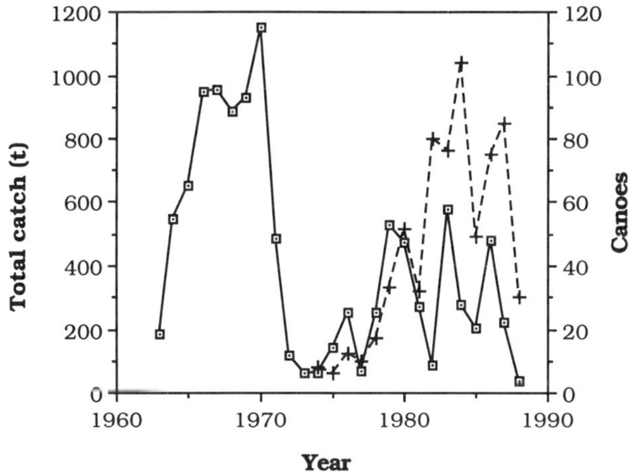

Fig. 14.1 The combined catch (solid line) of *T. zillii*, *O. leucostictus* and *M. salmoides* in Lake Naivasha between 1963 and 1989 together with the number of canoes in the fishery from 1974 (dashed line).

from 1974 onwards. The total catch and the lake water levels are plotted together in Fig. 14.2. The catch increased rapidly in the early years to a high of 1150 t. The early effort, the effects of which were made worse by the use

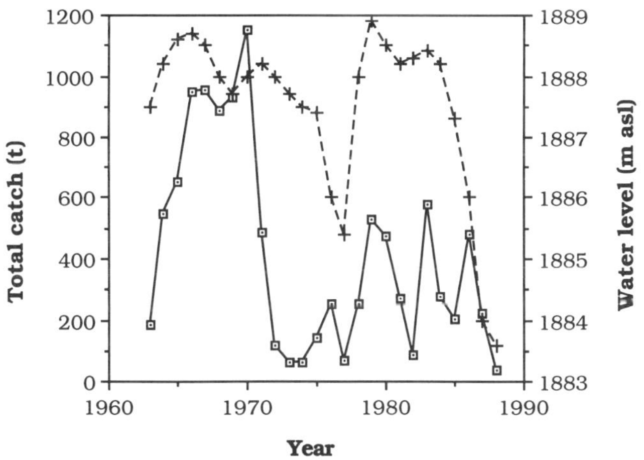

Fig. 14.2 The combined catch of *T. zillii*, *O. leucostictus* and *M. salmoides* in Lake Naivasha between 1963 and 1989 (solid line) together with the water levels (dashed line).

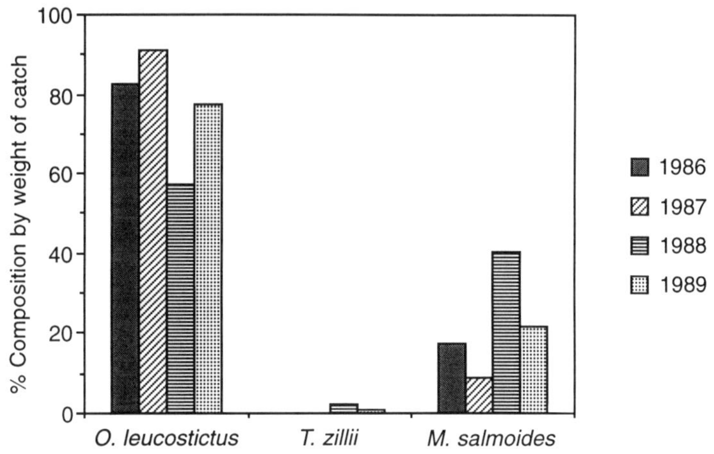

Fig. 14.3 Histograms showing the species composition of the catch for the last 4 years of the period under study.

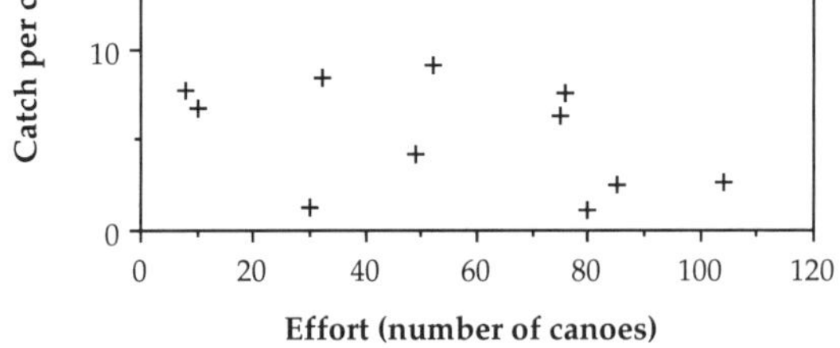

Fig. 14.4 Catch per unit effort, expressed as tonnes caught per canoe, plotted against the total effort expressed as the number of canoes. These data form the basis of the stock assessment given in the text.

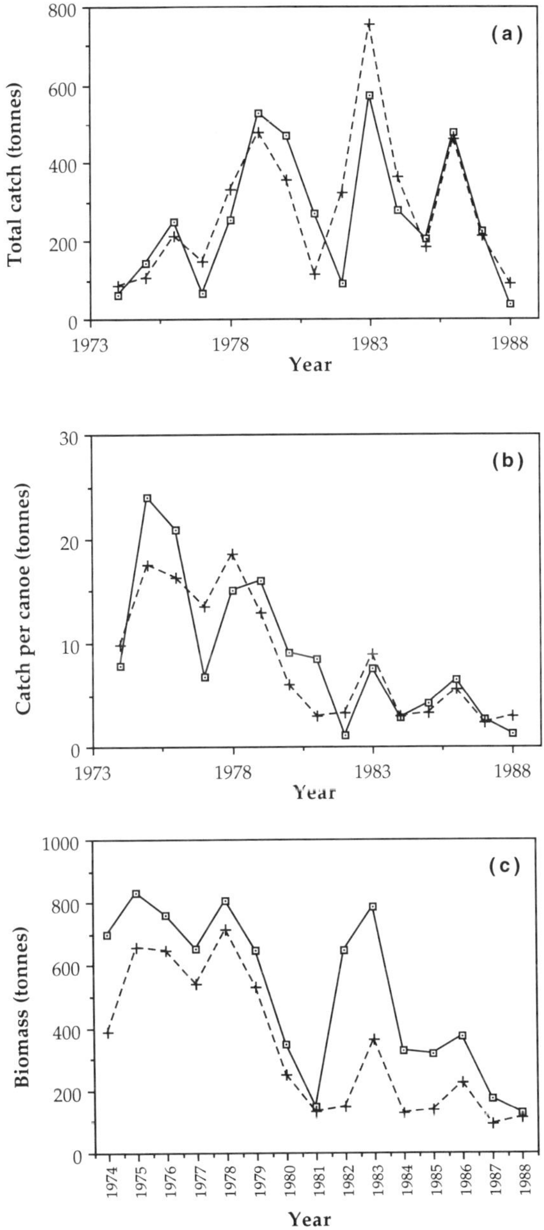

Fig. 14.5 Fit of the Gulland–Fox model to total catch from Lake Naivasha. (a) Actual (solid line) and predicted catch (dashed line), 1974–1988. (b) Actual (solid line) and predicted (dashed line) catch per unit effort for the same period. (c) Biomass estimates (solid line) for the 15 year period. This panel also shows biomass as calculated from the Schaeffer logistic model of population growth (dashed line).

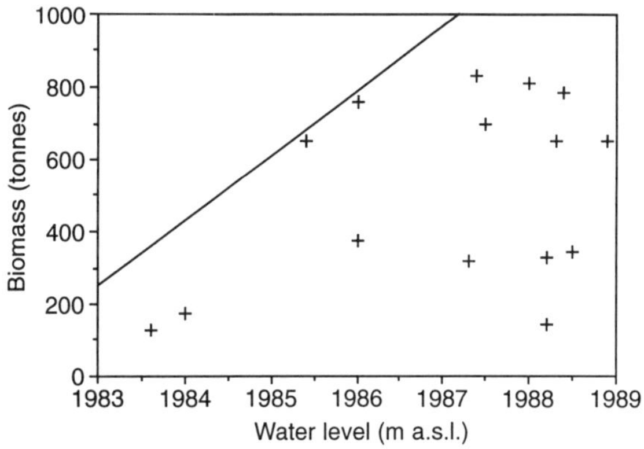

Fig. 14.6 Biomass of all three species plotted against the water levels over the period 1974 to 1988. The curve defines the empty area above which no biomass/water level combinations occur.

of small-mesh nets, led to a sharp decline in the early 1970s. The limnological changes already described may also have had an effect. Following the introduction of a 100 mm minimum mesh size in the early 1970s, the catch increased from a low of about 60 t to between 200 and 500 t, values which remained until the 1988 season when the catch fell to an all-time low for the period of 37 t.

The species composition of the catch between 1986 and 1989 is shown in Fig. 14.3. *O. leucostictus* dominated the fishery during this period. The fishery is mainly executed in shallow water less than 3 m deep. Survey gill netting showed that *O. leucostictus* occurs mostly in water of less than 3 m whilst *T. zillii* is more abundant in water deeper than 3 m (Muchiri, 1990).

The data on catch and number of canoes (Fig. 14.1) have been used to calculate catch per canoe as an index of abundance. Catch per canoe is plotted against the number of canoes in Fig. 14.4. The best fit to these data, using the CIDA software package (MRAG, 1992) is achieved with an exponential curve ($r^2 = 0.7$) and a Gulland–Fox model for population growth (Ricker, 1975). Using this, $B_\infty = 719$ t (90% confidence limits: 539–1040 t), maximum sustainable yield is 365 t and the number of canoes yielding this would be 37. The fit of the model to the data is shown in Fig. 14.5 (a, b, c).

Fitting a straight line to the catch per unit effort and effort plot (Fig. 14.4) leads to a logistic or Schaeffer model of population growth. Using this yields an estimate of $B_\infty = 892$ t, with an MSY of 446 t. For both these fits it should be remembered that the analysis assumes that fish population and fishery are in equilibrium, which they are probably not. The fit is also made to data aggregated from three species.

The estimate of biomass for the 15 years between 1974 to 1988 (Fig. 14.5(c)) has been plotted against water level in Fig. 14.6. The correlation of the two is not significant but it is interesting to note, as highlighted by the curve added to the figure, that high biomass never co-occurred with low water levels. The mechanisms determining this relationship are not clear, but there is a hint that low water levels reduce the areas available for spawning to the *Tilapia* and *Oreochromis* populations. At high water levels other mechanisms, such as increased fishing pressure, would have greater influence on biomass. It should also be remembered that this analysis is applied to the total catch of three species, each with its own population dynamics. The conclusions can only be preliminary.

14.5 WHY HAVE THE TILAPIA SPECIES PERSISTED IN THE FACE OF HEAVY FISHING PRESSURE AND HABITAT CHANGE?

Many species introduced into the lake have become extinct. The two tilapia species and the largemouth bass have persisted despite being the subject of fishing pressure. Lake Naivasha was not necessarily the most appropriate habitat for some of the introduced species and this may explain why they failed to establish themselves. A research programme was begun in 1987 to investigate aspects of the feeding ecology of the two tilapias. The results of the work (Muchiri, 1990) provide an insight into aspects of the trophic ecology of the two species that could explain their persistence in the face of a fluctuating environment and heavy fishing mortality. The results on the feeding habits of *T. zillii* and *O. leucostictus* will be summarized so that the reasons for persistence may be examined.

14.6 TROPHIC ECOLOGY OF THE TILAPIAS

Collection of fish samples

Fleets of multifilament gill nets of five mesh sizes (40, 48, 70, 100 and 125 mm stretched mesh) were used to collect monthly fish samples from three sampling sites (Fig. 14.7) from June 1987. Two sites were located at the main lake: (i) off Crescent Island, which at the beginning of sampling was devoid of submerged macrophytes, and (ii) off Safariland Lodge where, in contrast to the Crescent Island area, there was an extensive cover of submerged macrophytes. Sampling continued until June and July 1988 at sites (i) and (ii) respectively. The third site was in Oloidien Lake, which was confluent with Lake Naivasha before 1984 during high water. Sampling in Oloidien Lake started in August 1987 and continued until May 1988.

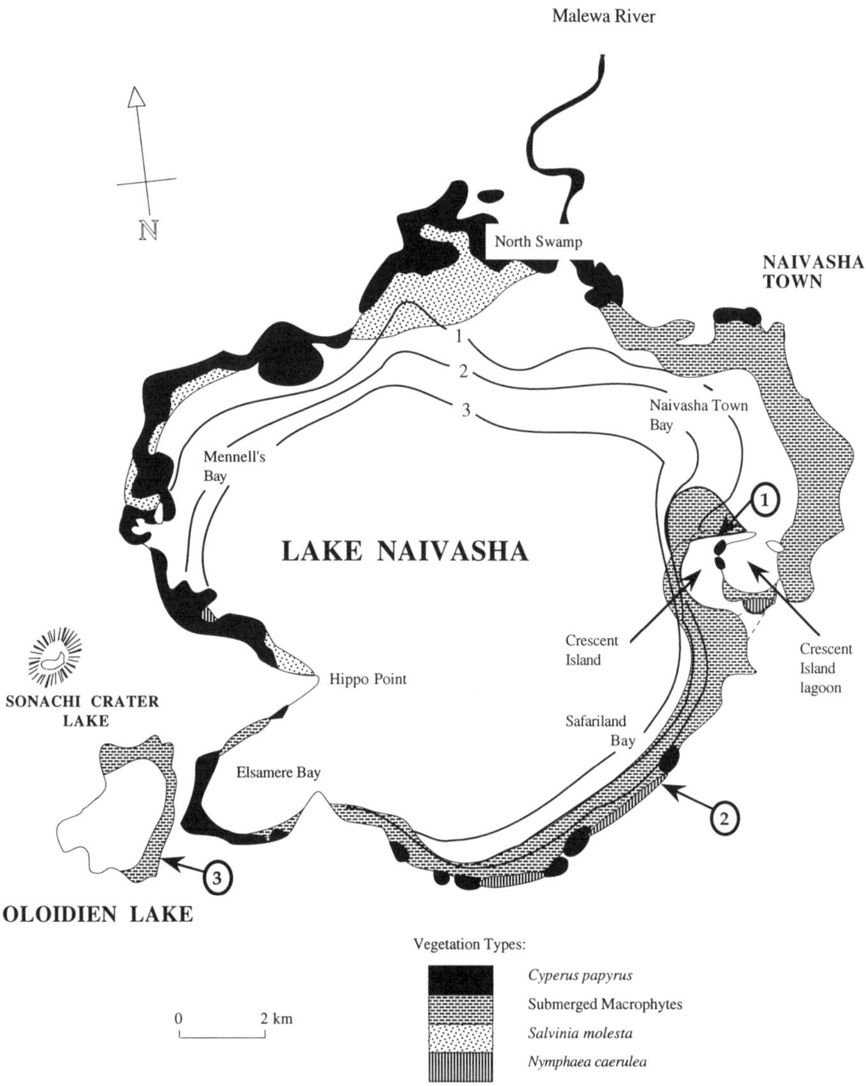

Fig. 14.7 Map of Lake Naivasha to show the sites (encircled numbers) used for sampling the populations of *O. leucostictus* and *T. zillii*: 1, Crescent Island; 2, Safariland Lodge; 3, Oloidien Lake. Additional information and other sites are mentioned in the text. Depth contours in m.

In 1989 further fish collection was done in January, March and April. In addition to the Oloidien site, gill netting was done in four habitat types in Lake Naivasha. These were: macrophyte zone, rocky shore, North Swamp and the open water. During the 1989 sampling sessions two smaller mesh sizes (22 and 30 mm stretched mesh) were introduced to sample the smaller fish sizes.

Description of the sampling sites

(i) Crescent Island At the beginning of sampling in June 1987 the Crescent Island site was devoid of submerged macrophytes and due to continued recession of the shoreline that began in March 1984, swamp vegetation was not established. The bottom substrate in this area was hard in the nearshore, gradually becoming soft flocculent mud along a transect line from Crescent Island towards the mouth of the River Malewa, on the north-east shore. A greater part of this area was less than 3 m in depth. By November 1987 submerged vegetation (mainly of *Potamogeton schweinfurthii*, *P. pectinatus* and *Najas pectinata*) was abundant. As the lake level began to rise again in the latter part of 1988 a fringing swamp of mainly *Cyperus* spp. formed along the shoreline.

(ii) Safariland site In contrast to the Crescent Island site, this station had beds of submerged vegetation up to 1000 m wide to a depth of 4 m. The shoreline had a thin cover of *Cyperus papyrus* swamp. As the lake level rose by about 1 m in early 1989, the swamps increased with lagoons forming behind them.

(iii) Oloidien Lake This is a small, saline, basin-shaped lake with a rim of water less than 3 m in depth stretching 200 m offshore, with a deeper central area going down to a maximum depth of about 6 m. Oloidien did not experience any loss of submerged plants (Harper, 1992). The substratum of the shallow zone of this lake has only a thin layer of organic mud and in some areas the bottom is hard and covered with volcanic gravel. The deeper parts of the lake have a flocculent iron-rich silt bottom.

(iv) Rocky shore Rocks are found only along the south-western shore of Lake Naivasha and the inner rim of Crescent Island. The shoreline of the south-western shore contains no submerged plants and falls steeply only a few metres offshore to 6 m, the deepest part of the lake.

(v) North swamp This is the northern part of the lake most influenced by fringing papyrus swamps and the inflowing rivers, which are often laden with silt and rotting organic debris brought in from the catchment area. It

is a shallow area (< 3 m deep) with a detritus-rich substrate and sparse vegetation cover.

(vi) Open water The open water is the most extensive of the habitats available to fish in Lake Naivasha, covering over 60% of the total surface area of the lake.

Laboratory treatment of fish samples

Fish caught in the gill nets were measured and weighed before being opened to remove their stomachs. The fish stomachs were preserved separately in 10% formaldehyde solution. The stomach contents were later examined with a low-power microscope. The larger food items were identified and their volumes estimated.

Because all the examined stomachs contained food items that were too small to be examined sufficiently under the low-power magnification, the stomach contents were mixed thoroughly to form a homogeneous mixture. Small amounts of this mixture were pipetted onto a microscope slide and examined under high-power magnification (up to × 400) using a Nikon compound microscope.

In both cases visual estimates of relative volumes were made by comparing the area occupied by each food item in the field of view. This meant that larger items contributed greater proportions (and therefore volume) of the stomach contents. The estimated volumes were allotted numbers between 1 and 10; 10 was allotted where the whole stomach content was made up of only one food item and 1 where a food item was estimated to constitute 10% of the contents.

The numbers of guts of each species in which a food item occurred were added up and expressed as a percentage (% occurrence) of the sum total of the non-empty stomachs for each of the food items. The proportion and occurrences were then used to estimate the relative importance of each food item consumed by employing Wilhm's (1967) prominence values (*PV*):

$$PV = \%\ \text{proportion} \times (\%\ \text{occurrence})^{\frac{1}{2}}. \qquad (14.1)$$

*PV*s range between 0.0 and 1000.0; 0.0 where a food item is not consumed at all and 1000.0 for a food item where only that one item is taken. This method was adopted by Hickley and Bailey (1987) in their analysis of the diets of fishes of the Sudd swamps of the River Nile.

The data on diet were analysed with two aims in mind: to examine the degree of diet overlap and to determine the food webs supporting each of the tilapia species. Dietary overlap was quantified by following Wilhm (1967) who used the *PV*s to calculate coefficients of similarity in populations of benthic organisms from different sampling stations. The scheme was employed in this study to examine similarity of food items eaten and hence diet overlap among the two tilapia species. The coefficient of similarity index, C, was calculated as follows:

$$C = 2 \times (\sum PV_{(\text{low})})/\sum PV_1 + \sum PV_2 \qquad (14.2)$$

where $PV_{(\text{low})}$ is the lower prominence value calculated for each of the food items taken in common by the two species to be compared, and PV_1 and PV_2 are the prominence values of each food item for species 1 and 2 respectively. C ranges between 0.0 and 1.0, where 0.0 shows no sharing of any food items and 1.0 occurs where there is total overlap in feeding. A 60% ($C = 0.6$) cut-off value was adopted in this study to imply competition in feeding: i.e. where 60% or more of the scores of food items eaten were shared by the two species as indicated by the coefficients of similarity, then competition for food would be a possibility (Zaret and Rand, 1971; Mathus, 1977, as quoted by Wallace, 1981).

The diet of *O. leucostictus*

Detailed results of stomach content examination for *O. leucostictus* during the sampling period 1987–88 and 1989 respectively are given in Muchiri (1990). Only the main features are summarized in this chapter. Throughout the sampling periods, detritus constituted the highest proportion in the diet of *O. leucostictus*. Detritus found in the guts of fish during the first few months of sampling in Lake Naivasha and in all sampling occasions in Oloidien Lake was mainly of coarse rotting material, in most cases associated with grit. This suggested bottom feeding. In later months the detritus found in the guts was generally of material that had a finer particle size.

Phytoplankton also constituted significant proportions of the gut contents collected from the Lake Naivasha sampling stations. *Closterium* spp., *Botryococcus* sp., and *Microcystis* spp. were the dominant species, in that order. Occasionally, in some stomach samples examined the filamentous diatom *Melosira* sp. constituted significant proportions. Other phytoplankton found in the stomachs of *O. leucostictus* included the bluegreen alga *Spirulina* sp., *Merismopedia* sp., *Ceratium* sp., *Pediastrum* sp., and *Chlorococcus* sp. During 1988 and 1989 phytoplankton constituted smaller proportions of the gut contents of *O. leucostictus*. This was particularly evident where there was extensive cover of submerged macrophytes.

Macroinvertebrates were also consumed, albeit in small proportions. In the guts of some individual fish, macroinvertebrates constituted significant proportions of the diet. This was particularly true for insects (mainly *Micronecta*) which are large food items and are therefore more prominent than other items.

Benthic macroinvertebrates are less significant, except in two habitat types in Lake Naivasha (macrophyte beds and rocky shore) in 1989, where oligochaete worms constituted notable proportions of the diet. It is, however, likely that fish picked up the worms which were attached to macrophytes and rocks rather than collecting them from the bottom mud. In Oloidien Lake, chironomid larvae were taken in significant proportions.

Zooplankton were consumed in very small quantities and were absent in most cases in both Lake Naivasha and Oloidien Lake.

The diet of *T. zillii*

Detailed results of the *T. zillii* stomach content analysis are given in Muchiri (1990). As with *O. leucostictus*, detritus constituted the most important food item in the diet of *T. zillii* in both Lake Naivasha and Oloidien Lake.

During the loss of submerged macrophytes in Lake Naivasha, *T. zillii* consumed mainly detritus and insects (dominated by the corixid *Micronecta scutelaris*). This was indicated by the feeding of *T. zillii* at the Crescent Island site during the months of August to November 1987. Macrophytes constituted the only other significant food item in Naivasha during the latter months of 1987 and 1988, while chironomid larvae were also consumed in significant proportions in Oloidien Lake.

In the habitat types sampled in 1989, chironomid larvae were consumed in significant proportions in the macrophyte zone whereas oligochaete worms were more significant in the rocky shore. It is unlikely that the detritus and macroinvertebrates consumed in Lake Naivasha this time were collected from the bottom because these items were not associated with grit (as they were in samples from Oloidien Lake). The detritus had a fine particle size and could have been of planktonic origin sedimented onto the macrophytes.

Other food items were found in small amounts in the guts of *T. zillii*, but these were presumably taken incidentally as fish fed on the preferred food items. Epiphytes and molluscs, for example, were ingested as the fish consumed macrophytes.

Feeding overlap

A summary of the feeding habits of the two tilapias in the main lake is given as Fig. 14.8. Coefficients of similarity calculated to assess the apparent sharing of food revealed a great deal of overlap in the diet of *O. leucostictus* and *T. zillii* in virtually all the habitats. By adopting 60% cut-off value to decide competitive sharing, the feeding overlap in the two species suggested competition for food. The coefficient of similarity at Crescent Island in 1988 was 0.60 while those of Safariland in 1987 and 1988 were 0.63 and 0.68 respectively. At Oloidien Lake the coefficients were 0.66 in 1987 and 0.63 in 1988. Out of the three habitats sampled in 1989, where sample sizes were

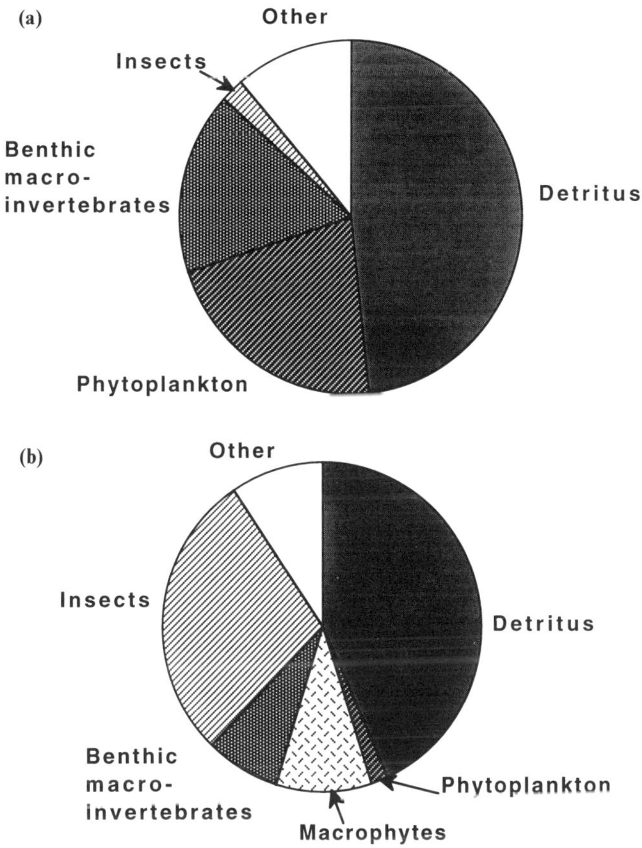

Fig. 14.8 Pie charts illustrating the diet composition for (a) *O. leucostictus* and (b) *T. zillii* in the main lake.

large enough for both aspects to be compared, two habitats, the macrophyte zone and rocky shore, had coefficients that suggested competition (0.67 and 0.60 respectively). Closer scrutiny of the data showed that the observed large coefficients of similarity were a result of the two tilapia species feeding on large amounts of detrital material. By recalculating the coefficients leaving out detritus, feeding overlap was shown to be very small in all habitats.

Food webs

Food webs for both tilapia species are shown in Fig. 14.9 (a and b) and are based on data given in Muchiri (1990). The only difference between the two is that *O. leucostictus* did not make use of macrophytes.

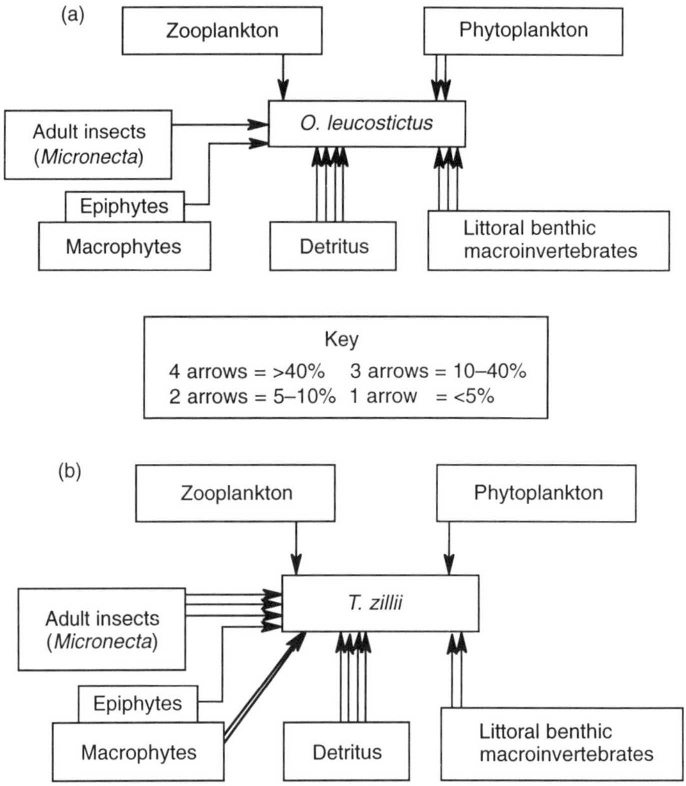

Fig. 14.9 Proposed food webs for the two tilapia species indicating the proportion of the diet, shown by the number of arrows, coming from the main sources of food. (a) *O. leucostictus*, (b) *T. zillii*.

14.7 DETRITUS AS A BUFFER AGAINST HABITAT VARIABILITY

Detritus was shown in this study to be the most important food item for both *O. leucostictus* and *T. zillii* in Lake Naivasha and the adjoining Oloidien Lake. In Lake Naivasha during the early phase of the study, sources of detritus were mainly from *Cyperus* spp. in the lake, from allochthonous organic material coming into the lake in river water, from surface run-off and from sedimenting plankton. Recovery of submerged macrophytes and rise in the lake level, thus flooding terrestrial vegetation in the drawdown zone, led to increased formation of detritus. The loss of submerged vegetation and the decline in the lake level did not therefore have great effects on the quantity of detritus. This meant that detritus was available at all times. The availability of this almost unlimited food item allowed the two tilapia species to coexist with minimum competition for food.

The importance of detritus as food for tilapias has been discussed in detail by Fryer and Iles (1972) and Bowen (1979, 1982). Bowen (1982) pointed out that detritivory in fish ensures that food quantity does not become the limiting factor. This is particularly important in Lake Naivasha because some of the other food items consumed by the tilapias are also shared with other animals in the ecosystem, although the quantitative aspects of the overlap are not known. The tilapias share insects and chironomid larvae with juvenile largemouth bass (Hickley and North, unpublished data). *T. zillii* also shares submerged macrophytes with crayfish and the large numbers of coot, *Fulica cristata*, the yellow-billed ducks, *Anas andulata*, and the African pochard, *Aythia erythrophalma*.

It would be interesting to know more about the fitness of the two species in this regime compared with one in which a more varied selection of food types was available. The effect of more choice on *O. leucostictus* may not be so significant because the species commonly feeds on detritus in other habitats. Earlier studies in Lake Naivasha have also shown that detritus is important to the species. Malvestuto (1974) and Siddiqui (1977) reported the diet of *O. leucostictus* in Lake Naivasha to consist of mainly chironomid larvae and detritus. The diet of *T. zillii* was, however, reported in the past to be 67.7% macrophytes (Siddiqui, 1977). *T. zillii* has been known to feed on other foods in other habitats. Abdel-Malek (1972) and Spataru (1978) have reported *T. zillii* as feeding on benthos in Lake Quarun, Egypt, and Kinneret, Israel, respectively. In Kinneret they also consumed phytoplankton. In the Sudd swamps of the River Nile, Hickley and Bailey (1987) found the diet of *T. zillii* to be 61% periphyton. These results may indicate that *T. zillii* is feeding on a poorer diet than it would choose if greater prey diversity was available.

Analysing the trophic relationship between the two tilapia species is handicapped without a knowledge of their population sizes. From the fishery data, it could be concluded that *O. leucostictus* is very much more abundant than *T. zillii*. As *T. zillii* is more likely to be caught in deeper water than is *O. leucostictus* (Muchiri, 1990), it is possible that *T. zillii* is underrepresented in the fishery.

Population stability of a species could be influenced by the complexity of the food web from which it is supplied (Pimm, 1991). In Lake Naivasha the dependence of both tilapia species on a large reservoir of organic detritus as their main food source seems to damp out the effect of any fluctuations in the populations of other food organisms. It is also the case that both species are dependent on nearly the same food web, so that both should be similarly affected by changes in food species. It is more likely that fluctuations in the populations result from reduced stocks of submerged plants diminishing the fish's breeding opportunities (Fig. 14.6). The nest-building *T. zillii* is most likely to be affected by these changes. Fishing pressure would have added further perturbations to the populations (Muchiri and Hickley, 1991).

Several major sources of fish food in Lake Naivasha remain underutilized. These include the benthos, especially in the offshore zones, and the limnetic zooplankton and phytoplankton. This is because all the existing fish species in Lake Naivasha live in the shallow littoral zones. Analysis of fish yields in relation to the lake's primary productivity has shown that Lake Naivasha could have a capacity for higher yields than are now realized (Muchiri and Hickley, 1991). This has led to suggestion for careful consideration of further introductions of fish species that would use the unexploited resources, to diversify the species composition of the fish community and increase the yield for the fishery (Muchiri and Hickley, 1991).

14.8 HABITAT VARIABILITY AND EXTINCTION RATES IN THE FISH COMMUNITY

It has been assumed earlier that the recent variability of the lake ecosystem may be related to the sparseness of the lake's fish fauna. With only five fish species now present and only ten species (excluding hybrids) known over the past 100 years, the fauna is atypical of many tropical lakes (this volume). This situation is likely to be a result of a number of factors, mostly historical. Lake variability is probably not, on its own, a deciding factor.

Lake Naivasha is not well connected to other lakes and rivers so that it is hard for immigrants to find their way to it. With no surface outlet, fish from other lakes or river systems cannot get into Lake Naivasha. The inflowing rivers are short and not connected to any other lakes (map, Chapter 1, Fig. 1.1). The nearest lakes in the Rift Valley, such as Nakuru and Elmenteita, are soda lakes and unlikely to have fish species that would thrive if they arrived in the fresh water of Lake Naivasha. In addition, the instability of the lake level over the past 150 years would have made it difficult for immigrants to establish themselves.

These remarks still leave untested the assumption that even if individuals of an immigrant species do arrive at the lake, the species will not last long because of the variability of the ecosystem. This variability is likely to have strong effects on the population abundance of immigrants, making it fluctuate widely in unison with the habitat changes.

Early studies on extinction rates of immigrants on islands did not account for population variability (MacArthur and Wilson, 1967) and concluded that extinction rate would be inversely correlated with population size. In general, large populations would not go extinct. As the carrying capacity increased (K), time to extinction would become infinitely long. Recent theoretical work by Mangel and Tier (1993) and empirical work with birds (Pimm *et al.*, 1988;

Tracey and George, 1992) and spiders (Schoener and Spiller, 1992) has shown that the early ideas were faulty because they did not account for variability in population abundance. The theoretical study has shown that if catastrophes are included in the model that predicts time to extinction, then the very large values disappear. The study also shows that if there are catastrophes, even large populations can become extinct.

The empirical work has produced conflicting results. Pimm *et al.* (1988), working with birds on British islands, found that large populations were less likely to go extinct, but the risk of extincton was greater for species the populations of which varied greatly with time. Reanalysing the data, Tracey and George (1992) found that there was no evidence for this idea. Extinction rates of spiders on Caribbean islands showed an inverse relation with temporal variability in population size (Schoener and Spiller, 1992). Schoener and Spiller also found that the coefficient of variation of population size was directly proportional to population size; large populations were more variable. These relationships are explained by arguing that a small population invading an island and going extinct several times over a short time span, would never have the chance to build up temporal variability in population size. These small populations would become extinct for three reasons: (1) demographic – stochastic accidents under 'normal' or 'average' conditions, such as a series of deaths before sufficient births had occurred, would cause the population to die out; (2) islands might have very poor conditions, so that births and deaths would be substantially worse than normal; (3) females might fail to reproduce because they would not encounter a male in time. The second condition might describe the situation in Lake Naivasha. The environmental variability makes it difficult for a new species to establish itself. The tilapia species have persisted because they can exploit a resource that is not affected by changes in water level and reductions in macrophyte communities.

14.9 SUMMARY

The tilapias, *Oreochromis leucostictus* and *Tilapia zillii*, have been introduced into Lake Naivasha, Kenya and form the basis for a commercial fishery. Lake Naivasha is characterized by frequent changes in water level and varying amounts of submerged vegetation. Changes in the lake ecosystem tend to occur rapidly and have significant ecological consequences. The endemic fish community in the lake has been sparse during recent times. Human additions to the fish fauna have led to the present community which is dominated by largemouth bass, *O. leucostictus* and *T. zillii*. The persistence of the two tilapia species in the face of rapid ecological change and heavy fishing is an interesting challenge to our understanding. To elucidate the problem, a study of the diet of the tilapias was made over a 3 year period at three contrasting sites in the lake. For each species, 40–50% of the food eaten was detritus. The rest of the diet of *O. leucostictus* was composed of 22% phytoplankton, 17% benthic macroinvertebrates and a small proportion of insects. Apart from detritus, the diet of *T. zillii* included 28% insects and then less than 10% each of macrophytes, benthic macroinvertebrates and non-specified material. The emphasis on detritus in both species indicates that competition for food between the two is unlikely. Their diets may also mean that changes in macrophyte or invertebrate abundances do not seriously influence the feeding ecology of the two fish species. The quantity of detritus is unlikely to change quickly in response to change in macrophytes and water levels, so damping out short-term ecosystem changes. Environmental variablility is more likely to influence the fish populations by changing the availability of spawning habitat.

ACKNOWLEDGEMENTS

This Chapter formed part of the joint research work of the University of Leicester, University of Nairobi and the Department of Fisheries, Kenya at Lake Naivasha in 1987–1989. The authors are grateful to the Office of the President, Kenya for research permission, and to the Director of Fisheries for giving time to SMM to undertake a higher degree for which this work was carried out. The work was funded by the British Council, Earthwatch, Elsa Wild Animal Appeal Fund and others. The authors are also grateful to the following for support in various ways: Phil Hickley, Frank Clark, Aggrey Oluoch, Rick North and Ken Mavuti for helping in work and many hours of discussions; Peter Kubai, John Dedge, Simon Orege, Paul Njomo and volunteers from Earthwatch helped with the sampling. Peter Magius, all the staff of the Department of Fisheries, Naivasha and staff at Elsamere Conservation Centre and others helped with logistics.

REFERENCES

Abdel-Malek, S.A. (1972) Food and feeding habits of some Egyptian fishes in Lake Quarun. Part I. *T. zillii* (Gerv.) A. According to different localities. *Bull. Inst. Oceanogr. Fish. Cairo,* **2,** 185–201.

Bowen, S.H. (1979) Determinants of the chemical composition of periphytic detrital aggregate in a tropical lake (Lake Valencia, Venezuela). *Arch. Hydrobiol.,* **87,** 166–77.

Bowen, S.H. (1982) Feeding, digestion and growth – qualitative consideration, *The Biology and Culture of Tilapias (ICLARM Conf. Proc.,* 7) (eds R.S.V. Pullin and R.H. Lowe-McConnell), International Centre for Living Aquatic Resource Management, Manila, Philippines, pp. 141–56.

Clark, F., Beeby, A. and Kirby, P. (1989) A study of the macro-invertebrates of Lake Naivasha, Oloidien and Sonachi, Kenya. *Revue Hydrobiol. trop.,* **22**(1), 21–3.

Fryer, G. and Iles, T.D. (1972) *The Cichlid Fishes of the Great Lakes of Africa: Their Biology and Evolution*, T.F.H. Publ., Neptune City, NJ. Also publ. by Oliver and Boyd, Edinburgh.

Gaudet, J.J. (1977) Natural drawdown on Lake Naivasha (Kenya) and the formation of papyrus swamps. *Aquat Bot.*, **3**, 1–47.

Harper, D.M. (1992) The ecological relationships of aquatic plants at Lake Naivasha, Kenya. *Hydrobiologia*, **232**, 65–71.

Harper, D.M., Muvuti, K.M. and Muchiri, S.M. (1990) Ecology and management of Lake Naivasha, Kenya in relation to climatic change, alien species introductions and agricultural development. *Env. Conserv.*, **17**, 328–36.

Hickley, P. and Bailey, R.G. (1987) Food and feeding relationships of fish in the Sudd swamps (River Nile, southern Sudan). *J. Fish Biol.*, **30**, 147–59.

MacArthur, R.H. and Wilson, E.O. (1967) *The Theory of Island Biogeography*. Princeton Univ. Press, Princeton, 203 pp.

Malvestuto, S. (1974) The fishes of Lake Naivasha: their biology, exploitation and management. Mimeo report to the Director of Fisheries, Kenya Government. 37 pp.

Mangel, M. and Tier, C. (1993) Dynamics and metapopulations with demographic stochasticity and environmental catastrophes. *Theor. Pop. Biol.*, **44**, 1–31.

MRAG (1992) Catch-effort data analysis. Unpublished software manual, MRAG Ltd, London and ODA, London.

Muchiri, M. (1990) The feeding ecology of tilapia and the fishery of Lake Naivasha, Kenya, PhD thesis, University of Leicester, UK.

Muchiri, S.M. and Hickley, P. (1991) The fishery of Lake Naivasha, Kenya, in *Catch Effort Sampling Strategies: Their Application in Freshwater Fisheries Management* (ed. I.G. Cowx), Blackwell Scientific Publications, Oxford, pp. 382–92.

Pimm, S.L. (1991) *The Balance of Nature? Ecological Issues in the Conservation of Species and Communities*, Univ. of Chicago Press, Chicago, 434 pp.

Pimm, S.L., Jones, H.L. and Diamond, J. (1988) On the risk of extinction. *Am. Nat.*, **132**, 757–85.

Richardson, J.L. and Richardson, A.E. (1972) History of an African Rift lake and its climatic implications. *Ecol. Monogr.*, **42**, 499–534.

Ricker, W.E. (1975) Computation and interpretation of biological statistics of fish populations. *Bull. Fish. Res. Bd Canada*, no. 191, 382 pp.

Schoener, T.W. and Spiller, D.A. (1992) Is extinction rate related to temporal variability in population size? An empirical answer for orb spiders. *Am. Nat.*, **139**, 1176–1207.

Siddiqui, A.Q. (1977) Lake Naivasha (Kenya, East Africa) fishery and its management together with a note on the food habits of fishes. *Biol. Conserv.*, **12**, 217–18.

Spataru, P. (1978) Food and feeding habits of *Tilapia zillii* (Gervais) (Cichlidae) in Lake Kinneret, (Israel). *Aquaculture*, **14**, 327–38.

Tracey, C.R. and George, T.L. (1992) On the determinants of extinction. *Am. Nat.*, **139**, 102–22.

Wallace, R.K. jun. (1981) An assessment of diet overlap indexes. *Trans. Am. Fish. Soc.*, **110**, 72–6.

Wilhm, J.T. (1967) Comparison of some diversity indices applied to populations of benthic invertebrates in streams receiving organic wastes. *J. Water Poll. Contr. Fed.*, **39**, 1673–88.

Zaret, D. and Rand, A.S. (1971) Competition in stream fishes: support for the competitive exclusion principle. *Ecology*, **52**, 336–42.

Chapter fifteen

Species changes in reservoir fisheries following impoundment: the case of Lake Itezhi-tezhi, Zambia

Ian G. Cowx and Cyprian K. Kapasa

15.1 INTRODUCTION

In Africa since the 1950s, a large number of dams have been constructed for electricity generation, improved water supply or for irrigation (Table 15.1). In all the major and many of the larger reservoirs impounded by these dams, commercial fisheries have developed. However, many of these larger water

Table 15.1 Relationship between surface area (km^2 at upper storage level, u.s.l.) and number of reservoirs in Africa

Reservoir category	Surface area (km^2 at u.s.l.)	Number	Examples
Major	>1 000	8	Kariba, Kainji, Volta
Large	100 to 999	32	Metra, Bamendjing, Itezhi-tezhi
Medium	10 to 99.9	74	Massinga, Nyumba ya Munga
Small	1.0 to 9.99	118	
Other	0.01 to 0.99	>623	

The Impact of Species Changes in African Lakes. Edited by Tony J. Pitcher and Paul J.B. Hart. Published in 1995 by Chapman & Hall, London. ISBN 0 412 55050 4.

bodies, and most of the medium and smaller systems, have not achieved their full fisheries potential in terms of predicted yield.

In many cases the newly developed fisheries are highly productive, but suffer from declining catches through overexploitation and marked shifts in the species composition of catches in later years (Petr, 1968; Lelek, 1973; Dadzie, 1980; Balogun, 1986). This chapter examines the changes that have occurred in the fisheries of Lake Itezhi-tezhi, central Zambia, following impoundment of the upper Kafue River. It evaluates the factors which have contributed towards these changes with a view to improving the management of fisheries in existing and newly impounded reservoirs.

15.2 MATERIALS AND METHODS

Lake Itezhi-tezhi (Fig. 15.1), built as an upstream storage reservoir for the Kafue Gorge Upper Hydropower Scheme, was completed in 1977 as an

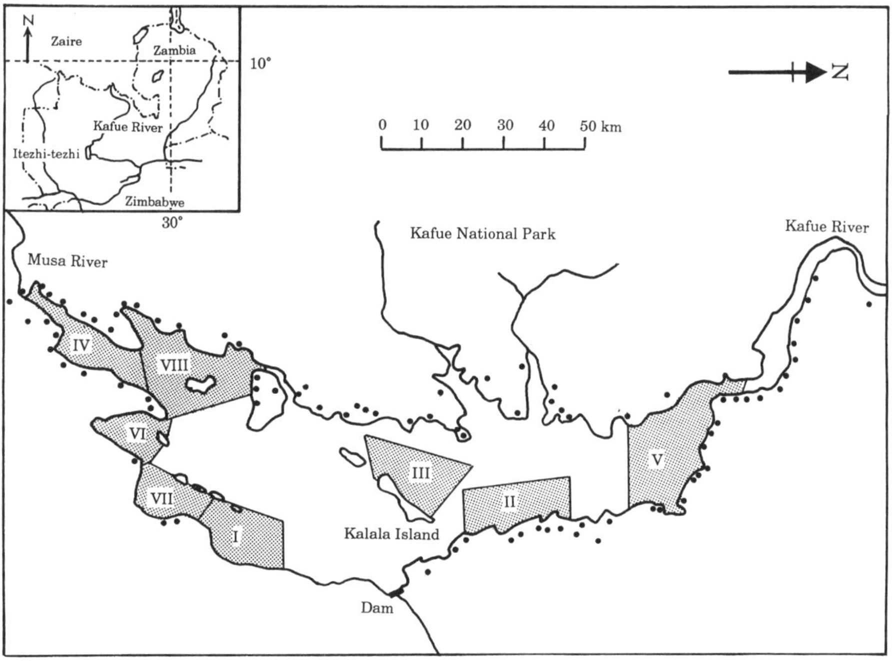

Fig. 15.1 Location of Lake Itezhi-tezhi in Zambia and details of its catchment and sampling sites. I–VIII, Sampling stations used in the present study; ●, landing stations used in catch assessment survey.

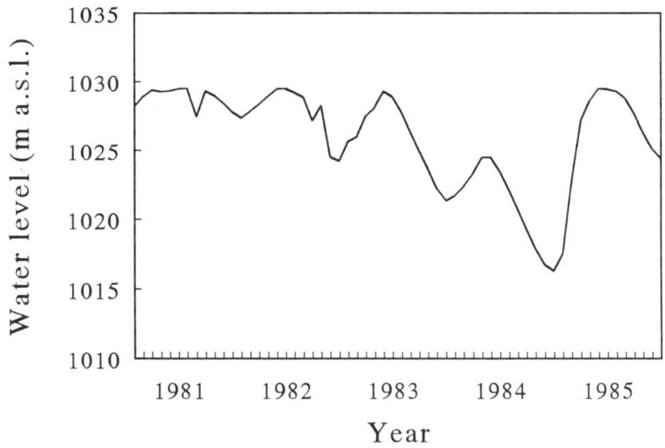

Fig. 15.2 Monthly variations in water level of Lake Itezhi-tezhi from 1981 to 1985.

extension of the Kafue Gorge Upper Power Station. At the upper storage level of 1029.5 m above sea level (a.s.l.) the reservoir has a surface area of approximately 370 km^2, an area 42 times greater than the river production area before inundation. However, as a result of the operational strategy practised at the reservoir, it is subject to large-scale drawdown (Fig. 15.2) which significantly reduces the surface area and volume of the impoundment.

A detailed description of the Kafue river system is given by Dudley and Scully (1980) and the hydrological and topographical features of Lake Itezhi-tezhi by Obrdlik *et al.* (1989) and Kapasa and Cowx (1991), respectively. Of particular note to this study is the flooding pattern of the lake, which varies both seasonally and annually (Fig. 15.2).

Post-impoundment changes in the fisheries of Lake Itezhi-tezhi were assessed in two ways. First, the total fish yield from the reservoir has been recorded by a simple catch assessment survey (Bazigos, 1974) since 1980. However, due to inconsistencies in the data collection, this information can only be used to indicate relative changes in the fish yield from the lake.

The second method was a fishery-independent survey carried out by the Zambian Fisheries Department to support the catch assessment survey. Eight locations, with differing habitat topography, were sampled regularly between January 1981 and December 1985 (Fig. 15.1).

All fish were caught in 25 m × 3 m gill nets ranging in mesh size from 25 to 178 mm. The nets of mesh size less than 50 mm were designed to catch *Alestes lateralis.* Between 1981 and 1983, nets were set in fleets of eight at each sampling station on two consecutive nights, at least twice a month. The number of fishing nights per sample was increased to three in January

1984 but in January 1985, owing to a national fuel crisis, only stations I, III, V and VI were sampled. All fish caught were identified to species, and the number and weight were recorded. Further details of the fishery-independent survey and the results obtained are provided by Kapasa and Cowx (1991).

15.3 RESULTS

Species diversity

Prior to any impoundment, the Kafue fishery system had over 56 species (Mortimer, 1965). The dominant group was the 'Kafue breams' (Bell-Cross, 1965). However, since inundation only 24 species have been recorded from Lake Itezhi-tezhi (Table 15.2; Kapasa and Cowx, 1991).

The decline in species diversity was primarily due to members of the families Cyprinidae, Mormyridae and Schilbeidae being unable to adapt to the new lacustrine environment, particularly in relation to spawning and feeding grounds, and being unable to cope with the fluctuating water levels in the lake. The increase in diversity in 1985 was due to the rise in water levels which brought in, from the Kafue and Musa Rivers, rheophilic species that have a preference for vegetation (Kapasa and Cowx, 1991). These included *Barbus, Mormyrus* and *Haplochromis* species.

Table 15.2 Pre- and post-impoundment diversity of fish fauna in Kafue River and Lake Itezhi-tezhi. Data are number of fish species

Family	Pre-impoundment 1971	Post-impoundment				
		1981	1982	1983	1984	1985
Mormyridae	4	2	3	3	0	3
Characidae	4	2	1	1	1	2
Cyprinidae	18	5	4	3	2	5
Schilbeidae	2	2	1	1	1	1
Clariidae	4	2	1	1	2	1
Mochokidae	3	1	1	1	2	1
Cichlidae	13	10	8	8	8	9
Cyprinodontidae	3	0	0	0	0	0
Kneriidae	1	0	0	0	0	0
Citharinidae	2	0	0	0	0	0
Anabantidae	1	0	0	0	0	0
Mastacembelidae	1	0	0	0	0	1
Total	56	24	19	18	16	23

Fish yield

Fish production in Lake Itezhi-tezhi (Fig. 15.3) exhibited the typical pattern of high yield, characteristic of most impounded waters in the early years after inundation, followed by a marked decline in catches (Kapetsky and Petr, 1984).

In addition to this initial decline in high productivity, several other factors contributed towards the pattern observed. Firstly, coincidental with the decline in catches was a mass immigration of fishers, from approximately 800 in 1979–1980 to more than 1250 in 1982–1984, following their immigration from the Kafue floodplains, which at the time were being overfished.

The fall in catches in 1987 was also the result of a management decision to impose a closed season on the fishery between November and March. This was accompanied by a ban on small-mesh nets (< 76 mm mesh size) which effectively excluded the catching of the characid, *A. lateralis*. These regulations were enacted in 1984. As this species previously accounted for more than 40% of the total commercial catch by weight, the ban caused fishers to change profession or move to other fisheries and their numbers dropped to fewer than 750 in 1985.

Finally, and perhaps most importantly, the fluctuating water level as a result of the drawdown procedure operated on Lake Itezhi-tezhi has a marked effect on catches. Experimental gill net catches (kg per 25 m gill net) were negatively related to reservoir water level (m a.s.l.) according to the following relationship:

$$\textit{Water level} = 0.334\ \textit{Catch} - 34 \qquad (r = 0.78,\ P < 0.05) \qquad (15.1)$$

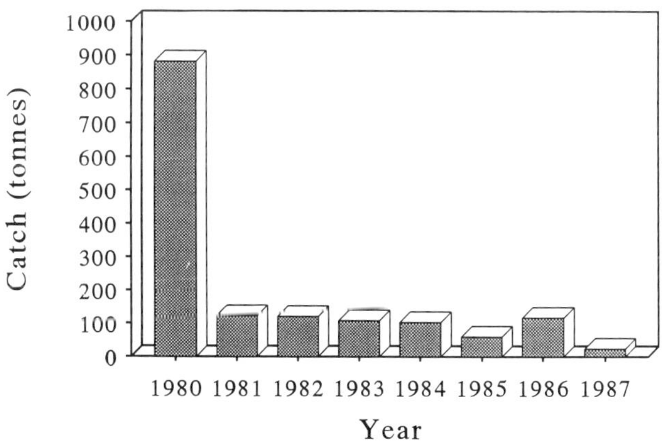

Fig. 15.3 Annual commercial catch from Lake Itezhi-tezhi from 1980 to 1987 (source: Department of Fisheries, Chilanga, Zambia). Potential annual yield based on morphoedaphic index is 1800 tonnes (Coopconsult-Propesca, 1982).

Irrespective of these factors, the lake has never achieved the predicted yield of 1800 t, based on the morpho-edaphic index (Coopconsult-Propesca, 1982).

Species composition of catches

Marked changes in the composition of the fish catches from Lake Itezhi-tezhi, as reflected in the experimental gill net catches, have occurred since impoundment (Figs 15.4 and 15.5). The most prominent change has been the shift from catches with a preponderance of the characid *A. lateralis*, which represented 49% of the fish caught by number in 1981 and declined to 21% in 1985, to one dominated by Cichlidae.

Ten members of the family Cichlidae were recorded in 1981, constituting 11% of the total catch by number and 31% by weight. By the end of 1985 this contribution had increased to 63% and 71% by number and weight, respectively. The changes were mainly due to a steady increase in the catches of *Oreochromis microchir*, a dramatic increase in numbers of *Haplochromis giardi* in 1985 and an increase in the weight of *Oreochromis andersonii* in 1985 (Kapasa and Cowx, 1991).

The family Cyprinidae accounted for the greatest number of species present in the Kafue river system, comprising 18 species (Mortimer, 1965). However,

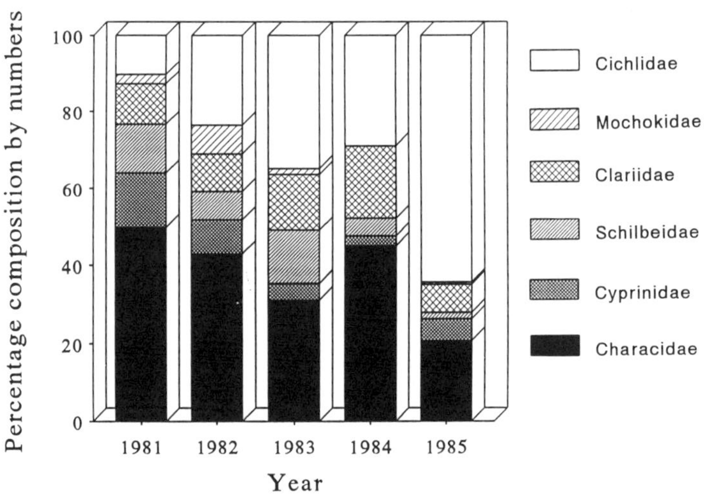

Fig. 15.4 Change in the species composition by numbers of experimental gill net catches from Lake Itezhi-tezhi between 1981 and 1985.

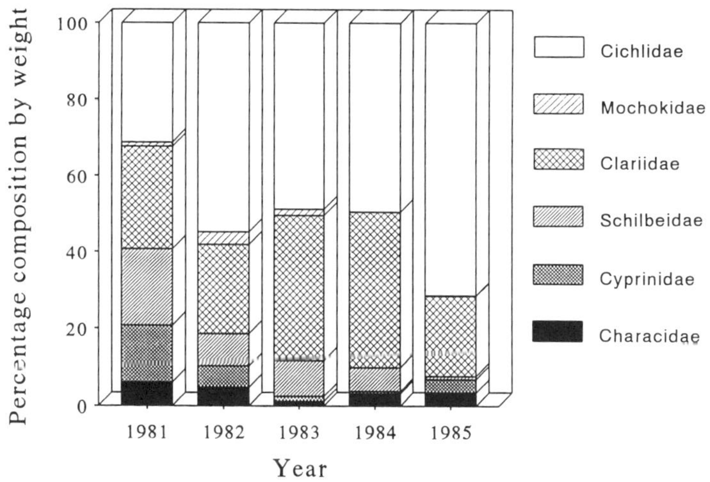

Fig. 15.5 Change in the species composition by weight of experimental gill net catches from Lake Itezhi-tezhi between 1981 and 1985.

only five, *Labeo cylindricus*, *Labeo molybdinus*, *Barbus poechii*, *B. paludinosus* and *B. marequensis*, were found in the commercial and experimental gill net catches. In all cases their contribution generally declined over the study period.

Two species of Schilbeidae were found in the lake. *Eutropius depressirostris* disappeared almost immediately after filling whilst *Schilbe mystus* declined both in numbers and weight over the study period but still remained an important component of the ichthyofauna.

Two other species important to the catches, particularly in terms of weight, were *Clarias gariepinus* and *Synodontis kafuensis*, the latter being the only member of the family Mochokidae to be caught. Although the importance of *Clarias* in terms of numbers landed was low (Fig. 15.4), it contributed more than 20% of the total catch by weight (Fig. 15.5).

The distribution of fish species in the lake, and hence composition of the catches, is strongly influenced by the topography of the sampling station. *A. lateralis* appears to be more abundant in open water areas of the shallow littoral region, particularly during low water levels, and cichlids (e.g. *Oreochromis microchir* and *Tilapia rendalli* were generally found near the shores amongst the vegetation. Similar distribution characteristics for cichlids were found by Lelek (1973), Dadzie (1980) and Balogun (1986) for other impounded waters in tropical Africa.

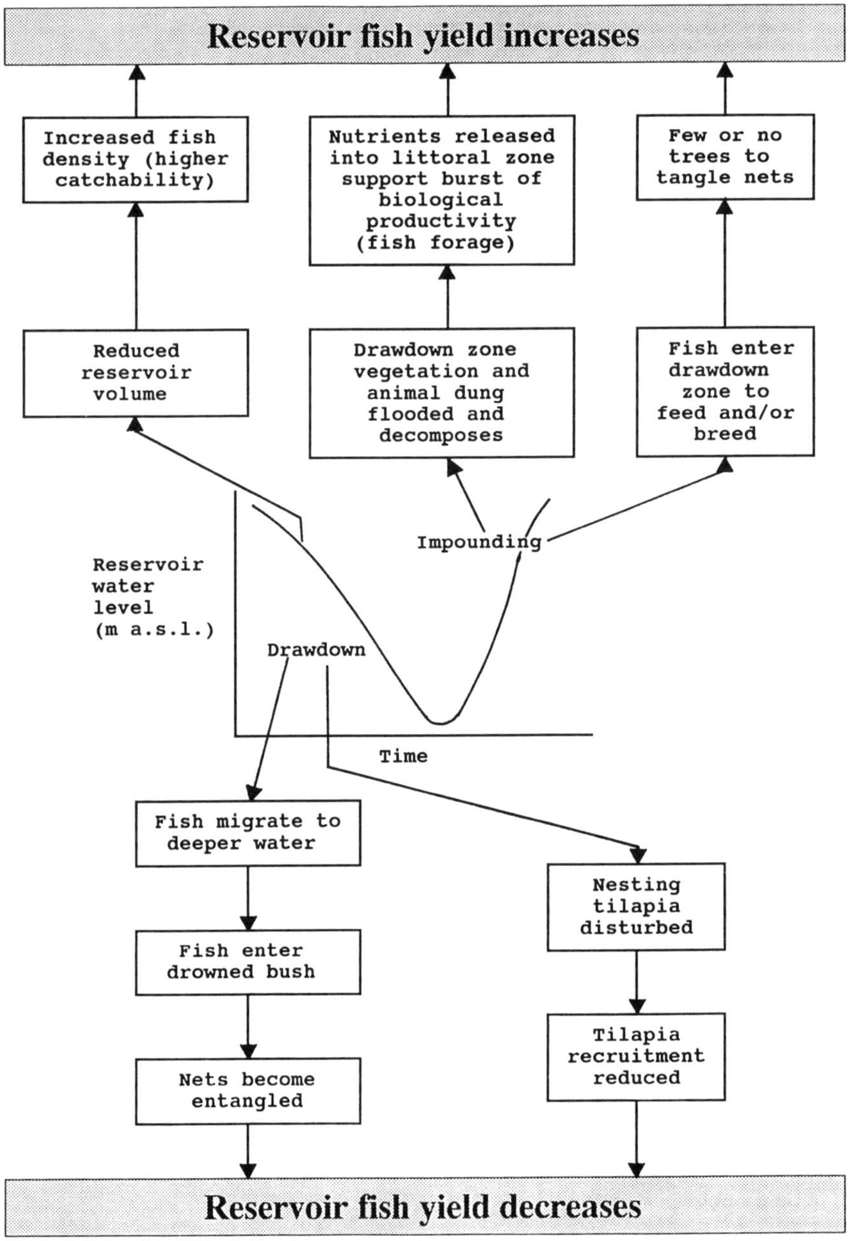

Fig. 15.6 Summary of the effects of water level fluctuations in reservoirs on fish yield. Adapted from Bernacsek (1984).

15.4 DISCUSSION

The most obvious changes in the fish community structure in Lake Itezhi-tezhi since its completion have been the decline in species diversity and the shift in species composition from a reasonably diverse community, with a preponderance of *A. lateralis*, to one dominated by cichlids (Table 15.2). The reduction in diversity suggests that some species, including the mormyrids, *Labeo* spp., *E. depressirostris* and *H. odoe* have not been able to adapt to the new environmental characteristics of the lake, e.g. a shift in spawning and feeding grounds. Similar observations have been made for Lake Kainji (Lelek, 1973; Balogun, 1986), Lake Kariba (Balon, 1972) and Lake Kamburu (Dadzie, 1980).

Fish production in Lake Itezhi tezhi followed the characteristic pattern of early high productivity in the first few years after flooding followed by a steady, but much lower yield (Petr, 1975), well below that predicted (1800 t). Several factors are considered to contribute towards the pattern observed. In the early inundation period, fish species that were able only to tolerate riverine habitats withdrew from the flooded zone. During the same period fish of lacustrine character, or those able to adapt, invaded the reservoir. These changes were coupled with the liberation of nutrients through bacterial decomposition of organic matter from flooded plants, animals and organic debris which subsequently resulted in high primary and secondary productivity, including that of the fish fauna (Entz, 1984). Once the nutrients became exhausted, the peak was followed by a sharp decline in production at all levels. Similar trends have also been reported by Balon and Coche (1984) in Lake Kariba, Lelek (1973) and Blake (1977) in Lake Kainji and Vanderpuye (1984) in Lake Volta.

Once the fish community had established in Lake Itezhi-tezhi it was apparent that no indigenous plankton-feeding species was present to exploit the open-water pelagic zone. A similar situation was observed in Lake Kariba, where the clupeid plankton feeder *Limnothrissa miodon*, a sardine locally known as kapenta, was introduced from Lake Tanganyika in an attempt to increase productivity. This introduction proved successful and the sardine now sustains a major part of the fishery (Marshall, 1988). In other lakes, such as Volta or Kainji, clupeids were present in the pelagic zone and such an introduction was not considered necessary (Entz, 1984). In an attempt to increase the sustainable fish yield from Lake Itezhi-tezhi, *Limnothrissa* was introduced in early 1992, and now appears to be well established throughout the lake (Chapters 13, 24).

Unfortunately, during the highly productive initial phase, the fishery of Lake Itezhi-tezhi was heavily exploited because of the immigration of fishers from the Kafue floodplain fishery, which itself was being overexploited. This accelerated the fall-off in fish yields as the reservoir stabilized.

Dams built for hydroelectric power generation control the water level in response to power supply requirements and to accommodate the flood water from their catchments at the start of the rainy season. This drawdown, and the way in which it is achieved may not be advantageous to the development of fisheries in reservoirs (Fig. 15.6; Chapter 12). Lowering the water level may concentrate the fish and allow higher catches to be taken. However, the rapid drawdown might have an adverse effect on fish that spawn in the littoral zone, such as nest-building tilapias, killing eggs and larvae and thus reducing future recruitment to the fishable stocks of these species. On the other hand, this drawdown may clear trees and brush in the drawdown zone over a period of years (Bernacsek, 1984), thus improving the fishing pitches and obviating the need for expensive clearing prior to inundation. Figure 15.6 summarizes the effects on fish yields of large-scale fluctuations in water level, a practice which in Lake Itezhi-tezhi, like many other reservoirs, appears to be detrimental to fish catches.

15.5 CONCLUSIONS AND RECOMMENDATIONS

This study clearly shows the potential of reservoirs for providing a valuable source of high quality protein to rural communities. However, there are a number of conclusions and recommendations which should be considered in the development of fisheries in new and existing reservoirs, in Africa and elsewhere, to maximize the yield and enable rational management of the resource.

Lake Itezhi-tezhi, like many other reservoirs in Africa, has not achieved its predicted yield. This may be for two reasons. First, the methods for predicting fish production are usually based on morpho-edaphic indices or by comparison with similar reservoirs; neither approach has proved particularly accurate. There is thus a need to focus attention on pre-impoundment studies aimed at predicting yield of future reservoirs and to develop more accurate reservoir-specific predications based on early catch data and morpho-edaphic indices (Chapter 23). Second, there is often a vacant niche for a plankton-feeding fish species in the open-water pelagic zone. Consideration should therefore be given to introducing a species, such as *L. miodon*, to exploit this niche and improve production (EIFAC, 1988).

There is a need to develop methods for improving reservoir and downstream fisheries through better dam design and reservoir operation (Bernacsek, 1984). This specifically relates to the reservoir drawdown procedure which often results in a rapidly fluctuating discharge regime in the river below the dam. This practice can be detrimental to the stability of fish stocks through impaired recruitment, particularly of those fish inhabiting the littoral zone of the reservoir. Research should be targeted at improving our understanding of the detrimental effects of reservoir operating procedures on fisheries and identifying mechanisms of minimizing the impact.

The literature on clearing endemic vegetation from the reservoir area prior to inundation is vague and contradictory. Evidence suggests that bush clearing reduces periphyton production which in turn lowers the fish production and consequent yield (Entz, 1984). However, the presence of trees and bushes, which can take many years to decay, can impede fishing and reduce access to the fishery. The importance of flooded bushes etc. for primary production and the impact that clearing has on fish production need to be evaluated. Consideration should be given to the benefit gained from clearing in relation to the high cost of the exercise and how drawdown can be utilized to establish fishing pitches (Bernacsek, 1984).

Finally, the fisheries of many of the medium and small reservoirs have received little attention. Action should be focused on increasing the sustainable yield from these reservoirs through improved management strategies, thus making available much-needed protein in rural communities.

15.6 SUMMARY

The many reservoirs in Africa offer the potential for increasing the supply of fish to rural communities but rational management is required to meet this objective. This Chapter examines the shifts in the fish community and fisheries of Lake Itezhi-tezhi, Zambia, following inundation to identify the problems associated with reservoir fisheries.

The fish fauna of the newly inundated Lake Itezhi-tezhi was observed between 1980 and 1985. Marked changes in the community structure and yield were identified. The most obvious were a decline in fish catches following early high productivity, a decline in species diversity and a shift in species composition from a reasonably diverse community with a preponderance of *Alestes lateralis* (Characidae), to one dominated by cichlids. These changes were primarily due to members of the families Cyprinidae, Mormyridae and Schilbeidae being unable to adapt to the new environmental conditions, such as spawning and feeding grounds, and their inability to cope with fluctuating water levels in the lake. Fish yield from the lake never achieved its predicted potential and the lack of a plankton-feeding species to exploit the open-water pelagic zone was considered contributory. General recommendations for improving the fish yield from Lake Itezhi-tezhi and other artificial lakes are discussed.

ACKNOWLEDGEMENTS

Thanks are expressed to the Government of Zambia through the Director of the Fisheries Department for providing a scholarship to CK and The World Bank for financial support to conduct this work. Thanks also go to all those who assisted in the data collection. The opinions expressed are those of the authors.

REFERENCES

Balogun, J.K. (1986) Fish distribution in Kainji Lake, Nigeria. *J. Fish Biol.*, **29**, 489–98.

Balon, E.K. (1972) Possible fish stock assessment and available production survey as development on Lake Kariba. *Afr. J. trop. Hydrobiol. Fish.*, **5**, 119–62.

Balon, E.K. and Coche, A.G. (eds) (1974) Lake Kariba: a man-made tropical ecosystem in Central Africa. *Monogr. Biol.*, **24**, 767 pp.

Bazigos, G.P. (1974) The design of fisheries statistical surveys – inland waters. *FAO Fish. tech. Pap.*, no. 133, 122 pp.

Bell-Cross, C. (1965) Additions and amendments to the check-list of the fisheries of Zambia. *Puku, Occ. Pap. Dept Game Fish., Zambia*, **3**, 29–43.

Bernacsek, G.M. (1984) Dam design and operation to optimize fish production in impounded river basins. *CIFA tech. Pap.*, no. 11, 98 pp.

Blake, B.F. (1977) Lake Kainji, Nigeria: a summary of the changes within the fish population since the impoundment of the Niger in 1968. *Hydrobiologia*, **53**, 131–7.

Coopconsult-Propesca (1982) Zambia inland fisheries development project. Interim report. Repub. Zambia, Min. Lands Nat. Res. Ed Coopconsult-Propesca S.p.A Inc., Il Nuovo Castro, Rome, January 1982. 59 pp., 7 Annex.

Dadzie, S. (1980) Recent changes in the fishery of a new tropical man-made lake, Lake Kamburu (Kenya). *J. Fish Biol.*, **16**, 361–7.

Dudley, R.G. and Scully, R.J. (1980) Changes in the experimental gillnet catches for the Kafue Floodplain, Zambia, since construction of the Kafue Gorge Dam. *J. Fish Biol.*, **16**, 521–37.

Entz, B.A.G. (1984) A synthesis and evaluation of activities of FAO/UNDP projects on five African man-made lakes: Kainji, Kariba, Kossou, Nasser-Nubia and Volta. *FAO Fish. Circ.*, no. 774, 34 pp.

EIFAC (European Inland Fisheries Advisory Commission) (1988) Code of practice and manual of procedures for consideration of introductions and transfers of marine and freshwater organisms. Rome, FAO, *EIFAC occ. Pap.*, no. 23, 23 pp.

Kapasa, C.K. and Cowx, I.G. (1991) Post-impoundment changes in the fish fauna of Lake Itezhi-tezhi, Zambia. *J. Fish Biol.*, **39**, 783–93.

Kapetsky, J.M. and Petr, T. (eds) (1984) Status of African reservoir fisheries. *CIFA tech. Pap.*, no. 10, 326 pp.

Lelek, A. (1973) Sequence of changes in fish populations of a new tropical man-made Lake Kainji, Nigeria, West Africa. *Arch. Hydrobiol.*, **71**, 381–420.

Marshall, B.E. (1988) A preliminary assessment of the biomass of the pelagic sardine *Limnothrissa miodon* in Lake Kariba. *J. Fish Biol.*, **32**, 515–24.

Mortimer, M.A.E. (1965) *Natural Resources Handbook. The Fish and Fisheries of Zambia.* Government of Zambia Publication, 27 pp.

Obrdlik, P., Mumeka, A. and Kasonde, J.M. (1989) Regulated rivers in Zambia – the case study of the Kafue River. *Regulated Rivers: Res. Manage.*, **3**, 371–80.

Petr, T. (1968) The establishment of a lacustrine fish population in the Volta lake in Ghana during 1964–1966. *Bull. Inst. Fr. Afr. noire*, **30A**, 257–69.

Petr, T. (1975) On some factors associated with the initial high fish catches in new African man-made lakes. *Arch. Hydrobiol.*, **75**, 32–49.

Vanderpuye, C.J. (1984) Volta Lake (Ghana), in Status of African Reservoir Fisheries (eds J.M. Kapetsky and T. Petr), *CIFA tech. Pap.*, no. 10, pp. 261–319.

Part Three

Evaluation of Species Changes in African Lakes without Introduced Species

Chapter sixteen

Changes in species composition and abundance of fish populations in Lake Turkana, Kenya

Jeppe Kolding

16.1 INTRODUCTION

Lake Turkana, a large holomictic endorheic Rift Valley lake between 2°27′ N and 4°40′ N in the arid northern part of Kenya (Fig. 16.1), is relatively little studied and understood. Besides a 1930–31 Cambridge University Expedition (Worthington and Ricardo, 1936), which worked for only a few months on the lake in a localized area, only one major investigation into the biology has been made. This was the Lake Turkana Fisheries Research Project sponsored by the Overseas Development Administration of the UK Government from 1972 to 1975, which submitted an exhaustive report on the limnology, biology and productivity of the lake (Hopson, 1982a). Probably due to its equatorial location, Lake Turkana has commonly been associated with environmental and biological stability (e.g. Beadle, 1981; Merona *et al.*, 1988). It appears, however, under closer scrutiny to be more in a state of perpetual change strongly governed by the hydrological regime (Kolding, 1992). The first account of unstable conditions was given by the District Commissioner, who wrote in 1954:* "The yields vary enormously from year

*KNA, Kenya National Archives, Nairobi. File name: DC/LODWAR/1/8/5 – Fishing on Lake Rudolf 1954–57.

The Impact of Species Changes in African Lakes. Edited by Tony J. Pitcher and Paul J.B. Hart. Published in 1995 by Chapman & Hall, London. ISBN 0 412 55050 4.

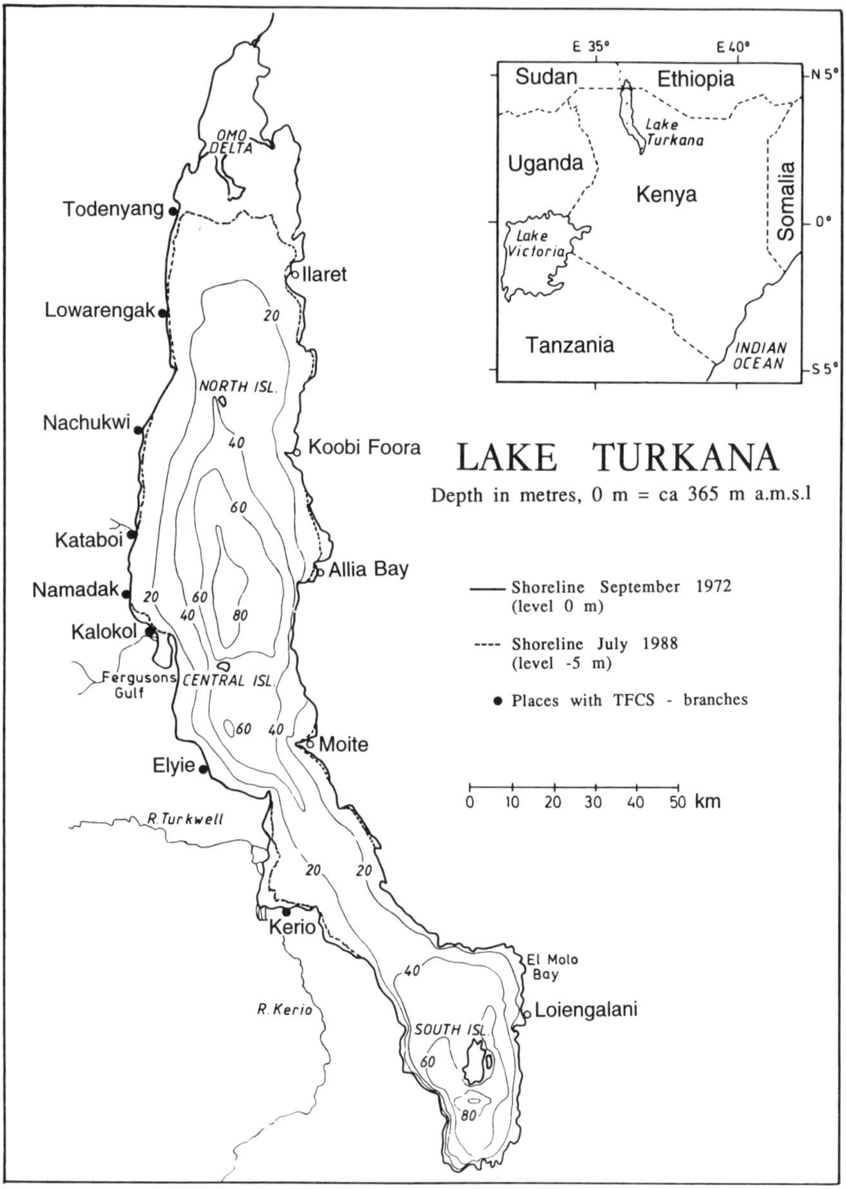

Fig. 16.1 Topography and bathymetry of Lake Turkana, Kenya. Turkana Fishermen's Cooperate Society (TFCS) branches correspond to experimental fishing stations. Depth contours refer to 0 m lake level in September 1972. Modified from Hopson (1982a) and Källqvist *et al.* (1988).

to year and from month to month and even from day to day – nobody seems to know why. If, as is claimed, the lake really does teem with fish it certainly does not teem close inshore for more than a few days at a time in favourable seasons ...". However, with the exception of Bayley (1977) from the Hopson team, who dealt with the changes in commercial landings during the initial period from 1962 to 1975, nothing empirical has been published on natural or fishery-induced changes in the fish of Lake Turkana during recent decades. This is presumably for several reasons: first, there has been a nearly complete cessation in the monitoring of biological parameters since the termination of the Hopson survey. Second, the scattered fishery statistics available have been subject to repeated criticism regarding accuracy and reliability (Bayley, 1977, 1982a; Gjerstad, 1983, 1986; NORAD, 1985), which has resulted in many changes. Lastly, a complete time series on catch composition, yield and effort did not exist until recently. A reconstructed time series of these variables from 1962 to 1988, based on all the reports, documents and information that could be traced, was compiled by Kolding (1989), who also analysed an experimental fishing survey conducted in 1986–87 under the auspices of the Norwegian Agency for Development Cooperation (NORAD). This unpublished material, together with the Hopson report, give the background information from which the results of this Chapter are derived. The period under discussion encompasses major changes in the fishery of Lake Turkana in terms of effort, yield and species composition as well as environmental changes in terms of fluctuating lake levels. The aim of this Chapter is to describe these changes and to discuss their relative impacts and importance for the observed changes in the fish populations of Lake Turkana.

16.2 BRIEF HISTORY OF THE FISHERY

The commercial fishery in Lake Turkana commenced around 1946 with one fishing concession at Kalokol. From the early 1960s it increased and expanded north and south on the western shore to reach Todenyang and Kerio by 1971 (Fig. 16.1). In 1968 the Kenya Government established the Turkana Fishermen's Cooperative Society Ltd. (TFCS) and by 1973–74 all fishing branches on the western shore had been amalgamated under this society, which gained a monopoly of the fishing and trade on the western side (the eastern side is much less populated and there are few permanent fishing settlements). The fishing effort grew nearly exponentially from 1962, with about 150 fishers and four boats, to more than 7000 fishers and 200 boats in 1978–1980. Since then the effort has rapidly declined to fewer than 2000

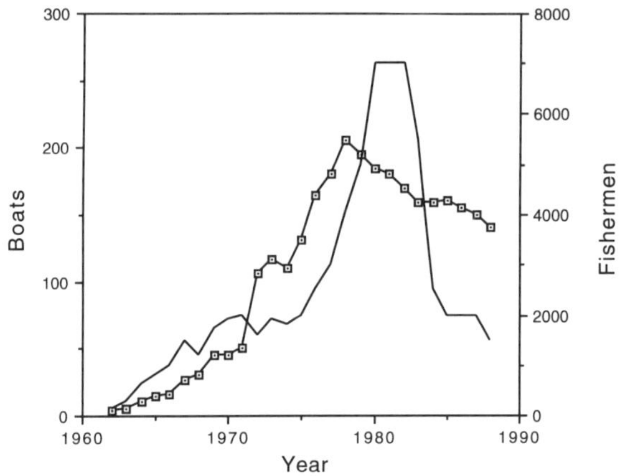

Fig. 16.2 Fishing effort in terms of number of fishers (solid curve) and number of boats (squares) on the western side of Lake Turkana, 1962–1988.

fishers with fewer than 150 boats in 1988 (Fig. 16.2). The fishery is entirely inshore (less than 10–15 m depth) and utilizes passive gear only, especially gill nets, as no fishing boats are motorized on Lake Turkana. Beach seining has been prohibited since 1976, but is still widely used at sheltered localities for catching tilapia.

The species composition in the landings has changed considerably over the years (Table 16.1). Initially it was virtually a single-species fishery (first beach seining for tilapia and later, when large-meshed 200 mm gill nets were introduced, for *Citharinus citharis*). Around 1972, after the collapse of this latter stock, a more diversified fishing pattern with a wider range in mesh sizes (minimum legal is 125 mm) and lines began, which also diversified the landings (Bayley, 1977). However, large natural fluctuations, especially intermittent peak tilapia catches (Kolding, 1993a), are reflected in the landing data.

Due to the high evaporation rate (about 2.5 m year^{-1}) and closed basin, the water level of Lake Turkana fluctuates widely following the general sub-Saharan periods of drought alternating with above-average rainfalls (Butzer, 1971; Ferguson and Harbott, 1982; Kolding, 1989). In all other climatic parameters (irradiation, wind pattern, temperature and circulation), Lake Turkana is remarkably stable compared with other large African lakes (Beadle, 1981). Between 1979 and 1987 the lake level receded almost linearly by 6 m and during the same period commercial landings went down some 80% (Fig. 16.3).

Table 16.1 Total commercial yield (tonnes) and relative species composition in per cent wet weight of landings on the western side of Lake Turkana. Based on Kolding (1989)

Year*	Yield (t)	Tilapia†	*Lates* sp.	*Citharinus*	*Distichodus*	*Labeo horie*	*Barbus bynni*	*Hydrocynus* sp.	Others‡	Unspecified
1951[a]	69	80	11							9
1952[a]	37	84	10							6
1953[a]	23	78	14							8
1954[a]	58	79	8							13
1955[a]	13	69	21							10
1956[a]	27	67	23							10
1963[b]	1 153	3	23	70	4					
1969[b]	1 805	2	2	90	5				1	
1970[b]	2 628	45		55						
1972[c]	2 101	5	14	24	17	21	3	8	8	
1973[c]	2 276	7	14	13	9	26	10	10	11	
1974[c]	2 250	13	14	3	3	29	12	9	17	
1975[c]	3 516	56		1	2	13	7			21
1976[c]	21 070	>80								<20
1979[c]	4 628	43								57
1980[c]	4 154	39								61
1981[c]	2 871	28								72
1982[d]	4 013	69	8	1	4	13	2	1	2	
1983[d]	3 220	34	14							52
1984[d]	2 018	14	11							75
1985[d]	1 146	16	32	3	5	23	3	2	3	
1986[d]	850	11	38	1	2	36	2	2	8	
1987[d]	906	25	31	1	1	31	2	1	8	
1988[d]	873	17	30	3	5	27	5	1	13	

* Gear used: [a]100 mm seine and few hooks; [b]200 mm nets; [c]125–200 mm nets, hooks and increasing seining; [d]100–150 mm nets, 75–100 mm seines and some hooks.

† After 1977 most of the tilapia is caught on the eastern side of the lake, except around the 1982 'boom' in Ferguson's Gulf.

‡ Consisting mainly of *Bagrus* sp., *Synodontis schall*, *Clarias lazera* and *Heterotis niloticus*.

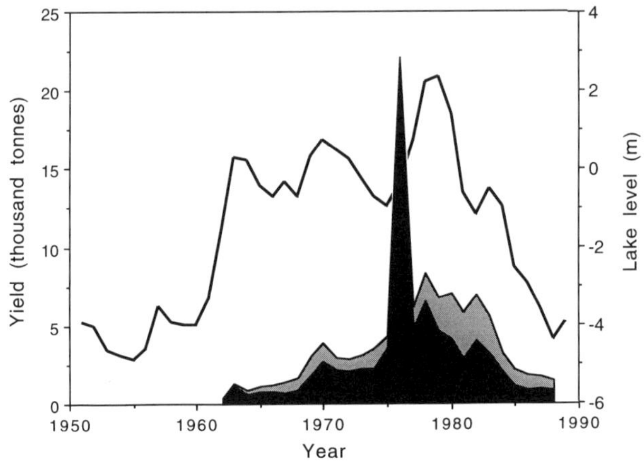

Fig. 16.3 Commercial annual yields (TFCS purchases, dark shading) and estimated local subsistence consumption (light shading) in tonnes wet weight on the western side of Lake Turkana together with mean annual lake levels (solid curve) over the last four decades.

16.3 EXPERIMENTAL FISHING SURVEYS

Only two extensive experimental fishing surveys have been carried out in Lake Turkana. The first was the Lake Turkana Fisheries Research Project (Hopson, 1982a), which is by far the most comprehensive including acoustic, trawl and gill net investigations over most parts of the lake. The second was the NORAD experimental gill net and longline fishery from September 1986 to September 1987 (Kolding, 1989), which covered the north-western side from Kerio to Todenyang out to the 25 m depth contour. The nets were mounted in fleets consisting of four mesh sizes (58, 78, 105 and 125 mm), and a total of 769 gill net settings, set both at the surface and on the bottom, and 107 longline settings were performed at seven sampling stations coinciding with TFCS branches (Fig. 16.1).

In the combined catch composition of all mesh sizes in the 1986–87 survey the contribution of the catfish *Synodontis schall* was by far the greatest (64% by number and 43% by weight). This picture changed when looking at the individual mesh sizes (Fig. 16.4). In the larger mesh sizes, 125 and 105 mm, the dominant species was the cyprinid *Labeo horie* with respectively 60% and 36% on average by weight. The third most important fish was the two species of the centropomid piscivore genus *Lates*, which were not taxonomically differentiated. In the longlines, the demersal predator *Bagrus bayad* dominated the catch with 42% by weight followed by large Nile perch, *Lates niloticus*,

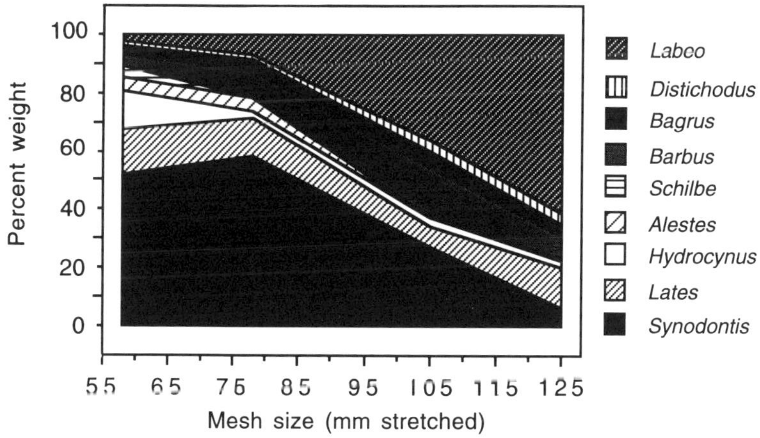

Fig. 16.4 Species composition of the dominant species in per cent by weight interpolated between the four mesh sizes 58, 78, 105 and 125 mm used in the NORAD experimental fishery survey.

with a mean weight of about 11 kg. Large *Clarias lazera*, a species not caught with the gill nets, was also important on the longlines, but was restricted to the northern stations.

Figure 16.5 shows a three-dimensional plot of the mean standard CPUE of each of the four mesh sizes at seven sampling stations from north to south while Fig. 16.6 shows the mean CPUE, all mesh sizes combined, for the five dominant species at the seven stations. An overall decline in CPUE from north to south, with a minimum in Kalokol, is apparent, particularly for the two smallest mesh sizes (78 and 58 mm). This diminishing gradient with distance from the north is in accordance with the primary productivity, which in the southern part of the lake is on average less than 10% of that in the north (Hopson, 1982a). For the largest mesh size (125 mm), which catches mainly the mud-sifting *Labeo horie*, highest values are found in the north (Todenyang) and the south (Kerio), with lower values in between. The plots show that the stations close to river inlets (Todenyang and Kerio) are the most productive from a commercial point of view (125 mm is the smallest legal mesh size and *Labeo horie* the most important commercial species). The River Kerio, however, flows only intermittently. The River Omo in the north, contributing about 90% of the water inflow, has the greatest overall abundance of 'small fish', whether they be small species or juveniles of larger species. This is consistent with the large proportion of species in Lake Turkana which are potamodromous spawners: *Labeo horie*, *Barbus bynni*, *Citharinus*

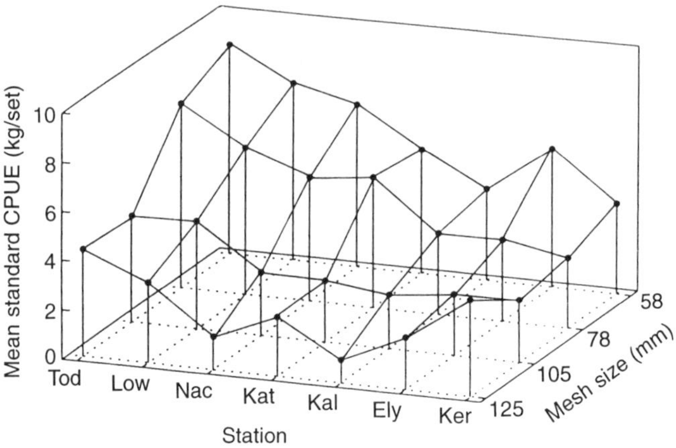

Fig. 16.5 Mean total catch per unit effort (kg per standard setting) for each of the four mesh sizes at the seven stations from north to south in the 1986–87 NORAD survey. Tod, Todenyang; Low, Lowarengak; Nac, Nachukwi; Kat, Kataboi; Kal, Kalokol; Ely, Elyie; Ker, Kerio; see Fig. 16.1 for locations.

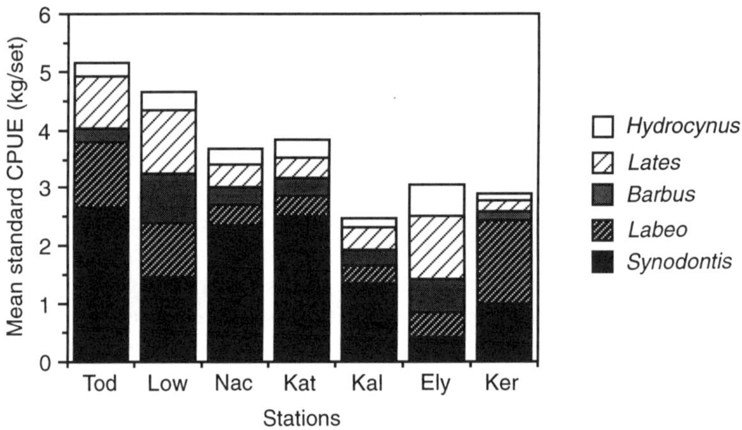

Fig. 16.6 Mean total catch per unit effort (kg per standard setting), all mesh sizes combined, for the five dominant species in the gill nets at the seven stations from north to south in the 1986–87 NORAD survey. Tod, Todenyang; Low, Lowarengak; Nac, Nachukwi; Kat, Kataboi; Kal, Kalokol; Ely, Elyie; Ker, Kerio; see Fig. 16.1 for locations.

citharis, *Distichodus niloticus*, *Alestes baremose*, *Hydrocynus forskalii*, *Clarias lazera*, *Synodontis schall* and *Schilbe uranoscopus*.

Comparison between the two surveys

The Hopson team conducted experimental fishing with coarse gill net fleets (100–230 mm) from November 1972 to November 1973 (Bayley, 1977, 1982a), and with surface fleets of fine nets (51–100 mm) in the period May 1974 to August 1975 (Hopson and McLeod, 1982). In four key areas, Todenyang, Kataboi, Kalokol and Kerio (Fig. 16.1), these surveys coincided with the NORAD survey and, because the net properties and timing of the sampling were almost identical between the two periods (Kolding, 1989), the catch per unit effort (CPUE) in the same mesh sizes should be comparable as indices of stock abundance because the catchability coefficient (q) can be assumed constant. In order to make comparisons between the surveys, all CPUE values are expressed as mean catch (kg wet weight or number of fish) per standard unit of setting. For gill nets the standard unit is a net size of 45.7 m (50 yards) long and 4 m deep, and for longlines the standard unit is 100 hooks. The CPUE distributions are strongly skewed and the application of statistical tests is not valid without normalization or without knowledge of individual observations. This has not been possible as the results available from the Hopson survey are in condensed tabulated form with little information of variance. Detailed data from the NORAD survey are tabulated in Kolding (1989).

Coarse gill nets

The comparison between the 1973 Hopson survey and the 1986–87 NORAD survey is summarized in Table 16.2. Concentrating on the 125 mm mesh size, which is the most common gear in the commercial fishery, Table 16.2 indicates an overall reduction in standard CPUE of 2.3 kg set^{-1} (48%) since 1973, but the decline is relatively greater in the south. Looking at the individual species, it appears that the main differences are due to *Labeo*, *Hydrocynus*, *Synodontis* and *Barbus bynni* but that the pattern is not uniform. The dominant species *Labeo* has increased in Todenyang but decreased at the two other stations, particularly at Kerio, although it is still by far the most important species there. The opposite feature is seen for *Barbus bynni* although much less pronounced. Both *Hydrocynus* and *Synodontis* have substantially decreased at both Todenyang and Kalokol but were intermittently present in Kerio in either of the two surveys. *Lates* is the only species that has increased in all areas. Apparently *Distichodus* has also increased, whereas *Citharinus* has declined. *Bagrus bayad* has changed relatively little in the north, but has decreased notably in Kerio. Although a general decline is apparent, the conclusion on the status of the individual species is more diverse.

Table 16.2 Comparison of experimental mean catch per effort for individual species of fish in kg fresh fish per standard net-setting between 1973 (Bayley, 1977, Tables 8, 9, 10, 11) and the 1986–87 NORAD survey (Kolding, 1989) at three stations: Todenyang, Kalokol and Kerio. The number of settings (*N*) of each mesh size is given in parentheses and ± indicates standard deviation. The Δ125 column gives the differences in the mean CPUEs in the 125 mm mesh size. All mesh sizes are stretched mesh (mm)

Station and species	Bayley 1973		NORAD 1986–87		
	100 mm	125 mm	105 mm	125 m	Δ125
Todenyang (inshore and offshore)	(*N* = 0)	(*N* = 59)	(*N* = 24)	(*N* = 24)	
Labeo horie		1.18	2.07	1.94	+0.76
Synodontis schall		0.64	1.25	0.16	−0.48
Barbus bynni		1.13	0.43	0.33	−0.80
Distichodus niloticus		0.12	0.14	0.24	+0.12
Hydrocynus forskalii		1.09	0.09	0.17	−0.92
Citharinus citharis		0.09	0.09	0.06	−0.03
Lates sp.		0.89	0.06	1.04	+0.15
Bagras bayad		0.22	0.07	0.26	+0.04
Mean total		5.36 ± 3.10	4.20 ± 3.27	4.20 ± 4.76	−1.16
Kalokol (inshore)*	(*N* = 2)	(*N* = 24)	(*N* = 21)	(*N* = 21)	
Labeo horie	1.39	1.21	0.82	0.42	−0.79
Synodontis schall	5.25	0.64	0.70	0.14	−0.50
Barbus bynni		0.05	0.21	0.07	+0.02
Distichodus niloticus			0.04		
Hydrocynus forskalii	3.89	0.65	0.13		−0.65
Citharinus citharis		0.32	0.02	0.06	−0.28
Lates sp.	0.25	0.03	0.04	0.13	+0.10
Bagrus bayad		0.04	0.09		−0.04
Mean total	10.78 ± 3.10	2.94 ± 1.80	2.05 ± 1.39	0.82 ± 0.98	−2.12
Kerio (inshore)	(*N* = 14)	(*N* = 18)	(*N* = 19)	(*N* = 19)	
Labeo horie	3.65	4.23	1.01	1.59	−2.64
Synodontis schall		0.08	0.14	0.05	−0.03
Barbus bynni	0.21		0.08	0.06	−0.06
Distichodus niloticus			0.14	0.10	+0.10
Hydrocynus forskalii					
Citharinus citharis		0.15			−0.15
Lates sp.		0.20		0.24	+0.04
Bagrus bayad		0.54	0.01	0.06	−0.48
Mean total	3.86 ± 1.52	5.24 ± 4.61	1.38 ± 1.54	2.09 ± 3.72	−3.15
Overall weighted mean		4.76 ± 2.04		2.47 ± 2.12	−2.29

*Bayley 1973 data at Kalokol include Ferguson's Gulf, which was dry in 1986–87.

Fine gill nets

Hopson and McLeod's 1974–75 survey with fine surface set nets all over the lake showed, like the NORAD survey (Figs 16.5, 16.6), a marked tendency for the mean catch rates to decrease along the lake from high values in the north to more moderate levels in the south. Two of the stations, Kataboi and Kalokol inshore, coincided with the NORAD survey and a comparison is summarized in Table 16.3. The difference between the two surveys is notable. Both *A. baremose* and *H. forskalii* have apparently declined substantially, the former having almost disappeared. *Synodontis* seems to have increased, at least at Kataboi, and *Distichodus* has reappeared, although still scarce. *Barbus bynni* and *Lates* sp. seem to have changed little. *Labeo horie* has declined, especially at Kalokol. However, as Ferguson's Gulf was an important nursery ground during 1974–75 (Hopson, 1982a), it is difficult to judge whether

Table 16.3 Comparison of experimental mean catch per effort for individual species of fish in kg per standard net-setting between 1974–75 (Hopson and McCleod, 1982, Table 8.25g, j) and the 1986–87 NORAD survey (Kolding, 1989) at two stations, Kalokol and Kataboi, fished with fine gill nets. The number of settings (*N*) of each mesh size is given in parentheses and the last column, Δ mean, gives the weighted mean differences between surveys. All mesh sizes are stretched mesh (mm)

Stations and species	Hopson 1974			NORAD 1986–87			
	64 mm	76 mm	100 mm	58 mm	78 mm	105 mm	Δ Mean
Kalokol, off Longesh Spit, depth 8–10 m	(*N*=12)	(*N*=14)	(*N*=4)	(*N*=21)	(*N*=21)	(*N*=21)	
Labeo horie	0.95	1.19	8.79	0.04	0.08	0.82	−1.79
Synodontis schall	1.43	2.90	7.93	2.11	2.06	0.70	−0.38
Barbus bynni	0.05	0.05	0.62	0.17	0.29	0.21	+0.10
Distichodus niloticus				0.03	0.01	0.04	+0.03
Hydrocynus forskalii	9.96	6.41	2.52	0.53	0.12	0.13	−7.05
Alestes baremose	9.17	10.59	0.05	0.02			−8.61
Lates sp.	1.28	0.10		0.77	0.32	0.04	−0.18
Schilbe uranoscopus	0.24	0.05		0.07	0.01		−0.10
Mean total	23.08	21.29	19.91	3.84	2.98	2.05	−18.87
Kataboi, 1 km offshore, depth 8–10 m	(*N*=4)	(*N*=4)	(*N*=4)	(*N*=21)	(*N*=21)	(*N*=21)	
Labeo horie	0.19	0.23	1.24	0.08	0.05	0.42	−0.13
Synodontis schall	0.19	2.33	4.75	4.30	5.00	1.44	+1.19
Barbus bynni		0.71		0.10	0.57	0.59	+0.18
Distichodus niloticus				0.04	0.09	0.16	+0.10
Hydrocynus forskalii	1.95	0.52	1.28	0.57	0.10	0.11	−0.87
Alestes baremose	13.72	14.29		0.05			−9.32
Lates sp.	1.33	0.19	0.24	0.40	0.36	0.11	−0.30
Schilbe uranoscopus	0.95	0.10		0.07			−0.33
Mean total	18.33	18.37	7.51	5.64	6.34	2.86	−9.79

this local decline is because of the desiccation of the Gulf or lower stock size. It is clear, however, that the pelagic stocks of *A. baremose*, *H. forskalii* and *Schilbe uranoscopus* have dramatically diminished. It should be noted that none of these stocks was subject to any significant fishing between the two survey periods.

Longlines

Hooks and lines are used by the Turkana fishers to a limited extent in shallow water, mainly to catch *Lates niloticus*. Longlining in deeper waters to exploit *Bagrus bayad* was introduced in 1972 but the effort declined after a few years. An experimental longline fishery was carried out in 1974 (Jan.–Sep.) from Namadak between 1 and 15 nautical miles from the shore (Østvedt, 1974; Lock, 1982). The catches consisted mainly of *Bagrus*, 40–90 cm TL, and a few large *L. niloticus*. A comparison between the 1974 and the NORAD survey (combined results) is summarized in Table 16.4.

Although the gill net catches (Table 16.2) indicate that the abundance of *Bagrus bayad*, except in Kerio, seemingly has changed little and *Lates* even increased, there is, according to Table 16.4, a notable decline in both CPUE and individual mean weight for both species in the longline catches. Especially *Bagrus* had declined by nearly 80% in kg set^{-1}, by about 60% in mean individual weight and 40% in numbers. *Lates* had reduced by half in mean individual weight but the CPUE by number had increased by 60% (Kolding, 1989). The results are, however, difficult to evaluate for this species considering the high coefficient of variation. Lowest catch rates were found at Kataboi, which is very close to Namadak (Fig. 16.1), whereas Lowarengak in the north produced a mean CPUE of 9 kg set^{-1} in the NORAD survey.

16.4 STATUS OF THE MAJOR TAXA

Citharinids

Citharinus citharis and *Distichodus niloticus* are two large benthic-feeding species which formed the backbone of the fishery from 1966 to 1972 when 200 mm gill nets were used almost exclusively. Both species, especially *Citharinus*, declined rapidly; by 1974 they formed only 6% by weight of the commercial yield and have remained at this low level since (Table 16.1). The largest catches were taken at the northern stations and limited reproductive data (Bayley, 1982b,d) showed that both species breed only in the River Omo and that the juvenile fish stay in the delta. However, the virtual absence of juvenile fish of both species during the 1972–1975 surveys was taken as an indication of recruitment failure since 1970 coinciding with reduced

Table 16.4 Comparison of experimental longline results in 1974 (Lock, 1982, Table 10.28 and Appendix E) and 1986–87 (Kolding, 1989) for the two dominant species *Bagrus bayad* and *Lates niloticus*

Year	*Bagrus bayad*			*Lates niloticus*		
	Mean CPUE*	Mean weight (kg)	Mean length (cm FL)	Mean CPUE*	Mean weight (kg)	Mean length (cm TL)
1974	15.58 ± 4.64 (118)	3.29	66.8	3.55 ± 1.89 (120)	24.11	119.1
1986–87	3.59 ± 4.30 (107)	1.20	45.5	2.90 ± 6.38 (107)	11.34	81.6

*Catch per unit effort, in kg per 100 hooks. Values are mean ± SD; numbers in parentheses are numbers of nights fished.

Table 16.5 Comparison between the Hopson survey and the NORAD survey on (1) length range of 10 largest fish caught in experimental gill nets, (2) mean length and standard deviation in 125 mm mesh size (sample size in parentheses) with significance levels, (3) size at first maturity and (4) size at first breeding of males and females (in parentheses). All units in cm fork length, except *Lates* sp. which is total length. Based on data in Hopson (1982a) and Kolding (1989)

Species	Range of 10 largest fish		Mean length, SD and sample size in 125 mm mesh tested for difference*		Size at first maturity ♂ (♀)		Size at first breeding ♂ (♀)	
	1973	1987	1973	1987	1973	1987	1973	1987
Labeo horie	73– 78	60– 65	58.6 ± 5.4 (236)	50.4 ± 4.3 (191) **	51 (56)	34 (28)	51 (56)	38 (46)
Synodontis schall	33– 37	32– 36	24.4 ± 3.5 (639)	22.7 ± 4.1 (180) **	16 (17)	14 (15)	19 (19)	18 (16)
Barbus bynni	63– 73	46– 54	41.9 ± 5.6 (135)	38.5 ± 5.2 (56) **	30 (28)	30 (24)	38 (48)	(34)
Lates sp.†	108–184	66–126	47.7 ± 11.9 (65)	30.6 ± 18.1 (96) **	18 (17)	18 (20)		
Citharinus citharis	62– 66	31– 36	49.1 ± 9.5 (14)	29.1 ± 3.3 (16) **	48 (54)	28 (32)	48 (58)	
Distichodus niloticus	75– 83	46– 59	53.6 ± 8.4 (9)	45.1 ± 7.4 (19) *	36 (36)	28 (36)	60 (66)	
Hydrocynus forskalii	78– 80	53– 65	64.8 ± 11.5 (64)	47.4 ± 16.9 (9) *	26 (28)	26 (26)	26 (28)	(26)
Bagrus bayad	76– 82	62– 80	46.6 ± 12.6 (41)	49.1 ± 14.0 (21) ns	26 (26)	30 (24)	36 (33)	40 (34)
Alestes baremose‡	43– 44	38– 42	35.4 ± 2.7 (1884)	35.0 ± 4.2 (123) ns	30 (31)	28 (30)	31 (32)	(30)

*Significance for the differences between means is tested with a modified *t*-test for independent samples when the two population variances cannot be considered equal (Snedecor and Cochran, 1989); ** indicates significance at the 1% level, * is significant at the 5% level and ns is non-significant.
† Largest fish are *Lates niloticus* but mean lengths in 125 mm mesh are *L. niloticus* and *L. longispinis* combined.
‡ Mean lengths are from 78 mm mesh size.

floods. During 1972–1975 commercial catches consisted mainly of fish from 40 to 60 cm in length and only very few fish below 30 cm were caught in the lake.

During the NORAD survey only 36 *Citharinus* were caught, which suggests that a further decrease in abundance has taken place since Bayley's survey (Table 16.2). The material is too sparse for biological interpretation. The only marked difference was the significant decrease in mean size of the fish (Table 16.5). *Distichodus*, however, appears to have remained stable or even to have increased over the last decade (Tables 16.1–16.3). The population structure appeared healthy with relatively large numbers of juveniles between 20 and 30 cm FL (Fig. 16.7). Large fish, however, were scarce compared with 1973 (Table 16.5) and only about 10% of the sample population was more than 40 cm long. Still *Distichodus* seems to have managed better than *Citharinus* during the last decade of falling lake level.

The abundance of citharinids, especially *Citharinus*, in the Lake Turkana fishery during the late 1960s after the considerable 4 m rise in the lake level in 1962–63 (Fig. 16.3) displays an interesting similarity to that of newly flooded reservoirs. Short-lasting 'booms' of citharinids after new inundations have been observed in artificial lakes Volta, Kainji, Nubia and Kariba (Petr, 1967; Lelek, 1973; Ali, 1984; Kenmuir, 1984; Chapter 12).

Cyprinids

Labeo horie and *Barbus bynni* are also benthic-feeding anadromous species with a similar niche to *Citharinus* and *Distichodus* but are slightly smaller and possibly less dependent on the River Omo for spawning (Bayley, 1982c). Exploitation of these species began around 1970–1972 when smaller mesh sizes than the initial 200 mm were introduced. *Labeo* soon became one of the most commercially important species in Lake Turkana. By 1974 it contributed about 30% in the landings but was already then considered heavily exploited (Bayley, 1977, 1982a). Potamodromous species are considered to be particularly vulnerable to recruitment overfishing on their spawning runs (Cadwalladr, 1965) so that, after the experience with the rapid decline of citharinids, a similar collapse of the *Labeo* stock was feared. Strict control of mesh sizes and prohibition of fishing near the mouths of and in rivers was therefore urgently recommended (Bayley, 1982a), but this has been hopeless to enforce in the wild and remote Omo delta. However, from the catch records and the results of the NORAD survey it seems that the feared collapse of the *Labeo* fishery has not occurred. It is still the most important commercial species and the proportion of landings has remained stable at around 30%, except for the years with exceptional tilapia booms (Table 16.1). An increased stress, whether natural or from fishing pressure, seems to have influenced the population structure as indicated from the marked reduction in the mean size of mature fish and the maximum size observed between the two studies

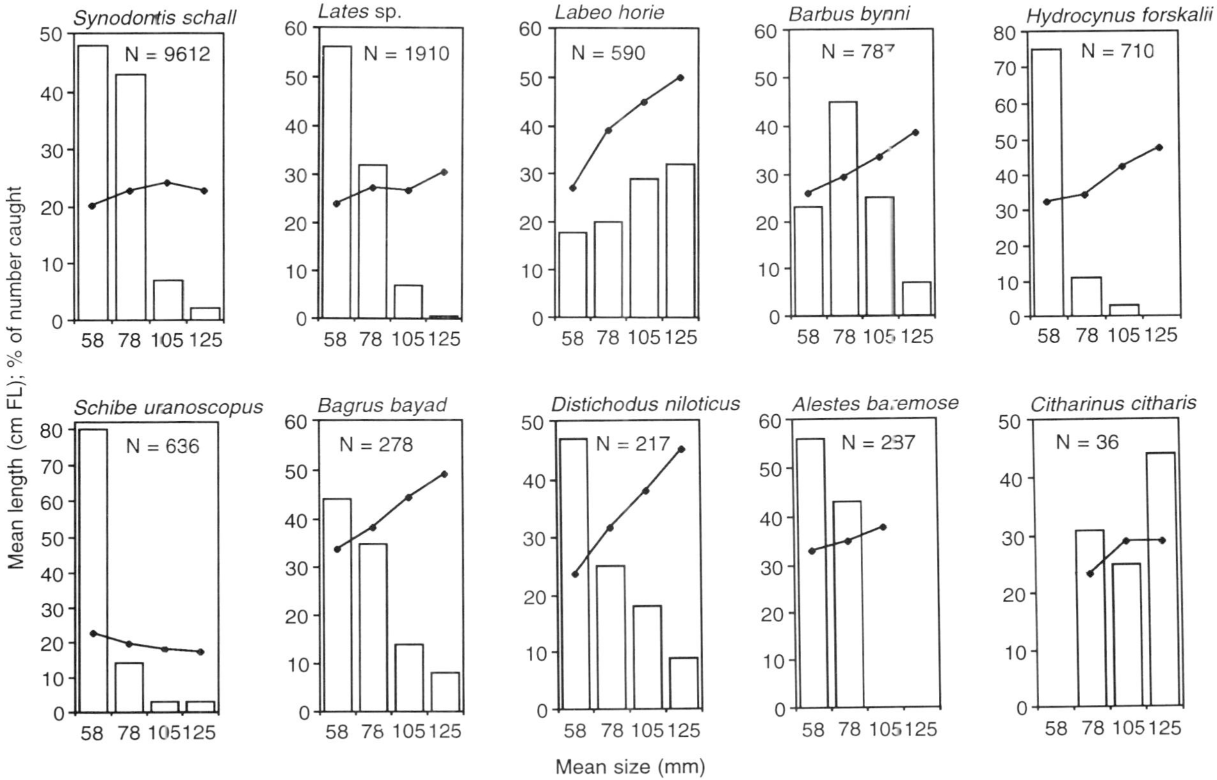

Fig. 16.7 Per cent numbers caught (columns) and mean length (dots) of the ten most dominant species in the four experimental mesh sizes (58, 78, 105, 125 mm) employed during the 1986–87 NORAD survey. Note differing vertical scales.

(Table 16.5). Another important feature is that the number of fish caught increased with increasing mesh size, indicating a low presence of juveniles in the fished areas (Fig. 16.7). Together with *Citharinus citharis* these are the only two species displaying this feature during the NORAD survey. Whether this is an indication of recruitment failure or whether the juveniles are localized in other areas is unknown. Compared with the distribution of the stock as described by Bayley (1982c), it is most likely the result of reduced recruitment prior to the time of the study.

Barbus bynni is the most widely distributed of the larger cypriniform fish in Lake Turkana (Mraja, 1982), but less numerous than *Labeo*. The proportion of the commercial yields has decreased from about 10% since the Bayley survey but has remained stable at 2–3% over the last decade with a slight relative increase in 1988 (Table 16.1). The results of the NORAD survey indicate that the stock of *Barbus bynni* in Lake Turkana has changed relatively little since the mid 1970s. A decrease in catch rates of larger fish is only observed at Todenyang and this might be due to seasonal differences and to a more even spatial distribution. The status of the juvenile population (Fig. 16.7) indicates, in contrast to *Labeo*, no recent recruitment failure. There are, however, similar indications of an increased stress from the observed decrease in mean length and in sizes at maturity (Table 16.5).

Characids

This family has shown some interesting changes. In the fine gill net survey in 1974 (Hopson and McLeod, 1982) *Alestes baremose* was by far the most important species, particularly in the northern sector, with a mean standard CPUE for the combined 64–89 mm fleet from all stations of 12.1 kg set^{-1}. High values of over 50 kg set^{-1} were recorded around North Island. Offshore it contributed some 80% of the catch but even inshore *A. baremose* frequently yielded more than 50% in these fine mesh sizes. *Hydrocynus forskalii* was usually the subdominant species with a mean CPUE of 5.5 kg set^{-1} in the same mesh sizes and at the same stations. Both *H. forskalii* and *A. baremose* belong to the non-exploited offshore pelagic community in Lake Turkana and the potential sustainable yield, especially of *Alestes*, was considered very high after the Hopson survey (at least 11 000 t year^{-1}). A regular offshore fishery with 76–100 mm surface nets was recommended to take the pressure off the heavily exploited inshore demersal fishery (Hopson, 1982a). However, due to the general mesh size limt of 125 mm and the lack of suitable boats, this fishery never materialized.

From the results of the NORAD survey it seems that these characid species have decreased dramatically since the mid 1970s. The result of 36 offshore settings of 78 mm mesh nets (Kolding, 1989) gave a mean of only 0.14 kg set^{-1} of *Hydrocynus* and 0.21 kg set^{-1} of *Alestes*, which both correspond to less than 2–3% of the mean CPUEs in 1973. Because both these stocks are hardly exploited ($F \approx 0.03$ for *Hydrocynus* and practically zero for *Alestes*) it must be concluded that the very substantial reduction of these stocks is entirely a result of natural causes, whether through recruitment failure or reduced carrying capacity of the environment.

There is indirect evidence that the latter plays a role. In the open pelagic zone of Lake Turkana two other characids, the small endemic *Alestes minutus* (3–4 cm FL) and *A. ferox* (6–8 cm FL), constitute the bulk of the pelagic fish biomass and are the most important link between zooplankton and higher trophic levels (Hopson, 1982a; Kolding, 1993b). Although few data exist of the changes, it seems that these species have also declined significantly in abundance. Hydro-acoustic surveys in 1984–1986 (Lindem, 1984, 1985, 1986) revealed that the amount of small pelagic fish was unexpectedly low compared with the results from the 1970s. Furthermore, the densities seemed to follow the changes in lake level over the 3 years of the surveys.

Centropomids

Two species of the predatory genus *Lates* are common in Lake Turkana; the large *L. niloticus* (up to 190 cm TL) which occurs throughout the Sudanian region of Africa, and the smaller endemic *L. longispinis* (up to 64 cm TL). Although their biology differs to some extent, morphologically they are very alike and are therefore not differentiated in most catch statistics (including the NORAD survey). *L. niloticus* is widely distributed both inshore and offshore, whereas *L. longispinis* is primarily pelagic and uncommon inside the 10–15 m depth contour (Hopson, 1982b). *L. niloticus* is therefore the most commercially important and has formed about one-third of the total yield in recent years (Table 16.1).

Except for the smallest mesh size (Table 16.3), *Lates* are the only species that seem to have increased in abundance at all stations since the Hopson survey. This is remarkable because, as it is a piscivorous top predator, its prey sources have seemingly all declined. The decrease in the smallest mesh sizes might therefore indicate an increased cannibalism as has been observed in Lake Victoria after the Nile perch depletion of small haplochromine cichlds (Ogari and Dadzie, 1988; Ogutu-Ohwayo, 1990; Chapters 5 and 8). Further evidence is a marked decline in relative abundance of *Lates* below 18–20 cm TL in the combined length frequencies (Kolding, 1989). Although Nile perch can attain very large size, its main quantitative impact on prey populations is on small fish between 3 and 17 cm SL (Gee, 1969; Ogari and Dadzie, 1988). However, as indicated in both the longline catches (Table 16.4) and the gill nets (Table 16.5), the larger specimens of this genus have also decreased since the 1970s. In the 1986–87 experimental gill net catches,

95% of the fish were less than 33 cm TL and only two fish exceeded 1 m (Kolding, 1989).

Silurids

In the 1972–1974 survey, the relatively small omnivorous mochocid catfish *Synodontis schall* proved to be the most plentiful of all species in the lake (Lock, 1982). Although *Synodontis* elsewhere is mainly a demersal species (Bishai and Gideiri, 1965; Lowe-McConnell, 1987), it is mainly pelagic in Lake Turkana. This species is very lightly predated upon (Hopson, 1982a; Kolding, 1993b), probably due to its long serrated locking pectoral and dorsal spines, and apparently the abundance has changed little since the mid 1970s. The abundance of large specimens has, like that of all other species, declined as seen from the coarse gill net data (Table 16.2), but the stock of smaller fish seems to have remained more or less stable (Table 16.3).

Commercially *Synodontis* has little value, especially since the smoked fish trade was abolished around 1982. As a bycatch it is consumed locally and is in high demand. Lock (1982) considered it the most common species eaten by the fishing people. Although both experimental surveys have shown that *Synodontis* is potentially one of the most important fishes in the lake, an expansion in exploitation has hitherto conflicted with policies for conservation of the larger commercial species. Both the experimental fishing surveys showed that 78 mm surface- and offshore-set gill nets catch juveniles of the larger species to only a minor extent. Provided legislation and education were provided to prevent these nets being set on the bottom or in shallow water, an offshore surface *Synodontis* fishery for subsistence and local markets would seem rational. Hopson (1982a) estimated the maximum sustainable yield (MSY) of this species to be 22 000 t year^{-1}, while Kolding (1993b) estimated the sustainable yield at 12 000 t year^{-1}. This is a lot considering the total commercial landings at present are less than 1000 t year^{-1}.

Bagrus bayad is a little-exploited, deep-water demersal predator in Lake Turkana (Lock, 1982). In both experimental surveys, mean densities increased with depth and reached a maximum in the deepest parts of the lake. A comparison of the gill net surveys (Table 16.2) indicates a decrease in abundance to the south (Keiro) but a slight increase to the north (Todenyang). This is the only species that does not display a marked reduction in mean size and maximum length in the experimental gill net fishery (Table 16.5). However, from the longline surveys (Table 16.4) it appears also that larger individuals of this species are becoming relatively more scarce.

Schilbe uranoscopus is a small (maximum length 34 cm FL), mainly pelagic predator. Although widely distributed in the lake, it is generally neither abundant nor commercially important. Although the basis for comparison is limited, it seems that its abundance has fallen (70–90%) since the last survey 1972–1974 (Table 16.3).

Cichlids

The littoral tilapiine cichlids, particularly *Oreochromis niloticus*, are one of the most important fish resources of Lake Turkana and periodically they dominate the commercial landings (Table 16.1). They are principally caught by beach seining and only a few specimens were caught in the experimental gill net surveys. The fishery, however, experiences enormous boom-and-bust cycles. For example, in 1976, these cichlids yielded a staggering 16 000 t from the small Ferguson's Gulf near Kalokol (Fig. 16.1). Smaller 'booms' took place in 1963, 1970 and 1982, and are the main reason for the peaks in Fig. 16.3. The reason for the dramatic but ephemeral booms in the tilapia populations may be closely related to the floodplain conditions of shallow sheltered areas in conjunction with changing lake levels (Harbott and Ogari, 1982; Kolding, 1993a).

16.5 GENERAL TRENDS IN CATCH RATES VERSUS LAKE LEVELS

Before the 1972–1975 survey (Hopson, 1982a), there is very limited information on catch per unit effort in Lake Turkana. The only previous experimental fishing was conducted by Mann (1964), who sampled a short while off Longesh spit at Kalokol. The results of the three experimental surveys in this area have been plotted in Fig. 16.8 to show trends over a 24 year

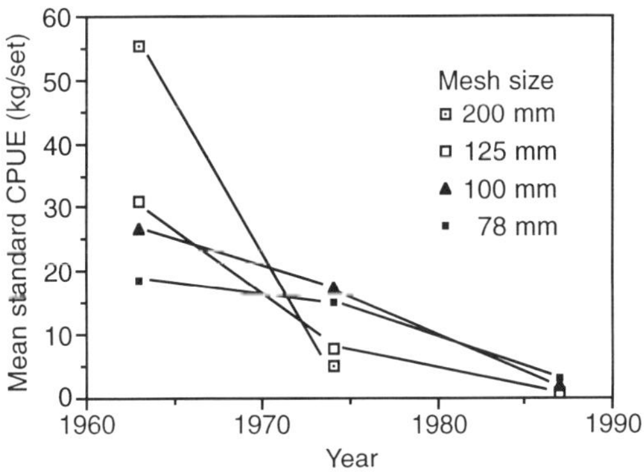

Fig. 16.8 Comparison of the mean standard CPUE (kg per standard setting) of three experimental fishing surveys (1963, 1974 and 1986–87) on the inshore (within 10 m depth) grounds off Kalokol.

period. Figure 16.8 clearly indicates that a strong decline in catch rates has taken place, but that catches in the larger mesh sizes have decreased much more rapidly than in the smaller. Reflecting this trend, the order of appearance of the various mesh sizes on the ordinate has reversed over the period, indicating the shift from fishing on an accumulated stock of old, large fish to a situation where the small or juveniles contribute the largest biomass. This was predicted by Mann (1964: 57) who pointed out: "The remarkable large size of many specimens taken is also due to the presence of old fish, the fishing effort being so low that many adult fish survive to a large size. These aged specimens will be eliminated as the fishing effort increases". A fall in CPUE is evident for all mesh sizes, even in the two smallest which are not part of the commercial gear. As these two mesh sizes (78 and 100 mm) are mostly catching small non-exploited species with a relatively short generation time, the decrease in catch rates for these nets must be considered to be influenced by factors other than the fishery alone.

The hydrological regime in Lake Turkana may have a strong impact on the ecological processes in the lake (Bayley, 1977; Harbott, 1982; Talling, 1986; Kolding, 1992). To obtain some evidence for this, the mean CPUE for 78 and 100 mm mesh size combined over the three experimental surveys is plotted against the mean annual lake level (Fig. 16.9). The closed basin of Lake Turkana means that water level is directly proportional to the influx of nutrients such as nitrogen and allochthonous organic matter.

The indication in Fig. 16.9 of a close association between fish production and the hydrology in Lake Turkana has been further tested using the independent data set from commercial CPUE (Fig. 16.10). There are two

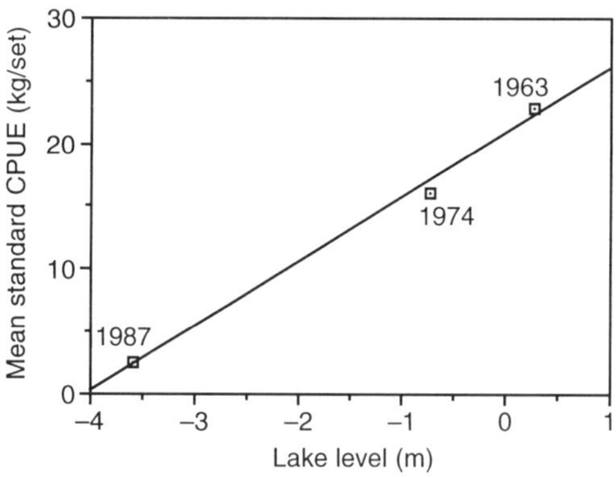

Fig. 16.9 Mean standard CPUE (kg per standard setting) of 78 and 100 mm mesh sizes combined, from the three experimental fishing surveys off Kalokol plotted against mean annual lake level of the same years. Least-squares regression ($r^2 = 0.99$, $P = 0.048$).

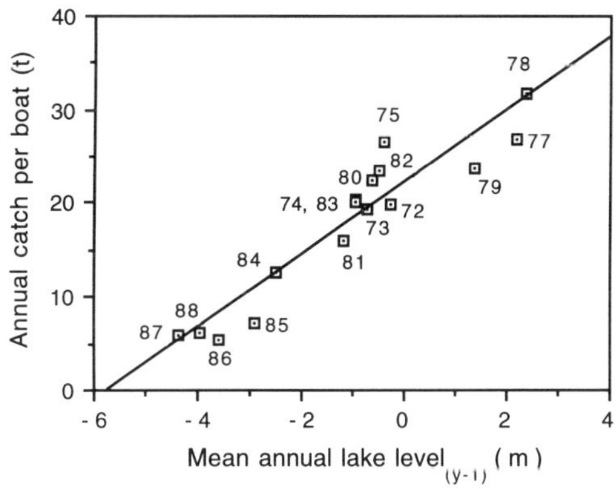

Fig. 16.10 Commercial annual catch per boat (year$_i$) plotted against mean annual lake level (year$_{i-1}$) for the period 1972 to 1988, except 1976. Catch per boat in tonnes wet weight and lake level in m. Least squares regression: $y = 22.15 + 3.87x$ ($r^2 = 0.89$, $P = 0.000$). From Kolding (1992).

different time trends in these data: (1) the expected decline of mean CPUE when initializing a fishery on virgin stocks, i.e. before equilibrium is reached between production and yield, and (2) the suggested independent effect from environmental variations. For those reasons, the correlation between commercial CPUE (mean annual catch per boat) and lake level is limited to cover only the period from 1972 on; 1972 was chosen as it is about the time when the first major stocks were depleted and the first reports of overfishing began. The atypical 1976 tilapia 'boom-year' is also omitted because this took place in Ferguson's Gulf, which is not representative of the rest of the lake (Kolding, 1989, 1993a).

Quantifying an environmental impact on a biological system is always risky and the determination coefficient in Fig. 16.10 could be the result of a spurious fit to a limited number of data points. Still it is reasonable to conclude that the plot not only strongly supports the hypothesis of a high positive correlation between biological production and the hydrological regime in Lake Turkana, but also indicates that the rate of change in the production is fairly rapid and resilient.

16.6 DISCUSSION

The drastic decline in the commercial yields of Lake Turkana over the last decade has almost unanimously been attributed to overfishing. Practically all reports after the Hopson survey and up to very recent times have expressed

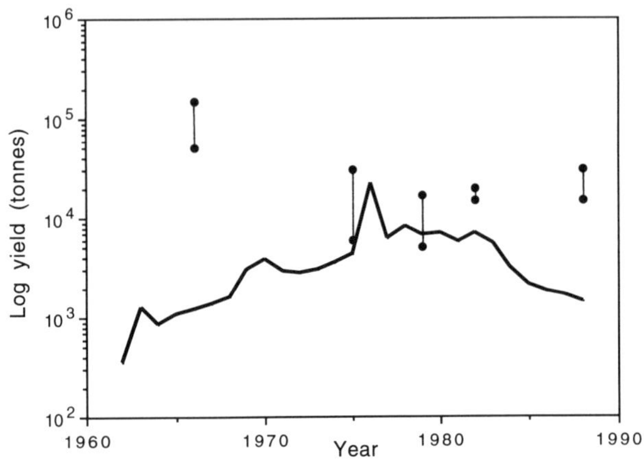

Fig. 16.11 Logarithmic plot of estimated maximum sustainable yield ranges (MSY) by various authors versus the obtained total yield (solid curve) of the western side of Lake Turkana in the period 1962–1988. Sources of MSY: 1966, Rhodes (1966); 1975, Hopson (1982a); 1978, Kjær (1979); 1982, Odero (unpublished data); and 1988, Källqvist *et al.* (1988).

concern about this issue (Kolding, 1989). Only one evaluation (NORAD, 1985) clearly states that there is no evidence of overfishing causing the decline. One reason for this widely held view was perhaps misleading or missing statistics (Kolding, 1989), which meant that the simultaneous strong decline in fishing effort (Fig. 16.2) was overlooked. The large discrepancy between the potential sustainable yield predictions (MSY), ranging from a minimum of 5000 to some 50 000 t year^{-1} (Fig. 16.11 and Chapter 23), and the actual landings has contributed to sustain the notion of overfishing. It should be noted that catches matching even the lowest of these forecasts were only attained for a short period between 1976 and 1983.

Several of the attempts to calculate potential yields were based on the classic equations of Gulland (1971), Henderson and Welcomme (1974) and Melack (1976). The long-term predictive power of these methods is subject to the inherent assumption of stability in the parameters. Hopson (1982a), for example, used the well-known Gulland (1971) expression

$$\mathrm{MSY} = \mathrm{a}M\mathrm{B}_0 \tag{16.1}$$

where a is normally set to 0.5, M is the natural mortality rate and B_0 is the virgin biomass. Apart from the general criticisms of this equation in overestimating MSY (e.g. Beddington and Cooke, 1983), and the difficulty in estimating the parameters M and B_0, the prerequisite of stable values for these parameters is far from satisfied in the Lake Turkana system. The results in this Chapter have shown that even the unexploited stocks fluctuate greatly in abundance. For example, Kolding (1993b), applying a multispecies trophic box model, found the natural mortality rate due to predation alone (*M2*) to vary by more than 100% in both exploited and unexploited stocks.

Clearly, under such circumstances, the whole concept of maximum sustainable yield as a management objective becomes invalid. As early as 1977, Bayley (p. 130) concluded that predictions in Lake Turkana were "at worst very mistaken, and at best very conditional", and this has been verified repeatedly by the history chronicled here. The MSY concept is evidently rooted in planners, administrators and research personnel (the overall objectives of both the ODA and the NORAD-sponsored surveys were to establish sustainable yields) but, if anything, MSY estimates have so far only created misconceptions and false expectations. Much hope was put into the unexploited stocks of *Alestes baremose* and *Hydrocynus* after the ostensible collapse of the inshore demersal species (NORAD, 1983). These pelagic stocks were predicted to be more stable than the inshore demersal fish and to have high potential yields, with MSY in the region of 11 000 tonnes a year (Hopson, 1982a). From the results presented here the opposite seems to be the case. In fact, the common denominator of most of the offshore pelagic species is that they have suffered the heaviest mortalities during the period of receding lake levels despite having the least exploited species.

The empirical relationship between the overall fish production index and lake levels (Fig. 16.10) strongly suggests that the environment, more than the fishery, is the dominant agent of change in the Lake Turkana ecosystem. Lowe-McConnell (1979) summarized some of the salient attributes of fish populations living under fluctuating seasonal conditions such as floodplains: generally, selection pressures favour high fecundity with rapid development and growth due to high natural mortality. As a result the populations contain relatively few age groups and generation times are short. The marked seasonal variation in primary and secondary productivity causes the higher trophic levels to adapt a more generalized feeding behaviour with little time or opportunity to develop complex patterns. The abiotic factors influencing spawning success vary from year to year, together with high predation pressure, lead to large fluctuations in numbers of particular groups or species. The overall pattern is thus a predominantly *r*-selective type of reproduction (MacArthur and Wilson, 1967; Pianka, 1970), aimed at rapid recolonization of habitats in an unstable fluctuating environment. I suggest that ecological theory predicts that such communities will not only respond markedly to changes in environmental factors but will also have a high degree of resilience.

These characteristics all seem appropriate to Lake Turkana with its basically riverine nilotic fish fauna (Hopson, 1982a). The diversity (48 species, of which 36 occur regularly within the actual lake) and endemism (10 species) is low for a large ancient lake. Moreover, most species have euryphagic habits;

either herbivore/detritivore or piscivore (Hopson, 1982a; Kolding, 1992). There is a marked seasonal variation in the primary and secondary productivity (Ferguson, 1982; Harbott, 1982; Talling, 1986; Källqvist *et al.*, 1988) and, as described here, great fluctuations in fish abundance. The results of Kolding (1993b) showed that the human exploitation rates ($E = F/Z$, where F is mortality occasioned by fishing and Z is total mortality rate) in 1987 of *Synodontis, Lates* and *Hydrocynus* ranged from 0.01 to 0.04, whereas for benthic fish (*Labeo, Barbus, Distichodus* and *Citharinus*) E averaged 0.27. Clearly, even for the most heavily exploited populations, natural mortality rates (M) are high compared with the fishing mortality. The general reduction in mean individual sizes (Tables 16.4, 16.5) indicates that the total mortality rate (Z) of all the sampled stocks has risen between the mid 1970s and the late 1980s (Kolding, 1993b). At least since the early 1980s this must be due mainly to natural causes, as the number of fishers has reduced by 80%. Increased predation would be consistent with *Lates* being the only stock that has increased in abundance. All in all, there seems ample justification to consider changes in the natural environmental as having the main impact on the Lake Turkana fish populations.

16.7 CONCLUSIONS

The predictability of fish populations in Lake Turkana is obviously very low, and the standard explanation, that a decline in landings is most likely due to overfishing, seems not to fit the evidence. The imporant lesson to be learnt from these results is that it may not be sufficient to focus on the variables of which we usually have most knowledge and control, i.e. fishing effort and landings. The general conclusion of this investigation, without incorporating the environmental changes, would most likely be in line with most previous reports; that Lake Turkana appears overfished and that consequently effort should be reduced (e.g. NORAD, 1983). I consider that this prevailing view is the direct outcome of an equilibrium theory, regulated by feedback controls, which forms the theoretical background for the logistic growth curve, 'surplus-yield modelling' and the MSY concept. The idea, that there is a balance in nature and that this state is maintained by interactions among the various components is a deeply rooted paradigm in science. Under these assumptions it is logical to reason that internal adjustments (e.g. fishing pressure) in the system structure can correct the imbalances and restore it to normal. However, an antithetical point of view highlights the role of external drivers at a level above the scale of observations as control mechanisms of a system (e.g. rainfall). In such a case the development of feedback controls within the focal system is less important (e.g. DeAngelis and Waterhouse, 1987; Ulanowicz, 1990). In fact, this very dichotomy has prompted famous scientific debates, such as realism versus empiricism, reductionism versus holism or the debate over density-dependent versus density-independent factors in ecology in the 1950s. The first (balance-of-nature) school has traditionally had the more pervasive role in fisheries science and leans heavily on deterministic models. The latter (external-driver school) has been more concerned with phenomenology and causality at larger scales but is often discredited (e.g. Kerr, 1982; Rigler, 1982).

While both views represent reality in different biological 'systems' (it is merely a matter of scale), it is evident that only the former legitimates the use of fixed standards, such as 'the carrying capacity' and the derived concepts of MSY and overfishing. As stochastic external forces become stronger, such as abiotic perturbations, the result is fluctuations and destabilization, and terms such as MSY progressiveley lose their practical meaning (Vetaas and Kolding, 1991). Under these circumstances, persistence is dependent on adaptation to unpredictable alterations in the environment and survival can be explained only by opportunistic and/or compensatory mechanisms. There are basically two strategies: either follow the fluctuations in limiting factors (resilience or 'boom and bust', e.g. small pelagic species and tilapias), or increase the spatial or temporal scale of responses to the system (endurance, e.g. large predators and human fishers). Large predators have a much lower but more stable productivity. Consequently an opportunistic and flexible fishing pattern with diversified gears seems required to follow the dynamic changes in Lake Turkana. On the other hand, as fishers are less adapted than the fish to short-term fluctuations, a relatively low standing effort is necessary if they are to survive economically (Kolding, 1992).

In this context it is doubtful whether overfishing has ever been a problem in Lake Turkana. Effort regulation as a management objective seems therefore not to be a key issue. Rather, the catch per unit effort will, in terms of an economic break-even, be a regulating factor by itself. The comparison between the two experimental fishing surveys in 1973–74 and 1986–87 showed an average decreased from 4.8 to 2.5 kg set^{-1} (nearly 48%) for the nets with 125 mm mesh size in the three areas Todenyang, Kalokol and Kerio combined (Table 16.2). The commercial landings have in the same period decreased from 20 to 6 t boat^{-1} year^{-1} (Fig. 16.10), representing a 70% reduction. All other things being equal, the commercial yields have decreased to half of what would be expected from knowledge of the resources themselves. This indicates that socio-economic factors may also play a role.

16.8 SUMMARY

Major changes in the relative composition, abundance and structure of the fish populations in Lake Turkana between 1973 and 1987 are evaluated and discussed. The investigation is based partly on two experimental fishing surveys in the north-western part of the lake and partly on long-term trends in the catch rates and composition of the commercial fishery. A general reduction in catch per unit effort since the mid 1970s was the result, but it is unevenly distributed among the geographic localities and among the various species. Most of the species of the demersal inshore community (*Labeo horie, Barbus bynni, Citharinus citharis* and *Distichodus niloticus*) have changed relatively little despite sustaining the heaviest fishing pressure. In contrast, some major species from the offshore pelagic community (*Hydrocynus forskalii, Alestes baremose* and *Schilbe uranoscopus*) have been reduced drastically in abundance. Interestingly the decrease is not caused by the fishery as these species are hardly exploited commercially. A few exploited species (*Lates* spp. and *Synodontis schall*) have increased in abundance. However, all the sampled stocks display increased stress symptoms indicated by a general decrease in mean individual size, and for some also there is evidence of a decreased mean size at first maturity. A general assessment of the major changes and trends in time is presented and the relative impact of the possible agents of change, environment or the fishery, is discussed. The work concludes that natural changes in the environment have the main impact on the fish stocks and that overfishing has probably never been a serious issue. This conclusion has important consequences for future management strategies and yield forecasts.

ACKNOWLEDGEMENTS

I am much indebted to Francis Mutungu for his diligence and stamina in the field during the trial fishery. Parts of this study were financed by the Norwegian Agency for Development Cooperation (NORAD) and I would especially like to thank Inger Stoll, NORAD, for her help and backing.

REFERENCES

Ali, M.el-T. (1984) Fishes and fisheries of Lake Nubia, Sudan. *Hydrobiologia*, **110**, 305–14.

Bayley, P.B. (1977) Changes in species composition of the yields and catch per unit effort during the development of the fishery at Lake Turkana, Kenya. *Arch. Hydrobiol.*, **79**, 111–32.

Bayley, P.B. (1982a) The commercial fishery of Lake Turkana, in *Lake Turkana* (ed. A.J. Hopson), ODA, London, Vol. 2, pp. 349–557.

Bayley, P.B. (1982b) The biology of *Citharinus citharis* (Geoffr) in Lake Turkana, in *Lake Turkana* (ed. A.J. Hopson), ODA, London, Vol. 3, pp. 789–802.

Bayley, P.B. (1982c) The biology of *Labeo horie* (Heckel) in Lake Turkana, in *Lake Turkana* (ed. A.J. Hopson), ODA, London, Vol. 3, pp. 803–15.

Bayley, P.B. (1982d) The biology of *Distichodus niloticus* (L) in Lake Turkana, in *Lake Turkana* (ed. A.J. Hopson), ODA, London, Vol. 3, pp. 843–7.

Beadle, L.C. (1981) *The Inland Waters of Tropical Africa – An Introduction to Tropical Limnology*, 2nd edn, Longman, London and New York, 475 pp.
Beddington, J.R. and Cooke, J.G. (1983) The potential yield of fish stocks. *FAO Fish. tech. Pap.*, no. 242, 47 pp.
Bishai, H.M. and Gideiri, Y.B.A. (1965) Studies on the biology of the genus *Synodontis* at Khartoum. 2. Food and feeding habits. *Hydrobiologia*, **26**, 98–113.
Butzer, K.W. (1971) Recent history of an Ethiopian delta – the Omo river and the level of Lake Rudolph. Univ. Chicago, Dept Geography, Res. Pap. no. 136, 184 pp.
Cadwalladr, D.A. (1965) The decline in the *Labeo victorianus* Blgr. (Pisces: Cyprinidae) fishery of Lake Victoria and an associated deterioration in some indigenous fishing methods in the Nzio River, Kenya. *East Afr. Agri. Forestry J.* (January), 249–56.
DeAngelis, D.L. and Waterhouse, J.C. (1987) Equilibrium and nonequilibrium concepts in ecological models. *Ecol. Monogr.*, **57**, 1–21.
Ferguson, A.J.D. (1982) Studies of the zooplankton of Lake Turkana, in *Lake Turkana* (ed. A.J. Hopson), ODA, London, Vol. 1, pp. 161–245.
Ferguson, A.J.D. and Harbott, B.J. (1982) Geographical, physical and chemical aspects of Lake Turkana, in *Lake Turkana* (ed. A.J. Hopson), ODA, London, Vol. 1, pp. 1–107.
Gee, J.M. (1969) A comparison of certain aspects of the biology of *Lates niloticus* (L.) in some East African lakes. *Revue Zool. Bot. afr.*, **80**, 244–62.
Gjerstad, D. (1983) Fish processing in Turkana: present and future methods, MSc thesis, Univ. Tromsø, Norway, 139 pp.
Gjerstad, D. (1986) Lake Turkana fisheries statistics, recent findings 1983–1985. NORAD, Oslo, 36 pp.
Gulland, J.A. (1971) *The Fish Resources of the Oceans*, Fishing News Books, West Byfleet, 255 pp.
Harbott, B.J. (1982) Studies on algal dynamics and primary productivity in Lake Turkana, in *Lake Turkana* (ed. A.J. Hopson), ODA, London, Vol. 1, pp. 109–61.
Harbott, B.J. and Ogari, J.T.N. (1982) The biology of the larger cichlid fish of Lake Turkana, in *Lake Turkana* (ed. A.J. Hopson), ODA, London, Vol. 5, pp. 1331–55.
Henderson, H.F. and Welcomme, R.L. (1974) The relationship of yield to morpho-edaphic index and numbers of fishermen in African inland fisheries. FAO, Rome, *CIFA occ. Pap.* no. 1, 19 pp.
Hopson, A.J. (ed.) (1982a) *Lake Turkana. A Report on the Findings of the Lake Turkana Project 1972–1975*. Overseas Development Administration, London, Vols 1–6, 1614 pp.
Hopson, A.J. (1982b) The biology of *Lates niloticus* (L) in Lake Turkana, in *Lake Turkana* (ed. A.J. Hopson), ODA, London, Vol. 5, 1258–1303.
Hopson, A.J. and McLeod, A.A.Q.R. (1982) Gillnet survey, in *Lake Turkana* (ed. A.J. Hopson), ODA, London, Vol. 2, pp. 619–23.
Källqvist, T., Lien, L. and Liti, D. (1988) Lake Turkana – limnological study 1985–1988. Norwegian Inst. Water Research (NIVA), Kenya Mar. Fish. Res. Inst. (KMFRI). NIVA Report No. 0-85313, ISBN 82-577-1510-7, Oslo, 98 pp.
Kenmuir, D.H.S. (1984) Fish population changes in the Sanyati Basin, Lake Kariba, Zimbabwe. *S. Afr. J. Zool.*, **19**, 194–209.
Kerr, S.R. (1982) The role of external analysis in fisheries science. *Trans. Am. Fish. Soc.*, **111**, 165–70.
Kjær, O. (1979) Specialist marketing mission to lake Turkana Fishermen's Cooperative Society. NORAD, Oslo, 77 pp.
Kolding, J. (1989) The fish resources of Lake Turkana and their environment. Cand. Scient. thesis, Univ. Bergen and final report of project KEN 043, NORAD, Oslo, 262 pp.
Kolding, J. (1992) A summary of Lake Turkana: an ever-changing mixed environment.

Mitt. int. Verein. Limnol., **23**, 25–35.

Kolding, J. (1993a) Population dynamics and life history styles of Nile tilapia (*Oreochromis niloticus*) in Ferguson'a Gulf, Lake Turkana, Kenya. *Env. Biol. Fishes*, **37**, 25–46.

Kolding, J. (1993b) Trophic interrelationships and community structure at two different periods of Lake Turkana, Kenya – a comparison using the ECOPATH II box model, pp. 132-139 in *Trophic Models of Aquatic Ecosystems* (eds V. Christensen and D. Pauly) (*ICLARM Conf. Proc.*, **26**), ICLARM, Manila, Philippines, pp. 116–23.

Lelek, A. (1973) Sequence of changes in fish populations of the new tropical man-made lake, Kainji, Nigeria, W. Africa. *Arch. Hydrobiol.*, **71**, 381–420.

Lindem, T. (1984) Results from the hydroacoustic survey in Lake Turkana, December 1984. NORAD, Oslo, 30 pp.

Lindem, T. (1985) Report from the hydroacoustic stock assessment survey in Lake Turkana, October 1985 (KEN 043). NORAD, Oslo, 27 pp.

Lindem, T. (1986) Results from the hydroacoustic survey in Lake Turkana – June 1986. NORAD, Oslo, 61 pp.

Lock, J.M. (1982) The biology of siluriform fishes in Lake Turkana, in *Lake Turkana* (ed. A.J. Hopson), ODA, London, Vol. 4, pp. 1021–1281.

Lowe-McConnell, R.H. (1979) Ecological aspects of seasonality in fishes of tropical waters. *Symp. Zool. Soc. Lond.*, **44**, 219–41.

Lowe-McConnell, R.H. (1987) *Ecological Studies in Tropical Fish Communities*, Cambridge Univ. Press, Cambridge, 382 pp.

MacArthur, R.H. and Wilson, E.O. (1967) *The Theory of Island Biogeography* (Monogr. Population Biology 1), Princeton Univ. Press, Princeton, NJ, 203 pp.

Mann, M.J. (1964) Report on a fisheries survey of Lake Rudolph, Kenya. Jinja, Uganda, *E. Afr. Freshwat. Fish. Res. Org. Annual Report 1962/63*, Appendix G:53–62.

Melack, J.M. (1976) Primary productivity and fish yields in tropical lakes. *Trans. Am. Fish. Soc.*, **105**, 575–80.

Merona, B., Hecht, T. and Moreau, J. (1988) Croissance des poissons d'eau douce Africains, in *Biologie et Ecologie des Poissons d'Eau Douce Africains* (eds C. Lévêque, M. Bruton and G. Ssentongo), ORSTOM, Paris, pp. 191–219.

Mraja, A.H.S. (1982) The biology of *Barbus bynni* (Forsk.) in Lake Turkana, in *Lake Turkana* (ed. A.J. Hopson), ODA, London, Vol. 3, pp. 817–27.

NORAD (1983) Summary of plenary discussions and papers presented during a seminar on the future of the Lake Turkana fisheries, Kalokol, September 20th–25th, 1982. NORAD, Oslo, March 1983, 91 pp.

NORAD (1985) Evaluation report on the Lake Turkana Fisheries Development Project. Evaluation Rep. no. 5.85, NORAD, Oslo, 95 pp.

Ogari, J. and Dadzie, S. (1988) The food of the Nile perch, *Lates niloticus* (L.), after the disappearance of the haplochromine cichlids in the Nyanza Gulf of Lake Victoria (Kenya). *J. Fish Biol.*, **32**, 571–7.

Ogutu-Ohwayo, R. (1990) Changes in the prey ingested and the variations in the Nile perch and other fish stocks of Lake Kyoga and the northern waters of Lake Victoria (Uganda). *J. Fish Biol.*, **37**, 55–63.

Østvedt, O.J. (1974) Assessment of fisheries resources in Lake Rudolph. NORAD, Oslo, 25 pp.

Petr, T. (1967) Fish population changes in the Volta Lake in Ghana during its first sixteen months. *Hydrobiologia*, **30**, 193–200.

Pianka, E.R. (1970) On *r* and K selection. *Am. Nat.*, **104**, 592–7.

Rhodes, D.H. (1966) Fisheries Development Possibilities, Report to the Government of Kenya. FAO, Rome, UNDP/FAO Rep. No. TA 2144, 77 p.

Rigler, F.H. (1982) The relation between fisheries management and limnology. *Trans. Am. Fish. Soc.*, **111**, 121–32.

Snedecor, G.W. and Cochran, W.G. (1989) *Statistical Methods*, 8th edn, Iowa State Univ. Press, Ames, Iowa, 503 pp.

Talling, J.F. (1986) The seasonality of phytoplankton in African Lakes. *Hydrobiologia*, **138**, 139–60.

Ulanowicz, R.E. (1990) Aristotelian causality in ecosystem development. *Oikos*, **57**, 42–8.

Vetaas, O.R. and Kolding, J. (1991) Environmental variability in the Sahel and its constraints on animal husbandry. *Forum for Utviklingsstudier, NUPI, Oslo*. **1991**(2), 215–28.

Worthington, E.B. and Ricardo, C.K. (1936) Scientific results of the Cambridge Expedition to the East African Lakes, 1930–31. 15. The fish on Lake Rudolph and Lake Baringo. *J. Linn. Soc. Zool.*, **39**, 353–89.

Chapter seventeen

Management, conservation and species changes of exploited fish stocks in Lake Malaŵi

George F. Turner

17.1 INTRODUCTION

Lake Malaŵi is home to at least 500 and possibly as many as 2000 fish species which are endemic or unique to the lake. These may account for as much as 2–5% of all existing vertebrate species. Almost all of them belong to one family, the cichlids, and apart from three tilapiines, all are closely related haplochromine species, possibly descended from a single, recent common ancestor (Meyer *et al.*, 1990, 1991). Understanding the mechanisms of the evolution of this diversity may provide one of the most critical tests of models of the origin of species (Mayr, 1984; Turner, 1994a). In addition to their great scientific importance, the fishes of Lake Malaŵi are an important component of the aquarium fish trade, a major tourist attraction and perhaps of considerable potential in aquaculture.

For the peoples of this region of Africa, however, these fish overwhelmingly represent a cheap supply of protein and a much-needed source of employment and cash income. It is estimated that 60–70% of the Malaŵian nation's protein consumption is supplied by fish, and that more than 70% of this comes from the Lake Malaŵi system (Mkoko, 1992). For these reasons, it is necessary to maximize long-term sustainable fish yields, while preserving the unique fauna: two needs which may be in conflict. In this Chapter, I review the current state of the major fisheries in Lake Malaŵi and associated water bodies and identify further research and management requirements.

The Impact of Species Changes in African Lakes. Edited by Tony J. Pitcher and Paul J.B. Hart. Published in 1995 by Chapman & Hall, London. ISBN 0 412 55050 4.

17.2 FISHING AREAS

The Lake Malaŵi system can be defined according to the distribution of the lacustrine endemic fauna and thus comprises Lake Malaŵi and associated lagoons, Lake Malombe and part of the single outflowing river, the Shire, which links the two lakes. Some 50 km south of Lake Malombe, the 80 km stretch of the Murchison Cataracts serves as a barrier to the Zambezian fishes which dominate the Lower Shire River (Tweddle *et al.*, 1979). The shore of Lake Malaŵi is bounded by Malaŵi, Tanzania and Mozambique. Little information is available from the latter two states, and fishing is generally assumed to be less intense than that in the Malaŵian sector. For the rest of this Chapter, discussion will be limited to the Malaŵian portion of the lake, unless otherwise stated.

In terms of limnology, fish distribution and exploitation patterns, the Lake Malaŵi system can be divided into five main areas (Table 17.1). Although comprising less than 10% of the total surface area, the southern areas (SE and SW Arms of Lake Malaŵi, plus Lake Malombe and the Upper Shire River, Fig. 17.1) are estimated to have provided 80% of the total fish catch between 1976 and 1990. The northern and central area employ almost half of the fishers, but produce less than one-quarter of the total yield. The greater fish yields from the south are probably due to limnological conditions: the broad shallow shelf areas support high benthic productivity and facilitate trawling and seining, while the shape of the lake and the direction of the prevailing winds means that seasonal upwelling of nutrient-rich water occurs mainly in the south (Eccles, 1974). By contrast, most of the rest of the lake is bordered by steeply shelving, often rocky, beaches and the fisheries are mainly dependent on zooplankton-feeding fish, which in turn are supported by low pelagic primary productivity.

17.3 FISHERIES STATISTICS

Mechanized fishing companies are required to provide catch returns, and these are generally reliable with regard to total yield and effort. Artisanal fisheries have been assessed since 1976 using a boat-based sampling system, the catch assessment system – CAS (Bazigos, 1974). This system contains a number of methodological flaws and, in Malaŵi, error checking seems not to have been rigorously performed. CAS provides reasonably good estimates of annual catch over large sampling areas, but does not accurately reflect catches by gear and provides very poor effort estimates. From June 1990, a gear-targeted system introduced in the SE Arm of Lake Malaŵi, Lake Malombe and the Upper Shire River (Alimoso *et al.*, 1991) has produced more accurate estimates. Neither system adequately records small-scale subsistence fishing

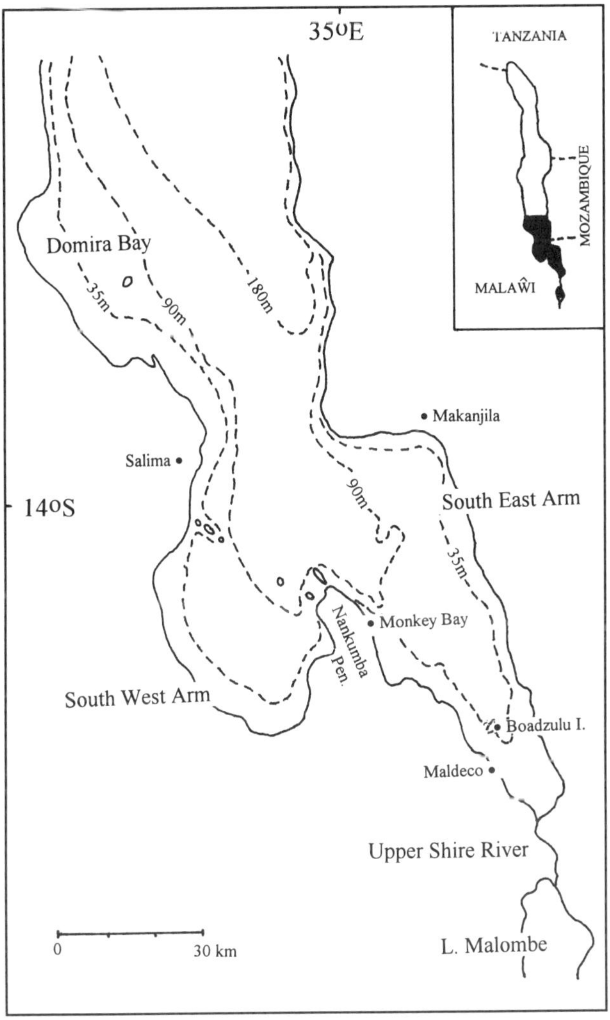

Fig. 17.1 Map of southern areas of Lake Malaŵi, showing places mentioned in the text. Inset, location of map on whole lake.

activities, such as handlining and trapping; these are major fisheries in some areas (Turner *et al.*, 1992). Nor does either system adequately cover the nocturnal inshore pelagic fishery for *Engraulicypris sardella* (a small cyprinid, locally known as usipa). Lewis and Tweddle (1990) estimated that only 10% of the usipa catch was recorded in 1985–86, and that the true annual yield could have been as high as 50 000 tonnes (t) for this species alone.

Table 17.1 Summary of limnology and exploitation of the fishing area of the Lake Malaŵi system (Malaŵian waters)*

	Lake Malombe	Upper Shire River	SE Arm Lake Malaŵi	SW Arm, Lake Malaŵi
Limnology				
Depth	V. shallow (max. 7 m)	V. shallow (max. 7 m)	Shallow (max 120 m)	Shallow (max. 120 m)
Shoreline length (km)	72	17	217	155
Shore type	Reeds, mud	Reeds	Sand, reeds	Reeds, sand
Bottom type	Mud	Mud	Diatom ooze	Diatom ooze
Productivity	Very high	High	High	High
Exploitation				
Main artisanal gears	Nkatcha and kambuzi seines	Chambo and kambuzi seines	Chirimila and chambo seines	Chirimila seine and Gill net
Artisanal craft	593	103	1 725	1 036
Artisanal fishers	2 077	1 125	7 306	3 563
Artisanal catch (t)	8 400	700	6 800	6 200
Mechanized gears	–	–	Trawls, ringnet	Pair trawl
Mechanized units	0	0	11	7
Mechanized catch (t)	0	0	6 700	1 100
Main taxa	Kambuzi, chambo	Chambo, kambuzi	Chambo, kambuzi	Kambuzi, utaka
Catch (t)	8 400	700	13 600	7 300
% Total catch	22	2	35	19

	North and Central Lake Malaŵi	Totals
Limnology		
Depth	Deep (max. 700 m)	
Shoreline length (km)	637	
Shore type	Rock, sand	
Bottom type	Sand, rock	
Productivity	Low	
Exploitation		
Main artisanal gears	hirimila seine and Gill net	
Artisanal craft	4 382	7 839
Artisanal fishers	12 912	26 983
Artisanal catch (t)	8 500	30 600
Mechanized gears	Pair trawl	
Mechanized units	1	19
Mechanized catch (t)	125[a]	7 925
Main taxa	Usipa, utaka	
Catch (t)	8 600	38 600
% Total catch	22	100

* Data from Malaŵi Fisheries Department. Catch figures, in tonnes, are means of 1976–1990, except [a] 1990 only. Numbers of artisanal craft and fishers from Tweddle, Alimoso and Sodjapanza, unpublished data. See Table 17.3 for explanations of local names for fish taxa.

17.4 STOCK IDENTIFICATION AND MANAGEMENT

None of the catch-recording systems in use accurately identifies fish to species, and the broad marketing categories used are not consistently applied. Assessment of Lake Malaŵi's fisheries is handicapped by the enormous species diversity of the fish (more than 200 species are exploited for food in the SE Arm alone), lack of adequate guides to their identification (the majority of exploited species are still undescribed) and the variety of fishing techniques used (Table 17.2).

In particular, before an assessment of stocks can be made, it is necessary to determine which fishing techniques exploit the same stocks. At first approximation this means the same species. The dominance of haplochromine cichlids, variously recorded according to size and the gear employed as kambuzi, utaka, ncheni, ndunduma, mbaba and chisawasawa (Table 17.3) means that fisheries statistics based on marketing categories are of little assistance in this regard. The high-value tilapiine species, recorded as chambo when adult, are distinguished by fishermen as kasawala when young, but not recorded separately from haplochromines. Until the present study, no records of the species composition of commercial catches had been made since the surveys of Lowe (1952) and Jackson *et al.* (1961).

17.5 GENERAL TRENDS IN THE FISHERIES

Since 1976, total annual fish landings in Malaŵi from the Lake Malaŵi system have been estimated to be between 29 000 t and 47 000 t. In 1983, landings in Mozambique were estimated at 9 000 t (Ssentongo and Dampha, 1992) and it seems unlikely that Tanzanian catches are higher.

Between 1976 and 1990, total fish production from the Malaŵian sector of Lake Malaŵi and associated water bodies remained fairly stable, and catches have been almost constant since 1986 (Table 17.4). However, since 1980, there has been an increase of almost 60% in the total fishing fleet, and of 100% in the mechanized fishery. There has been a switch from large-meshed to small-meshed gears: in the artisanal fishery from gill nets and chambo seines (mainly 76–90 mm stretched mesh) to kambuzi, chirimila and nkatcha seines (mainly 6–20 mm); in the mechanized fishery from 102 mm to 38 mm or less. The total catch value has declined, so that the value of the catch per craft has fallen to half of the 1983 figure. It is noticeable that catches of the larger fishes (tilapias and catfish) have declined. Although catches of the smaller species have shown an increasing trend, yields of usipa and demersal haplochromines seem to have stabilized, while the utaka catch has been in decline since 1987. This overall picture should alert fishery managers to the

Table 17.2 Principal fisheries of the Lake Malaŵi system

Fishing sector and gear type	Explanation	Principal catch*	% Total catch by weight*
Mechanized			18 (total)
Midwater trawl	Semipelagic trawl with bottom-set otter boards	Chambo, ndunduma, chisawasawa, ncheni, utaka	3
Bottom trawl	Demersal trawl, single boat	Chisawasawa, ndunduma	3
Pair trawl	Demersal trawl, two boats	Chisawasawa, ncheni, utaka	9
Ring net	Manually operated purse seines	Chambo	2
Commercial artisanal			82 (total)
Chirimila seine	Open-water seine/lift net	Utaka, usipa	33
Chambo seine	Large-meshed beach seine	Chambo	2
Kambuzi seine, nkatcha seine	Small-meshed beach and offshore demersal seines respectively	Kambuzi, usipa, immature chambo	29
Gill nets	Passive or active gill nets	Chambo, catfish, anadromous spp.	18
Subsistence			Not recorded
Longline	Deep-water baited hooks usually left unattended	Catfish	–
Handline	Active angling with baited hooks	Utaka, ncheni, kambuzi	–
Fish trap	Reed basket traps	Kambuzi, catfish	–
Scoop net	Netting between two poles	Ncheni, utaka	–
Cast net	Weighted margins, thrown	Kambuzi, ncheni	–

*Catch statistics from Malaŵi Fisheries Dept, 1987–1989.

Table 17.3 Commercial importance of different taxa in the Lake Malaŵi system

Taxon	Local name	Number of exploited species	Number of exploited endemics	Total catch (tonnes)	Total catch (%)
Haplochromine cichlids					
Demersal spp.	Kambuzi, chisawasawa, mbaba	150+	150+	15 000	32
Pelagic piscivores and zooplanktivores	Ncheni, ndunduma	25+	25+	1 600	3
Semipelagic zooplankton feeders	Utaka	25+	25+	11 000	24
Totals		200+	200+	28 000	59
Tilapiine cichlids	Chambo	5	3	8 000	17
Engraulicypris	Usipa	1	1	7 000	15
Catfish	Various	15+	12	3 500	7
Others	Various	15+	2	1 000	2

possibility of widespread overfishing, although more detailed investigations of individual fisheries are required to confirm this.

17.6 TILAPIINE FISHERIES

Stock identificaion

Five tilapiine species are found in southern Lake Malaŵi, Lake Malombe and the connecting Upper Shire River. *Tilapia rendalli* and *Oreochromis (Oreochromis) shiranus* are of little commercial importance in Lake Malaŵi. The bulk of the commercial tilapiine catch consists of three endemic *Oreochromis (Nyasalapia)* species (Lowe, 1952; Trewavas, 1983; Turner *et al.*, 1989): *O. lidole, O. squamipinnis*, and *O. karongae*, collectively known as 'chambo'.

Oreochromis karongae is a morphologically variable species. On coarse sand substrates (Karonga, the Nankumba Peninsula and Makanjila), they have expanded lower pharyngeal bones, while some individuals on rocky coasts (Nankumba Peninsula) have wide bands of jaw teeth. Specimens taken over soft muddy bottoms (Lake Malombe, the SW Arm and the southern part of the SE Arm) have reduced jaw and pharyngeal dentition, and were described as *Tilapia saka* by Lowe (1953). All of these forms overlap morphologically and are now considered to be conspecific (Turner and Robinson, 1991), and they share a number of ecological and behavioural characteristics: black ripe

Table 17.4 Summary of trends in the fisheries of the Lake Malawi system

Year	Catches (t × 1000)						Catch value* (MK × 10^6)	No. of craft	Catch wt per craft (t)	Catch value per craft (MK × 10^3)
	Tilapiines	Demersal haplo-chromines	Utaka	Catfish	Usipa	Total				
1976	16	9	6	5	2	38	51	5 000	7.7	10.3
1977	16	8	6	4	2	37	51	5 100	7.2	10.0
1978	15	5	5	4	2	33	47	5 100	6.4	9.1
1979	13	5	5	3	1	29	41	5 000	5.7	8.1
1980	13	6	7	3	1	30	41	5 000	5.9	8.1
1981	17	7	6	3	1	35	51	5 200	6.7	9.9
1982	19	8	7	3	2	40	58	5 400	7.4	10.9
1983	21	8	6	3	2	41	63	5 700	7.2	11.1
1984	20	8	7	3	2	41	61	5 900	6.9	10.3
1985	19	9	8	3	3	42	59	6 300	6.8	9.4
1986	15	11	11	3	5	46	54	6 500	7.1	8.3
1987	12	13	13	3	5	47	49	6 800	7.0	7.3
1988	10	13	12	3	7	47	46	7 300	6.4	6.3
1989	10	14	9	3	5	44	44	7 700	5.8	5.7
1990	9	14	8	3	7	45	44	7 900	5.7	5.5

* All catch values at 1991 equivalent prices. MK denotes Malawian Kwacha; approximately US$ 1 = 3.5MK.

male colour; facultative brownish colour in females, juveniles and non-ripe males; reproduction beginning at the end of the cold season; nesting areas extending into shallow water; small nest size; demersal habits and benthic diet (Turner and Robinson, 1991; Turner *et al.*, 1991a,b; Turner and Mwanyama, 1992). *O. lidole* is a more pelagic species, recognized by its elongated shape, dark grey colour, larger head and mouth and reduced dentition. Huge nests, up to 3 m diameter, are excavated in deep water, the shallowest nest being found at 17 m (Turner *et al.*, 1991b). Ripe males are black. In its distribution and trophic morphology, *O. squamipinnis* is intermediate between the other species. Ripe males are silvery, with a blue or white head and a dark ventral surface. Reproduction begins later than the other two species (Fig. 17.2) and like *O. karongae*, it sometimes breeds in shallow water (Turner *et al.*, 1991b). Reproduction in all three species is seasonal, with a marked decline during the cold season months of April to June. Reproduction in Lake Malombe appears to begin earlier than in Lake Malaŵi.

From October 1990 to September 1991, monthly samples of tilapiine catches were collected from the semipelagic trawl catch in the north of the SE Arm of Lake Malaŵi, and from gill nets and beach seines at six locations throughout Lake Malombe and the SE Arm of Lake Malaŵi. Inconsistency in effort or low catches meant that sporadic samples were obtained from purse seines and demersal trawls in Lake Malaŵi and from seine net catches from the Upper Shire River. It was found that all three species are taken, in varying proportions, by mechanized and artisanal fisheries in the SE Arm of Lake Malaŵi (Table 17.5). Thus there seems little justification for considering that the mechanized and artisanal fishing sectors exploit different tilapiine stocks, and I consider that no attempt should be made to perform separate stock assessments on these fisheries.

From experimental fishing, gonad analysis and morphometric studies from 1990 to 1991, Seisay *et al.* (1992a) concluded there was little migration of chambo between Lake Malombe and Lake Malaŵi. However, with the collapse of the Lake Malombe stock (Chapter 19), the density of chambo in the Upper Shire River was shown to have suffered an almost 30-fold decline since 1973. It is possible that migration between the two lakes was formerly much more important than now, but on the basis of recent data, it seems appropriate to assess the chambo stocks of Lake Malaŵi and Malombe as separate units.

Assessment

Chambo fisheries in the SE Arm have generally been assessed using the purse seine ('ring net') pull as the standard unit of effort (FAO, 1976; Tweddle and Magasa, 1989). The total catch of all gears is combined with the equivalent effort level calculated by dividing the purse seine CPUE. Because the 38 mm semipelagic trawl had a low effort level in some years because the artisanal

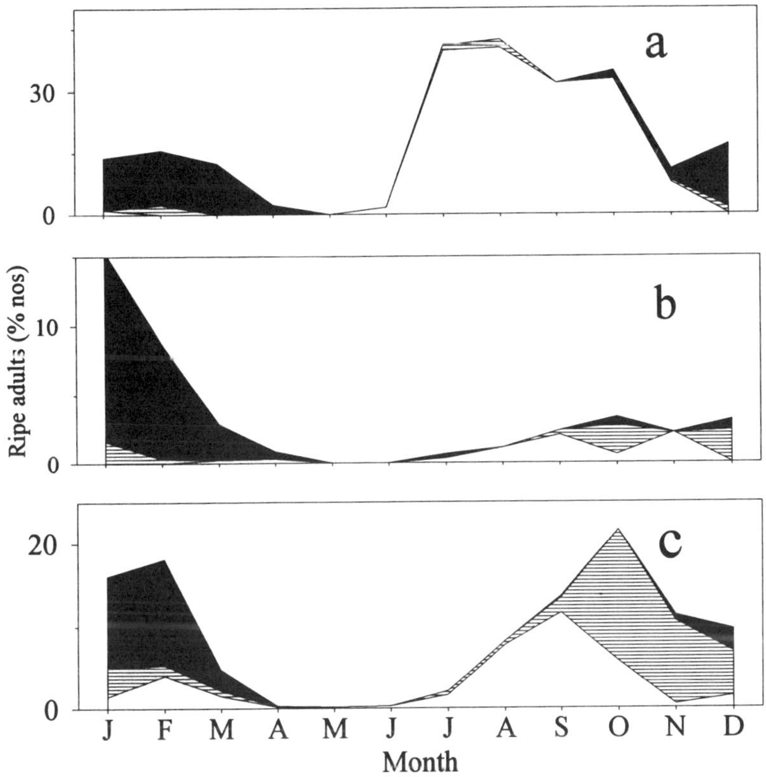

Fig. 17.2 Breeding seasons of endemic tilapiines *Oreochromis karongae* (white polygon), *O. lidole* (light shading) and *O. squamipinnis* (black) expressed as percentage by number of ripe adults taken from all gears sampled from: (a) Lake Malombe; (b) SE Arm of Lake Malaŵi, South of Boadzulu Island; (c) SE Arm of Lake Malaŵi, North of Boadzulu Island. High levels of breeding activity were recorded from all months, except April to June. Reproduction begins earlier in Lake Malombe than in Lake Malaŵi. North of Boadzulu Island *O. karongae* has a longer, and possibly bimodal, breeding season. The main breeding area for *O. lidole* is north of Boadzulu Island.

fisheries estimates are less reliable than the mechanized fisheries returns, it seems wise to continue to use the purse seine as a standard effort unit.

Between 1982 and 1986, high catch rates were recorded by the three main chambo-exploiting gears, the semipelagic trawl, purse seine and gill net (Fig. 17.3). There is no indication that this was due to a decline in fishing effort in the SE Arm and extraneous causes must be invoked. Similarly high catches were observed in Lake Malombe and the Upper Shire River at approximately the same time, although the SW Arm was not greatly affected

Table 17.5 Species composition (percentage weights) of tilapiine catches in the SE Arm of Lake Malaŵi, 1990–91

Gear	*Oreochromis lidole* Trewavas 1941	*O. squamipinnis* Gunther 1864	*O. karongae* Trewavas 1941
Purse seine	75.3	20.5	4.2
Semipelagic trawl	60.4	25.7	13.8
Bottom trawl	20.2	72.7	7.1
Pair trawl	1.3	70.2	28.5
Chambo seine	33.7	32.3	34.0
Gill net	23.3	40.8	35.9

(Fig. 17.4). This increase in stocks was perhaps caused by a reduction in gill net effort in Lake Malombe in the late 1970s (shown by both gear census and CAS surveys), leading to an increase in the tilapiine abundance in Lake Malombe and the SE Arm of Lake Malaŵi. Alternatively it may be a reflection of recruitment variability. Irrespective of cause, such years of exceptional stock density should not be used in an equilibrium calculation of MSY, as this may lead to a temptation to try to aim for an increased yield. It is notable

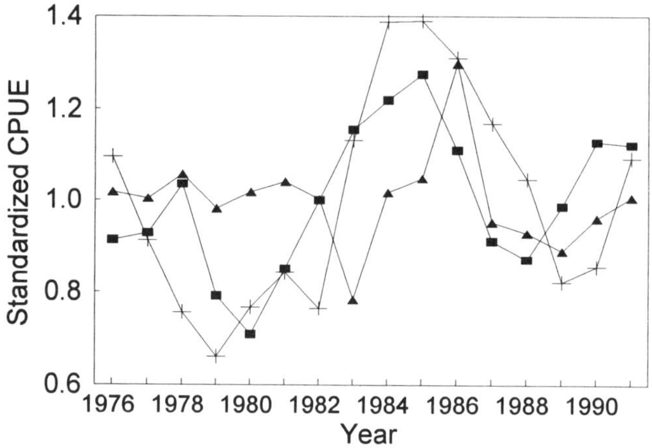

Fig. 17.3 Trends in the CPUE of tilapiine fisheries in the SE Arm of Lake Malaŵi. Catch rates by all three main gears were high in the mid 1980s and declined from 1986 to 1989. Since then there has been a substantial recovery of stocks. Artisanal (■) and purse seine (+) CPUEs have shown similar trends throughout, but that of the semipelagic trawl (▲) has only been related to the others since the mid 1980s, suggesting that this gear formerly exploited a different stock. CPUE for all gears has been standardized as the ratio of each year to that of the mean of 1976 to 1991.

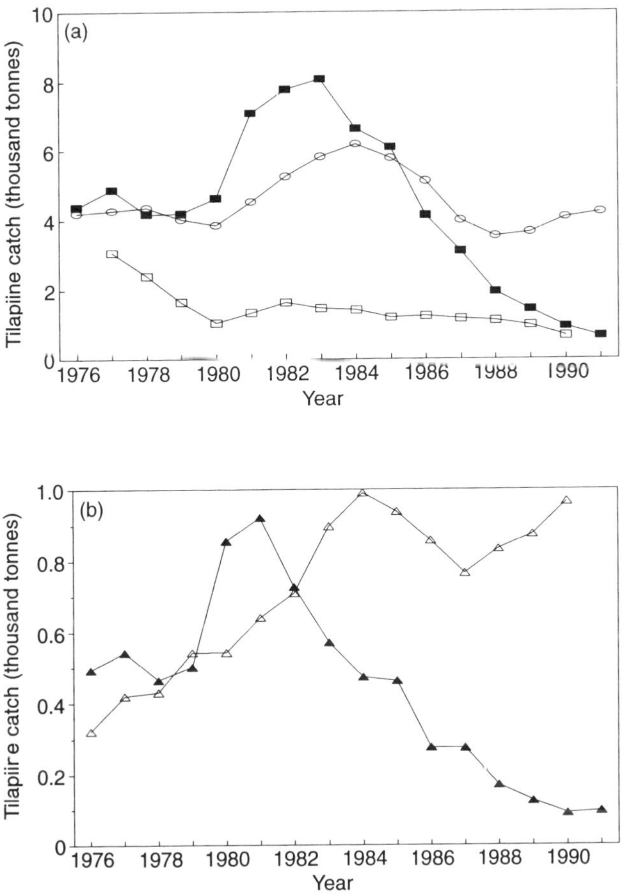

Fig. 17.4 (a) Total tilapiine catches in Lake Malombe (■) and the SW Arm of Lake Malaŵi (□) have declined substantially, while that of the SE Arm (○) has remained relatively stable. (b) The tilapiine catch in the Upper Shire River (▲) declined dramatically at the same time as that in the adjacent Lake Malombe. Yields in northern and central Lake Malaŵi (△) have steadily risen, although they still constitute a very small proportion of the total fish catch in these areas. Note that a temporary increase in catches occurred in all areas in the mid 1980s. Data are from commercial fishing returns and three-year running means for artisanal catches. SW Arm data not collected in 1976 and not available for 1991.

that the inclusion of these years in the calculation does not lead to a higher estimate of F_{MSY}. Should such high stock densities ever re-occur they should be regarded as a bonus whereby 50% higher yields can be achieved at approximately the same effort level.

From Fig. 17.3, it can be seen that purse seine and artisanal fisheries show similar trends in CPUE over the 13 years to 1991, while the semipelagic trawl and purse seine CPUEs were negatively correlated between 1976 and 1984 ($r = -0.77$). From 1985 to 1990, however, there was a strong positive correlation ($r = 0.86$). The CPUE of the purse seine fishery was substantially elevated in 1991, so that if this year is included the correlation drops to 0.40. The explanation for this is that, under new management, the overall effort of the purse seine fishery was reduced, and fishing concentrated on the months that have traditionally had the highest CPUE (Fig 17.5). Thus, for stock assessment, the 1991 purse seine CPUE was estimated by multiplying the 1991 semipelagic trawl CPUE by the mean of the ratios of the purse seine to semipelagic trawl CPUE for 1985 to 1990.

Although it might appear that the high effort level and low CPUE for 1979 might have an undue influence on the calculation of MSY, sensitivity analysis indicates that this is not the case; leaving out this year has only a minor effect (1–2% change in MSY, 4–6% in F_{MSY} and 7–8% reduction in the correlation coefficient), and so the data for 1979 have been retained. Artisanal fisheries catches were averaged over 3 years to reduce noise resulting from the CAS system.

Fitting of a Schaefer or Fox surplus-production model made little difference

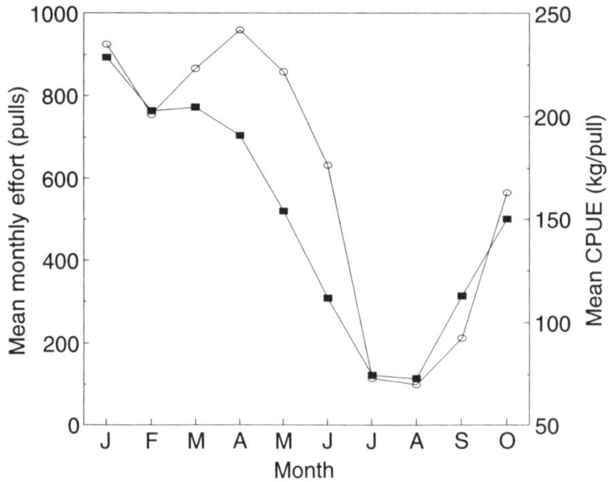

Fig. 17.5 The purse seine fishery for tilapiines is strongly seasonal. CPUE (■) and effort (○) are low during the cold-season months of June to August. Shown are means of CPUE and effort for the years 1976 to 1991. Mean monthly CPUE and effort are significantly correlated: $r = 0.893$, 9 d.f., $P < 0.001$. The fishery is closed during November and December.

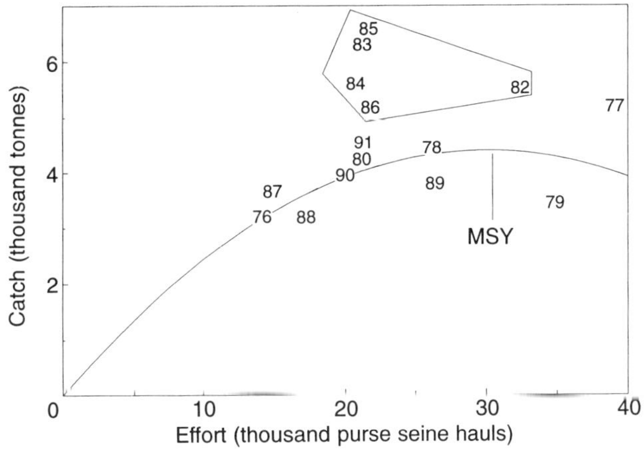

Fig. 17.6 The maximum sustainable yield of the tilapiine fishery in the SE Arm of Lake Malaŵi is estimated to be approximately 4400 t, at an effort level equivalent to around 30 000 pulls. The fit shown is the Schaefer model: $r = 0.899$, 9 d.f., $P < 0.001$. The unusually high catch rates of 1982 to 1986 were excluded from the analysis. Due to changes in fishing practices, purse seine CPUE in 1991 was estimated from semipelagic trawl CPUE, using the relationship between the CPUEs of the two gears from 1985 to 1990.

to the results: MSY was estimated at 4300–4400 tonnes at an effort level of 30–33 thousand ring net hauls ($r = 0.89$ in each case, 9 d.f.).

Thus, it appears that, except in years of exceptional recruitment, the SE Arm could yield no more than about 4000 t of chambo. Because F_{MSY} is only marginally above the effort estimate for 1991 (Fig. 17.6), the data suggest that there would be no benefit in increasing the fishing fleet targeting chambo.

From an analysis of length data, Seisay *et al.* (1992b) found that the mortality rate (Z) of immature *Oreochromis lidole* was much lower than those of *O. squamipinnis* and *O. karongae* (2.8, 7.4 and 7.5 respectively), although there was little difference in exploitation rates of adult fishes (0.85, 1.09, 0.81). Difficulty in distinguishing young of these species, as well as the unexplained scarcity of 1 + *O. lidole*, mean that these results must be treated with some caution.

Given that these species appear to be susceptible to stock collapse through excessive exploitation of immature fish (van Zalinge *et al.*, 1991; Chapter 19), any future expansion of effort should be prevented or confined to those gears which preferentially catch mature fishes, particularly *O. lidole* (i.e. mechanized fisheries employing large meshes) and only if there is a concomitant reduction in gears that catch immature fish, particularly of the other two species; that is, mainly small-meshed seines.

Mechanized fisheries

Few tilapiines are caught by demersal trawling, either by the deep-water single-boat trawlers or the pair trawlers that operate further inshore. Until recently, the mechanized tilapiine fishery was dominated by purse seining and semipelagic trawling, both using 102 mm mesh. Effort by purse seines has declined since the late 1980s, and the gear appears to be less economical than trawling. The 102 mm mesh semipelagic trawl, intended to fish for tilapiines, has not been used since 1988, being replaced by a 38 mm mesh net supposedly targeted on pelagic (*Diplotaxodon* spp.; ndunduma) and semipelagic haplochromines (*Copadichromis* spp.; utaka). Expansion of the semipelagic trawl fishery is under consideration.

Length-frequency samples, taken thrice monthly from October 1990 to September 1991, indicate that 66.6% of all individual tilapiines taken by the 38 mm net are immature (47.3% of catch weight). A substantially smaller proportion of *O. lidole* are taken when immature: 55.4%, compared with 77.9% for the other two species. The high proportion of immature fish taken is likely to be detrimental to the tilapiine stock. A reduction of effort by this gear, and a return to the use of 102 mm net would appear to be advisable, particularly as, in the past, the tilapiine catch per unit effort of the 102 mm net generally equalled or exceeded that of the 38 mm (Turner and Mdaihli, 1992). However, the 38 mm gear also produces a substantial bycatch of smaller haplochromine cichlids which are not available to the larger-meshed gear. Thus, assessment of the optimum effort for the small-meshed midwater trawl requires consideration of this bycatch.

17.7 HAPLOCHROMINE TRAWL FISHERIES

Stock identification

Haplochromine cichlids are at present exploited by three trawling techniques: semipelagic trawling, demersal single-boat trawling and demersal pair trawling (1, 2 and 20 units respectively in 1991). Under the present licensing system, units are allocated to eight fishing areas: maximum levels of exploitation are based on tentative biomass estimates made by experimental trawling between 1971 and 1973 (FAO, 1976), no distinction being made between vessels using different trawling techniques.

Before assessing the state of the haplochromine fisheries, it is necessary to know which vessels are exploiting the same stocks. The numbers and weights of each species were recorded from 54 commercial trawl catches (Turner, 1992).

The haplochromine catch of the semipelagic trawl is dominated by a single previously undescribed pelagic cichlid (Table 17.6), *Diplotaxodon limnothrissa* (Turner, 1994b). This species is a zooplanktivorous cichlid of comparable

Table 17.6 Major fish taxa from trawl catches in the SE Arm of Lake Malaŵi (% weights, excluding tilapiines and catfish)

Taxon*	Maximum total length (cm)	Habitat	Diet	Semi-pelagic trawl	Demersal single-boat trawl	Demersal pair trawl
Diplotaxodon limnothrissa Turner 1994	18	Pelagic	Zooplankton	62	18	5
Rhamphochromis spp.	30–50	Pelagic	Fish	11	8	15
Lethrinops microdon Eccles & Lewis 1977	20	Demersal	Sediment	3	28	–
Lethrinops 'blue-orange'	8	Demersal	Benthic invert.	P†	–†	15
Lethrinops 'dark'	10	Demersal	Benthic invert.	P	5	8
Otopharynx 'argyrosoma b'	14	Demersal	Benthic invert.	1	P	11
Placidochromis 'cf. *longimanus*'	12	Demersal	Benthic invert.	P	–	11
Aulonocara 'orange'	10	Demersal	Benthic invert.	1	P	8
Lethrinops longipinnis Eccles & Lewis 1978	21	Demersal	Benthic invert.	2	5	2
Copadichromis 'cf. *virginalis*'	11–15	Demersal	Zooplankton	4	P	4

* Taxa designated with 'cf.' before specific name comprised a number of related taxa which could not be adequately distinguished from the described species.
† P, present; –, not recorded.

size and body shape to *Limnothrissa miodon* from Lake Tanganyika: it is numerous throughout the pelagic zone of the lake from the surface to at least 120 m depth. Demersal stern trawl catches also contained large proportions of these pelagic species, but were dominated by large demersal species, particularly *Lethrinops microdon*. An entirely different community of species is exploited by pair trawlers. South of Boadzulu Island, catches were mainly composed of small undescribed species characteristic of relatively shallow water of about 40 m or less. All techniques made substantial catches of the pelagic *Rhamphochromis*: there are probably at least 15 species in this genus and it is likely that different species are caught preferentially by different gears.

This analysis indicates that there is probably little direct competition between these three trawling techniques. Unlike the tilapiine cichlids, the principal deep-water haplochromine cichlids (except possibly *Rhamphochromis* spp.) do not have inshore spawning or nursery areas, and are thus not affected by seining.

Assessment of the semipelagic trawl haplochromine fishery

Using the Schaefer model, MSY is estimated to be about 720 t at an effort of 390 boat days ($r = 0.72$, 13 d.f., Fig. 17.7), indicating little or no room for expansion of this fishery in the SE Arm.

Assessment of the demersal stern trawl fishery

Demersal stern trawling in deep water (50–70 m) has been carried out by a variety of vessels since 1972, but only Maldeco's *Dennis Sanudi* has operated every year. It was thus taken as the standard unit of effort (Turner, 1992). This fishery is severely overexploited (Fig. 17.8). Total catches averaging around 1500 t had been maintained through a range of effort from 400 to 700 boat days, although the licensing of a new vessel in 1983 coincided with a sharp decline in CPUE (Fig. 17.9). From 1987 to 1989, a further increase in effort led to a second decline in CPUE, and a decline in overall catch. Despite a reduction of effort in 1990 and 1991, the fishery has yet to recover. Units added from 1983 onwards were nominally fishing in license area C, while previously fishing was concentrated to the south in area B. From the clear effects of fishing in area C on the CPUE of the *Dennis Sanudi* (area B), it is apparent that this arbitrary division does not reflect the boundary of the stock. Because the landing and refuelling sites for all these vessels are to the south, it is probable that economic considerations mean that fishing by these units has been concentrated in the southern part of area C, or (illegally) in area B.

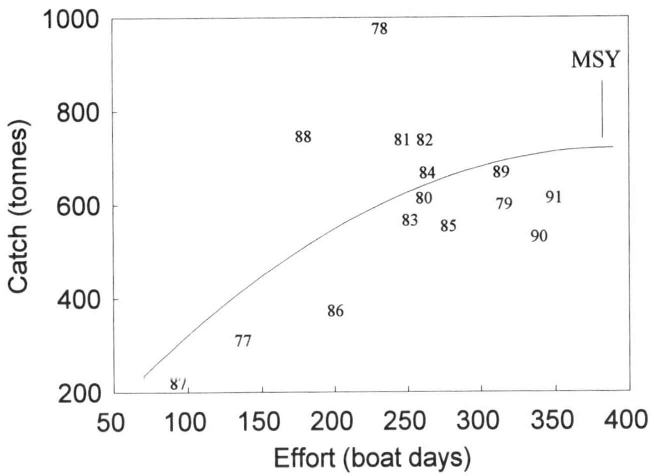

Fig. 17.7 The maximum sustainable yield of the haplochromine catch of the semipelagic trawl fishery of the SE Arm of Lake Malaŵi is estimated at 720 t at F_{MSY} of 390 boat days, and is thus more or less fully exploited. Fit: Schaefer model, $r = 0.721$, 13 d.f., $P < 0.01$.

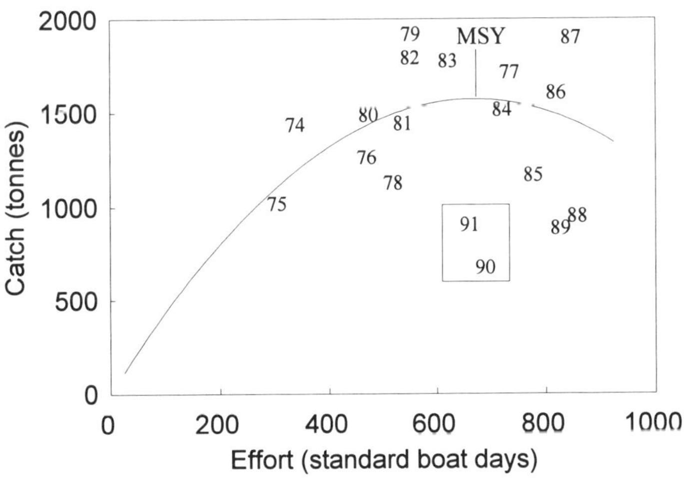

Fig. 17.8 The single-boat demersal trawl fishery in the SE Arm of Lake Malaŵi is seriously overfished. MSY is estimated at about 1570 t at an effort of 670 standard boat days. Effort was standardized to the CPUE of the *Dennis Sanudi* by calculating the ratios of the CPUEs of each vessel to that of the *Dennis Sanudi* over the whole data series. Fit is Schaefer model, $r = 0.826$, 14 d.f., $P < 0.001$. Data from 1990 and 1991 (box) were excluded from the analysis as they clearly deviated from the yield curve, perhaps as a result of recruitment overfishing.

Assessment of pair trawl fisheries

South of Boadzulu Island, there was an initial period of instability in pair trawl CPUE as exploitation of the virgin stock was begun, and a temporary decline in CPUE following an increase in mesh size from 25 mm to 38 mm in 1977. Since 1978, CPUE has remained fairly stable, varying by no more than 50% across this time period, while effort has shown changes of more than 150% (Fig. 17.10).

MSY was estimated by FAO (1976) and Tweddle and Magasa (1989) to be around 1400 t. Both analyses used Gulland's modification of the Schaefer model (Gulland, 1961), which is no longer considered valid (Pitcher and Hart, 1982). Reanalysis of the data, including subsequent years, indicates that there is no significant correlation between CPUE and effort if the early years of the fishery (1968–1971) are left out (Fig 17.10). Equilibrium versions of the Schaefer and Fox models cannot be used on non-equilibrium situations, such as the first few years of exploitation (Pitcher and Hart, 1982). Closer analysis of trends since 1978 reveals that three separate catch–effort relationships can be tentatively identified (Fig. 17.11). These changes are probably the result of changes in species composition (Chapter 18), mesh

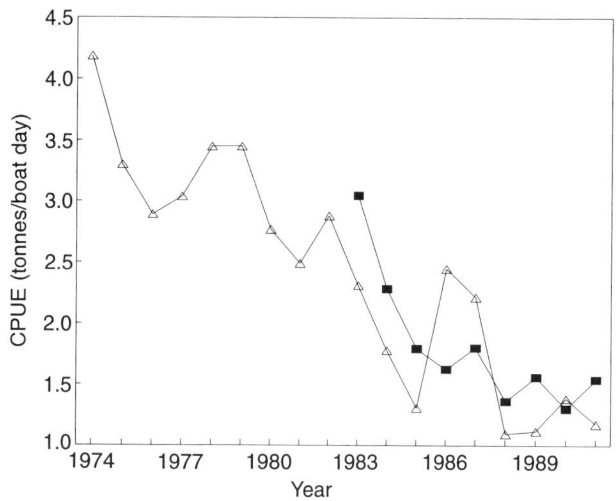

Fig. 17.9 CPUE of the demersal single-boat fishery in the SE Arm of Lake Malaŵi has declined since the inception of the fishery in 1972. Commercial fishing was initiated by Maldeco Fisheries (△): shown is the CPUE for the two smaller boats operated by this company. Licensing of another company (■) was accompanied by a sharp decline in CPUE.

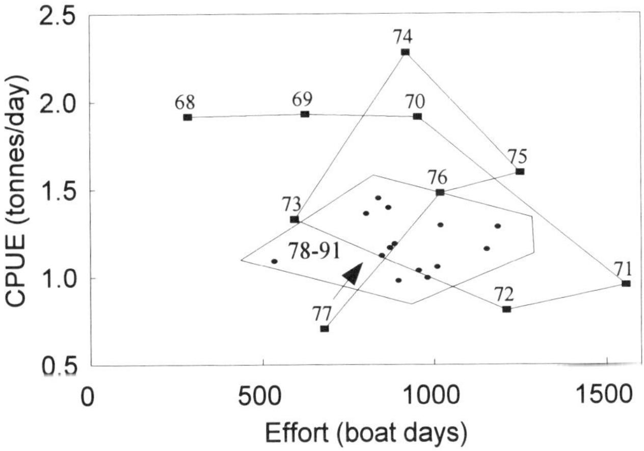

Fig. 17.10 From its inception in 1968, the pair trawl fishery exhibited large fluctuations in CPUE before stabilizing in 1978, since when CPUE has varied by less than 50% despite changes in effort of more than 150%. There was no statistical relationship between CPUE and effort in this fishery unless the first year of exploitation is included. Because large changes in age- and species composition are likely to occur during the early stage of exploitation, the fishery is likely to be far from an equilibrium condition.

size and depth fished. The current phase is characterized by high catches at low effort levels, but with a rapid decline when fishing pressure is increased. It appears to be highly unstable and requires close monitoring, although from the data available, it would be unwise to attempt to estimate MSY or optimum fishing effort.

The species composition of catches in the SW Arm has not yet been investigated, but the total yield of the fishery appears to show a similar resilience to that south of Boadzulu Island. The tentative MSY prediction of FAO (1976) has been exceeded on several occasions, but with no indication of a decline in CPUE, which again shows no correlation with fishing effort. Pair trawling was recently initiated in Domira Bay (area G). CPUE has so far been less than half that of the long-exploited area A.

An economic analysis of the pair trawl fisheries (Turner and Mdaihli, 1992) indicates that in the south of the lake in 1991, fishing would be profitable even at a relatively low activity level: this would have been 93 days in area A (south of Boadzulu Island), 115–116 days in the SW Arm and the SE Arm north of Boadzulu Island. Concern has been expressed that low catch rates have been due to declining mechanical efficiency of the pair trawl units

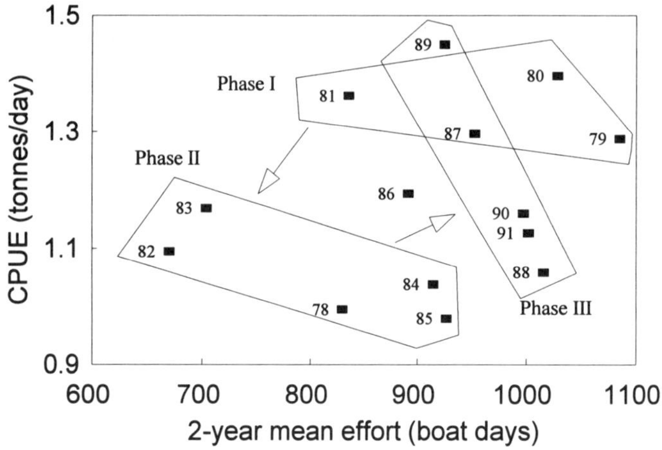

Fig. 17.11 Since 1978, the relationship between CPUE and effort in the pair trawl fishery south of Boadzulu Island has moved through three phases. From 1979 to 1981, changes in effort produced moderate changes in CPUE. From 1982 to 1985, despite a decline in effort, CPUE was lowered but changes in effort had a similarly moderate effect on CPUE. A third phase has lasted from 1987 to 1991, where high CPUE is maintained despite high effort levels, but small increases in effort have led to large declines in CPUE. Effort given as two-year means.

(Tweddle and Magasa, 1989), leading to suggestions that development aid should be provided to refit the vessels. However, from the 1991 estimates, I estimate that any unit operating for 260 days per annum in the southern part of the lake could generate sufficient profits to pay for the complete replacement of the machinery of both boats within a single year.

Nevertheless, at present price levels, pair trawling is uneconomical in Domira Bay. Because experimental fishing (FAO, 1976) indicated that catch rates in areas north of Domira Bay are approximately the same or lower, it would be inadvisable to introduce pair trawling into the northern and central parts of the lake.

17.8 SEINE FISHERIES

Assessment of the haplochromine seine net fisheries is hindered by the weakness of artisanal fisheries data, particularly because there may have been considerable (unrecorded) changes in the mean headline lengths of the nets, which currently vary from 50 to 1000 m. CPUE figures for these gears cannot be expected to be comparable.

Southern Lake Malaŵi

Detailed biological investigation of the catches of chirimila (open-water seine) nets were last performed in the 1950s (Jackson *et al.*, 1961), primarily in the north of the lake. Catch/effort returns from 1990–1991 indicate that semipelagic and demersal haplochromine cichlids (day fishing) and pelagic cyprinids (night fishing) dominate catches. Systematic sampling of this gear has not been performed and is urgently required. Preliminary examination of catches taken around the Nankumba Peninsula (G.F. Turner, pers. obs., L. Smith, pers. comm.) has indicated that zooplanktivorous cichlids of the genus *Copadichromis* constitute the bulk of the catch. Small catches are regularly made of a group of inshore-living utaka: *Copadichromis chrysonotus* (*sensu* Konings, 1990), *C. borleyi*, *C. jacksoni*, and *C.* cf. *prostoma*. These species were never recorded from trawl catches, which, however, often contained the related species *C. pleurostigma*, *C. pleurostigmoides*, *C. quadrimaculatus* and especially *C. virginalis* (two or more species). This second group of species are sporadically caught, in large numbers, by chirimila fishers, suggesting a possible competition between the two fisheries. Over the last 16 years, the chirimila fishery has greatly expanded in the SE Arm of Lake Malaŵi and there has been a substantial decline in CPUE. Many of the exploited *Copadichromis* spp. have restricted distributions (Konings, 1990) and may be vulnerable to local overfishing. The state of this major fishery requires more investigation.

In the SE Arm of Lake Malaŵi, small-meshed 'kambuzi' beach seines catch around 1000 tonnes per annum. Catch–effort returns indicate that the great bulk of the catch is 'usipa' (*Engraulicypris*) in the northern part of the SE Arm, while small demersal haplochromines (kambuzi) dominate in the southern part of the arm. Detailed biological analysis (of a few samples) suggests that, in the southern part of the Arm, *Engraulicypris* and immature *Oreochromis* each account for around 25% of the catch by weight. Haplochromines make up most of the rest. This is mainly composed of adults and juveniles of small permanently inshore-living species which form around 60% while the rest consists of large numbers of juveniles of many of the large species which are important in gill net, handline and pair trawl fisheries. Because recent developments in Lake Malombe have indicated that expansion of kambuzi seining can be seriously damaging to other fisheries, it would be advisable to restrict or ban this gear in the rest of the lake. This has been approved in principle by the Malaŵi Fisheries Department, but the practicality or time scale for its implementation is by no means clear.

Northern Lake Malaŵi

The commercial artisanal fisheries north of Salima are largely based on chirimila fishing for haplochromine cichlids and *Engraulicypris*, while gill net fisheries for catfish are of lesser and declining importance.

While it is generally considered that fisheries in the north of the lake are underexploited, this view is outdated. From 1976 to 1986, the fishing fleet north of Domira Bay expanded from 2500 to 3200 craft while yields increased from 7600 to 10 000 t. However, between 1986 and 1990, the number of fishing craft increased from 3200 to 4600, while total yields have remained constant at 11 000–12 000 t. This suggests that further expansion using existing fishing practices should be avoided.

17.9 PROBLEMS WITH FISHERIES ASSESSMENT

Although there is evidently a general problem with overexploitation of Lake Malaŵi's fisheries, it should be emphasized that the figures given for maximum sustainable yields and optimum fishing effort should not be taken too seriously for individual fisheries. For the most part, historical data series consist entirely of catch and effort figures, which have been analysed using production models with the assumption of equilibrium conditions (Pitcher and Hart, 1982). In some cases, changes in species composition or in fishing practice may have rendered this assumption invalid. The rapid change in population structure at the beginning of exploitation may also pose problems, as with the pair trawl fishery south of Boadzulu Island. However, leaving out the early years of low exploitation may mean there is insufficient contrast in the range of effort values for adequate model fitting (Hilborn and Walters, 1992). Furthermore, these models can really only give a good idea of the maximum sustainable yield after the fishery has declined; false plateaux may lead to an overly pessimistic estimate of MSY (Hilborn and Walters, 1992). This may perhaps be a problem for both the tilapiine and pelagic haplochromine fisheries in the SE Arm. Perhaps the use of non-equilibrium production models (Holden and Bravington, 1992) may be more appropriate, but initial attempts to use them with the pair trawl fishery have so far proved unpromising.

Analytical models of population structure would seem to offer the way forward to a more satisfactory understanding of Lake Malaŵi's fisheries. As yet, there have been only four published studies of growth and mortality rates in Lake Malaŵi cichlids (Lowe, 1952; Jackson *et al.*, 1961; Tweddle and Turner, 1977; Seisay *et al.*, 1992b). All have suffered from being conducted over short time periods and from difficulties with species identification which had to be worked out (with varying degrees of success) in parallel with the population studies rather than prior to them. Unfortunately none of these population monitoring programmes has been continued on a routine basis after the end of the externally funded project. A further difficulty lies in the sensitivity of these techniques to the assumed natural mortality rate. Empirical formulae such as that of Pauly (1980) were established largely for fecund marine species which are likely to continue to suffer substantial predation pressure throughout their life histories. It is not clear how appropriate such relationships are for low-fecundity freshwater species, some of which, like the tilapiines, are probably almost free from predation as adults.

17.10 PROSPECTS

Fisheries

Throughout the period 1976 to 1991, there has been a trend towards the increased use of smaller-meshed gears: chirimila seines, kambuzi seines, nkatcha nets and trawls of 38 mm mesh or less. Large-meshed gears, such as gill nets, chambo seines and purse seines have shown reductions in effort, while the 102 mm midwater trawl has been withdrawn. Malaŵi has lost three of its four principal tilapiine fishing grounds: Lake Malombe, Upper Shire River and the SW Arm of Lake Malaŵi. The anadromous cyprinid, *Labeo mesops*, which formerly supported the second largest fishery in the lake (FAO, 1976), is now of little commercial importance except in the vicinity of the closed fishing area of Liwonde National Park to the south of Lake Malombe. Many of the larger haplochromine species have been eradicated by pair trawling south of Boadzulu Island (Chapter 18), whereas expansion of this fishery into the north of the lake would seem to be constrained by economic factors. Few large fish species remain in Lake Malombe, where the biological diversity is now low. Deep-water demersal haplochromine stocks in southern Lake Malaŵi are already heavily overexploited. Thus, there is no evidence that Lake Malaŵi's demersal or semipelagic fisheries could benefit from expansion.

Prospects for the expansion of the offshore pelagic fishery must await the findings of the ODA/SADC Pelagic Fish Resource Assessment Project, in progress (see Chapter 23 for preliminary findings). However, even if a substantial stock is discovered, it should be emphasized that any plan for its exploitation should take into consideration that tradition, economics, personal safety and time may all dictate that fishing will continue to be concentrated on inshore areas until stocks have been depressed to very low levels by overfishing. Biological interactions between inshore and offshore stocks are poorly known. Therefore, before any attempt is made to expand offshore fisheries there should be: (1) investigation of an appropriate technology to harvest these stocks, (2) consideration of the social and economic factors relevant to decisions about where and how to fish, (3) a thorough investigation of the systematics and life histories of the major pelagic species, and (4) an assessment of the competing exploitation by inshore fisheries.

While Malaŵi's population continues to grow at more than 3% per annum, decision makers must accept that the fish catch cannot be expanded to keep pace with this. The continuing profitability of fishing, even in fisheries that are heavily overexploited (Mdaihli and Donda, 1992), coupled with the lack of alternative employment, means that there will continue to be a strong expansionary pressure within all fisheries for the foreseeable future. On the other hand, the enforcement and extension capacities of the fisheries department are inadequate to prevent overexploitation, and as the present Chapter has demonstrated, little is known of the majority of important fisheries, so that overfishing is unlikely to be detected until the effects have become severe and perhaps irreversible (Hilborn and Walters, 1992). Malaŵi's fisheries are expanding at a rate incommensurate with the available monitoring and management capabilities. Overhasty government or external intervention aimed at further accelerating the expanding fishing effort on Lake Malaŵi will only exacerbate this process, having a transient effect of increasing yields and bringing closer the day when overexploitation leads to decline in the economic value, biological diversity and total yield of Lake Malaŵi's unique fish community.

Possible introductions

Lake Malaŵi contains not only more species of fish, but also a wider variety of different trophic adaptations than all of the fresh waters of Europe combined (see also Chapter 23). It is therefore most improbable that species introductions could increase the total fish yield of Lake Malaŵi, as has been suggested. For example, the introduction of the Lake Tanganyika clupeid *Limnothrissa miodon* was mooted by FAO in 1982, to exploit the large numbers of chaoborid larvae said to be found in the pelagic zone. Chaoborids were rarely found in the stomachs of *Engraulicypris*, and the zooplanktivorous *Copadichromis* are largely inshore species, because of their demersal spawning habits. In fact FAO (1982) did not present any data to indicate that chaoborids form a large part of the pelagic biomass, apart from ad hoc observations of large swarms of adults. The discovery that *Diplotaxodon* species are large, numerous, truly pelagic, zooplankton feeders (Turner, 1994b) indicates that there is no empty niche waiting to be filled by *Limnothrissa* (this is now confirmed by A. Menz ODA/SADC project, pers. comm. and Chapter 23 – *Eds*).

The introduction of predatory species is generally justified as a way of converting small 'trash' species into large fish fit for human consumption. This is not relevant to Malaŵi, where all sizes of fish are in high demand as human food. It is not possible to increase the total ecological production of any water body through switching a fishery from primary or secondary consumers to a top predator, any more than it would be possible to increase the yields of European or American meat production by switching from production of cows to husbandry of lions (Fryer, 1973), as there is a huge loss of energy in the conversion of animal protein at each trophic level.

Conservation of biological diversity

Fishing, of the intensity currently found in Lake Malombe and in Lake Malaŵi south of Boadzulu Island, has led to great changes in community structures, with a reduction in diversity and increasing dominance of smaller species. If such practices were to spread throughout the lake, species extinctions may be inevitable. Many, if not most, of the endemic species seem to have restricted distributions within the lake, although at present little is known about fish communities outside the SE Arm.

Tilapiine cichlids and non-cichlid species have wide distributions within the lake, large population sizes and relatively high fecundities. Species extinctions in these taxa seem unlikely. The uniqueness of and need to preserve the rocky-shore haplochromines (mbuna) is well appreciated both in Malaŵi, where they are protected by the Lake Malaŵi National Park, and internationally. However, at present there is no threat to these species. They are not highly esteemed as food, and their behaviour means that they are difficult to catch with intensive commercial fishing techniques.

The demersal and pelagic haplochromine cichlids, the species of greatest interest to evolutionary biologists, are the species most at risk. These fishes have low fecundities, and many have small population sizes and restricted distributions. They can be exploited by small-meshed, unselective gears. Conservationists must realize that preservation of biological diversity may not be compatible with the optimization of fish yields; higher yields may be obtained from a simplified community of a few small-maturing species. As there is very little difference in the prices of different sizes or species of haplochromine cichlids and little local awareness of their diversity, there is unlikely to be interest in conservation among the fishing communities. The deeper-water species are often drab, inaccessible to divers and snorkellers and few are exported as aquarium fish. Thus, awareness of these species among tourists and aquarists is likely to remain low. It is likely that, unless vigorous intervention is initiated by the scientific community, the last chance to investigate the possibility of large-scale rapid sympatric speciation in vertebrates (Turner, 1994a) will shortly disappear.

Future research priorities

The present work has indicated the great utility of species identification in characterizing stocks and thus in assessing their state of exploitation. Species identification is also an essential prerequisite for framing conservation strategies in that knowledge of the extent of species distributions and the influence of exploitation on stocks is impossible if species are not identifiable. The first priority for research on Lake Malaŵi must therefore be the inventory of species and the preparation of a field guide for their identification. For this purpose, formal descriptions are not necessary, and informal 'cheironyms' could be used; this approach has been successfully adopted by Ribbink *et al.* (1983) for rocky shore species. Subsequently distribution, abundance and exploitation should be investigated. Basic biological parameters can then be studied, such as size at maturity, growth rate, diet and mortality rates, all of which are poorly known in Malaŵian species. Determination of natural mortality rates in unexploited stocks should be attempted while these still exist, because yield-per-recruit, VPA and other assessment techniques are sensitive to the natural mortality rate assumed.

Much more information about the species composition of commercial catches is required, and should be collected on a regular basis. For example, the chirimila fishery has not been adequately investigated and no biological work has yet been carried out in the SW Arm, which is also a major fishing ground.

17.11 SUMMARY

The biological bases and state of exploitation of the major fish stocks of the Lake Malaŵi system are reviewed. Prior to 1986, endemic tilapiine cichlids (chambo) dominated the fisheries. Beach seine, gill net, trawl and purse seine fisheries were found to compete for the same three species. Of four formerly important fishing grounds, only the SE Arm of Lake Malaŵi still retains a viable tilapiine fishery, but this appears to be fully exploited. Little is known about the endemic haplochromine cichlids which now dominate Malaŵi's fisheries, and attempts to assess these stocks will continue to be unsatisfactory while the present situation persists where few of the major commercial species can be identified in the field. It is tentatively concluded that the three main trawl fisheries can be assessed independently. Of these, the deep-water bottom trawl fishery is heavily overexploited while the midwater trawl stocks are fully exploited in the SE Arm. The pair trawl fishery could not be assessed using classical equilibrium-based production models, almost certainly because of changes in fishing practices and in species caught. However, since 1987, pair trawl stocks in the heavily exploited waters south of Boadzulu Island appear to have become less stable. Pair trawling was found to be highly profitable in all areas to the south of Salima, but uneconomical to the north. Although there is good evidence that the decline of the tilapiine fisheries has been the result of catches of juveniles by seines, it is currently not possible to assess the likelihood of interactions between the seine and trawl haplochromine fisheries, which together account for almost 60% of the total yield. Even in the remote and poorly studied northern and central regions of the Malaŵian sector of Lake Malaŵi, the artisanal fishing fleet is expanding rapidly, although yields remain roughly constant. Until the mid 1980s, the main priority for Malaŵi's fisheries was investment in the expansion of the fishing fleet, coupled with an increase in its efficiency through technological improvement. This phase seems to have been a wise expansion but, as far as the artisanal and demersal fisheries are concerned, this phase is now at an end. Future investment must be directed towards the improvement of stock-assessment techniques and effort-limitation mechanisms. There is no reason to believe that species introductions could increase fish yields, which may, in any case, be much higher than present statistics indicate.

ACKNOWLEDGEMENTS

Most of this work was carried out while I was employed on the FAO/UNDP/Government of Malaŵi Chambo Fisheries Research Project. I thank my colleagues Steve Alimoso, Moses Banda, Marina Mdaihli, Nelson Mwanyama, Mohammad Seisay, and Niek van Zalinge for help and discussion. Additional material was taken from the ODA-funded *Oreochromis lidole* project carried out by the University of Wales at Bangor, and I thank Andrew Grimm, Tony Pitcher and Rosanna Robinson for their collaboration on this work. All the data on artisanal fisheries from 1976 to 1989 were taken from the unpublished working papers of the ODA-funded Traditional Fisheries Assessment Project, and I thank Denis Tweddle for sharing this information, as well as other aspects of his knowledge of Malaŵian fish and fisheries. I also thank Lance Smith of the Department of National Parks and Wildlife, and Eddie Allison and Ben Ngatunga of the ODA/SADC Pelagic Resources Project for their collaboration and discussion of their work in progress. My efforts to come to terms with the systematics and identification of the Lake Malaŵi cichlids were greatly aided by David Eccles, Ad Konings, Ro Lowe-McConnell, Lameck Phiri, Jay Stauffer, and in particular, the late Ethelwyn Trewavas. I acknowledge the Departments of Fisheries and of Research and Environmental Affairs of the Government of Malaŵi for the provision of facilities and permission to conduct this research.

REFERENCES

Alimoso, S.B., Seisay, M.B.D. and van Zahlinge, N.P. (1991) An efficient method for catch-effort sampling of the artisanal chambo fisheries of the south east arm of Lake Malawi, the Upper Shire River, and Lake Malombe. FAO, Rome, *FI:DP/MLW/86/013 Field Doc.*, 6, 26 pp.

Bazigos, G.P. (1974) The improvement of the Malawian fisheries statistical system. FAO, Rome, *FAO/MLW/16*, 23 pp.
Eccles, D.H. (1974) An outline of the physical limnology of Lake Malawi (Lake Nyasa). *Limnol. Oceanogr.*, **19**, 730–42.
FAO (1976) Promotion of integrated fishery development, Malawi. FAO, Rome, *FI:DP/MLW/77/516 tech. Rep.* 1, 73 pp.
FAO (1982) Fishery expansion project, Malawi. Biological studies on the pelagic ecosystem of Lake Malawi. FAO, Rome, *FI:DP/MLW/75/109, Tech. Rep.* 1, 176 pp.
Fryer, G. (1973) Hopeful empiricism versus scientific method: the Nile Perch controversy revived. *Biol. Conserv.*, **5**, 222–3.
Gulland, J.A. (1961) Fishing and the stocks of fish at Iceland. *Fishery Invest. Lond. Ser. 2*, **23**, 1–52.
Hilborn, R. and Walters, C.J. (1992) *Quantitative Fisheries Stock Assessment: Choice, Dynamics and Uncertainty*, Chapman & Hall, New York, 570 pp.
Holden, S. and Bravington, M.V. (1992) *The CEDA Package User Manual*, MRAG, London, 90 pp.
Jackson, P.B.N., Iles, T.D., Harding, D. and Fryer, G. (1961) *Report on the Survey of Northern Lake Nyasa*, Government Printer, Zomba, Nyasaland, 171 pp.
Konings, A. (1990) *Cichlids and All the Other Fishes of Lake Malawi*, TFH Publications, Neptune City, New Jersey, 495 pp.
Lewis, D.S.C. and Tweddle, D. (1990) The yield of usipa (*Engraulicypris sardella*) from the Nankumba Peninsula, Lake Malawi (1985–1986). *Coll. Rep. Fish. Res. Malawi*, **1**, 57–66.
Lowe, R.H. (1952) Report on the Tilapia and other fish and fisheries of Lake Nyasa, 1945–1947. *Col. Off. Fish. Publs*, **1**, 126 pp.
Lowe, R.H. (1953) Notes on the ecology and evolution of Nyasa fishes of the genus *Tilapia*, with a description of *T. saka* Lowe. *Proc. zool. Soc. Lond.*, **122**, 1035–41.
Mayr, E. (1984) Evolution of fish species flocks: a commentary, in *The Evolution of Fish Species Flocks* (eds A.A. Echelle and I. Kornfield), Univ. Maine at Orono Press, Orono, pp. 3–11.
Mdaihli, M. and Donda, S. (1992) Profitability of fishing in Lake Malombe, the Upper Shire River and the South East Arm of Lake Malawi. FAO, Rome, *FI:DP/MLW/86/013, Field Doc.*, 17, 22 pp.
Meyer, A., Kocher, T.D., Basasibwaki, P. and Wilson, A.C. (1990) Monophyletic origin of the Lake Victoria cichlid fishes suggested by mitochrondrial DNA sequences. *Nature, Lond.*, **347**, 550–53.
Meyer, A., Kocher, T.D. and Wilson, A.C. (1991) African fishes. *Nature, Lond.*, **350**, 467–8.
Mkoko, B.J. (1992) The fishing industry with actual and pipeline programmes, in Ssentongo and Dampha (eds) (1992), q.v., pp. 43–52.
Pauly, D. (1980) On the interrelationships between natural mortality, growth parameters, and mean environmental temperature in 175 fish stocks. *J. Con. int. Explor. Mer*, **39**, 175–92.
Pitcher, T.J. and Hart, P.J.B. (1982) *Fisheries Ecology*, Croom Helm, Beckenham, 414 pp.
Ribbink, A.J., Marsh, B.A., Marsh, A.C., Ribbink, A.C. and Sharp, B.J. (1983) A preliminary survey of the cichlid fishes of the rocky habitats of Lake Malawi. *S. Afr. J. Zool.*, **18**, 149–310.
Seisay, M.B.D., Turner, G.F. and van Zalinge, N.P. (1992a) The relationships between the chambo stocks of Lakes Malawi and Malombe: studies of migration through the Upper Shire River. FAO, Rome, *FI:DP/MLW/86/013 Field Doc.*, 18, 20 pp.
Seisay, M.B.D., van Zalinge, N.P. and Turner, G.F. (1992b) Population of dynamics

and stock estimates of chambo (*Oreochromis* spp.) in the south east arm of Lake Malawi and Lake Malombe – length-based approach. FAO, Rome, *FI:DP/MLW/86/013 Field Doc.*, 19, 34 pp.

Ssentongo, G.W. and Dampha, N.F. (eds) (1992) Report of the technical consultation between Malawi and Mozambique on the development and management of the fisheries of Lake Malawi, Chilwa and Chiuta. FAO, Rome, *RAF/87/099-TD/33/92*, 89 pp.

Trewavas, E. (1983) *Tilapiine Fishes of the Genera* Sarotherodon, Oreochromis *and* Danakilia, British Museum (Natural History), London, 583 pp.

Turner, G.F. (1992) The mechanised fisheries of Lake Malawi. FAO, Rome, *FI:DP/MLW/86/013 Field Doc.*, 23, 24 pp.

Turner, G.F. (1994a) Speciation mechanisms in Lake Malawi cichlids: a critical review. *Adv. Limnol.*, **44**, 139–60.

Turner, G.F. (1994b) A description of a commercially important new pelagic species of the genus *Diplotaxodon* (Pisces: Cichlidae) from Lake Malawi, Africa. *J. Fish Biol.*, **44**, 799–807.

Turner, G.F. and Mdaihli, M. (1992) A bioeconomic analysis of the industrial and semi-industrial fisheries of southern Lake Malawi. FAO, Rome, *FI:DP/MLW/86/013 Field Doc.*, 22, 21 pp.

Turner, G.F. and Mwanyama, N.C. (1992) Distribution and biology of chambo, *Oreochromis (Nyasalapia)* spp. in Lakes Malawi and Malombe. FAO, Rome, *FI:DP/MLW/86/013 Field Doc.*, 21, 28 pp.

Turner, G.F. and Robinson, R.L. (1991) Ecology, morphology and taxonomy of the Lake Malawi *Oreochromis (Nyasalapia)* species flock. *Ann. Mus. roy. Afr. Centr. Sc. Zool.*, **262**, 23–8.

Turner, G.F., Pitcher, T.J. and Grimm, A.S. (1989) Identification of Lake Malawi *Oreochromis (Nyasalapia)* spp. using multivariate morphometric techniques. *J. Fish Biol.*, **35**, 799–812.

Turner,G.F., Grimm, A.S., Mhone, O.K., Robinson, R.L. and Pitcher, T.J. (1991a) The diet of *Oreochromis lidole* (Trewavas) and other chambo species in Lake Malawi and Malombe. *J. Fish Biol.*, **39**, 15–24.

Turner, G.F., Witimani, J., Robinson, R.L., Grimm, A.S. and Pitcher, T.J. (1991b) Reproductive isolation and the nest sites of Lake Malawi chambo, *Oreochromis (Nyasalapia)* spp. *J. Fish Biol.*, **39**, 775–82.

Turner, G.F., Seisay, M.B.D. and van Zalinge, N.P. (1992) An evaluation of the Malawi traditional fisheries catch assessment survey. FAO, Rome, *FI:DP/MLW/86/013 Field Doc.*, 14, 26 pp.

Tweddle, D. and Magasa, J.H. (1989) Assessment of multispecies cichlid fisheries of the Southeast Arm of Lake Malawi, Africa. *J. Cons. int. Explor. Mer.*, **45**, 209–22.

Tweddle, D. and Turner, J.L. (1977) Age, growth and natural mortality rates of some cichlid fishes of Lake Malawi. *J. Fish. Biol.*, **10**, 385–98.

Tweddle, D., Lewis, D.S.C. and Willoughby, N.G. (1979) The nature of the barrier separating the Lake Malawi and Zambezi faunas. *Ichthyol. Bull. J.L.B. Smith Inst.*, **39**, 1–10.

van Zalinge, N.P., Alimoso, S.B., Donda, S.J., Mdaihli, M., Seisay, M.B.D. and Turner, G.F. (1991) Preliminary note on the decline of chambo catches in Lake Malombe. FAO, Rome, *FI:DP/MLW/86/013, Field Doc.*, **9**, 9 pp.

Chapter eighteen

Changes in demersal cichlid communities as a result of trawling in southern Lake Malaŵi

George F. Turner, Denis Tweddle and Ramsey D. Makwinja

18.1 INTRODUCTION

Many of the endemic cichlids of Lake Malaŵi have low fecundities and restricted distributions within the lake (Fryer and Iles, 1972; Ribbink *et al.*, 1983). Consequently, they may be particularly vulnerable to local overfishing. Many of the fisheries on Lake Malaŵi catch a large number of species with different life history characteristics (Chapter 17). Most species have similar economic values. Thus, populations of more vulnerable species may decline long before fishing becomes uneconomical. The two fishing techniques most likely to pose such a threat are small-meshed beach seines and demersal pair trawls. At present, almost nothing is known of the biological basis of the seine fisheries. However, recent studies on trawling grounds in the south of Lake Malaŵi have shown that fishing has had considerable effects on the populations of many cichlid species.

Commercial trawling on Lake Malaŵi was introduced in 1968. Since then the fishery has undergone steady expansion. It now yields more than 6000 tonnes per year, about 13% of the total fish production of the lake (Turner, 1992). In collaboration with the Malaŵi Government Department of Fisheries

The Impact of Species Changes in African Lakes. Edited by Tony J. Pitcher and Paul J.B. Hart. Published in 1995 by Chapman & Hall, London. ISBN 0 412 55050 4.

(MFD), the Food and Agriculture Organisation of the United Nations (FAO, 1976) carried out experimental trawling from 1971 to 1975. They divided the southern portion of the lake into seven fishing areas and estimated total fish biomass and tentative maximum sustainable yields (MSY). To date, these estimates have been used to determine the number of trawling licences issued by the MFD. Subsequent developments in the fishery were analysed by Tweddle and Magasa (1989) and Turner (1992). In all areas north of Boadzulu Island (map, Chapter 17, Fig. 17.1), effort levels have remained below those recommended and yields have been lower than the MSY estimates. South of Boadzulu Island (area A), levels of effort and yield have been near to or above those recommended for the last 15 years. Unexpectedly, there has been no decline in CPUE (Turner, 1992). Tweddle and Magasa (1989) suggested that yields of cichlid fisheries are highly resilient.

Between 1971 and 1974, experimental trawls revealed substantial decreases in populations of larger haplochromine cichlids in this area (FAO, 1976). On the recommendation of FAO, the minimum legal trawl cod-end mesh size was increased from 25 to 38 mm. Pair trawl CPUE declined for 1 year, but thereafter returned to previous levels (Tweddle and Magasa, 1989).

Until 1989, no attempt was made to see if the hoped-for recovery of the large cichlid species had occurred. From 1989 to 1992, experimental trawling was used to estimate total fish biomass and the species composition of the catch was investigated in 1991 and 1992. The results were compared with those of surveys made between 1971 and 1975 to determine whether the populations of these larger species had recovered or continued to decline.

18.2 SURVEY METHODS AND ANALYSIS

Survey design

Between September 1989 and May 1992, eight experimental trawl surveys were carried out. All procedures were designed to replicate the surveys of 1971–1975 (Turner, 1977a,b). An otter trawl of 25 mm cod-end mesh, 17 m headrope and 20 m footrope was hauled from the 88 hp stern trawler RV *Ethelwynn Trewavas*. Each tow was made for 30 min at a speed of approximately 3.7 km h^{-1} (2 knots). Thirty-five stations in the SE Arm of Lake Malaŵi were sampled. Although this gave uneven coverage of different areas and depths (Table 18.1), the survey design was retained for the purpose of comparability. The total catch was weighed on board. To reduce bias, large specimens of tilapiine cichlids and catfish were discarded. A subsample of 6–15 kg was preserved in 10% formalin for the determination of species composition. In the laboratory, the formalin was washed out and the samples

Table 18.1 Experimental trawl survey design

Location and depth	Fishing area code	Surface area (km^2)	Number of trawl stations	Area (km^2) per station
South of Boadzulu Island				
0–50 m	A	221	7	31.6
Boadzulu Island – Monkey Bay				
0–50 m	Bi	231	10	22.1
50–100 m	Bii	233	7	33.3
North of Monkey Bay				
0–50 m	Ci	257	3	85.7
50–100 m	Cii	537	6	89.5
>100 m	Ciii	263	2	131.5

were sorted according to species or species group. For each sample, the number of individuals and total weight was determined for each taxon.

Estimation of standing stock

Standard stock estimation employed the standard swept-area method, and followed the methods used by J.L. Turner (1977a,b) for the previous surveys, except that mean catch rates were calculated for depth ranges of 0–50 m, 51–100 m and 101+ m. This is because although J.L. Turner (FAO, 1976; J.L. Turner, 1977a,b) based his calculation on depth strata of 0–29 m, 30–59 m and 60–100 m, the data were presented (FAO, 1976) according to the depth ranges used in the present work.

In publications reporting the results of the previous surveys (FAO, 1976; J.L. Turner, 1977a,b), the 50 m contour had been drawn well to the north of Boadzulu Island, whereas in fact it extends further south to lie off the east of the island. Because this could bias the calculation of areas, and thus yields, all areas were recalculated. Comparison of these figures with the unpublished calculations used for the FAO (1976) project indicated that the surface area over depths of 0–50 m in area B had been overestimated by 62%. Although other subareas were also miscalculated, the total estimated surface area for the SE Arm was not significantly different. Furthermore, it had previously been ascertained (Tweddle, 1981) that calculation of swept areas had used the metric equivalent of the statute mile instead of the correct nautical mile, and had used the full headrope width instead of the generally accepted 1/2 headrope length.

Therefore, to make comparisons between standing stocks in the 1970s and 1990s, all biomass estimates from the 1970s were recalculated from the raw data.

Estimation of MSY

FAO (1976) presented tentative estimates of the MSY for each trawled area. First, they calculated the MSY for area A from catch and effort returns using Gulland's modification of the Schaefer model. Then, they calculated the ratio of the MSY to the standing stock biomass obtained from the swept-area method. To obtain an MSY for other fishing areas, the MSY:biomass ratio for area A was multiplied by the biomass estimated by trawling surveys in that area. There are grounds for believing that the MSY calculations are unreliable (see Chapter 17 for references and a discussion of the drawbacks of this analysis). Therefore, we feel it would be unwise to revise the recommended MSYs. However, the FAO estimates should not be ignored, as they have continued to form the basis for licensing the trawl fleet.

As a result of our recalculations of the standing stocks in the 1970s, the maximum sustainable yield calculated for area A by FAO (1976) has been shown to have been 32.5% of the 1970s biomass and not 45% as stated. All of the FAO estimates for MSY have been recalculated accordingly using this revised MSY: biomass ratio.

Species identification

Identification of Lake Malaŵi fishes is exceptionally difficult and is complicated by the fact that most species are undescribed. In contrast to the shallow-water species on rocky shores (Ribbink *et al.*, 1983), there are not even temporary cheironyms in the literature. In all, 177 taxa were distinguished: 3 siluroids (6 species), 4 cyprinids, 1 tilapiine cichlid (3 species) and 169 haplochromine cichlids. Of the haplochromine cichlids, 79 taxa could be unambiguously referred to described species (Eccles and Trewavas, 1989) while 53% represented unknown or unidentifiable species. The undescribed species were recognized from: (1) comparison with preserved specimens in the field museum at Monkey Bay Fisheries Station, (2) clear-cut differences in meristics or body shape and, (3) differences in male breeding colour, which are thought to indicate reproductive isolation (Ribbink *et al.*, 1983). It is likely that these methods underestimate the number of undescribed species.

In the 1970s, extensive use was made of unpublished names for undescribed species. Although many of these unpublished names were used to label preserved specimens in the field museum in the Monkey Bay Research Station, it was not possible to recover the provenance of all such names. In addition, several 'catch-all' categories had been employed to cover a number of species with similar meristics, but often different sizes at maturity and habitat preferences. The original categories are retained in the present analysis for the sake of comparability with the 1970s.

In addition, there was evidence of some inconsistency in nomenclature in the 1970s records. For example, sample records from the Chekopa I station showed that the dominant species was '*Haplochromis ericotaenia*', '*Haplochromis sphaerodon*', and '*Haplochromis mollis*'. All these species are superficially similar. In the 1990s samples, the first two species were very rare and the third absent. In every sample from the same station in the 1990s, the dominant species was a fourth, undescribed form, superficially similar to the previous three species. In such cases, the taxa were combined for both 1970s and 1990s data.

Rare species were recorded, but excluded from the quantitative analyses of the present study. The more accurate identifications from the 1990s have been retained (Turner, unpublished data).

Guild structure

To gain an insight into the effects of trawling on fish community structure, the 40 principal taxa were grouped into ten ecological guilds. These were based on diet, adult size and degree of association with the substrate (Turner, unpublished data). These guild allocations are tentative and are often based only on functional morphology, as dietary studies have not been performed on many species. As an overall measure of change in community structure, an index of the change in guild proportions ∂g was calculated as:

$$\partial g = \sum_{x=1}^{n} \sqrt{(G_{x,i} - G_{x,j})^2} \qquad (18.1)$$

where G is the percentage of the total fish biomass belonging to guild x; i and j are the two sampling periods; n is the number of guilds present.

Statistical analysis

Total catch rates for all species combined were analysed using analysis of covariance tests, with catch rate as the dependent variable, trawling depth as the covariate, and sampling area and date as factors (Sokal and Rohlf, 1981).

When abundance of each taxon was expressed as percentage sample weights, the standard errors were invariably large in comparison with the means. Furthermore, there were many samples in which species were not recorded. The large number of zeros rules out the use of even non-parametric statistical tests. Changes in the proportion of samples in which a given species occurred were analysed using contingency tests (G-test). Significance was taken at 1% level, to reduce the risk of type I error, because a large number of independent comparisons were performed (Sokal and Rohlf, 1981).

18.3 COMPARISON OF SURVEY RESULTS

Biomass estimates 1989 – 1992

Mean catch rates for each area and fishing depth are shown on Table 18.2. Catch rates were significantly influenced by fishing area and by date of survey (Table 18.3). There was also a significant interaction between these factors. Water depth had little influence on catch rates. However, if the single 1989 survey is excluded, there is no significant influence of area or survey date, although the interaction term remains. Comparison of the experimental trawl surveys with commercial fishing returns (based on a much larger sample) made over the same period indicates that the surveys systematically overestimated the decline in stock density in all areas between 1989 and 1990. There was more consistency between the figures for 1990 – 1991 (Table 18.4). Thus, the 1989 survey was exclued from further analysis.

There was no significant change in catch rates of the experimental trawler in areas A and B throughout the surveys of 1990 to 1992 (analysis of variance: Table 18.5). The decline in areas A between 1991 and 1992, although large (Table 18.2; it becomes more striking if only the May survey is considered) is not statistically significant. An exception was area C. In this

Table 18.2 Mean catch rates in experimental trawl surveys, 1989–1992, in kg per 30 min haul. Number of sample hauls in parentheses

	1989	1990	1991	1992
Area A	237 (7)	110 (14)	135 (21)	89 (14)
Areas B and C				
0–50 m	183 (16)	100 (22)	74 (32)	97 (22)
>50 m	154 (16)	115 (31)	95 (49)	76 (32)

Table 18.3 Analysis of covariance of experimental trawl catch rate by area, quarterly survey and depth

	1989–1992			1990–1992		
	F	d.f.	*P*	*F*	d.f.	*P*
Area	3.96	2	0.020	2.61	2	0.076
Quarterly survey	8.73	7	<0.001	1.58	6	0.153
Area × survey	2.33	14	0.005	2.29	12	0.009
Depth	0.07	1	0.792	0.36	1	0.551
Error	–	246		–	215	

Table 18.4 Comparison of changes in experimental and commercial trawl CPUE, 1989–1991

		1989–1990	1990–1991
Area A	Experimental	−54%	+23%
	Commercial	−22%	+1%
Areas B and C	Experimental	−45%	−26%
0–50 m	Commercial	−12%	−57%
>50 m	Experimental	−25%	−17%
	Commercial	+24%	−15%

Table 18.5 Analysis of variance of experimental trawl catch rate from 1990 to 1992 for each area

Area	*F*	d.f.	*P*
South of Boadzulu Island (A)	1.59	6,42	0.174
Boadzulu Island to Monkey Bay (B)	1.55	6,105	0.170
North of Monkey Bay (C)	2.36	6,69	0.039

case the small sample size, large heterogeneous area covered and marginal significance level, suggested that a type II error was possible. A Scheffe range test indicated that the variation in catch rate in this area was the result of a transient decline in the middle of 1991. There was no directional trend across the whole time period. Thus, standing stock estimates were calculated from the mean of all surveys from 1990 to 1992.

Comparison of biomasses in 1970s and 1989–1992

Trawlable fish biomass in all areas has declined substantially since the 1970s (Table 18.6). The largest declines have been in areas A and B, which are the more heavily fished areas. Over the whole of the SE Arm at depths of less than 100 m, the total trawlable fish density declined from 12.2 to 6.1 $t\,km^{-2}$. In the productive area A, fish density has fallen from 18.8 to 7.3 $t\,km^{-2}$.

Table 18.6 indicates that trawl yields in some areas are substantially below the maximum sustainable yield estimated by FAO (1976). However, it should be remembered that this could be due to overfishing, rather than underexploitation. For example, the stock at depths of 50–100 m collapsed in the late 1980s (see analysis of the bottom trawl fishery in Chapter 17). Yields have fallen to around half of the mid-1970s figure of more than 1500 t $year^{-1}$.

Table 18.6 Standing stock estimates and yields (tonnes). 1990s biomasses given as means and 95% confidence intervals

Depth (m)	Fishing area	FAO 1976*		1990–1992 survey		Yield (mean 90–91)
		Biomass (t)	MSY (t)	Biomass (t)	Decline (%)	
0–50	A	4 160	1 350	1 620 ± 350	61	1 112
	B	3 200	1 020	1 240 ± 180	61	807
	C	2 780	890	1 700 ± 600	39	
51–100	B	3 340	1 070	1 560 ± 260	53	778†
	C	4 450	1 420	2 980 ± 340	33	
101–150	C‡	–	–	1 450 ± 360	–	–
0–100	Total	17 930	5 750	9 100	49	6 024§

* Biomass and MSY calculated following FAO (1976), but corrected for revised area estimates (see text).
† Stock overexploited: catch in 1980s exceeded 1 500 t.
‡ Not calculated in the 1970s.
§ Including artisanal fishery haplochromine catch.

Little is known about the possible interactions between the artisanal and trawl fisheries. If artisanal fisheries are included, the total yield of haplochromine cichlids in the SE Arm of Lake Malaŵi is at present above the MSY predicted from the trawl surveys of the 1970s.

Changes in species composition

Particularly to the south of Boadzulu Island, significant changes in abundance of a number of taxa occurred between the mid 1970s and the 1990s (Table 18.7). *Lethrinops macracanthus* and *L. mylodon*, two of the larger species already in decline by 1975, appear to have become locally extinct. *Lethrinops longimanus, Ctenopharynx nitidus, Taeniolethrinops furcicauda* and *T. praeorbitalis* have also declined substantially.

To the north of Boadzulu Island, these large shallow-water species have been largely unaffected (Table 18.8), but there has been a substantial decline in the population of *Copadichromis* cf. *virginalis* (= *Haplochromis mloto* of FAO, 1976; J.L. Turner, 1977a,b; Tweddle and J.L. Turner, 1977), a species taken in large numbers by the artisanal 'chirimila' fishery. However, the most intensive trawling in this area has been in the 40–60 m depth range. Here, the percentage biomass of the formerly dominant *Lethrinops microdon* (= part of *L. stridei* in earlier papers) has declined sharply, although there has been no decline in the proportion of samples in which it occurs.

Table 18.7 Changes in fish communities south of Boadzulu Island. Taxa are listed by guild. Statistical comparisons, using *G*-tests, were made between the number of samples (*N*) in which the taxon occurred in consecutive sampling periods. Because of the large number of comparisons made, statistical significance was taken at the 1% level. Percentage weights given are mean sample percentages for the year, area and depth range indicated

Survey years:	1971–1973			1975			1991–1992	
Number of samples:	32			48			30	
Mean depth fished (m):	28.2			26.4			30.4	
	% Wt	*N*		% Wt	*N*		% Wt	*N*
Large piscivores								
Buccochromis spp.	2.65	24		1.23	41		0.78	19
Nimbochromis spp.	0.36	9		0.25	27		0.40	10
Otopharynx speciosus	0.49	5		0.24	17		0.30	8
Medium piscivores								
Corematodus taeniatus	0.81	22		0.14	31	**	0.12	10
Sciaenochromis spp.	0.39	9		0.09	15	***	0.97	26
Taeniochromis holotaenia	0.00	0		0.03	5		0.02	2
Molluscivores								
Chilotilapia rhoadesii	1.30	18		0.66	32	**	0.69	11
Lethrinops mylodon	0.30	5		0.01	1		0.00	0
Mylochromis spp.	2.99	21		1.31	41		1.19	21
Trematocranus spp.	0.78	10	***	2.06	36	**	0.73	13
Large arthropod feeders								
Ctenopharynx intermedius	1.71	13		0.85	28		0.73	12
Lethrinops longimanus	1.52	8		0.83	15	**	0.01	1
Lethrinops macracanthus	3.12	10	**	0.11	4		0.00	0
Taeniolethrinops furcicauda	10.17	9	**	1.21	28	***	0.08	4
Taeniolethrinops praeorbitalis spp.	3.41	17		1.72	35	**	0.63	12
Medium arthropod feeders								
Ctenopharynx nitidus	0.71	12		0.35	29	***	0.00	1
Lethrinops lethrinus	1.00	8		0.48	22		2.28	18
Lethrinops longipinnis	5.58	19		1.82	21		2.15	11
Lethrinops parvidens spp.	8.53	20		5.30	39		4.03	19
Placidochromis subocularis	0.25	11	***	0.74	41		0.67	21
Small arthropod feeders								
Aulonocara spp.	7.21	24		3.31	42		8.20	29
Lethrinops altus	1.78	9		0.38	16		0.89	5
Otopharynx argyrosoma spp.	12.33	25		11.26	37		14.77	25
Synodontis njassae	1.28	20	**	1.24	42	***	1.04	15
Large sediment feeders								
Lethrinops stridei/micrentodon	8.05	9		2.90	14		3.88	11
Otopharynx auromarginatus spp.	0.57	3	***	4.70	27	***	0.47	5
Small sediment feeders								
Aulonocara brevirostris spp.	0.89	8		1.68	10	**	2.65	15
Lethrinops 'oliveri' spp.	1.01	7		2.51	23		1.68	7
Lethrinops small spp.	4.43	21		7.68	43		6.09	25
Placidochromis cf. *longimanus*	1.86	3	**	6.76	19		9.23	19
Pseudotropheus elegans	1.54	14	***	3.59	44		8.84	28
Pseudotropheus livingstonii	5.00	23	**	17.16	45		15.31	28
Zooplankton feeders (benthic)								
Copadichromis cf. *eucinostomus*	0.47	1	**	3.65	12		0.07	4
Copadichromis large spp.	0.29	8	***	0.49	31	***	0.27	7
Copadichromis cf. *virginalis*	3.15	26		3.13	33		1.55	15
Pelagic zooplankton/fish eaters								
Diplotaxodon spp.	0.76	14		1.39	28		1.17	9
Rhamphochromis spp.	0.89	21	**	0.33	44		3.05	29

**, $P < 0.01$; ***, $P < 0.001$ (see caption).

Table 18.8 Species composition changes north of Boadzulu Island, for selected species owing to small sample sizes, and absence of data for 1975, deep-water (>60 m) samples from 1973 are combined with those of 1971–1972. Percentage weights given are mean sample percentages for the year, area and depth range indicated. Statistical comparisons made as in Table 18.7

	1971–1972		1973–1975			1991–1992	
	% Wt	*N*	% Wt	*N*		% Wt	*N*
DEPTH LESS THAN 40 M							
Molluscivores							
Chilotilapia rhoadesii	2.80	29	0.64	10		0.97	13
Mylochromis spp.	11.28	37	6.63	20		8.58	36
Medium arthropod feeder							
Ctenopharynx nitidus	3.39	36	1.94	18	***	0.84	11
Benthic zooplankton feeder							
Copadichromis virginalis spp.	11.08	22	11.43	10		1.56	14
DEPTH 40–60 M							
Large arthropod feeders							
Lethrinops macracanthus	2.64	11	1.26	2		0.00	0
Small arthropod feeders							
Otopharynx argyrosoma spp.	6.49	15	6.30	9		15.61	33
Synodontis njassae	4.70	30	3.88	13		12.05	43
Large sediment feeders							
Lethrinops microdon	16.23	13	13.27	4		1.85	10
Small sediment feeders							
Aulonocara brevirostris spp.	0.52	4	1.41	5		1.39	13
Lethrinops 'oliveri' spp.	2.07	12	4.20	8		7.29	21
Lethrinops small spp.	1.80	21	1.33	6		7.45	25
Pseudotropheus elegans	0.03	1	0.42	5		0.23	20
Pseudotropheus livingstonii	0.05	3	0.09	4		0.06	10

Alteration of community structure

Changes in the abundance of individual taxa have also been paralleled by changes in community structure. Overall, the greatest changes in guild proportions have occurred in areas B and C in the 40 – 60 m depth range, and in area A, the shallow water to the south of Boadzulu Island (Table 18.9). These are the areas where most trawling has taken place. Most of the changes in area A had already occurred by the second sampling period in 1975. North of Boadzulu Island, in the 40 – 60 m depth range, the principal changes occurred after 1975, although changes during the mid 1970s may have been obscured by the necessity of pooling the samples from 1973 to

Table 18.8 (Continued)

	1971–1973			1991–1992	
	% Wt	*N*		% Wt	*N*
DEPTH GREATER THAN 60 M					
Small piscivores					
Sciaenochromis spp.	2.65	3	***	3.47	32
Molluscivores					
Lethrinops mylodon	0.40	4		0.01	1
Small arthropod feeders					
Lethrinops altus	2.88	9		5.28	31
Otopharynx argyrosoma spp.	0.02	1		7.28	15
Synodontis njassae	5.79	14		10.70	43
Small sediment feeders					
Aulonocara brevirostris spp.	0.04	2		0.84	13
Lethrinops 'oliveri' spp.	9.01	9		13.86	30
Lethrinops small spp.	2.23	9		3.09	19

**, $P < 0.01$; ***, $P < 0.001$ (see caption).

Table 18.9 Changes in community structures, given as the sum of differences between sampling periods in the weights of each ecological guild. In all cases, sampling period III is 1991–1992. For earlier data, surveys were less regular and so had to be combined differently: south of Boadzulu Island: I, 1971–1973; II, 1975; north of Boadzulu Island 0–60 m: I, 1971–1972; II, 1973–1975; >60 m: I, 1971–1973

Area	Sampling periods compared		
	I, II	II, III	I, III
South of Boadzulu Island	53.4	30.4	61.4
North of Boadzulu Island			
0–39 m	30.9	25.6	41.3
40–60 m	21.3	62.4	66.5
>60 m	–	–	48.2

1975. In areas B and C, communities have been more stable in the less heavily fished waters of less than 40 m or greater than 60 m depth.

South of Boadzulu Island, large and medium-sized benthic arthropod feeders, molluscivores, benthic zooplanktivores, large sediment feeders and large piscivores have all declined (Fig. 18.1). Although the proportion of small sediment feeders and pelagic species has increased, the total standing stocks of these guilds have remained largely unchanged (Fig. 18.1). There

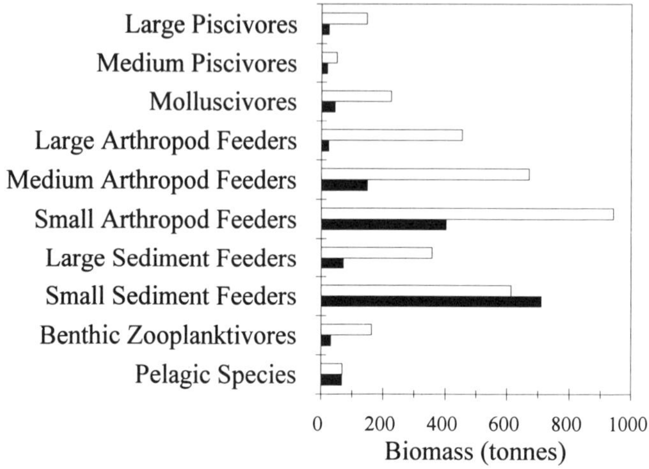

Fig. 18.1 South of Boadzulu Island, biomasses of eight of the ten guilds of haplochromine cichlids declined between the surveys of 1971–1973 (open bars) and 1991–1992 (filled bars). Only small sediment feeders showed a slight increase. There is little evidence that fishing has enabled the smaller species to profit by the decline of their larger predators or competitors.

is no indication, therefore, that smaller, faster-maturing species have become more abundant as the larger species have declined.

18.4 DISCUSSION

State of exploitation

The fish biomass in areas B and A has declined by 50–60% since the mid 1970s. At that time these stocks were already heavily exploited. The decline in area C of 30–40% suggests that these stocks are less depleted. However, this picture is a simplification.

The majority of benthic haplochromine cichlids in Lake Malaŵi have relatively restricted depth preferences (Eccles and Trewavas, 1989). Vertical migration appears to be limited to movements associated with reproduction, particularly by benthio zooplanktivores ('utaka') and by species found at depths of 30 m or less (Turner, unpublished observations). Horizontal migration is most likely to be constrained by substrate preferences. Many littoral species have strong preferences for rocky, sandy or reedy shores. However, there is at present no indication that deeper-water species are similarly constrained by substrate type. The steeply shelving rocky shores at the extreme north of the SE Arm may be a barrier to the movements of all species, except those found in the unexploited extreme deep waters, or the pelagic zone. Thus, the SE Arm can tentatively be considered a natural unit for stock assessment.

There is no natural barrier to movements of fish between areas B and C, nor between the shallow waters of areas A, B and C. The western shore of areas B and C is probably largely isolated from the eastern shore. Shallow-water species with poor depth equilibration abilities (Ribbink *et al.*, 1983) are unlikely to cross the deep water separating the two shores. High fishing mortality in area A is likely to inhibit movement of large fish through the southern part of the lake. It appears more appropriate, therefore, to subdivide the haplochromine stocks according to depth, rather than fishing area. In practice, this means that pair trawlers and single-boat bottom trawlers should be assessed separately, as these mainly fish at 0 – 50 m and 50 – 70 m respectively.

Given the absence of barriers to horizontal movement, it is quite possible that the present high levels of exploitation in area A have been sustained only by immigration from the less heavily exploited north. Furthermore, until the relationships between artisanal and trawl fisheries have been clarified, it remains possible that these fisheries are competing for the same stocks.

The current state of exploitation of the SE Arm stocks may be viewed in two ways.

1. Excluding the artisanal fishery, and assuming little horizontal migration, there is considerable room for expansion in the northern part of the arm.
2. Including the artisanal fishery, and assuming considerable horizontal migration, the present yield is almost half of the 1970s biomass, and would have been higher had not the deep-water bottom trawl fishery been exploited in the late 1980s. Using Gulland's (1971) formula, and assuming a natural mortality coefficient of 1.0 (FAO, 1976), this would mean that the stock is fully exploited. However, this assumes that the mid 1970s biomasses were equal to the virgin biomass, which is unlikely, as the stocks had been exploited heavily for several years.

We conclude, therefore, that the stocks are probably overexploited and if the intention is to optimize yields, no increase in the exploitation of the haplochromine stock in the SE Arm of Lake Malaŵi can be recommended until more information is available.

Changes in community structure

It is clear that heavy fishing pressure has markedly different effects on different ecological guilds of haplochromine cichlids. The decline in large piscivores is unsurprising and probably beneficial to the overall yield of the fishery, although regrettable for handline fishermen for whom these 'saguga' (*Buccochromis* spp.) represent a high-value component of the catch.

Although the small sediment feeders have increased as a proportion of the community, their total biomass has remained largely unchanged. The probable explanation for this is their inability to exploit all available food resources.

Declining populations of molluscivores and large arthropod feeders may have led to the diversion of production from fish to benthic invertebrates. Deep-burrowing invertebrates, such as bivalves and some Ephemeroptera, were probably exploited by large long-snouted haplochromines such as *Taeniolethrinops praeorbitalis*, while specialized mollusc-crushers such as *Mylochromis anaphyrmus* and *Lethrinops mylodon* were capable of taking large tough-shelled snails. As these items are likely to be inaccessible to small unspecialized fish species, populations of larger benthic invertebrates may have risen in heavily fished areas. This could have undesirable consequences: (1) increasing snail populations may lead to an increase in the incidence of human and animal schistosomiasis in the lake; (2) energy may be diverted away from fish production.

FAO (1982) suggested that the existence of large populations of chaoborid larvae in the pelagic zone of Lake Malaŵi would lead to a loss of energy and nutrients from the aquatic food web, as the adult insects are often consumed by terrestrial or aerial predators. They recommended that consideration be given to introducing the Lake Tanganyika clupeid *Limnothrissa miodon* to exploit this resource. It is now believed that this is unlikely to be beneficial because there are numerous indigenous pelagic cichlids which eat chaoborid larvae (Chapter 17). Although it is unlikely that a large food resource such as chaoborids should remain unexploited by such a diverse fish community, the argument does seem valid for a community severely disrupted by human intervention.

Although no surveys of benthic invertebrates have been made on Lake Malaŵi, the existence of relatively unfished areas to the north could allow a comparative study to be made. It is notable that the small cichlid species that have increased most rapidly in area A are *Pseudotropheus livingstonii* and *P. elegans*, both of which use gastropod shells as a refuge for at least part of their life history. Although there are anecdotal reports that large beds of empty gastropod shells have long been found in southern Lake Malaŵi, their abundance and locality have not been recorded.

Conservation status of haplochromine populations

It is apparent that the increase in mesh size recommended by FAO (1976) has been ineffective. Perhaps yet larger mesh sizes, such as 50 mm, might prevent local extinctions, but the consequent loss in catches of small species would probably be so great as to render fishing uneconomical (FAO, 1976), and so this cannot be recommended. It is probable that the 38 mm minimum mesh size restriction was unsuccessful partly because it was not adequately enforced. Furthermore, pair trawl units (the most numerous trawlers) typically made pulls of 8–12 h duration. This would lead to clogging of the meshes of almost any practical size.

It is our impression that all endemic species found in areas A and B are also present in area C. Thus we recommend that the fisheries of area A and B be managed to maximize long-term sustainble yields, irrespective of the effects on the taxonomic composition of the fish community.

We suggest that species conservation is probably best achieved through the closure to trawling of the north-eastern part of area C. At present, economic considerations mean that little fishing takes place in this area. This situation is unlikely to change provided no onshore support facilities for trawlers are constructed on the eastern shore of area C. Thus, the limitation of effort would be self-enforcing. The present low level of fishing in the southern part of area C could be tolerated. It is strongly recommended that no expansion of the trawl fishery in area C be permitted until a feasibility study into setting up a sanctuary from trawling has been carried out.

18.5 SUMMARY

As a result of the recommendations of FAO (1976), the minimum legal mesh size for demersal trawlers on Lake Malaŵi was increased from 25 to 38 mm. This was intended to reverse the decline in populations of larger haplochromine cichlid species. Experimental trawl surveys from 1989 to 1991 have shown that this measure has not been successful. Species, such as *Lethrinops mylodon* and *Lethrinops macracanthus*, which had declined substantially in the 1970s, have become locally extinct. Other large species have also declined, such as *Taeniolethrinops furcicauda*, *T. praeorbitalis*, and *Ctenopharynx* spp. The continued presence of these large species in waters to the south of Boadzulu Island may depend on immigration from lightly fished areas to the north. In the northern part of the SE Arm of the lake, there has been a substantial decline in the population of *Lethrinops microdon*, formerly the mainstay of the demersal fisheries at depths of 40–70 m. It is probable that some species or races are confined to the SE Arm of Lake Malaŵi, although there is no indication that any species are restricted to the most heavily fished southern parts of the arm. Expansion of the trawl fleet in the northern portion of the SE Arm to achieve levels of exploitation currently seen to the south of Boadzulu Island could lead to species extinctions. It is recommended that no expansion of trawl fishery or associated support facilities be permitted in the north-eastern part of the SE Arm of Lake Malaŵi, pending a study into the feasibility of establishing a refuge from trawling in that area. To the south of that area, it is recommended that the trawl fishery be managed to optimize yields rather than species diversity.

ACKNOWLEDGEMENTS

The trawl survey was carried out by the Malaŵi Government Department of Fisheries, assisted by the Overseas Development Administration (UK). GFT was employed on the FAO/UNDP/Malaŵi Government Chambo Fisheries Research Project. DT was an ODA-funded OSAS officer in the Malaŵi Government Fisheries Department. RDM is a Fisheries Research Officer in the Malaŵi Department of Fisheries. Quantification of species abundance was carried out with the assistance of D. Mandere, L.M. Phiri, O. Mponda, and A. Phiri. L. Mzunzu assisted with data entry. We are grateful to D.H. Eccles for guiding our initial attempts to identify the species.

REFERENCES

Eccles, D.H. and Lewis, D.S.C. (1977) A taxonomic study on the genus *Lethrinops* Regan (Pisces: Cichlidae) from Lake Malawi. Part 1. *Ichthyol. Bull. J.L.B. Smith Inst. Ichthyol.*, **36**, 1–12.

Eccles, D.H. and Trewavas, E. (1989) *Malawian Cichlid Fishes: The Classification of Some Haplochromine Genera*, Lake Fish Movies, Herten, Germany, 334 pp.

FAO (1976) Promotion of integrated fishery development, Malawi. *FAO tech. Rep.* FI:DP/MLW/71/516, 76 pp.

FAO (1982) Fishery expansion project, Malawi. Biological studies on the pelagic ecosystem of Lake Malawi. *FAO tech. Rep.* FI:DP/MLW/75/109, 176 pp.

Fryer, G. and Iles, T.D. (1972) *The Cichlid Fishes of the Great Lakes of Africa*, Oliver & Boyd, Edinburgh, 641 pp.

Gulland, J.A. (1971) *The Fish Resources of the Ocean*, Fishing News (Books) Ltd, West Byfleet, Surrey, UK.

Ribbink, A.J., Marsh, B.A., Marsh, A.C., Ribbink, A.C. and Sharp, B.J. (1983) A preliminary study of the cichlid fishes of the rocky habitats in Lake Malawi. *S. Afr. J. Zool.*, **18**, 149–310.

Sokal, R.R. and Rohlf, F.J. (1981) *Biometry*, 2nd edn, W.H. Freeman & Co., New York, 859 pp.

Turner, G.F. (1992) The mechanised fisheries of Lake Malawi. FAO, Rome, *FI:DP/MLW/86/013 Field Doc.*, 23, 24 pp.

Turner, J.L. (1977a) Some effects of demersal trawling in Lake Malawi (Lake Nyasa) from 1968 to 1974. *J. Fish Biol.*, **10**, 261–72.

Turner, J.L. (1977b) Changes in size structure of cichlid populations of Lake Malawi resulting from bottom trawling. *J. Fish. Res. Bd Can.*, **34**, 232–8.

Tweddle, D. (1981) A preliminary assessment of the bottom-trawling potential of northern Lake Malawi. *Luso J. Sci. Technol. (Malawi)*, **2**, 3–13.

Tweddle, D. and Magasa, J.H. (1989) Assessment of multispecies cichlid fisheries of the Southeast Arm of Lake Malawi, Africa. *J. Cons. int. Explor. Mer.*, **45**, 209–22.

Tweddle, D. and Turner, J.L. (1977) Age, growth and natural mortality rates of some cichlid fishes of Lake Malawi. *J. Fish Biol.*, **10**, 385–98.

Chapter nineteen

Changes in species composition and abundance as a consequence of fishing in Lake Malombe, Malaŵi

Denis Tweddle, George F. Turner and Mohammed B.D. Seisay

19.1 INTRODUCTION

The Shire River flows from the southern tip of Lake Malaŵi for about 12 km before widening to form Lake Malombe, a lake 30 km long by 15 km wide with a maximum depth of only 5 m (Fig. 19.1). At times of low water level in Lake Malaŵi, such as the first 35 years of this century, Lake Malombe does not exist and its bed is rich farmland. The lake is fed by the most eutrophic water in Lake Malaŵi, the South East Arm, and is further enriched by streams flowing into the lake from its highly populated catchment area and by recycling of nutrients from sediments as a result of the shallowness of the lake. Malombe is consequently much more productive than Lake Malaŵi, though detailed limnological investigations have not yet been made.

Fishing greater than at a subsistence level began in the 1960s after the destruction of the large crocodile population (A.J.P. Mzumara, pers. comm.). In the 1940s the lake was heavily overgrown with weeds (R.H. Lowe-McConnell, pers. comm.). By the 1970s, catches reached over 5000 t (tonnes) annually, with the bulk of the catch consisting of high-value chambo (*Oreochromis* spp.) (Chapter 17).

The Impact of Species Changes in African Lakes. Edited by Tony J. Pitcher and Paul J.B. Hart. Published in 1995 by Chapman & Hall, London. ISBN 0 412 55050 4.

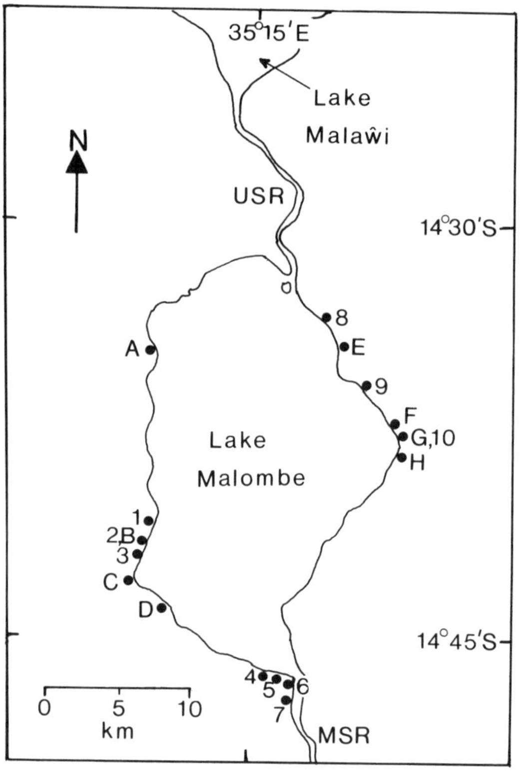

Fig. 19.1 Lake Malombe, showing fishing sites selected for sampling by the Fisheries Department's catch assessment survey, indicated by letters, and by the seine net survey conducted in the Traditional Fisheries Assessment Project, indicated by numbers. Beaches sampled are as follows. A, Chiwaula; B and 2, Mtanga; C, Changali; D, Chapola; E, Masame; F, Gola; G and 10, Kadewere; H, Magombo; 1, Kausi; 3, Katapwito; 4, Daitoni; 5, Paisoni; 6, Banaba; 7, Mvera; 8, Matache; 9, Sili. USR and MSR are Upper and Middle Shire River.

Seining was hampered by dense weed-beds and shoreline reeds and so gill nets were the main gear used and chambo were frightened into these nets by the fishers beating the water with clubs. Experimental trawl tows in the early 1970s produced high yields dominated by the small cichlid species *Placidochromis longimanus*, but further development of trawling was inhibited by weeds blocking the nets (J. Tarbit, pers. comm.). In the 1970s and 1980s, however, seining grew to dominate the fishery as weeds were cleared. In this chapter, the impact of rising levels of fish, particularly seining, on the fish populations is documented.

19.2 METHODS

Statistical data on the fisheries are collected and analysed using a system based on raising factors derived from the proportion of craft fishing from sampled beaches in relation to the number of craft in the entire sampling area. Lake Malombe is divided into two sampling strata. In each of these, catches from four randomly selected beaches are sampled for 4 days each month. Catches and fishing effort are estimated by combining data obtained in these monthly catch assessment surveys (CAS) and in annual 'frame surveys' of the total number of fishers, fishing craft and fishing gears (and see Chapter 17).

To assess the validity of data collected by the routine CAS, other surveys have been conducted. Two aerial surveys were conducted in 1987 to count fishing craft, while weekly surveys of the seine fisheries were conducted for a year from February 1988. In that programme, a frame survey covering all the fishing sites was carried out on Lake Malombe from October, 1987. At each site, details were taken of all seine nets and associated fishers. From February 1988, ten fishing sites were selected from which samples from the different kinds of seine nets operating in the lake could be recorded: samples comprised catch and fishing effort as well as data on species composition. The ten fishing sites were visited weekly and seine net fishers found there were interviewed on catches made during the previous week. Estimates of catches were raised by the total number of nets on the lake to obtain estimates for the lake as a whole. Samples of observed catches, taken one to four times a month, were sorted into species categories before being measured and weighed. A revised survey system was later implemented to obtain fuller information on the fisheries in the area from June 1990. All gears in use were covered, with raising factors based on numbers of gears recorded by the frame survey. The new frame system is designated as 'MTF', and has been run in parallel to the old CAS system in an attempt to cross-validate the results of both surveys. The differences between the two systems reflect the targeting of fishing gear (MTF), or fishing vessels (CAS), and these have effects on the accuracy, validity and error structure of the consequent fishery data.

MTF is gear-targeted, and beach recorders specifically aim to get samples every month from all gears. CAS is not gear-targeted and beach recorders aim to get samples from a proportion of craft on the beach according to a pre-defined rule for the sequence of landing of the boats. As a result of this vessel-order protocol, CAS can miss, or overestimate, the catch from rare gears, and this is important if, like 1 km long chambo seines, they are used with low effort yet have a high catch/effort ratio. MTF uses gear-based raising factors such that CPUE from the sample is raised by the proportion of days fishing by the gear, which is estimated independently, and by the number of gears in the sampling stratum. In contrast, CAS uses beach- and craft-based raising factors such that the sampled catches are taken as representative of all craft on the beach. This can bias results: for example, if a single chambo seine happens to be among four fishing units sampled out of 12 on the beach, then it is assumed that three (0.25 × 12) craft were using chambo seines, even if the beach recorder can clearly see that this is untrue. This catch is then raised by the ratio of craft recorded on the beach to the total number in the sampling stratum, and the monthly catch estimate is an unweighted average of all beaches sampled. This procedure gives equal weight to the catch from small and large beaches, the small ones contributing a lot of noise to the estimates. The large weighting given to small beaches leads to the magnification of recording and calculation error in CAS. MTF, on the other hand, gives equal weighting to all samples.

The result is that CAS data are unbiased, in the sense that positive or negative estimation errors are equally likely, but imprecise, in that the magnitude of errors is large. MTF fishery data are less noisy and are suitable for catch and effort analysis by gear and species. MTF's only bias is that sampling tends to be concentrated on large beaches and may miss catches at small landing sites. Both methods produce roughly the same total annual catch, but MTF provides more precise catch and effort data over shorter time periods.

19.3 RESULTS

The number of fishing craft recorded on the lake in annual frame surveys has fluctuated from year to year, probably as a result of fishers moving freely back and forth through the Shire River to Lake Malaŵi, but nevertheless showed a steady upward trend in numbers until 1986, since when it has stabilized (Fig. 19.2). In 1991, 620 craft, of which only 21 were dugout canoes, were operating on the lake. No outboard engines were in use, a remarkable change from the 221 engines present in 1980.

The total estimated annual catch in the period covered here (1976–1991) ranged from 2917 to 12 936 t (mean = 8569 t) (Fig. 19.3). In general, catches doubled from the 1970s to the 1980s but are now showing signs of decline. Chambo (*Oreochromis* spp.) have declined dramatically in importance, while catches of kambuzi (small haplochromines) increased steadily from 1981. Catches of minor species also improved in the 1980s.

A new method of fishing known as the nkacha net was developed in the 1970s. This is a small-meshed seine used in open water. After setting in a circle, the net is pulled together at the footrope by divers and the weights are tied together to 'purse' the net. This gear was not differentiated from small-meshed shore seines (kambuzi seines) until very recently, and thus has been combined with the kambuzi seines in the analysis presented here.

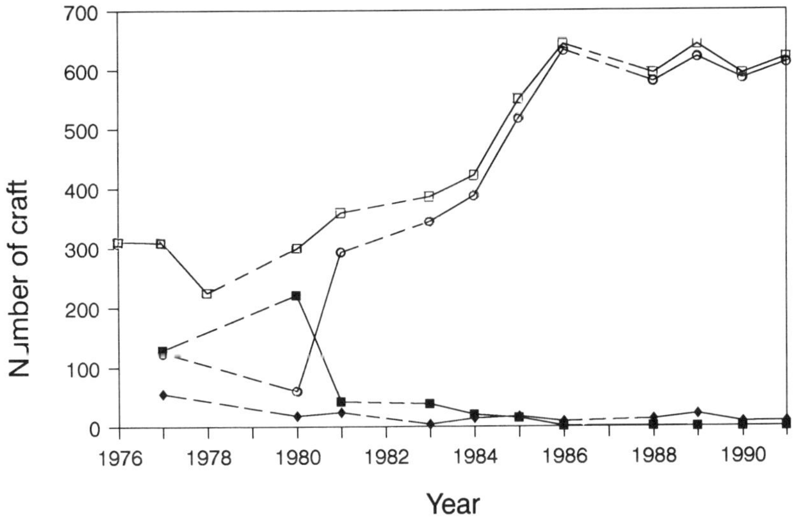

Fig. 19.2 Changes in the number of fishing craft operating on Lake Malombe, recorded in annual frame surveys: □, total fishing craft; ○, boats without engines; ■, boats with engines; ◆, dugout canoes. No frame surveys were conducted in 1982 and 1987. Some data from the 1970s are missing.

Gill nets and seine nets caught 99% of the total estimated catch, hence other gears are of negligible importance for stock assessment.

The data in Fig. 19.3 are divided into four species categories: chambo (three *Oreochromis* spp.), other tilapia (*O. shiranus* and *Tilapia rendalli*), kambuzi (numerous small cichlid species), and other species. The last category includes mbaba, which are larger cichlids than kambuzi, the catfishes mlamba (*Clarias gariepinus*) and kampango (*Bagrus meridionalis*), and the cyprinids usipa (*Engraulicypris sardella*) and nchila (*Labeo mesops*). Nchila used to be a very important commercial species but catches crashed in the 1960s and never recovered (Chapter 17). Nevertheless, in both the 1988 and 1991 rainy seasons, nchila catches in excess of 200 t were recorded in Lake Malombe.

With the expansion of small-meshed seine netting in the 1980s, catches of kambuzi increased greatly, while chambo catches declined from 5000 t to 500 t annually. The decline in chambo has resulted in a decline in profitability of the gill net fishery with a resulting drop in gill net ownership (Fig. 19.4). In contrast, the number of small-meshed seines in the fishery has increased sevenfold (Fig. 19.4). The result has been a big overall increase in kambuzi catches (Fig. 19.3) but a decline in the catch per unit effort (CPUE) and hence profitability of the seine nets (Fig. 19.5). Overall the fishery of Lake Malombe has changed from one based on large-meshed gears for high-value chambo

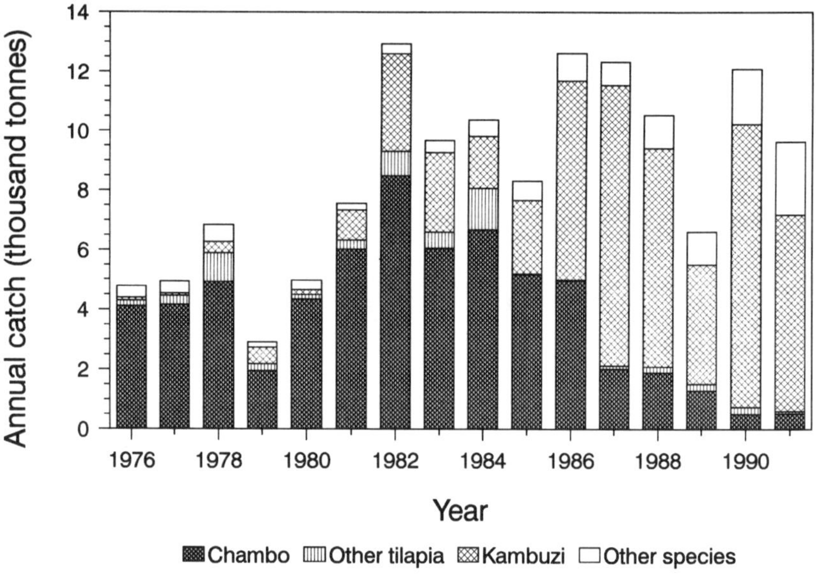

Fig. 19.3 Annual catches from Lake Malombe, based on the catch assessment survey.

to one based on intensive fishing with small-meshed nets for low-value kambuzi.

The data presented are calculated from the CAS as described above but there are problems with this system, particularly for irregularly distributed but important gears such as chambo seines, where apparent catches can fluctuate wildly and spuriously depending on whether such gears are included in the samples or not. However, the two parallel surveys have shown that although apparent catches do fluctuate to some extent from month to month in the CAS, over the course of a year the different systems produce consistent results for the annual catches of the commoner gears. For example, the parallel seine net survey suggested an overall yield of 9817 t from seines for the period February 1988 to February 1989, compared to the CAS estimate of 9067 t. These estimates are not significantly different (paired t-test on monthly estimates from the two systems; $P = 0.324$). On a monthly basis, the estimated catches of kambuzi seines and nkacha nets were correlated using the two systems (linear regression; $r = 0.748$; $P = 0.008$). The more recent survey shows a reasonable agreement between the estimated catches of the CAS system and the new MTF system (Fig. 19.6). In 1991, CAS and MTF estimates for total yield of Lake Malombe differed by about 10%, a reasonably good agreement, but monthly variations in CAS estimates were

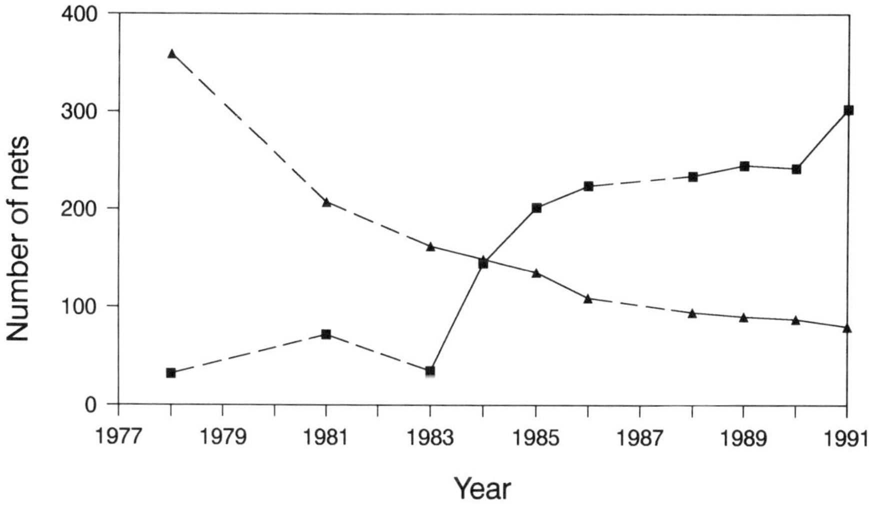

Fig. 19.4 Changes in ownership of gill nets (▲) and small-meshed seine nets (■), based on annual frame surveys. The data for seines show kambuzi seine and nkacha net data combined. Kambuzi seines are shore operated, nkacha are offshore encircling seines. N.B. gill net data are presented in multiples of five, i.e. one unit on the y-axis represents five 91 m (100 yd) long nets.

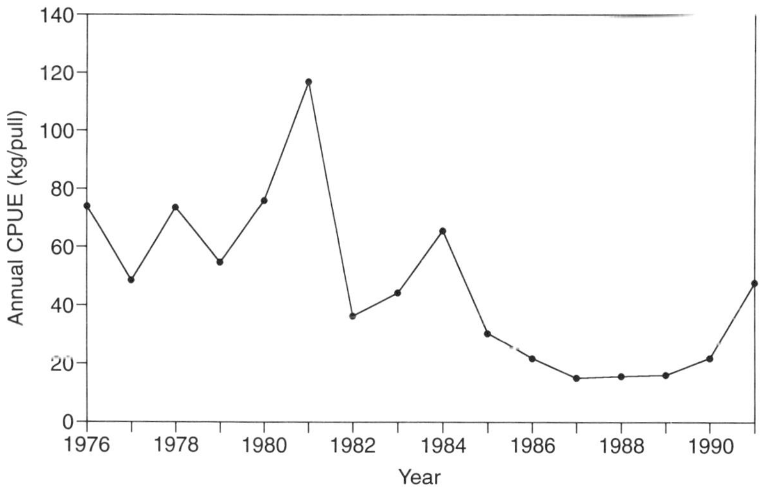

Fig. 19.5 Changes in the catch per unit effort (CPUE) for the small-meshed seine fisheries. The figures represent the mean for both nkacha nets and kambuzi seines combined.

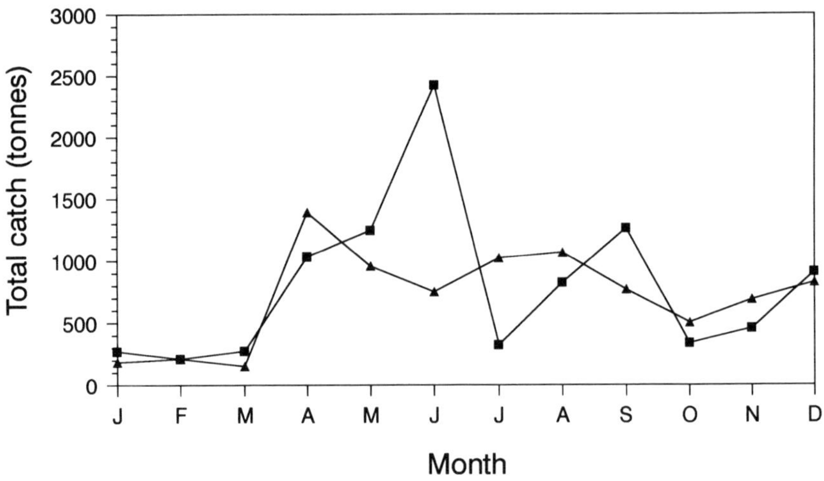

Fig. 19.6 Comparison of estimates for the Lake Malombe total catch using the catch assessment survey system (CAS) and the new gear-based frame system (MTF): ■, CAS estimate; ▲, MTF estimate. The total annual catch estimated by each system is 9625 t (CAS) and 8785 t (MTF).

substantially higher during the period when all gears were operating (from April to October). Although there is some 'noise' resulting from the older sampling system, particularly for rare but important gears such as the large chambo seines, the indications of long-term trends in the Lake Malombe fishery can be treated with confidence.

The seines currently in use have very small meshes and thus catch large quantities of immature fish. The species composition of several samples from both kambuzi shore seines and nkacha nets has now been investigated. The results are shown in Tables 19.1 and 19.2 and Fig. 19.7. They show that the fishery is currently based almost entirely on immature specimens of small cichlid species. The shore-based kambuzi seines are individually more damaging than the offshore nkacha nets in this respect (Fig. 19.7), although the larger numbers of nkacha nets mean that collectively they also have a considerable impact. The data also show that the greater concentration of fishing craft on the western shore has resulted in a much higher proportion of the smallest species in the catches. Catches are very much lower on the western shore.

Table 19.1 Mean percentage composition by weight of samples from Lake Malombe small-meshed seines

Species	Nkacha net (4 samples)	Kambuzi seine (5 samples)
Oreochromis karongae	0.60	0.09
Oreochromis shiranus	4.49	0.08
Tilapia rendalli	0.70	0.00
Aristochromis christyi	0.47	0.00
Astatotilapia calliptera	0.00	0.05
Aulonocara 'furcicauda'	0.21	0.25
Buccochromis spp.	0.40	0.40
Copadichromis cf. *chrysonotus*	2.28	3.59
Copadichromis cf. *virginalis*	18.01	17.50
Corematodus taeniatus	0.00	0.01
Ctenopharynx intermedius	1.05	0.42
Dimidiochromis strigatus	0.00	0.34
Dimidiochromis compressiceps	0.00	0.00
Hemitilapia oxyrhynchus	0.03	0.22
Lethrinops lethrinus	0.65	1.94
Lethrinops macrochir	0.69	1.36
Lethrinops 'pink head'	42.35	34.49
Maravichromis cf. *balteatus*	0.34	0.00
Maravichromis labidodon	0.14	0.00
Nimbochromis livingstonii	0.19	0.00
Otopharynx argyrosoma	6.70	30.56
Otopharynx cf. *tetraspilus*	1.53	0.00
Otopharynx tetrastigma	6.25	3.08
Placidochromis longimanus	3.06	0.00
Placidochromis subocularis	0.18	0.89
Protomelas cf. *labridens*	1.09	0.94
Pseudotropheus elegans	0.00	0.02
Pseudotropheus livingstonii	0.23	0.53
Rhamphochromis spp.	1.46	0.41
Tramitichromis lituris	0.00	0.42
Trematocranus placodon	0.29	0.00
Bagrus meridionalis	0.37	0.54
Barbus arcislongae	0.00	0.07
Barbus paludinosus	0.07	0.00
Brycinus imberi	0.00	0.29
Clarias gariepinus	3.02	0.00
Engraulicypris sardella	1.50	0.90
Labeo cylindricus	0.54	0.00
Aethiomastacembelus shiranus	0.36	0.00
Opsaridium microcephalum	0.55	0.54
Opsaridium microlepis	0.20	0.09

Table 19.2 Lake Malombe seine net fishery species composition (per cent by weight), showing differences between the West Malombe and East Malombe catches in relation to the number of seine nets in use. KS, kambuzi seine; NK, nkacha net; number of samples in parentheses

Species	West Malombe		East Malombe		Maximum total length (cm)
	KS (3)	NK (3)	KS (2)	NK (1)	
Lethrinops 'pink head'	34.8	52.8	34.0	11.0	12
Otopharynx argyrosoma	43.9	4.3	10.6	13.8	15
Otopharynx tetrastigma	3.7	8.1	2.2	0.6	14
Other small haplochromines	2.6	7.6	1.6	0.0	12–15
Small haplochromines total	84.9	72.9	48.4	25.4	
Copadichromis cf. *chrysonotus*	2.3	2.1	5.5	2.9	18
Copadichromis cf. *virginalis*	4.8	0.1	36.6	71.7	17
Other medium haplochromines	4.8	3.2	3.9	0.0	16–20
Medium haplochromines total	11.9	5.4	45.9	74.6	
Large haplochromines	0.9	5.2	2.6	0.0	20–40
Tilapiines	0.3	7.7	0.0	0.0	35–40
Large non-cichlids	1.1	6.7	1.3	0.0	20–100
Small non-cichlids	0.9	2.1	1.8	0.0	10–15
Number of nets	47	117	9	69	
Annual catch per net (1991) (tonnes)	12.7	24.5	82.6	46.3	

19.4 DISCUSSION

Dramatic changes have occurred in the fishery of Lake Malombe. The stocks of high-value chambo have been drastically reduced. The chambo fishery was based mainly on two inshore species, *Oreochromis karongae* (formerly known as *O. saka*; Chapter 17) and *O. squamipinnis*. Morphometric analysis suggests that the stocks are distinct from those of Lake Malaŵi (unpublished data) and thus may be managed separately. The third, more offshore, chambo species, *O. lidole*, does not live or breed in Lake Malombe, but brooding females from Lake Malaŵi enter the lake via the Shire River to utilize the lake as a nursery area. Excessive fishing for the females and fry may therefore have an adverse impact on the *O. lidole* population of Lake Malaŵi, although the species is of negligible importance (currently <10% of the total chambo catch) in the Lake Malombe fishery.

The chambo fishery has been almost totally replaced by an intensive fine-meshed seine fishery for kambuzi. The beach price for kambuzi is 30%

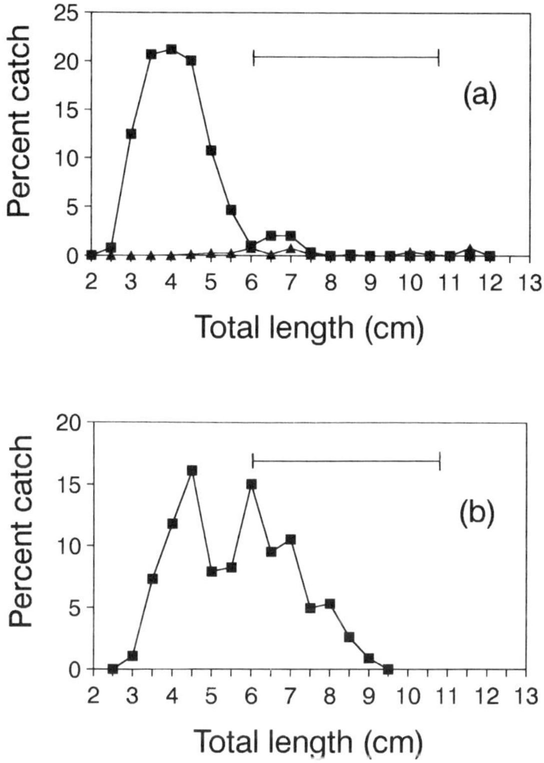

Fig. 19.7 Length distribution of samples of (a) kambuzi seines and (b) nkacha nets; ■, small haplochromine species; ▲, other species. The kambuzi seine in particular catches almost entirely juveniles of small species. The size range at which the small species mature is shown by the bar at the top of each figure.

of the chambo price (M. Mdaihli, pers. comm.). The number of fishing craft and, by implication, fishers has doubled since the mid 1970s. The total weight of fish caught in 1991 was about 60% greater than the late 1970s average. In real terms, each fishing craft is therefore now earning only one-quarter as much as in the 1970s. The decline in outboard engines from 221 to zero may be taken as evidence of economic decline but because the decline in engines began while the fishery was still apparently healthy, the reasons remain unclear.

Without the information on the species changes presented here, it might be argued that a fishery producing 60% more fish than before is healthy and gives no cause for concern. The kambuzi are, after all, less expensive and therefore more accessible to the poorer sections of the community. However, the data on the species composition and size of fish caught, and also the differences between the catches of the easily accessible and hence heavily fished western shore and those of the eastern shore, show that the current catch rates cannot be sustained as the fishery is increasingly reliant on tiny, immature fish. Urgent action is now necessary to avert a collapse of the fishery. Such a collapse would have a much more severe impact on the fishing communities than the reductions in catch which will temporarily occur if the fishery is restored to health by a carefully controlled blend of education, extension and law enforcement.

ACKNOWLEDGEMENTS

The analysis of the catch assessment survey data was carried out under the ODA-funded Traditional Fisheries Assessment Project. ODA funded the attendance of Mr Tweddle at this symposium. The species composition data for the seines were collected and analysed through the UNDP/FAO Chambo Research Project.

Chapter twenty

Effects of exploitation on the pelagic fish community in the south of Lake Tanganyika

Martin J. Pearce

20.1 INTRODUCTION

Lake Tanganyika has a simple but abundant pelagic fish community which in the south has been subject to a well-documented commercial fishery for 30 years. Fishing has, of course, been pursued for generations by the riparian inhabitants of Lake Tanganyika. Old accounts of fishing indicate that beach fires and charcoal braziers attached to boats were the traditional methods of attracting fish prior to capture. These methods are unlikely to have had any great effect on the pelagic fish populations.

In 1959, pressure paraffin lights were introduced to the local fishermen in the Zambian part of Lake Tanganyika, and the District Commissioner began fishing trials with a prototype purse seine in the same year. Significant exploitation can be considered to have started from this year.

At the time that exploitation began, there were six important species in the pelagic catch. These consisted of two species of sardine, locally called kapenta (*Limnothrissa miodon* and *Stolothrissa tanganicae*, Clupeidae), and four predators all of the genus *Lates*, Family Centropomidae. *L. mariae* and *L. angustifrons* are nominally benthic and together known as Nile perch; the other two species are *L. microlepis* (lake salmon) and *L. stappersi* (nvolo).

Changes in the fishery immediately after commercial exploitation began

The Impact of Species Changes in African Lakes. Edited by Tony J. Pitcher and Paul J.B. Hart. Published in 1995 by Chapman & Hall, London. ISBN 0 412 55050 4.

are well documented by Coulter (1970). The composition of the catch by weight was about 50% kapenta and 50% the *Lates* predators. This high proportion of predators fell rapidly between 1962 and 1966, while the catch rates of sardine increased. There was a linear relationship between the catch rates of predators and of sardines.

Since this early change in the pelagic community there have been several other changes in composition and abundance of these fish, and this chapter seeks to document and explain these changes.

20.2 SUMMARY OF FISH BIOLOGY

This brief account of the relative characteristics of the important species derives from Coulter (1966, 1970, 1976), Matthes (1968, unpublished), Ellis (1971), Chapman and van Well (1978a,b), Roest (1978, 1988), Pearce (1985, 1988, unpublished) and Kendall (unpublished). Table 20.1 gives a summary of the information relevant to this Chapter.

Lates angustifrons is the largest species and is normally benthic, feeding on cichlids, but when kapenta, especially *S. tanganicae*, is common the perch rises in the water column to feed (Coulter, 1970). *Lates mariae* is smaller and is more frequently near the bottom, often in very deep water close to the oxygen limit. *Lates microlepis* is pelagic and migratory and feeds principally on kapenta. All three species spend their first year in weed-beds in the lake margins. These three similar species will be collectively referred to as 'large *Lates*'. *Lates stappersi* is a medium-sized predator, truly pelagic, migratory and is never associated with shallow water during its life history. In the north of the lake the juveniles swim with kapenta in the first year, but later separate and prey on the kapenta. In the south the juveniles are uncommon, although the adults probably spawn in Zambian waters.

Limnothrissa miodon is the larger of the two clupeids, feeds on plankton and is not specialized. Mature adults are reported to be piscivorous (Chapter 13). Maturity is thought to be at 1 year of age and very few survive to 2 years. The fish is said to breed inshore throughout the year, but there is a seasonal peak in abundance of juveniles in May and June each year. Juveniles move offshore as they grow, so there is a pronounced size zonation in distribution. The largest fish (> 10 cm) can be found swimming in the offshore area with *L. stappersi*.

Stolothrissa tanganicae is the smaller sardine species. It eats copepods and pelagic prawns, with prawns becoming more important as the size of the fish increases. Juveniles appear in coastal waters at the beginning of the dry season often in May, but the bulk of the population does not appear in catches until half grown in July or August. The population drifts slowly offshore until

Table 20.1 Summary characteristics of the pelagic species of Lake Tanganyika

Species	Max. length (cm)	Max. weight (kg)	Maturity (years)	Life span (years)	Spawning	Juveniles	Adults	Presence
Lates mariae	80	5	4	10?	?	Inshore	Demersal and pelagic	Continuous
Lates angustifrons	180	60	4	10?	?	Inshore	Demersal and pelagic	Continuous
Lates microlepis	110	7	4	10?	?	Inshore	Pelagic	Erratic
Lates stappersi	50	0.8	2	4?	Pelagic?	Pelagic	Pelagic	Seasonal Oct.–May
Stolothrissa tanganicae	11	0.01	0.5	1	Pelagic	Pelagic	Pelagic	Seasonal June–Oct. variable
Limnothrissa miodon	16	0.05	1	2	Inshore	Inshore	Pelagic	Continuous

about October, when fish are fully grown, and by the end of the year the bulk of the population has disappeared.

20.3 ON THE RELATIONSHIPS OF THE PELAGIC FISHERIES

Nature of fisheries

Two fisheries, industrial and artisanal, exploit the pelagic fish community in southern Lake Tanganyika. The industrial fishery started in 1962, based on a ring net set from a boat which previously fished in Burundi. A second boat began fishing in 1963. Complete records begin in 1964 but sufficient data from 1962 exist to produce a time series over 30 years. This fishery catches all the pelagic species. As the industrial fishery expanded during the 1980s the fished area increased considerably.

The artisanal fishery, which catches only the two sardine species and is currently based mostly on beach seines, is less well documented. During the 1960s the dominant gear was a scoop (lusenga) net, but these were totally replaced by beach seines during the 1970s. From the late 1980s lift nets (Chapter 13) have been adopted by a small number of fishers. It was not until 1960 that the first estimates of catch and effort in the artisanal fishery were available. Though copious data on catch per unit effort, length-frequency distributions and market sales records exist, these have not been analysed to a level where total catches and effort can be inferred, despite the existence of official catch statistics.

Stocks exploited

Both fisheries exploit the same migratory stock of *S. tanganicae*, though because the fish tend to move towards the pelagic zone as they grow, the artisanal fishing season is slightly shorter, and the artisanal catch consists of shorter fish (Pearce, 1985) than the industrial catch.

The majority of fish caught in the artisanal fishery on *L. miodon* are very small (15–45 mm), often in the stage prior to metamorphosis. The industrial fishery only catches this species when the fish have moved offshore and are substantially bigger (> 50 mm). Hence, there is little overlap in the size range of fish that are caught by the two fisheries (Pearce, 1985). Thus the offshore *L. miodon* stock consists of fish that have passed through the artisanal fishery and are no longer affected by it. For the purposes of analysis of the status of the pelagic stock, the artisanal fishery can be ignored, as the industrial fishery is the only one exploiting the pelagic phase.

The two Nile perch species were also caught extensively in the demersal gill net fishery, and in artisanal beach seines. These fisheries are rather less well documented than the pelagic fishery. Most of the artisanal-caught fish are juveniles (unpublished data), though in the 1960s there was an artisanal fishery on the adults. Only adults appear in the pelagic industrial catch (Coulter, 1970). Therefore the same arguments used above for *L. miodon* apply, in that the artisanal fishery does not operate on the pelagic part of the stock.

There was only a small (handline) artisanal fishery for *L. stappersi* until October 1991, when a very poor kapenta season combined with the gradual introduction of the lift net into the artisanal fishery produced a sudden change in fishing patterns. Consequently, the industrial fishery records represent virtually 100% of the total catch and effort for this species.

Contribution of artisanal fishery

Because this Chapter presents a description of the changes which have occurred in the pelagic fish community, without making a formal assessment, I will use only the time series of catch and effort data from the industrial fishery, which represents fairly accurately, the position of all species except *S. tanganicae*.

An approximate assessment of the contribution of the artisanal fishery to the total catch and effort on *S. tanganicae* can be quantified as follows. The artisanal pelagic fishing fleet in 1964 consisted of 650 lusenga (scoop) nets and lights (Bowmaker and Mubanga, 1965). In 1990 there were 195 beach seines with 1436 lights (unpublished data). There is no evidence to assess the efficiency of a light used in conjunction with a lusenga net as compared with a beach seine, but the latter is probably somewhat more efficient. It may be reasonable to assume that the 1436 lights used in conjunction with beach seines had at least three times the catching power of 650 lights used with lusenga nets. The ratio of catch per night between commercial purse seine and artisanal beach seine is 24.1 (SD = 34.9, $n = 10$) between 1982 and 1991, with an average number of lights associated with each beach seine of 10.1. This produces an artisanal equivalent effort of 2.7 purse seine units in 1964, and 8 in 1990. The artisanal proportion of the catch of *S. tanganicae* has fallen from about 57% to about 32% during the last 30 years. Due regard to this contribution, as well as the magnitude of possible errors, needs to be made when considering industrial data on *S. tanganicae*.

20.4 INDUSTRIAL CATCH AND EFFORT DATA FROM THE SE ARM

The geography of the south of Lake Tanganyika (Fig. 20.1) makes it possible to demarcate two distinct areas, here referred to as the SE and SW Arms. Most of the industrial fishing has occurred in the SE Arm, and this area has

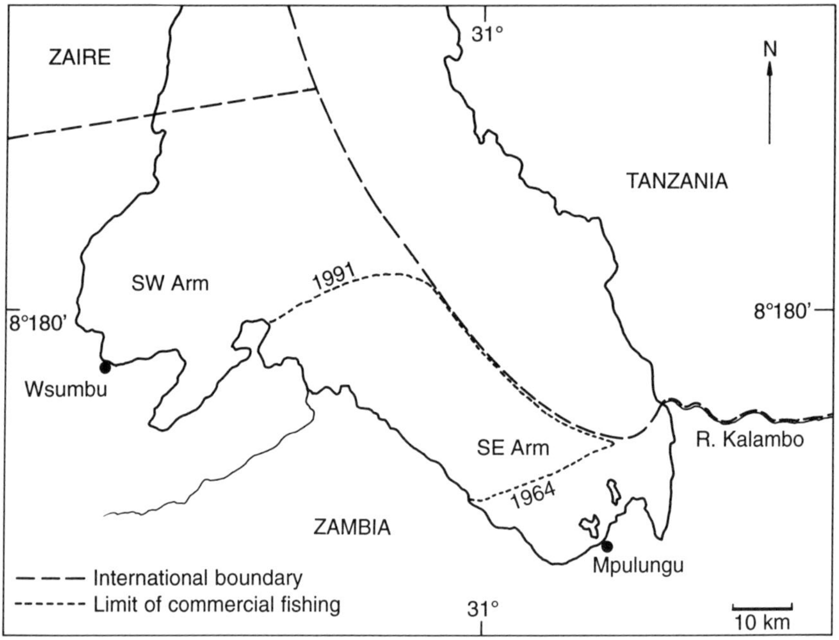

Fig. 20.1 Sketch map of southern Lake Tanganyika showing the principal fishing grounds of the commercial fleet.

the longest and most accurate series of data. This account largely limits itself to a consideration of what has occurred in the SE Arm of the lake.

One immediately noticeable characteristic of the industrial fishery is the large expansion through the 1980s (Fig. 20.2). There was a small increase in the number of boats beginning in the late 1960s, and although by 1978 this had reverted to two boats, from 1980 to the end of 1991 the fleet increased to 17. At the same time the fishing area, which was approximately maintained from 1962 to 1980, expanded from about 300 km^2 to about 1200 km^2 by the end of 1991. Total catches rose in line with the increase in effort up to 1985, but have subsequently declined a little. Since 1988 the effort has not increased any further. Data on fishing area are not precise and further analysis may necessitate a modification of the figures.

This increase in fishing area occurs only seasonally, and corresponds with the time that *L. stappersi* appears in abundance, the fishery then being directed towards only this species. Because the other main target species, *S. tanganicae*, is closer inshore when abundant, there is no increase in fishing area during this season.

It is often assumed that CPUE (catch per unit effort) is a fairly good index of stock abundance (e.g. Gulland, 1983). More strictly it is an index of local

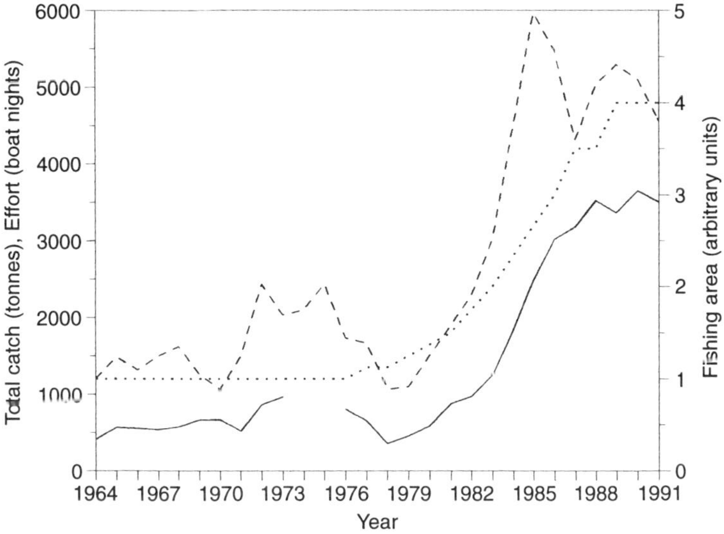

Fig. 20.2 Catch (dashed curve), effort (solid curve) and relative fishing area (dotted curve) for the commercial fleet in the SE Arm of southern Lake Tanganyika, from 1964. Effort data for 1974 and 1975 are missing.

density (Shepherd, 1988). For pelagic species there is evidence that CPUE is not a good index of abundance because the catchability coefficient can change with stock size and with stock distribution. For pelagic species, an index of effort that takes into account factors such as search time and seasonal changes in abundance can provide a better index of abundance (Csirke, 1988). *L. stappersi* may be one such shoaling species that maintains a constant local density because CPUE shows no trend with increased effort despite periodic fluctuations. The total distribution of this species during seasonal appearance is unknown, but is larger than the area fished. For these reasons I have applied a correction factor to the fishing effort data, to reflect both seasonal changes and searching effort, measured as area fished. Both species show marked seasonal changes in catch rates. The corrected CPUE was calculated by:

$$\textit{corrected CPUE} = \textit{catch}/(\textit{effort} \times \textit{fished area}) \qquad (20.1)$$

where *catch* is the total weight landed during the seasonal peak, *effort* is the number of boat nights in the same period, and *fished area* is proportional to the pre-1980s area. Throughout this chapter I have assumed that my CPUE index is a reliable index of population size.

The development of the fishery can be divided into three phases, corresponding to the amount of fishing. These are: low effort (1964–1980); increasing effort (1981–1988); and high effort (from 1988).

The influence of the greater number of boats on catch began to be apparent from about 1983, when the fleet started to increase the fishing area to avoid physical interference and to maintain catch rates, but by 1985 catch rates were falling despite the increased area.

During the period of industrial exploitation a number of changes in the species composition of the pelagic community have taken place, and the population sizes of the various species have fluctuated. These changes are shown in Fig. 20.3.

When industrial fishing started, the predator/prey catch ratio was approximately 1:1 and all four species of predators were regularly found in the catch. The large *Lates* rapidly disappeared, and the abundance of kapenta increased (Coulter, 1970). Some 10 years after the beginning of industrial fishing, the predator population increased again and kapenta abundance correspondingly began to fall. The composition of the predator population had changed. The abundance of large *Lates* species remained low, so that the predators now consisted almost entirely of the medium-sized *L. stappersi*, which had previously been unimportant. By 1977 the increase in *L. stappersi*

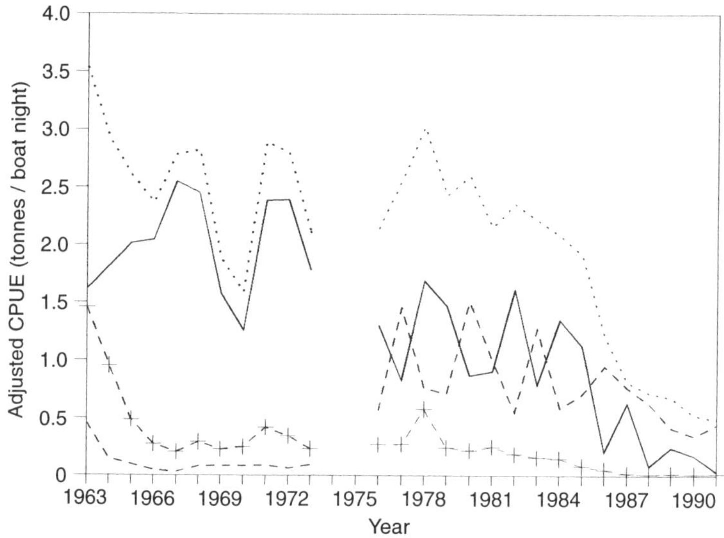

Fig. 20.3 Adjusted CPUE for major species caught by the commercial fishery in the SE Arm of southern Lake Tanganyika: dotted curve, total catch; solid curve, kapenta; dashed curve, *L. stappersi*; dashes with +, large *Lates*. Missing effort data for 1974–1975 preclude estimates of CPUE for these years.

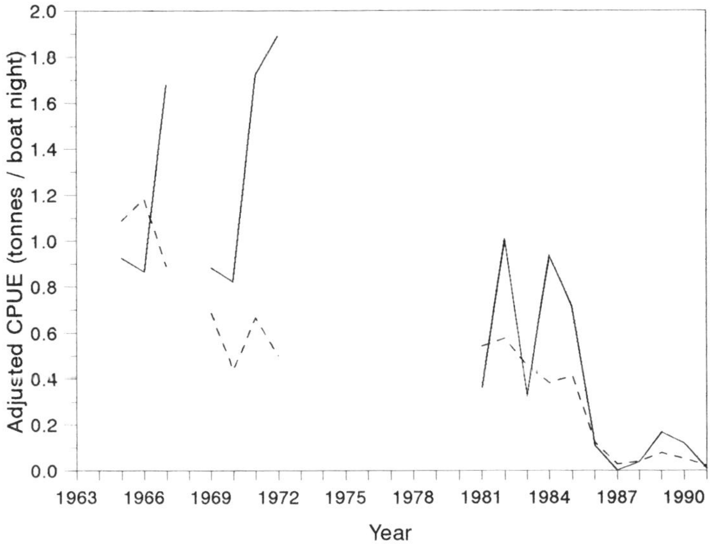

Fig. 20.4 Adjusted CPUE of *S. tanganicae* (solid curve) and *L. miodon* (dashed curve) from the commercial fishery of southern Lake Tanganyika. Missing values are for years when no samples were taken to assess the contribution of each species to the total kapenta catch.

population had brought the predator–prey ratio back to about 1:1. Throughout this period, which occurred with low fishing effort, the total catch rates remained almost constant except for cyclical fluctuations in the abundance of *S. tanganicae*.

During the period of increasing effort, there was a general fall in total adjusted CPUE for all species, although this was not noticeable in the daily landings of the boats. The 1:1 predator/prey catch ratio was maintained up to 1986, when there was a large decrease in the abundance of both species of kapenta (Fig. 20.4).

The population of *L. miodon*, which had previously been relatively unaffected by the changes in the composition and abundance of the other species, also declined markedly from 1986. This might be ascribed to the expansion of fishing in the pelagic zone. From about 1980 large specimens of *L. miodon* ($>$10 cm FL), which were not previously caught in large numbers (Coulter, 1970), were regularly caught in the pelagic zone during the rainy season (October to May) together with *L. stappersi*. From about 1986 these large adult fish again became scarce in the catch.

The period of high exploitation has only recently been in existence, and perhaps it is too soon to state with any assurance what are the major features of this phase, but it is probable that the observed very low kapenta populations and a predator–prey biomass ratio greater than 1 are likely to prove characteristic.

20.5 RELATIONSHIP BETWEEN *S. TANGANICAE* AND *L. STAPPERSI*

One notable feature of the pelagic community is the marked oscillations of kapenta and *L. stappersi* abundance from 1977, which show an inverse cyclical relationship with a periodicity of 3–4 years. Figure 20.4 indicates that *S. tanganicae* is the sardine species which is almost entirely responsible for these fluctuations.

Figure 20.5 shows the typical seasonal pattern of abundance of these two species. They do not coexist to a great extent, their migrations being so timed that one species disappears in the fishery as the other arrives.

A correlation between the catch rates of *L. stappersi* and *S. tanganicae* was obtained by regressing the mean CPUE of *S. tanganicae* during the peak season (the three contiguous months that produced the highest catch, normally July to September) against that of *L. stappersi* during the previous season (variable from October to May). Data used are given in Table 20.2. During the peak

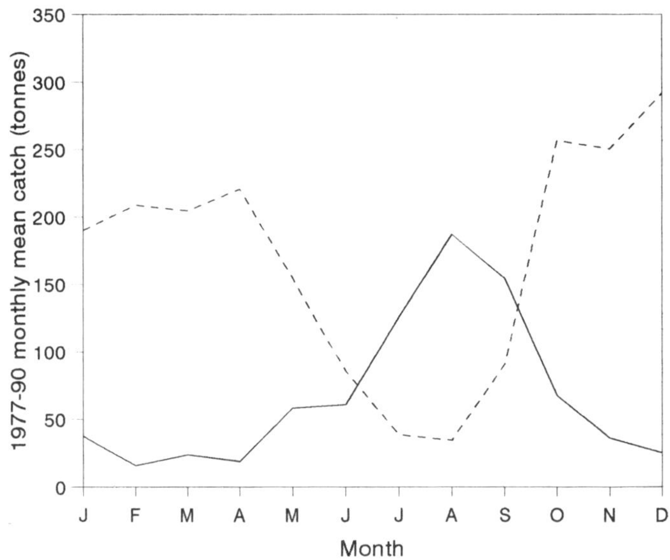

Fig. 20.5 Monthly mean change in tonnes of *S. tanganicae* (solid curve) and *L. stappersi* (dashed curve) from 1977 to 1990, showing the mutually exclusive seasons of abundance of these two species.

Table 20.2 Data used in correlation analysis between peak seasonal abundances of *S. tanganicae* and *L. stappersi*

Year	Seasonal peak CPUE (tonnes per boat night)	
	S. tanganicae	*L. stappersi*
1979–80	0.4448	1.9729
1980–81	0.6981	1.6513
1981–82	2.3680	1.1506
1982–83*	0.6529	3.1727
1983–84	2.2622	1.0363
1984–85	1.5952	1.6542
1985–86	0.3504	2.2379
1986–87	0.0037	2.1441
1987–88	0.2461	1.7760
1988–89	1.1189	2.1054
1989–90	0.9424	1.6336
1990–91	0.073	2.0972

*Data from this year were not used in calculating the regression shown in Fig. 20.6.

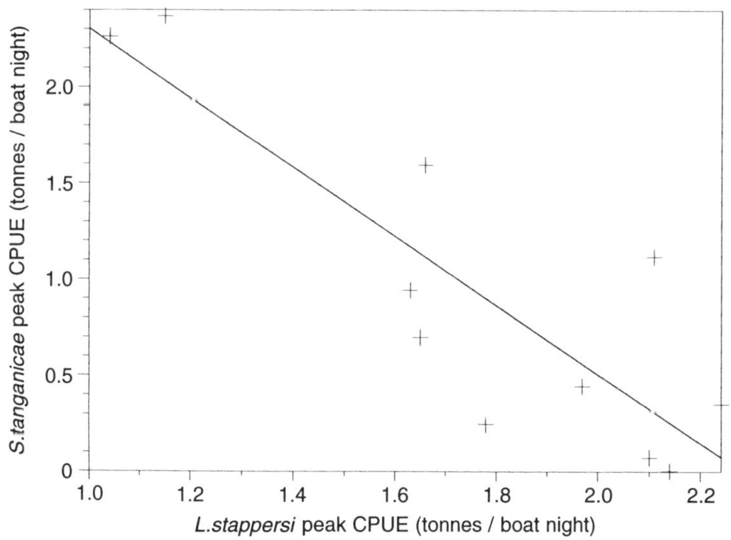

Fig. 20.6 Regression of the mean peak CPUE of *S. tanganicae* against the mean peak CPUE of *L. stappersi* from the previous season. Actual data points are given by +. Data for the year 1982–83 were excluded from this figure.

Table 20.3 Correlation coefficients obtained by regressing peak CPUE of *L. stappersi* against peak CPUE of *S. tanganicae* with different time lags*

Regression parameters	Peak CPUE of *L. stappersi* regressed against peak CPUE of *S. tanganicae* from:			
	Two seasons before	The previous season	The following season	Two seasons ahead
r	0.08	0.44	0.86	0.17
n	9	10	11	10

*No correlation is presented for zero time lag because the species do not occur together.

seasons these fish are the primary target of the industrial fishing fleet, and estimates of abundance during that period are more realistic than annual means. Figure 20.6 shows this inverse relationship between the two species, which has not been adjusted for changes in fishing area. There is no other significant cross-correlation between the abundance of *S. tanganicae* and that of *L. stappersi* (Table 20.3). So the analysis suggests that *L. stappersi* affects the abundance of *S. tanganicae*, but not vice versa.

From 1986 the abundance of *S. tanganicae* has been much reduced, and minimum values during the course of the cyclical fluctuations have reached virtually zero. At the same time the amplitude of the fluctuations in *L. stappersi* abundance has been decreasing. The proportion of *L. stappersi* in the total annual catch since 1986 has varied between 62% and 94%.

20.6 COMPOSITION OF THE PLANKTON

During 1989 regular plankton samples were taken at night under the lights of fishing boats. Vertical hauls from 50 m were made at intervals through the night using a plankton net of 30 cm diameter and mesh size which varied from 40 to 100 μm. A simplified summary of the results is given in Table 20.4.

Of particular interest in the context of this Chapter is that clupeid (*S. tanganicae*) larvae were present in considerable numbers from October to May. Ellis (1971) and Coulter (1970) also quote this period of the year as the major spawning period for *S. tanganicae*. The period during which these fry were common corresponds exactly with the period when *L. stappersi* is abundant.

Pelagic shrimps were also common in the plankton samples throughout the year. For the most part these consisted of unidentified larvae, and adults of the small species *Limnocaridina parvula*, Atyidae. Adults of the larger pelagic shrimp *Palaeomon moorei* were rarely caught in the plankton nets.

Table 20.4 Composition of the plankton during 1989 in the SE Arm of Lake Tanganyika. Data are numbers per m^2, with 95% confidence limits

Month	Copepods (thousands)	Shrimps	Jellyfish	Clupeid larvae
Jan.	65 ± 28	480 ± 190	40 ± 35	14 ± 20
Feb.*				
Mar.	40 ± 12	250 ± 90	24 ± 21	34 ± 22
Apr.	83 ± 22	550 ± 450	18 ± 9	64 ± 51
May	120 ± 37	520 ± 290	190 ± 180	54 ± 52
Jun.	93 ± 77	1100 ± 1000	1000 ± 1000	0
Jul.	24 ± 60	100 ± 300	110 ± 250	0
Aug.	116 ± 51	3800 ± 2000	190 ± 120	1 ± 2
Sep.	114 ± 26	2800 ± 1500	62 ± 41	6 ± 7
Oct.	84 ± 40	4100 ± 5900	500 ± 120	31 ± 15
Nov.	41 ± 38	740 ± 620	99 ± 65	10 ± 10
Dec.	48 ± 30	1800 ± 2900	5 ± 20	33 ± 54

*No data available.

20.7 FOOD OF *L. STAPPERSI*

Ellis (1978) reported that the food of *L. stappersi* in the north of the lake consisted of 95% *S. tanganicae*. This is not so in the south. Pearce (1985) reported the results of a qualitative analysis of *L. stappersi* stomach contents on a large number of fish over 3 years. He found that the shrimp *P. moorei* constituted about 50% of the diet and that most of the rest was *S. tanganicae*. He did not separate adult and larval specimens.

A second quantitative study was made in 1989 and a simplified summary of the results is shown in Table 20.5. It is apparent that whilst adult or subadult *S. tanganicae* are always important in the diet, shrimps are equally so. Nearly all the shrimps found in stomachs were adult and subadult *P. moorei* in the size range 15–25 mm (length, tail to base of rostrum). Clupeid larvae, which generally fall into the same size range, were important food items from October to March, and in one month were the principal item.

20.8 DISCUSSION

During the period of industrial exploitation the number of abundant species in the pelagic community has been progressively reduced from six to one. The first major change was the reduction in the abundance of the large *Lates* species and their replacement by the smaller predator *L. stappersi*. This was followed by a reduction in the abundance of the two kapenta species. The third change was the reduction in the total pelagic fish biomass.

It is not suggested here that the industrial fishery was entirely responsible

Table 20.5 Occurrence of food items in stomachs of *L. stappersi*, Dec. 1988–Nov. 1989, in the SE Arm of Lake Tanganyika

Mean	No. of fish	Mean wt fish (g)	Mean wt food (g)	% by wt of total stomach contents				
				*Stolo.**	*Stolo.*†	Shrimp	*Limno.*	Other
Dec.	80	173	2.0	48.0	0.6	38.5	0.0	12.0
Jan.	176	157	1.8	32.8	23.5	32.9	0.0	10.8
Feb.	146	138	1.7	23.1	43.1	27.1	0.0	6.7
Mar.	93	265	2.4	49.1	23.4	16.0	0.0	11.5
Apr.‡								
May	55	247	2.4	20.8	0.0	43.1	15.7	20.4
Jun.	75	211	1.9	35.6	7.6	44.9	1.2	10.7
Jul.‡								
Aug.	45	273	14.8	83.6	0.0	2.8	8.9	4.7
Sep.	120	248	9.0	30.3	3.9	51.4	1.1	13.3
Oct.	159	304	1.7	19.2	14.2	45.0	0.0	21.6
Nov.	106	333	1.2	6.8	13.8	49.5	4.3	25.6
Mean		228	3.1	35.3	12.9	34.5	3.2	13.1

* Adults/subadults.
† Larvae.
‡ No data available.

for all these changes, but in conjunction with the artisanal fishery it seems likely that fishing pressure was a likely contributor to many changes.

It is an axiom in fisheries management that large, long-lived species are more susceptible to fishing pressure than small, short-lived species, and are the first to disappear or be reduced after fishing starts or expands. This certainly seems to have been the case in southern Lake Tanganyika. Coulter (1970) has discussed this, together with the subsequent increase in the kapenta populations. Coulter made the point that there was now a vacant niche for a pelagic kapenta predator and predicted that some species would increase in abundance and fill that niche. *L. stappersi* duly fulfilled this prediction. Reasons for this are not hard to identify; an increased food supply and reduced predation pressure from the large *Lates* would be sufficient to explain this adaptation of the pelagic community.

From about 1977 to 1985, although there was a population of coastbound *L. miodon*, the great majority of the offshore community consisted of only the two species *L. stappersi* and *S. tanganicae*. The populations of these two species fluctuated markedly in an inverse cyclical way with a periodicity of about 3 years. The evidence that *L. stappersi* abundance greatly affects the abundance of *S. tanganicae* through predation on the larvae present at the spawning grounds of the latter species is considerable. Roest (1988) discusses a possible mechanism whereby abundant *S. tanganicae* populations can reduce *L.*

stappersi populations by competition with juvenile *L. stappersi* for the same food. Juvenile *L. stappersi* are rare in Zambian waters so there is no evidence for or against this hypothesis. There was no significant relationship between the abundance of *S. tanganicae* populations and future *L. stappersi* abundance, so it is unlikely that poor *S. tanganicae* seasons in Zambian waters can effect a reduction in *L. stappersi* abundance. In any case, while the few *L. stappersi* present in Zambian waters do feed heavily on adult and subadult *S. tanganicae* when the latter are abundant, the bulk of the nvolo population is elsewhere at this time. The reasons for the fluctuations in *L. stappersi* populations therefore remain unexplained. The *S. tanganicae* populations have been fluctuating in this way from before the time that the abundance of *L. stappersi* increased, and so larval predation pressure from *L. stappersi* cannot be a full explanation of this phenomenon.

In the south of Lake Tanganyika, *L. stappersi* feeds to a large extent on the shrimp *Palaeomon moorei*. Any reduction in the abundance of *S. tanganicae* is likely to be compensated for by increased feeding on this shrimp. It seems possible that the present situation, where *L. stappersi* is the only major pelagic species, is sustainable. The reduction in *L. stappersi* abundance since the kapenta species have been reduced can be ascribed either to high fishing pressures or to reduced food availability, and probably to a mixture of both causes. The reduced amplitude in the flucuations in the abundance of *L. stappersi* in recent years may also be attributed to either cause. However, a final understanding of the relationship between these two species must await data on *L. stappersi* during the time it is absent from Zambian waters, and on the discovery of the juveniles. These are presumably spawned in Zambian waters during seasonal visits by the adults.

The last question to be addressed is why the two kapenta species collapsed in 1986 and 1987, and why they did not recover to the levels seen in previous years. The answer is likely to lie directly with the increase in fishing mortality. Pearce (1985) made a stock assessment of the two species and found that they were being fully exploited or were overexploited in the period up to 1983.

L. miodon was, and is, subject to enormous fishing mortality by the artisanal kapenta seine fishery, and the abundance of this species in the pelagic area was only a small fraction of what it potentially could have been. This assessment may well have underrated the effect of artisanal mortality, because it took account of the fish that were landed. Rossiter (pers. comm.) followed an artisanal kapenta seine, using underwater gear whilst it was in operation, and he observed a large quantity of tiny pre-metamorphosic *L. miodon* fry passing through the meshes of the seine. All these fry died after passage. Moreover, another source of problems for *Limnothrissa* is that during the expansion of the industrial fishery into the offshore zone, large adult *L. miodon* began to be caught regularly for the first time. These fish were all fully mature (unpublished data). It may well be that these adult pelagic fish represent the previously unexploited breeding stock of the species. Ellis (1971) studied the gonad development of both species of kapenta, and noted that in no sample from the then catch were as many as 50% of the *L. miodon* specimens mature.

Pearce (1985) found that the *S. tanganicae* population was not as large as generally believed, and was already being fully exploited. The effect of the expansion of the industrial fleet on *S. tanganicae* would have been greater than the data presented here indicate, because in 1983 an expansion of industrial fishing took place in the SW Arm as well as the SE Arm. The stock of *S. tanganicae* which moves into the SW Arm is thought to derive from the same offshore larvae which supply the SE Arm with maturing juveniles each year. The stock of fish in the SW Arm, which had previously been lightly fished, was probably the major source of mature adults before exploitation began.

In the case of both these kapenta species, therefore, it would appear as if the reason for their decline lies in the expansion in fishing effort, particularly by the industrial fishery.

Solutions for recovery of the fishery, and to prevent it from becoming a seasonal fishery on one species only, may lie in reducing the inshore fishing effort on juvenile *L. miodon*, and in maintaining a high fishing pressure on *L. stappersi* without a concomitant fishing mortality on *S. tanganicae*.

20.9 SUMMARY

Catch and effort data from the commercial pelagic fishery of southern Lake Tanganyika indicate that the pelagic stocks have undergone several changes in composition since exploitation began. At present only two species are important, *Stolothrissa tanganicae*, a clupeid, and its major predator *Lates stappersi*. For the past 5 years *L. stappersi* has constituted 62–94% of the annual catch. Movement of one species into the fishing zone coincides with the departure of the other. There is a significant relationship between the CPUE of *S. tanganicae* and that of *L. stappersi* in the preceding season. A hypothesis is advanced that predation by *L. stappersi* is a major factor controlling the size of the *S. tanganicae* population. Supporting evidence is presented that shows that the *L. stappersi* fishing season and grounds coincide with a larval nursery area of *S. tanganicae*, and that these larvae form an important part of the diet of *L. stappersi* at this time, whereas adults are important at other times of the year. Equally important in the *Lates* diet is a pelagic shrimp, *Palaeomon moorei*, so it may be possible that the present situation is sustainable.

ACKNOWLEDGEMENTS

I wish to acknowledge the contribution of earlier researchers based at Mpulungu, notably G.W. Coulter and R. Kendall, who were responsible for the collection of data prior to 1974, and without whose efforts no long time series would be available. I wish also to acknowledge the contribution of the Department of Fisheries, Zambia, under whose auspices the data have been collected, and the many fish scouts who measured the catches.

REFERENCES

Bowmaker, A.P. and Mubanga, J. (1965) Results of a survey of gear and manpower on Lake Tanganyika and notes on the dagaa trade. *Fish. Res. Bull. Zambia*, **2**, 42–51.

Chapman, D.W. and van Well, P. (1978a) Growth and mortality of *Stolothrissa tanganicae*. *Trans. Am. Fish. Soc.*, **107**, 26–35.

Chapman, D.W. and van Well, P. (1978b) Observations on the biology of *Luciolates stappersi* in Lake Tanganyika (Tanzania). *Trans. Am. Fish. Soc.*, **107**, 567–73.

Coulter, G.W. (1966) Hydrological processes and the deep water fish community in Lake Tanganyika, PhD thesis, Queens University, Belfast, 204 pp.

Coulter, G.W. (1970) Population changes within a group of species following their exploitation. *J. Fish Biol.*, **2**, 329–53.

Coulter, G.W. (1976) Biology of *Lates* spp. (Nile perch) in Lake Tanganyika and the status of the pelagic fishery for *Lates* and *Luciolates stappersi*. *J. Fish Biol.*, **9**, 235–9.

Csirke, J. (1988) Small shoaling pelagic fish stocks, in *Fish Population Dynamics, the Implications for Management*, 2nd edn (ed. J.A. Gulland), John Wiley and Sons, Chichester, pp. 271–302.

Ellis, C.M.A. (1971) The size of maturity and breeding seasons of sardines in southern Lake Tanganyika. *Afr. J. Hydrobiol. Fish.*, **1**, 59–66.

Ellis, C.M.A. (1978) Biology of *Luciolates stappersi* in Lake Tanganyika (Burundi). *Trans. Am. Fish. Soc.*, **107**, 557–66.

Gulland, J.A. (1983) *Fish Stock Assessment. A Manual of Basic Methods*, John Wiley and Sons, Chichester, 223 pp.

Matthes, H. (1968) Preliminary investigations into the biology of Lake Tanganyika clupeidae. *Fish. Res. Bull. Zambia*, **4**, 39–45.

Pearce, M.J. (1985) A description and stock assessment of the pelagic fishery in the south east arm of the Zambian waters of Lake Tanganyika. Dept Fish. Res. Rep., P.O Box 350100, Chilanga, Zambia. Unpubl. 59 pp.

Pearce, M.J. (1988) Some effects of *Lates* spp. on pelagic and demersal fish in the Zambian waters of Lake Tanganyika, in *Predator–prey relationships, population dynamics and productivities of large African lakes* (ed. D. Lewis), FAO, Rome, *CIFA Occ. Pap.*, no. 15, pp. 69–87.

Roest, F.C. (1978) Stolothrissa tanganicae, *Population Dynamics, Biomass, Evolution and Life History in the Burundi Waters of Lake Tanganyika*, FAO, Rome, CIFA Symp. no. 27.

Roest, F.C. (1988) Predator prey relationships in Northern Lake Tanganyika and fluctuations in the pelagic fish stocks, in *Predator–prey relationships, population dynamics and productivities of large African Lakes* (ed. D. Lewis), FAO, Rome, *CIFA Occ. Pap.*, no. 15, pp. 104–29.

Shepherd, J.G. (1988) Fish stock assessments and their data requirements, in *Fish Population Dynamics, the Implications for Management*, 2nd edn (ed. J.A. Gulland), John Wiley and Sons, Chichester, pp. 35–62.

Chapter twenty-one

Changes in the pelagic fisheries of northern Lake Tanganyika during the 1980s

Philippe Petit and Antoine Kiyuku

21.1 INTRODUCTION

In the Burundi sector of Lake Tanganyika (Fig. 21.1), pelagic fisheries operate at night, mainly catching the centropomid *Lates stappersi* and its chief prey, two endemic species of sardine (*Stolothrissa tanganicae* and *Limnothrissa miodon*). *Lates mariae* and other *Lates* spp. are of secondary importance. Industrial fisheries use purse seines, the technology of which has not changed since they began during the 1950s. Artisanal fisheries use lift nets, the technology of which was improved during the 1980s; in consequence, artisanal fisheries operate in open waters in competition with industrial fisheries. The aims of this chapter are to describe recent changes in the artisanal fisheries, and to analyse and discuss the changing interrelationship of the pelagic fishery sectors during the 1980s.

21.2 INVESTIGATIONS INTO THE FISHERY

Information available on the artisanal fisheries for the 1980s is scattered and incomplete, so from January to June 1990 we carried out several investigations to evaluate recent changes in the artisanal fishery.

The Impact of Species Changes in African Lakes. Edited by Tony J. Pitcher and Paul J.B. Hart. Published in 1995 by Chapman & Hall, London. ISBN 0 412 55050 4.

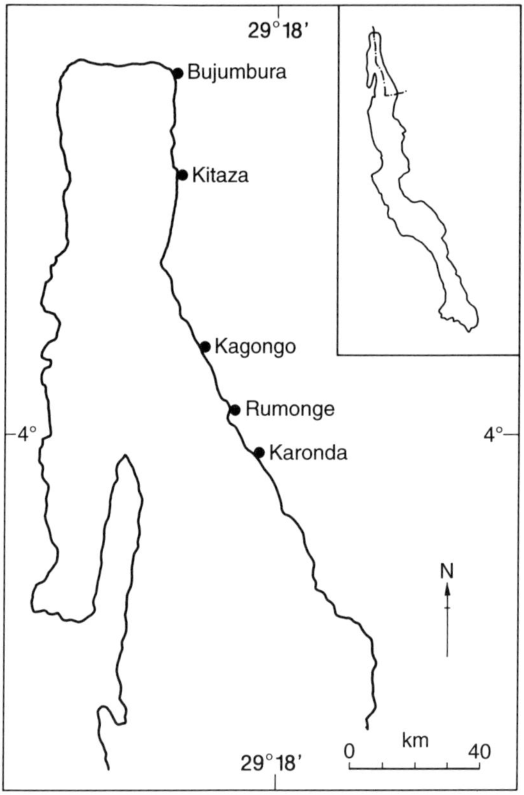

Fig. 21.1 Sketch map of northern end of Lake Tanganyika, showing the principal beaches where artisanal catches were analysed.

Interviews with fishers

The Burundi fishers were asked to describe their activity and current problems as well as the changes which they had observed since the fishery began. All the fishers were asked the same questions concerning their fishing methods and habits: when they go fishing and what gear they use. One hundred and fifty fishers, in 20 villages along the coast, were interviewed. Most of the interviews were conducted in Kiswahili, the most commonly used language in this area. Interviews were not constrained by specific short questions but encouraged fishers to talk about anything they wanted to mention on a free basis. This allowed the consistency of answers to be checked.

Study of fishing gear

Simultaneously with the above interviews, more detailed interviews were held to study the fishing gears on landing beaches at Kitaza, Kagongo, Rumonge and Karonda (Fig. 21.1). The beaches are situated in different ecological zones and are important places of commerce.

Interviews took place in the morning, on the beaches and at the homes of the fishers. Fishers were asked details of their fishing boats and gears (lamps, ropes, lines, lift nets), how these were employed, and their own activity levels. All questions are detailed in Petit (1990).

Additional data

To get as much information as necessary to try to understand changes in the pelagic fisheries, we obtained four types of additional data.

First, fisheries statistics obtained during the 1980s by the Department of Fisheries of Burundi have been analysed: these comprise actual catches, daily, monthly and annual sales. The official data accurately describe the industrial fishery, because all the catch coming from this fishery must be sold on the central market of Bujumbura.

Second, because in contrast, artisanal fishery catches are sold in several different places along the coast, the total number of artisanal fishery units has been estimated using information collected between 1977 and 1982, the interviews with fishers and discussions with local experts.

Third, from January to June 1990, fish samples were obtained from the artisanal catch (both *Lates* and sardines), on the same dates and using the same method as that used by the Department of Fisheries when sampling the industrial fishery.

Fourth, with the cooperation of the Immigration Police of Burundi, the increasing impact of piracy on fishing has been analysed.

21.3 RESULTS

Number of artisanal fishery units

About 700 catamarans (Fig 21.2) were operating between 1980 and 1982. At this time, an international artisanal fishery development project contributed to the increase in the number of catamarans. New boats with outboard engines started to operate, simultaneously with old boats that had not been abandoned or destroyed. Most likely, about 1000–1100 catamarans were effectively operating between 1983 and 1986. This number probably decreased between 1986 and 1990, because many old units are known to

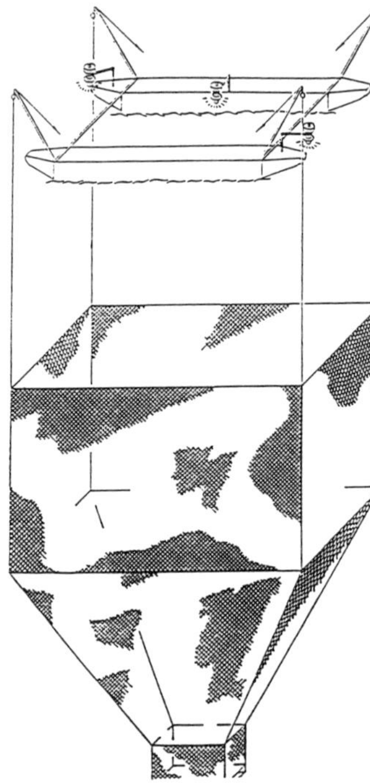

Fig. 21.2 Drawing of an artisanal unit: catamaran with lamps and lift net. The length of one boat is 10–12 m; the size of the net corresponds (length of rope is minimized for clarity).

have ceased fishing. However, 750 catamarans, most of them now motorized, were still active in 1990.

In fact, because of several constraints, such as piracy, failures, age and abandonment of the old units, it can be assumed that during the 1980s, the equivalent of 750 artisanal fisheries units were permanently operating in the Burundi sector of Lake Tanganyika.

Analysis of artisanal boats and fishing gears

Of the fishing boats, 70% have an outboard engine. This holds true all along the coast. Only 3% of these engines existed before 1987. It is usual for one boat with an engine to tow two catamarans to open water several kilometres from the shore. During fishing operations, this motorized boat remains apart from the catamarans in the dark, to avoid pirates who may steal the engine. This is a serious problem because more than 30% of the artisanal

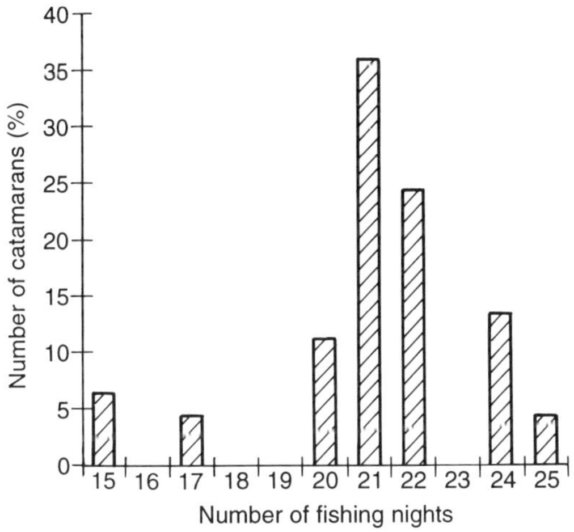

Fig. 21.3 Activity of the artisanal fishermen during one lunar month.

fishery units have been attacked at least once by pirates. We know one fisher who was seriously injured in 1988, and another who was killed in 1989.

Sixty per cent of fishing units started to operate since 1987. This shows the dynamism of artisanal fishers, and the level of activity is now the maximum allowed by the lunar cycle (Fig. 21.3). Fishing is inefficient during the full moon. The main constraint is the expected probability of attack by pirates. This makes it difficult to recruit crew, especially in the southern part of the Burundi sector where most recorded attacks have taken place.

With motorized boats, artisanal fishers are able to reach open waters and target *L. stappersi*, which is rarely caught near the coast and commands a high price. The selling price of a box of mukeke (e.g. *L. stappersi* adults) is higher than that of a box of ndagala (mixed *Stolothrissa*, *Limnothrissa* and *L. stappersi* juveniles). The average difference in price per box (40–50 kg) was between US$ 4 and US$ 12 in 1988 and US$ 20–35 in 1990.

The artisanal units now use new and more numerous lamps (Fig 21.4), called Anchor lamps, that are more powerful than the old lamps. Their light can reach deeper levels, where the predators live. It is now also common for a fishing unit to be assisted by a team (the 'renfort') which moves close to it with numerous very powerful lamps, to attract pelagic fishes which are expected to concentrate just above the lift net. This system is effective in increasing the catch and leads to severe competition with the industrial fishing boats which operate in the same areas.

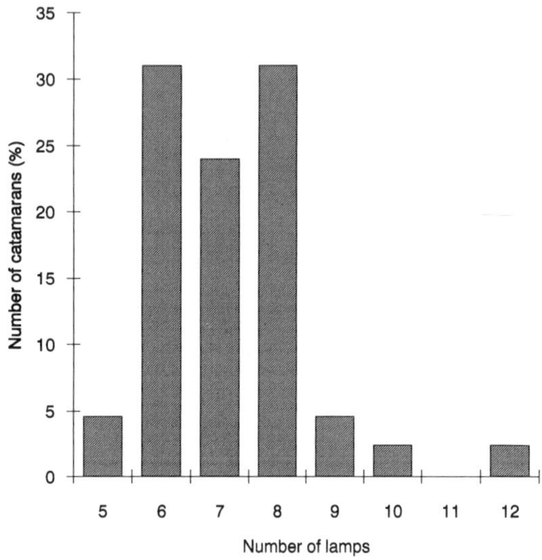

Fig. 21.4 Number of lamps used per catamaran (1990).

Generally, lift nets have 10 mm stretched meshes (Fig. 21.5), except for the purse itself which has 8 mm stretched meshes. The lift net is usually 22 m high and the opening is a square 10 m wide (Table 21.1). According to the fishers, the size of lift nets has recently and progressively increased up to a maximum compatible with the necessity of quickly raising the net during fishing operations.

Until 1980, lift nets were usually set at a depth of 30–35 m, enough for a quick catch of sardines; the main depth of operation is now at least 80 m (Fig. 21.6), so as to catch small *L. stappersi* (70–170 mm long), locally known as nyamu-nyamu (Coulter, 1970). Larger predators are seldom caught, because they can escape quickly from the lift net, while it is being raised. Figure 21.7 shows the difference in catchability of predators between artisanal and industrial fisheries.

Present situation of the pelagic fisheries

Total catch and catch per unit effort (CPUE)

Figure 21.8 shows the variation in the relative importance of the artisanal and industrial sectors in the total catch of Burundi. It shows clearly that, since 1980, the industrial fishery, long considered the main fishery of Burundi, has become of secondary importance.

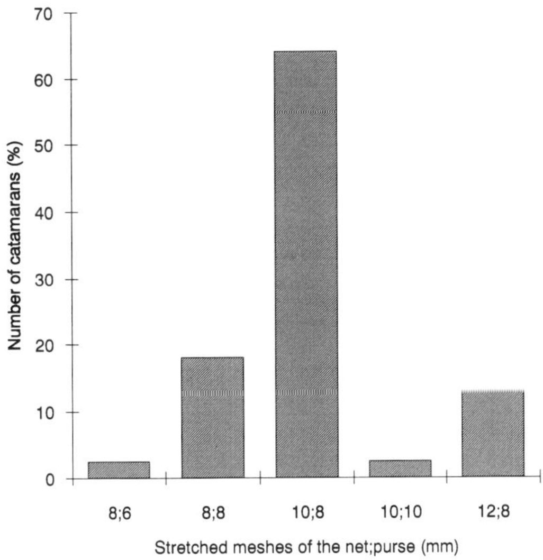

Fig. 21.5 Stretched meshes used by catamarans in Burundi (1990).

Because of improvement to their fishing gear, artisanal fishers are now able to exploit all the Burundi sector of Lake Tanganyika (Petit, 1990; Petit and Kiyuku, 1990). To some extent, this can explain the present crisis in the industrial fishery. The total catch of the industrial fishery decreased from 6000 t (tonnes) in 1980 to 2000 t in 1988 and 1990. Simultaneously, the artisanal fishery's catch increased (Fig. 21.8). Industrial fishers could catch 12.5 boxes per night of big *L. stappersi* (mukeke) in 1980 (around 500 kg)

Table 21.1 Main sizes of lift nets used in Burundi by the artisanal fishery. Number of units is expressed as percentage of the total number. 'Others' refers to net sizes unusual in Burundi, as used in Tanzania or Zaïre

Lift net sizes: upper width (m) × height (m)	% of units
8– 9 × 16–21	5
10–12 × 12–16	16
10–12 × 18–25	51
16–20 × 20–35	12
10–20 × 50–80	9
Others	7

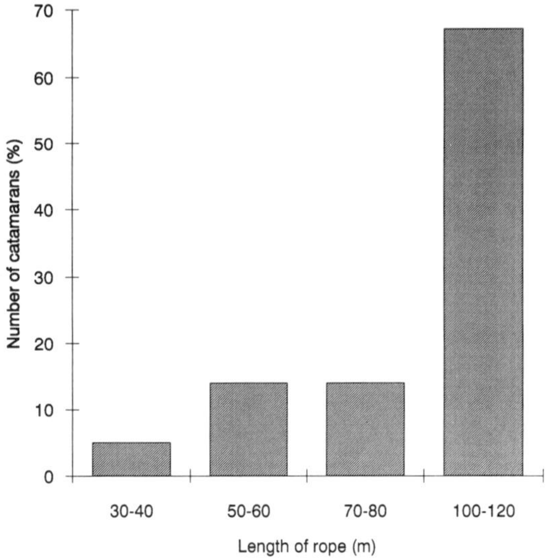

Fig. 21.6 Length of ropes used in Burundi artisanal fishery (1990).

but their catch per night decreased to only 2 boxes in 1990 (around 100 kg) (Fig. 21.9). Simultaneously, the catch per unit of effort (one fishing unit for one year) of adult *L. stappersi* in the artisanal fishery has nearly doubled since 1981 (Fig. 21.10).

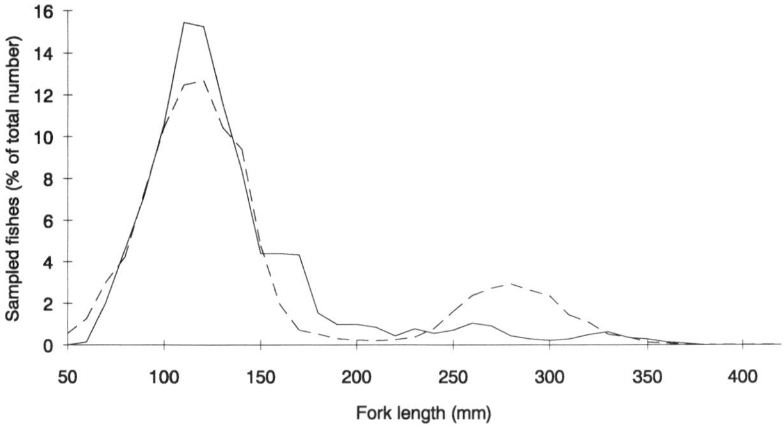

Fig. 21.7 Length frequencies of *Lates stappersi* caught by artisanal (solid curve) and industrial (broken curve) fisheries in Burundi, 1990.

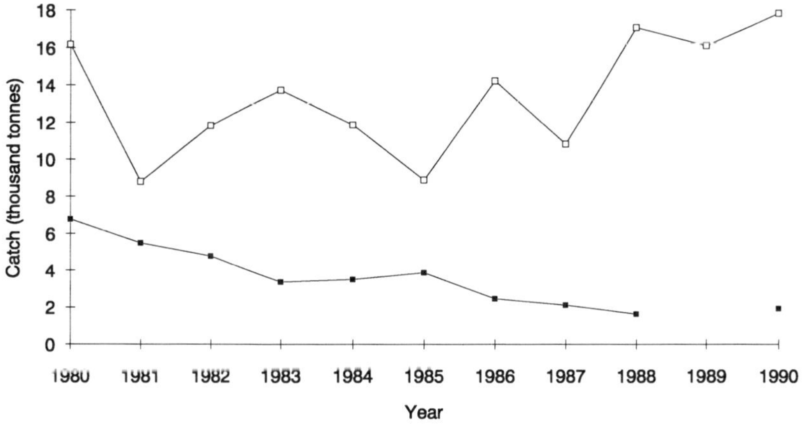

Fig. 21.8 Pelagic fisheries catches, Burundi, 1980–1990: □, artisanal fishery; ■, industrial fishery.

Impact of fishing effort on stocks

It is difficult to quantify the impact of the fisheries on the stocks, firstly because of the great variations in the annual artisanal catch and secondly because of large and sudden immigrations of predators into Burundi waters from the south.

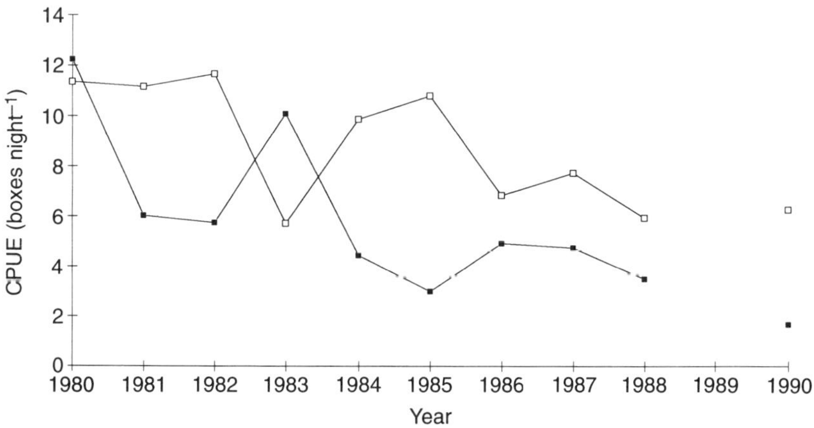

Fig. 21.9 Catch per unit of effort in the industrial fishery, 1980–1990: □, mukeke (adult *Lates stappersi*); ■, ndagala (mixed *Stolothrissa tanganicae*, *Limmothrissa miodon* and juvenile *Lates stappersi*).

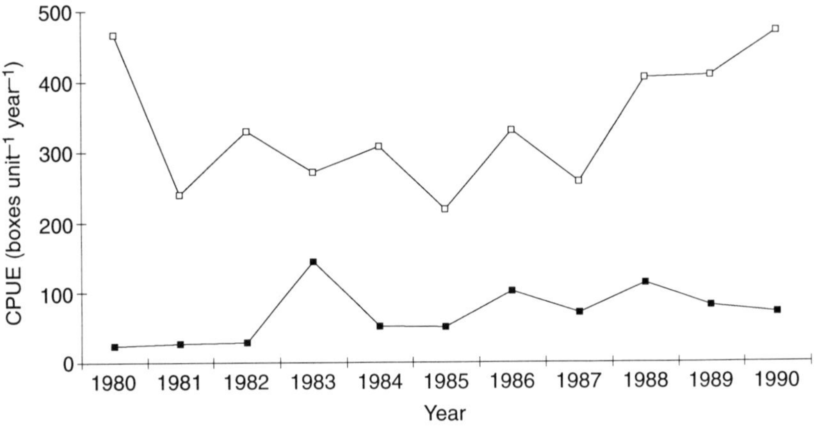

Fig. 21.10 Catch per unit of effort in the artisanal fishery, 1980–1990. Conventions as in Fig. 21.9.

We cannot explain why *L. stappersi* from 170 to 230 mm long are absent from the catch (Fig. 21.7) although many larger fish are present (Roest, 1988).

The very large purse seines likely have very low selectivity. We have used samples of this industrial fishery to estimate total mortality Z (Fig. 21.7). Z obtained for the younger fish (fork length between 100 and 170 mm) ranged from 7.3 to 8.3. These results are not compatible with the expected longevity of such a fish, which is more than 5 years (Lévêque *et al.*, 1977). The estimated Z cannot be assumed to reflect only mortality and we think, like Ellis (1978), that emigration may be an important factor.

Previous studies on the whole lake seem to show that sardines are rather sedentary: for example *Limnothrissa* populations display different demographic characteristics according to geographic zones (Henderson, 1976), and intensive fishing in one area seems to have no consequences in other ones (B. Nyakageni, fisheries officer, pers. comm.). Furthermore in the northern part of Lake Tanganyika, fishers of Zaïre, Burundi and Tanzania noticed that young *Stolothrissa*, which had been abundant close to the shore during the 1970s, disappeared progressively during the 1980s. According to the fishers, this disappearance of *Stolothrissa* from littoral areas occurred mainly where the artisanal fishery became very active.

21.4 CONCLUSION

During the 1980s, the pelagic fisheries of Burundi have been characterized by considerable changes, notably an increase in the relative importance of the artisanal fishery and the development of its fishing technology. At present, the industrial fishery is threatened with collapse.

It is important to evaluate the consequences of these changes on the sustainability of the fishery in the northern part of Lake Tanganyika because of the economic importance of the fish in the region.

ACKNOWLEDGEMENTS

We thank Dr T.J. Pitcher and the referees for their advice and we are grateful for the facilities provided by the Département des Pêches de la République du Burundi.

REFERENCES

Coulter, G.W. (1970) Population changes within a group of fish species in Lake Tanganyika following their exploitation. *J. Fish Biol.*, **2**, 329–53.

Ellis, C.M.A. (1978) Biology of *Luciolates stappersii* in Lake Tanganyika (Burundi). *Trans. Am. Fish. Soc.*, **107**, 557–66.

Henderson, H.F. (1976) Sur la grande taille de *Limnothrissa* dans les captures des pêcheries au filet tournant en Tanzanie. FAO, Rome, FAO/FI:DP/BDI/73/020/13, B, 5–10.

Lévêque, C., Durand, J.R. and Ecoutin, J.M. (1977) Relation entre le rapport P/B et la longévité des organismes. *Cah. ORSTOM, Sér. Hydrobiol.*, **11**, 17–31.

Petit, P. (1990) Place de la pêche artisanale dans les pêcheries du secteur nord du lac Tanganyika. ENSA, Institut National Polytechnique de Toulouse, 56 pp. (mimeo)

Petit, P. and Kiyuku, A. (1990) Les Pêcheries du Secteur Burundais du Lac Tanganyika. Evolution des captures pour les espèces principales entre 1980 et 1988. Contribution Laboratoire d'Ichtyologie de l'ENSA Toulouse/Dépt. des Pêches du Burundi, 7 pp.

Roest, F.C. (1988) Predator–prey relations in northern Lake Tanganyika and fluctuations in the pelagic fish stocks, in *Predator–prey relationships, population dynamics and fisheries productivities of large African Lakes* (ed. D. Lewis), FAO, Rome, *CIFA Occ. Pap.*, no. 15, pp. 104–29.

Part Four

Overviews of Fish Introductions in African Lakes

Chapter twenty-two

Genetic impacts of fish introductions: a perspective on African lakes

Gary R. Carvalho and Lorenz Hauser

22.1 INTRODUCTION

Preview

Human activities have had profound impacts on almost all freshwater ecosystems in the world. These activities include mechanical changes such as damming or drainage as well as chemical effects like eutrophication and pollution, all with some effects on fish communities. Against this background of massive, worldwide and direct human impacts on freshwater ecosystems, the effects of fish introductions may seem to have only very modest effects on fish communities (Fernando, 1991). However, the increasing concern about the loss of global biodiversity and the recent mass extinction of haplochromine cichlids in Lake Victoria associated with the introduction of the piscivorous Nile perch (Witte *et al.*, 1992; Chapters 2–11) has led to renewed research efforts on the effects of species introductions, especially in relation to commercial exploitation.

Major efforts have so far been directed towards examining the effects of fish introductions on species diversity and ecosystem structure (e.g. Zaret and Paine, 1973; Witte *et al.*, 1992). Effects on genetic diversity, especially at the intraspecific level, have only recently received specific attention (Soulé, 1986; Oldfield, 1989; Seitz and Loeschcke, 1991; Loeschke *et al.*, 1994). This review

The Impact of Species Changes in African Lakes. Edited by Tony J. Pitcher and Paul J.B. Hart. Published in 1995 by Chapman & Hall, London. ISBN 0 412 55050 4.

focuses on genetic aspects of species changes, particularly those leading to a loss of genetic diversity resulting from fish introductions. With special consideration of African lakes, we concentrate on changes in the intraspecific component of genetic diversity, typically the domain of population genetics; the decline in species diversity in African fresh waters has been the subject of many other reviews (e.g. Witte *et al.*, 1992). The aims here are to identify those species changes that result from introductions, and briefly to consider the tools available to detect their consequences in relation to management.

Biodiversity and fish resources: a genetic perspective

Components of biodiversity

Although the concept of biodiversity is not new (Elton and Miller, 1954; Elton, 1958), it is only in recent years that it has become central to the management of natural resources (IUCN *et al.*, unpubl. data; ODA, 1991; Solbrig, 1991a,b; Loeschke *et al.*, 1994), especially in the tropics. The continuous loss of animal and plant species, and a reduction in the abundance of many more species, has necessitated the development of a unified global approach to describing and monitoring the components of biodiversity. The value of such a concept in empirical terms does, however, depend largely on the provision of an unambiguous meaning, together with an appreciation of how natural and artificial changes in the environment may modify its components. A broad definition of biodiversity is that it encompasses 'the variety of plant and animal species on our planet'. A more valuable working definition (ODA, 1991) which shifts attention to the interacting hierarchical levels of organic diversity, is that biodiversity describes: (1) the genetic variation present within species (and the number of individuals within a species), (2) the variety of species within a habitat or ecosystem, and (3) the variety of habitats on Earth. Implicit in such a definition is that the diversity found within species is the ultimate source of biodiversity at higher levels, and that any changes in the amount (heterozygosity and genetic polymorphism) or distribution (within versus between populations) of genetic variability are likely to modify the interaction between a species and its environment. The nature and extent of such interactions can ultimately determine the abundance and persistence of a particular species, and so are an integral focus of attention for studies on biodiversity (Solbrig, 1991a).

Most interest has focused on the species diversity component of biodiversity because it is usually easier to distinguish species than populations. The usual rarity of obvious discriminative features for separating populations has caused major problems both in describing and in managing the intraspecific component of biodiversity.

Genetic changes and fish resources

When applying the concept of biodiversity to the management of fish it becomes imperative to consider genetic issues; a sustainable resource for the future must be ensured. Perpetuation of a natural resource requires continuity over time which necessarily involves a study of the hereditary mechanism and microevolutionary forces (Carvalho and Pitcher, 1994a). Because differential survival and reproduction of genotypes may affect the abundance and biological characteristics of a fish population, especially in changing environments, short-term benefits through heavy exploitation, for example, may lead to an irreplaceable loss of biodiversity (Nelson and Soulé, 1987). Indeed, the notion that fish constitute a renewable resource is a dangerous one, and is true only in part. The loss of a locally adapted population or an endemic species clearly cannot be reversed, though a diminution in genetic diversity can, at least in principle, be enhanced through carefully designed supplementary stocking programmes (Krueger *et al.*, 1981). A decrease in genetic variability within a population can, in the absence of migration, only be compensated through mutations on an evolutionary time scale. Such reductions in genetic diversity may render a population less able to adapt to artificial or natural changes in the environment, and thus more likely to undergo severe population fluctuations or eventual extinction (Nelson and Soulé, 1987; Carvalho, 1993). By failing to distinguish between reversible and irreversible genetic changes in fisheries management, dramatic species changes have sometimes occurred, which in the extreme have led to the loss of entire species flocks (Ribbink, 1987; Witte *et al.*, 1992). More common consequences, however, include a reduced adaptation to localized conditions (Hindar *et al.*, 1991; Carvalho, 1993) and changes in life history characters (Ricker, 1972).

Practices such as exploitation (Ward and Grewe, 1994), supplementary stocking (Ferguson, 1994) and introductions may modify a species' gene pool by exposing genotypes to new selection pressures, or through the interbreeding of previously isolated reproductive units. The effects of the latter may be especially rapid and far-reaching, not only through affecting individual populations, but also through interactions between populations, and as a consequence, the balance between species (Seitz, 1991).

22.2 FISH INTRODUCTIONS IN AFRICA: DEFINITIONS, PURPOSES AND EXTENT

Introduced species include any species intentionally or accidentally transported and released by humans into an environment where it was previously absent. This definition of introduction includes those fish moved to outside their geographic range, as well as so-called 'transfers' or translocations where a species is moved within its geographic range (Shafland and Lewis, 1984). Supplementary stocking is sometimes included in the definition of species introductions, but because at present it does not play a major role in African lakes it will not be considered here. Implicit in the above broad definition is the fact that an introduced species will now inhabit an environment which is different in some way from the native habitat. Introduced populations and resident species may thus be exposed to new selection pressures, as well as to a new source of potential mates if a related species is already present. For example, marked morphological changes have been recorded in Nile perch since its introduction to Lake Victoria (Harrison, 1991); these changes may in part be due to different environmental conditions.

There are several reasons for introductions, including recreation, aquaculture, improving the productivity of natural waters and control of undesirable organisms (Welcomme, 1988). In addition to enhancing fishery production, many species have been transferred within (e.g. tilapiine species) and introduced into Africa (e.g. rainbow trout and Chinese carps) for aquaculture, as well as for sport fisheries (e.g. trout and Nile perch; Ogutu-Ohwayo, 1992). Attempts at biological control of malarial and bilharzia vectors and water weeds through fish predation have further enhanced the number of introductions, including such species as *Gambusia affinis*, *Poecilia reticulata* and *Ctenopharyngodon idella* (FAO/CIFA, 1985).

A synopsis of worldwide fish introductions (Welcomme, 1988) documents a total of 1354 introductions of 237 species into 140 countries dating from the middle of the 19th century. Africa has been perhaps less affected overall by introductions than other continents, and may be viewed as a net exporter of species. Nevertheless, numerous fish introductions, many involving non-native species, have been made in Africa (FAO/CIFA, 1985), with both beneficial and undesirable consequences.

Approximately 50 fish species have been introduced or transferred within Africa (Table 22.1), though many species did not establish themselves. Despite the relatively small number compared with other continents (Welcomme, 1988), there has been increasing concern about the implications of fish introductions, especially within the context of their effects on biodiversity (FAO/CIFA, 1985; ODA, 1991; Ogutu-Ohwayo, 1992). Because the primary aim of African fish introductions has been to maintain or increase yields, it has proven particularly difficult to balance the short-term benefits of increased food production with long-term conservation of the intra- and interspecific components of biodiversity. For example, the important commercial fisheries of the tilapias *Oreochromis esculentus* and *O. variabilis* in Lakes Victoria and Kyoga were severely depressed by overfishing by the 1960s. Furthermore, the important riverine species, *Labeo victorianus* in the Lake Victoria basin, and *L. altivelis* in River Lualaba were depleted, resulting in a call for

Table 22.1 Reported fish introductions in Africa (after Ogutu-Ohwayo, 1992). Figures in parentheses show the number of countries to which each species has been introduced. The reasons for the introductions are given as: aquaculture (72), sport (angling and recreation) (26), to create fishery (23), forage for other fish (9), control weeds (6), control mosquitos (6), control snails (2), ornamental purposes (1), accidental transfer (1), no reason given (1)

Astatoreochromis alluaudi (5)	*Heterotis niloticus* (6)	*Oreochromis hornorum* (1)
Astronotus ocellatus (1)	*Hypophthalmichthys molitrix* (3)	*Oreochromis leucostictus* (3)
Bagrus meridionalis (2)	*Ictalurus punctatus* (1)	*Oreochromis macrochir* (13)
Barbus barbus (1)	*Labeo rohita* (1)	*Oreochromis mossambicus* (13)
Barbus kimberlevensis (1)	*Lates niloticus* (2)	*Oreochromis niloticus* (11)
Barbus natalensis (1)	*Lepomis cyanellus* (8)	*Oreochromis shiranus* (1)
Basilichthys bonarienses (1)	*Lepomis gibbossus* (2)	*Oreochromis spilurus* (6)
Carassius carassius (1)	*Lepomis macrochirus* (8)	*Osphronemus gourami* (3)
Carassius auratus (3)	*Lepomis microlophus* (2)	*Poecilia reticulata* (3)
Catla catla (2)	*Limnothrissa miodon* (5)	*Salmo trutta* (7)
Cichla ocellaris (1)	*Micropterus dolomieu* (2)	*Salvelinus fontinalis* (2)
Clarias lazera (5)	*Micropterus punctulatus* (2)	*Serranochromis robustus* (1)
Ctenopharyngodon idella (6)	*Micropterus salmoides* (15)	*Stolothrissa tanganicae* (5)
Cyprinus carpio (16)	*Oncorhynchus mykiss* (16)	*Tilapia rendalli* (12)
Esox lucius (3)	*Oreochromis andersoni* (1)	*Tiapia zillii* (7)
Etroplus suratensis (1)	*Oreochromis aureus* (1)	*Tinca tinca* (4)
Gambusia affinis (7)	*Oreochromis esculentus* (1)	*Xiphophorus maculatus* (1)

introductions of alien species into the lakes (Worthington, 1973). At an early stage there was resistance to such measures (Fryer, 1960, 1972), especially because some of the lakes contained the world's most diverse endemic fish fauna (Fryer and Iles, 1972). Nevertheless, fish introductions proceeded, including the much-debated introduction of Nile perch into Lake Victoria (Barlow and Lisle, 1987; Chapters 2–11).

Impacts of fish introductions include predation, competition, introduction of diseases, genetic changes in native fish, and disruption and degradation of the environment (Welcomme, 1984). All such impacts have been recorded in Africa (FAO/CIFA, 1985). The introduction of Nile perch, *Lates niloticus*, into Lake Victoria has caused global concern because of the associated extinction of many endemic cichlids (Barel *et al.*, 1985; Balon and Bruton, 1986; Witte *et al.*, 1992). The cichlid species within Lake Victoria have not co-evolved with the Nile perch which may explain their vulnerability to *L. niloticus* predation. However, the compatibility of cichlids and large centropomid predators is seen by the coexistence of several native *Lates* species and a diverse cichlid species flock in Lake Tanganyika (Coulter, 1970, 1976; Coulter *et al.*, 1986).

Here, we concentrate on the genetic consequences of introductions for both introduced and native fish populations, primarily within the African Great Lakes. Before considering these, however, it is appropriate to examine some pertinent characteristics of African fresh waters in relation to fish introductions.

22.3 FISH INTRODUCTIONS IN THE AFRICAN GREAT LAKES: SOME SPECIAL CHARACTERISTICS

Several features of the African Great Lakes render them particularly susceptible to irreversible genetic changes resulting from introductions. First, the Great Lakes are typically very large: Lake Victoria for example is the largest freshwater body in the tropics, with a surface area of 68 680 km^2, a mean depth of 40 m and a maximum depth of 80 m. The lake crosses national borders, with 51% in Tanzania, 43% in Uganda and 6% in Kenya (Acere, 1985), which in itself has produced serious delays in the implementation of a common management strategy regarding introductions (Ogutu-Ohwayo, 1992). It has thus been difficult to monitor the consequences of introductions, or to develop protocols for future planned releases.

Second, within such expansive water bodies are a highly diverse and complex fish fauna, with a high degree of endemism especially among the cichlids (Fryer and Iles, 1972). Although several lakes throughout the world support species flocks, the diversity of species within the African Great Lakes far exceeds the number of endemics recorded from elsewhere (Fig. 22.1). Such high species diversity and marked endemism in mature lake communities

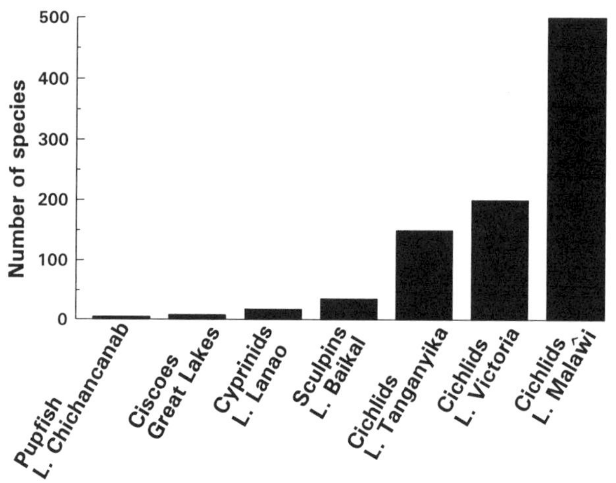

Fig. 22.1 Number of endemic species in species flocks from several lakes. Redrawn from Miller (1989). The Great Lakes referred to are in North America.

renders them particularly susceptible to perturbations, especially any rapid changes caused by humans, such as fish introductions which typically affect several trophic levels (Zaret and Paine, 1973). Stability has been essential for the evolution of species communities (Fryer, 1984), and once disrupted, the complex ecosystem dynamics and species interactions may recover only slowly, if at all (Margalef, 1968; May, 1975). Further, the loss of an endemic cannot be reversed by recruitment from elsewhere. Indeed, the African Great Lakes may be regarded as inverted islands, with their faunas having the same vulnerability to extinction as island faunas (Ribbink, 1987; Bruton, 1990).

Third, there is a preponderance of cichlids in the African Great Lakes. Despite the adaptive radiation occurring within this group (Fryer and Iles, 1972), many species are very similar and hybridize readily where previously allopatric species have come into contact (Ogutu-Ohwayo and Hecky, 1991), producing genetic changes in the surviving species. Many cichlids have a wide variety of trophic specializations (Fryer and Iles, 1972; Coulter *et al.*, 1986), but a narrow range of reproductive guilds (*sensu* Balon, 1975; Bruton, 1990). Most cichlids are mouthbrooders and invest much parental care into few offspring (Ribbink, 1987), which increases their competitive abilities in stable, predictable environments, but because of slow population growth, they are especially vulnerable to overexploitation (Coulter *et al.*, 1986; Ribbink, 1987), both by humans and by introduced predators. A further decrease in naturally small populations (Ribbink, 1987) typically leads to a loss of genetic diversity as described later. The similarity in ecological requirements of some tilapiines (Trewavas, 1983) may lead rapidly to competitive exclusion and the elimination of one or more species following an introduction. Thus, the fragile reproductive barriers and ecological requirements of such fish are particularly susceptible to introductions, so producing marked species changes.

22.4 GENETIC IMPACTS OF FISH INTRODUCTIONS

Genetic studies on the impact of introductions in Africa are scarce (but see Carvalho and Pitcher, 1994b; Hauser *et al.*, 1995), and to illustrate certain general points, examples will be drawn from elsewhere. Genetic impacts may be broadly divided into two groups: direct and indirect (Fig. 22.2). Direct impacts include those that operate on a species by initiating changes in gene flow (hybridization and introgression). Indirect effects are primarily caused by a reduction in population size, either in the introduced fish through the release of a small number of founders, or in the indigenous species through ecological processes such as competition, predation or new diseases or

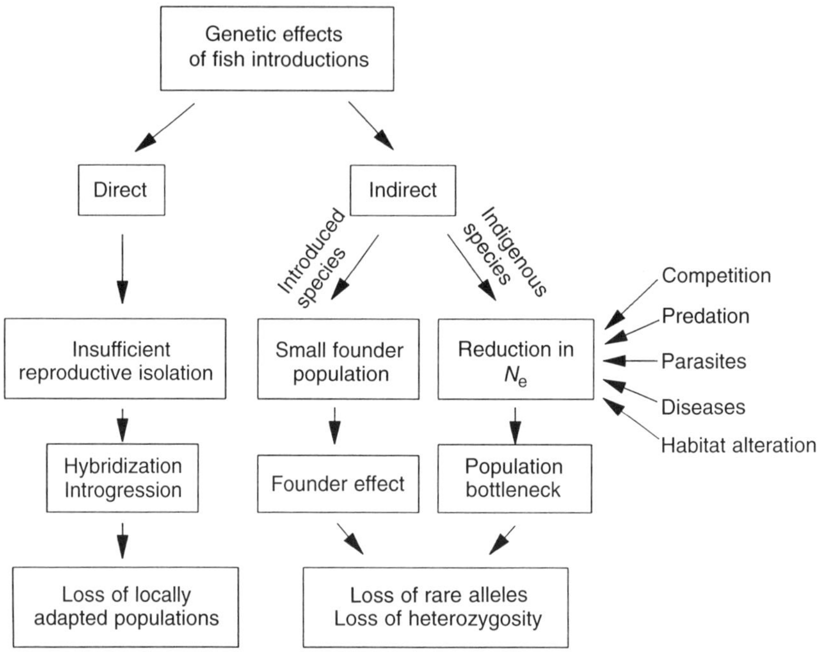

Fig. 22.2 Genetic effects of fish introductions. Introductions can affect both (i) indigenous species, through reducing the effective population size (N_e) and allowing hybridization between genetically distinct populations or species, and (ii) introduced species, through hybridization and founder effects. Irrespective of the genetic mechanisms involved, a decay in genetic diversity or loss of locally adapted populations may lead to a reduction in individual fitness (viability and fecundity) and thus the evolutionary potential of a species. The rate of population growth following a founder event or population bottleneck will affect the overall reduction in genetic diversity. See text for further details.

parasites. Such genetic effects may lead to a loss of locally adapted populations and genetic diversity through the microevolutionary forces described below.

Direct effects – hybridization and introgression

Hybridization and introgression can occur between species as well as between populations of the same species. The effects are very similar and are thus discussed together.

If reproductive isolation between two species or populations is maintained by geographic separation, and they are brought together by humans or natural causes, they may hybridize. In some cases the hybrids are sterile, but where sufficient genetic compatibility exists, introgression may occur, that is, the incorporation of genes of one species into the gene pool of another. Useful characteristics of the native stock or species for a fishery or aquaculture may be lost through such crosses, and only restored through time-consuming genetic selection programmes.

Hybridization can lead to new and unexpected ecological tolerances, leading to a spread of the introduced species to ecosystems previously unsuitable for them. In Australia for example, two strains of the common carp, *Cyprinus carpio*, were introduced and hybridized to give rise to the new, more vigorous and ecologically more tolerant 'Boolara' strain, which became far more widespread and problematic than the parental stocks (Arthington, 1991). For the same reason, interspecific hybrids between *Oreochromis mossambicus*, *O. hornorum* and *O. niloticus* occurring in North-East Australia (Bluhdorn *et al.*, 1990) give cause for concern (Arthington, 1991; Mather and Arthington, 1991).

Usually, however, hybridization and introgression lead to a reduction in fitness (Gregorius, 1991; Carvalho, 1993), through a reduction either in survivorship or in fecundity; the 'hybrid vigour' known to aquaculturists and plant breeders refers to hybrids between domesticated, highly inbred strains (FAO/CIFA, 1985). In natural populations, however, the number of parental lines is usually sufficiently large to avoid inbreeding with a subsequent loss of fitness. Even in hybrids between inbred lines, the benefits may be rather short-term. Initially the F_1 hybrids (that is, hybrids of the first generation) may exhibit increased fitness, as coadapted gene complexes of the parental groups remain intact, providing substantial benefit to the progeny (Krueger and May, 1991; Carvalho, 1993). However, recombination in subsequent generations will disrupt these gene complexes and cause fitness to drop below that of either parental group, a process known as outbreeding depression (Emlen, 1991; Krueger and May, 1991).

The total amount of genetic variability is adversely affected by hybridization in any case. Generally, the effect of gene exchange between previously isolated groups is to increase the variance within groups, decrease the variance between groups, and decrease the total genetic variance (Nelson and Soulé, 1987). Hybridization often means that unique qualities of individual species or populations are lost and are not available, for example, as sources of 'pure' strains for aquaculture, a problem well recognized in the tilapiine cichlids of Africa (FAO/CIFA, 1985). From a more scientific viewpoint, it is unfortunate that due to the mixing of gene pools within this group resulting from introductions and farm escapees, elucidation of the evolutionary relationships between tilapiine species may now be impossible (FAO/CIFA, 1985).

In practice, it may be difficult to separate the effects of hybridization from competition following an introduction (Ogutu-Ohwayo and Hecky, 1991). For example, *Oreochromis spilurus* was an abundant introduced species in

Lake Naivasha (Kenya) during the 1950s and 1960s (Siddiqui, 1977; Chapter 14, Table 14.1). In 1956, *O. leucostictus* was introduced and hybridized with *O. spilurus*, resulting in the subsequent disappearance of *O. spilurus*, and later also of the hybrids. Siddiqui (1977) attributes the disappearance of *O. spilurus* to the loss of spawning sites due to a rising lake level, but competition with *O. leucostictus* may also have been involved (Ogutu-Ohwayo and Hecky, 1991). Similarly, in Lakes Victoria and Kyoga, the endemic *O. esculentus* and *O. variabilis* disappeared, probably due to hybridization and competition with the introduced *O. niloticus* and *Tilapia zillii* (Ogutu-Ohwayo and Hecky, 1991). In the absence of any genetic investigations, the relative role of hybridization and competition in eliminating the indigenous species cannot be evaluated. Also, the extent to which introgression may have affected the genetic integrity of the surviving introduced species is not clear. An urgent need exists to undertake genetic studies to examine the extent and implications of hybridizations resulting from introductions.

If the introduced fish have been reared in a hatchery, possibly for many generations, the effects of hybridizations on the genetics of indigenous species or population may be even more severe. Most experience with effects of domesticated fish on wild populations stems from salmonids in North America and Europe (Berst and Simon, 1981; Billington and Hebert, 1991a; Ferguson and Thorpe, 1991; Hindar *et al.*, 1991; Taylor, 1991). However, the problem of releasing domesticated strains may also become important in Africa because of the increase in aquaculture, which inevitably leads to more farm escapees.

The main difference between introducing wild and domesticated fish is that the genetic constitution of the latter may have been changed by domestication and selective breeding, with a loss of genetic diversity due to inbreeding and genetic drift (Allendorf and Ryman, 1987; Garcia-Marin *et al.*, 1991). Such artificially reared fish often have extremely high post-release mortalities (Waples, 1991), but in many cases the survival is high enough to allow hybridization with the wild population, leading to a dilution of locally adapted gene pools and thus a reduction in population performance (Nelson and Soulé, 1987; Hindar *et al.*, 1991). For example, Reisenbichler and McIntyre (1977) demonstrated decreased survival rates in hybrids between wild and hatchery-bred steelhead trout (*Oncorhynchus mykiss*) and predicted an overall reduction in reproductive success in populations contaminated with hatchery fish.

A problem that is not yet serious, but may become so with advances in genetic engineering, is the release or escape of transgenic fish, i.e. fish in which novel genes have been inserted using gene technology. Although hatchery fish will also have been genetically altered over many generations, new molecular techniques make the integration of genetic information from very distantly related organisms into fish possible, allowing the production of genotypes that no previous technique could have produced (Tiedje *et al.*,

1989). It should shortly be possible to improve single performance traits by genetic engineering (Kapuscinscy and Hallerman, 1991) without the need to maintain populations for generations of selective breeding. However, the production of transgenic fish is perhaps better described as genetic 'tinkering' rather than engineering, because it is currently impossible to predict or explain the effects of inserted genes on the phenotype (Kapuscinscy and Hallerman, 1991). The possibility of introducing such genes into the gene pool of wild populations opens a new area of concern regarding the conservation of genetic resources. Policies controlling the production, rearing and release of transgenic fish have been formulated in North America (Kapuscinscy and Hallerman, 1991), and are now required in Africa, before releases of transgenic fish cause unpredictable damage to local gene pools.

Indirect effects – population bottlenecks

Ecological processes and reductions in population size

Successfully introduced fish often cause major changes in the abundance and species composition of the native fish fauna. Introduced predatory fish may reduce the population sizes of their native prey, often down to virtual disappearance. The best-known example for such an effect is the Nile perch, *Lates niloticus*, in Lake Victoria, which may have contributed to the disappearance of native haplochromine species (Ribbink, 1987; Ogutu-Ohwayo, 1990a; Chapters 2–11). The elimination of indigenous fish by introduced predators is a widespread phenomenon and well described in the literature (e.g. Jackson, 1960; Zaret and Paine, 1973; Siddiqui, 1977; Welcomme, 1984, 1988), and even small fish may have a profound effect on native species by eating their eggs and fry (Arthington, 1991). In addition, the introduced species may compete with indigenous species for food, spawning sites or other resources, again reducing the abundance of native fish (Welcomme, 1988), as seen commonly in the tilapiine cichlids (Siddiqui, 1977; Ogutu-Ohwayo, 1990b).

In addition to such direct ecological interactions, the indigenous fish community may also be affected by diseases and parasites introduced together with the fish (Hoffman, 1970; Arthington, 1991; Stewart, 1991). In Africa, the Asiatic ectoparasitic branchiuran, *Argulus japonicum*, was transferred to Lake Naivasha along with fish (Fryer, 1968) but fortunately does not seem to have persisted (Harper, pers. comm.). In Europe, North America and Australia, introduced diseases have sometimes eliminated native species and ruined aquaculture operations (Arthington, 1991; Stewart, 1991).

The introduction of an alien fish species can also have effects on the whole ecosystem, which can happen directly as in carp, the rooting behaviour of which may decrease water clarity and destroy beds of aquatic plants (Moyle *et al.*, 1986; Arthington, 1991), or indirectly through habitat alterations, especially by modifying trophic relationships. Zooplanktivorous fish can affect the nutritional status of a freshwater ecosystem by eliminating large zooplankton species and thus modifying the quality and quantity of the phytoplankton (Bartell, 1981; Vanni, 1986; Chapter 24). This seems to have occurred in Lake Kivu following the introduction of freshwater sardines (Chapters 13 and 24).

Predation, competition, diseases and ecosystem changes may cause the complete elimination of species or populations, which in turn causes an obvious and irreversible loss in biodiversity. However, even if the indigenous population escapes total extinction, its size may be drastically reduced, possibly resulting in a severe loss of genetic diversity within the population.

To appreciate the genetic consequences of such reduced population sizes, it is necessary to consider some population genetics theory before discussing examples in African lakes. To predict the loss of genetic variability in a natural population of a certain size, the concept of the 'effective population size' was developed.

Effective population size

Models on the effects of a small population size on genetic diversity necessarily assume ideal populations. In this context, the most important features of an ideal population are that the sex ratio is 1:1, the family size is randomly distributed among breeders, and each generation has the same number of individuals (Gall, 1987). Obviously these criteria are seldom met. Thus, to obtain valid predictions, the actual population size has to be converted to the 'effective population size' (N_e) according to the deviations from idealized conditions (Falconer, 1989). In simple terms, N_e can be seen as the size of an ideal population undergoing the same changes in genetic variability as the population observed.

In practice, N_e is very difficult to estimate for wild populations (Ryman *et al.*, 1981; Nelson and Soulé, 1987), but for a full appreciation of species changes resulting from a reduced population size it is important to consider how, and to what extent, N_e deviates from the actual numbers of fish observed in the field. Below, we emphasize in qualitative terms that N_e depends heavily on mating patterns and the lifetime reproductive investment of a species.

First, all juveniles and non-breeding adults do not contribute to N_e, an especially significant factor in highly fecund species with density-dependent recruitment; many fish may in fact be the progeny of very few adults (Cushing, 1973; Nelson and Soulé, 1987). N_e is decreased by an unequal sex ratio, which can be seen intuitively by considering a male spawning with several females (Frankel and Soulé, 1981). All the offspring of these females will be half- or full-siblings and much more closely related and genetically more homogeneous than had the females each mated with a different male. When a large proportion of the offspring are sired by only a few males, some of the original genetic variability will be lost.

N_e is also inversely related to the variance of the lifetime family size (Nelson and Soulé, 1987). A large variance in lifetime family size has a similar effect as skewed sex ratios; large families contribute a high proportion to the next generation. The lifetime family size is especially important in iteroparous fish (Nelson and Soulé, 1987) for two reasons. First, the enormous fecundity of many fish and associated density-independent mortality result in orders-of-magnitude variation in cohort strength (Cushing, 1973). Second, in iteroparous species, individuals surviving longer have more offspring than those dying shortly after maturity. In fish, this effect is enhanced by the fact that the fecundity increases with body size (Bagenal, 1973) and therefore with age. Thus a fish surviving twice as long after maturity has much more than twice the number of offspring.

If the size of the population varies largely over time, the overall N_e is more affected by the smaller generations than by the large ones. The breeding survivors of a population 'crash' contain only a sample of the original genetic variation; all future generations will have a corresponding deficit in genetic variability (Frankel and Soulé, 1981).

The combination of the above factors mean that, except in heavily managed hatchery populations, N_e is often an order of magnitude smaller than the census number of individuals (Nelson and Soulé, 1987). In practice, there is not much point in trying to estimate the N_e of natural populations, because usually too many of the required parameters are not known. In fact there are very few trustworthy estimates (e.g. Ryman *et al.*, 1981), and even these usually consider only some of the factors reducing N_e. N_e is mainly used for modelling genetic processes (Crow and Kimura, 1970; Frankel and Soulé, 1981; Falconer, 1989). The message here for managers is not to make judgements based on the number of fish counted, but rather to consider what proportion contributes to the gene pool. For natural populations, a feasible approach is to monitor changes in heterozygosity, or to compare the genetic diversity of undisturbed populations with that of the population in question. However, the large difference between N_e and census population size is an important point to consider in the following sections, where we deal with the theoretical effects of a reduced N_e on genetic variability.

Population bottlenecks

Periods of time when there is a reduced population size resulting in a loss of genetic variability are called population bottlenecks. In the following section we will illustrate and quantify this loss with as little mathematics as possible. For a more detailed description of the statistics and their derivation, the reader is referred to textbooks of theoretical population genetics (Crow and Kimura,

1970; Nei, 1987; Falconer, 1989) and its application to fish populations (Gall, 1987).

There are two main aspects of genetic diversity in bottlenecks: the qualitative and quantitative. The qualitative aspect considers the loss of rare alleles, whereas the quantitative considers both the number of alleles at a locus and their frequency. The distinction between qualitative and quantitative aspects is important because the loss of rare alleles has almost no effect on the heterozygosity, but nevertheless may be detrimental if the population has to adapt to new conditions.

At low population sizes, each generation can be seen as a small sample of the previous generation. Statistical theory shows that small samples are less representative than large ones because the sample mean may differ considerably from the population mean. Thus in small populations, the allele frequencies may differ greatly between generations, which leads to wide and random fluctuation of the allele frequencies in time, a process called random genetic drift. Under genetic drift, a population is likely to lose rare alleles just by chance – the larger the frequency of an allele, the larger the chance that it becomes fixed and all alternative alleles are lost (Fig. 22.3). Thus, in the long run, the prognosis for rare alleles is poor even at moderate population sizes, unless some form of selection elevates the fitness of the individuals carrying such genes. However, at small population sizes, selective forces have relatively

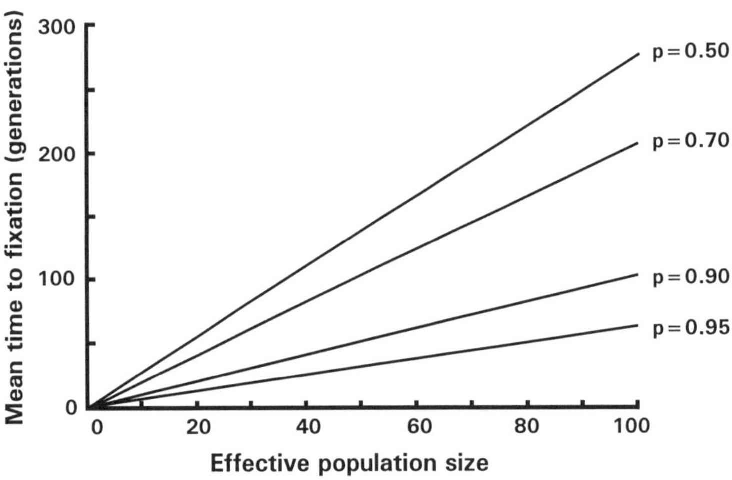

Fig. 22.3 Loss of rare alleles in population bottlenecks. The mean time to fixation is shown in relation to the effective population size (N_e) and the allele frequency (p). The higher the frequency of an allele, the faster it will become fixed (i.e. p = 1) with the loss of all other alleles at that locus. In this model, selection and mutation are not considered. See Frankel and Soulé (1981) for calculations.

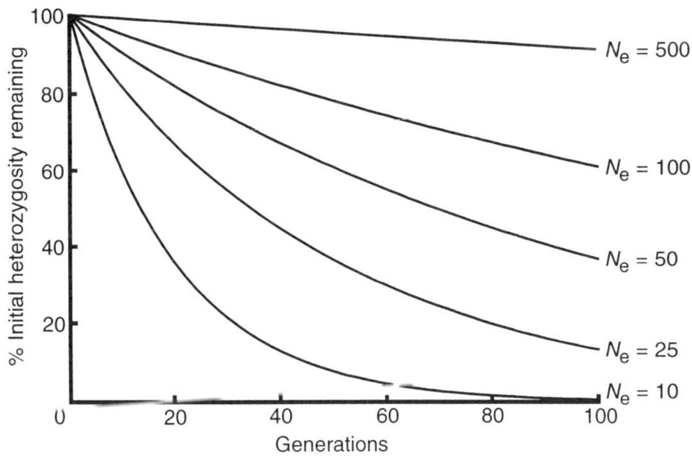

Fig. 22.4 Loss of heterozygosity in population bottlenecks. Heterozygosity is lost more rapidly in populations with smaller effective population size (N_e). It is assumed that there is no selection of mutation. See Frankel and Soulé (1981) for calculations.

little effect compared with genetic drift (Falconer, 1989), and thus even a large selective advantage may not secure the persistence of an allele. For the same reason, in small populations, deleterious alleles have roughly the same chance of being lost as beneficial ones (Frankel and Soulé, 1981). The fast and random loss of rare alleles may be detrimental to the adaptability of a population and thus the evolutionary potential of a species.

If a small population size persists for an extended period, even alleles with relatively high frequencies will be lost (Fig. 22.3), and the locus in question will become monomorphic. If this happens at progressively more loci, the quantitative aspect of genetic diversity, that is, the mean heterozygosity, will be affected as well. Clearly, the smaller the population size, the faster the decline in heterozygosity (Fig. 22.4). However, in the absence of selection and mutation, even the largest populations will slowly lose genetic variability. So, what is the minimum population size required to maintain genetic diversity?

Basic rule of conservation genetics

At present it is almost impossible to predict the reduction of genetic variability in real populations quantitatively; too many assumptions about effective population size, demographic and environmental stability, selection, mutation and migration have to be made. In addition, the interaction between different genes in the genome is usually not well understood. Such uncertainty means that only rough approximations of real situations are possible, which are useful to describe the processes leading to the loss of genetic variability, but insufficient for providing reliable recommendations of minimum population sizes. However, as an empirical rule of thumb used by animal breeders it has been postulated that, in the short term, selection can maintain genetic variability above an N_e of about 50 (Franklin, 1980; Frankel and Soulé, 1981; Simberloff, 1988).

This 'basic rule of conservation genetics' is the basis for calculating the absolute minimum N_e allowing the short-term preservation of genetic diversity. To protect the long-term adaptive potential of a population, however, more stringent rules are necessary. Long-term adaptability is especially important when considering the genetic effects of a reduced population size of a native species resulting from a fish introduction. Introduced fish, once successfully established in an ecosystem, are almost impossible to remove (Chapters 23, 24); the native fish may be kept at low population sizes for an extended period. Considering genetic drift, selection, mutation and gene interaction, Franklin (1980) suggests a minimum N_e of 500. It should be emphasized again at this point that this is an effective population size, and that a population of 5000 breeding individuals may be at risk of ultimately losing some of their genetic variability (Nelson and Soulé, 1987).

However, both these estimates of 50 and 500 individuals give only a rough order of magnitude of the minimum viable population size. The N_e required to maintain genetic diversity also depends on the traits studied; for quantitative traits (usually depending on a large number of loci) an N_e in the hundreds may be sufficient, whereas single-locus variation will be severely eroded if N_e is not in the hundreds of thousands (Lande and Barrowsclough, 1987). However, the evolutionary significance of single-locus variation is unclear, because many natural populations have sizes less than hundreds of thousands (Ribbink, 1987; Simberloff, 1988). The conclusion is that although it is very difficult to establish a minimum N_e for any given population, a useful strategy would be to monitor temporal and spatial patterns of genetic variability with one of the molecular methods described later (Nelson and Soulé, 1987).

Empirical examples and importance for African lakes

Although the theory of population bottlenecks is well developed (e.g. Nei *et al.*, 1975; Simberloff, 1988; Falconer, 1989) there are few emperical studies convincingly demonstrating its genetic effects. Usually an inference is made that if a population has low levels of genetic variability there must have been a bottleneck in the past (e.g. Bernatchez *et al.*, 1989), often without much supporting evidence. One example where it was possible to document a bottleneck was the elephant seal (Bonnell and Selander, 1974), which was hunted to near extinction in the 19th century and, probably as a consequence of the bottleneck, now shows no detectable polymorphism.

As discussed before, there is no threshold population size above which genetic variability can be eternally maintained and below which absolute homozygosity is inevitable. Rather, theory predicts that larger populations should have higher levels of genetic variability than smaller ones, other parameters like mutation rate, selection and population structure being equal. Indeed, such a relationship is seen across a wide range of taxa (Soulé, 1976; Kirpichnikov, 1992), indicating a strong correlation between population size and heterozygosity.

The correlation between population size and heterozygosity may render the cichlids in African lakes especially vulnerable to a loss of genetic variability. Many of the cichlid species in the African Great Lakes have small natural population sizes, estimated at fewer than 200 in some Lake Malaŵi cichlids (Ribbink, pers. comm.). Cichlids are often extremely stenotopic and thus the entire species is often restricted to tiny habitats (Ribbink *et al.*, 1983). Fishing pressure in most lakes, predation by the introduced Nile perch, *Lates niloticus*, and competition by the Nile tilapia, *Oreochromis niloticus*, in Lakes Victoria and Kyoga have reduced population sizes even further (Turner, 1977; Barel *et al.*, 1985; Hughes, 1986; Ogutu-Ohwayo, 1990b; Witte *et al.*, 1992) to levels where a loss of genetic diversity seems inevitable.

Prospects for the conservation of the intraspecific genetic diversity of the African cichlids, especially in Lake Victoria, are bleak (Ribbink, 1987). It may be that some species will adapt to the new predation pressure and their population sizes will recover. Such a recovery has happened, for example, in Gatún Lake, Panama, where the introduced predator *Cichla occelaris* reduced the population sizes of some indigenous fish species by more than 99% and caused the local extermination of 13 out of 17 species (Zaret, 1979). However, recently all but one of these species have recolonized the lake from adjacent refuges (Welcomme, 1988; pers. comm. in Witte *et al.*, 1992). Similarly, a recovery of some cichlids in Lake Victoria has been reported (Ogutu-Ohwayo, 1990b; Witte *et al.*, 1992). So far, however, population recovery in Lake Victoria cichlids has been restricted to shallow-water species (Witte *et al.*, 1992). The majority of species are still very rare or have totally disappeared from the catches. Even if these species are not completely exterminated, an extended bottleneck would reduce their genetic variability to levels sufficiently low to compromise their ecological adaptability and evolutionary potential. Genetic studies are needed to determine present levels of genetic diversity, followed by management measures ensuring a rapid increase in effective population sizes, either *in situ* by reducing the Nile perch population and closure of the trawl fishery (Bruton, 1990) or *ex situ* by captive propagation (Ribbink, 1987).

Founder effects

The same processes affecting native species after an introduction can also compromise the genetic variability of the introduced fish. In an introduction, only a relatively small sample of the source population is taken and transferred to a new site. Furthermore, probably only a proportion of the introduced fish survive and spawn in the new habitat. The founder effect can be seen as a one-generation bottleneck, and sometimes population bottlenecks and founder events are treated almost as synonyms (e.g. Nei *et al.*, 1975; Leberg, 1992). However, in contrast to population bottlenecks, the founder effect is an important mechanism in speciation (reviews: Barton and Charlesworth, 1984; Barton, 1989).

Similar to the events in bottlenecks, rare alleles are likely to be lost in founder events (Fig. 22.5; Frankel and Soulé, 1981). Rare alleles do not contribute much to the value of heterozygosity, but in a new environment the presence of certain allelic combinations may determine whether a species becomes established (FAO/UNEP, 1981).

In contrast to allelic diversity, heterozygosity is not strongly affected by a founder event, unless the founder population is extremely small (Fig. 22.6). Such a difference between allelic diversity and heterozygosity as indicators for founder events has been shown empirically in mosquitofish, *Gambusia holbrooki* (Leberg, 1992).

The effects shown in Figs 22.5 and 22.6 concern the loss of genetic variability in only one generation. However, even under ideal conditions it will take a few generations before the introduced fish establish a viable population. Events during the period of initial population growth can be viewed as a population bottleneck. Thus heterozygosity will decline to a certain minimum and then start to increase as the population becomes sufficiently large for mutations to accumulate (Fig. 22.7; Nei *et al.*, 1975). The faster the rate of population growth, the smaller the loss of heterozygosity after the founding event. Therefore, under ideal conditions for population growth, the minimum heterozygosity depends not only on the size of the initial founder population, but also on the intrinsic rate of population increase 'r' (Fig. 22.7).

Figure 22.7 also shows another important property of genetic heterozygosity; once it has been reduced to low levels it takes a long time to recover. In fact, the time in generations is roughly the reciprocal of the mutation rate (Nei *et al.*, 1975). Thus if the mutation rate is 10^{-8} per locus per generation, it takes 10^8 generations to reach the original level of heterozygosity. Such a time span is far beyond the time scale of fishery management and thus the minimum heterozygosity may well be regarded as the new heterozygosity in the introduced population. Figure 22.8 shows the percentage of the initial heterozygosity, which is retained after a founder event, and the subsequent

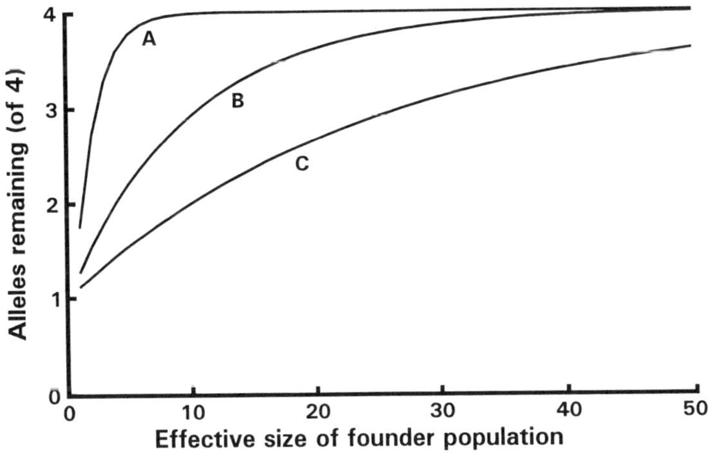

Fig. 22.5 Loss of rare alleles in founder events. The number of alleles remaining at a locus with initially four alleles is shown in the absence of selection and mutation. Initial allele frequencies are: A, four alleles with equal frequency ($p_1 = 0.25, p_2 = 0.25, p_3 = 0.25, p_4 = 0.25$); B, three moderately rare alleles ($p_1 = 0.85, p_2 = 0.05, p_3 = 0.05, p_4 = 0.05$); C, three rare alleles ($p_1 = 0.94, p_2 = 0.02, p_3 = 0.02, p_4 = 0.02$). In small founder populations, rare alleles have a high probability of being lost. See Frankel and Soulé (1981) for calculations.

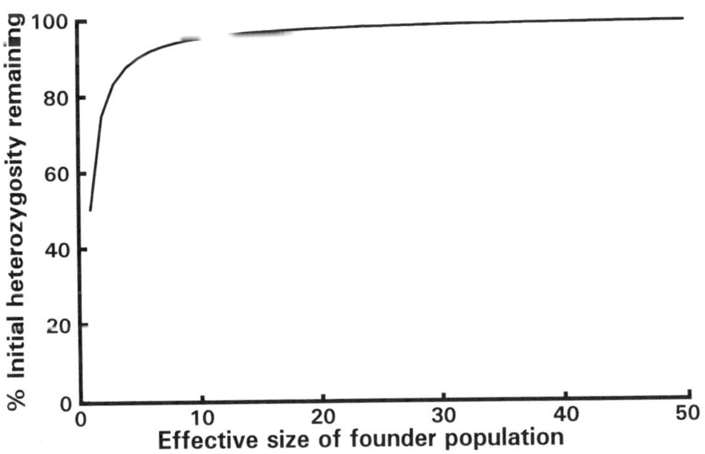

Fig. 22.6 Loss of heterozygosity in founder events with varying effective population size. Unless the founder population is very small most heterozygosity is retained. See Frankel and Soulé (1981) and Falconer (1989) for calculations.

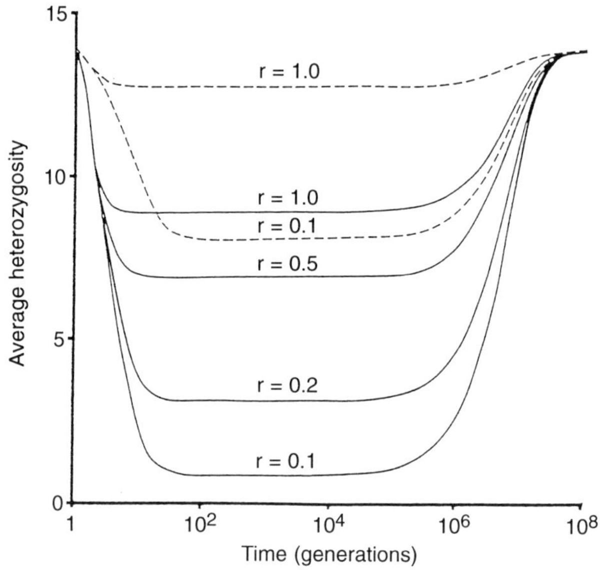

Fig. 22.7 Changes in average heterozygosity in a founder event. Solid curves refer to a founder population of 2, broken curves of 10 at generation 1. Population growth is assumed to be logistic with a rate of r. Initial and eventual population size is 4×10^6, the mutation rate is 10^{-8} per locus per generation. Heterozygosity drops after the founder event to a certain minimum because of the initially small population size; the level of this minimum depends on r. It takes a long time for heterozygosity to recover to the original level. Note the logarithmic scale of generations. Reproduced with permission from Nei *et al.* (1975).

bottleneck in relation to N_e of the founder population and the intrinsic rate of population growth. Again it should be emphasized that N_e of the founder population may be much smaller than the number of fish actually released. In addition to the factors determining N_e in native populations, N_e of introduced fish may be further reduced by a high initial mortality and low spawning success. Using estimates of survival and homing ability of pink salmon, *Oncorhynchus gorbuscha*, the number of fish surviving to breed following an introduction of 21 000 fry was estimated to be only 42 individuals (Gharrett and Thomason, 1987).

By definition, the intrinsic rate of population growth is only achieved under ideal conditions. However, the introduced fish may take some time to adapt to the new environment and may initially not achieve their potential population growth rate. Thus the period of small population size following the introduction will be extended, leading to a loss of genetic variability. A

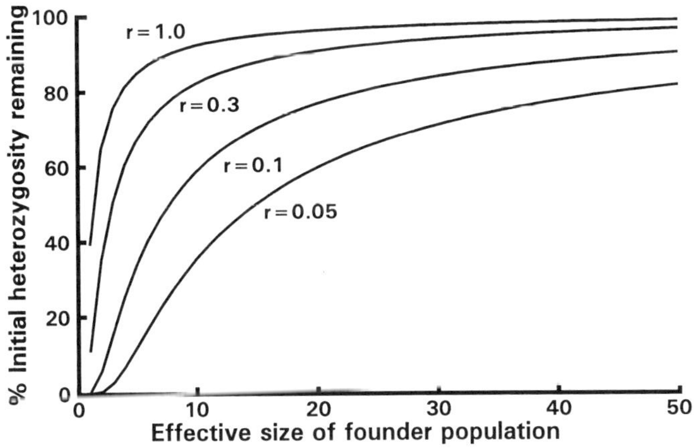

Fig. 22.8 Loss of heterozygosity in a founder event with a subsequent population bottleneck during population growth; r, intrinsic rate of population growth. At lower values of r, the population remains small for longer periods, resulting in a greater loss of heterozygosity. See Nei *et al.* (1975) for calculations.

good example for a slow increase in population size following an introduction is the Nile perch in Lake Victoria. Nile perch was introduced in the late 1950s and early 1960s, but significant numbers were caught only from the early 1980s onwards (Achieng, 1990; Chapters 2–11). The reason for this time lag is unclear, but it is likely to have caused a loss of heterozygosity, especially as only about 400 fish were originally introduced (Achieng, 1990).

In contrast to population bottlenecks there are numerous empirical studies on founder effects (e.g. Baker and Moeed, 1987; Knight *et al.*, 1987; St. Louis and Barlow, 1988; Leberg, 1992; Shaw *et al.*, 1992), probably because species introductions are usually better documented than periods of reduced population size, and also because founder events can be simulated easier and faster in laboratory experiments. All of these studies report a reduction in allelic diversity and often to a lesser degree also in heterozygosity, as predicted by theory. In a study on guppies (Shaw *et al.*, 1992), the size of the founder population was 200 fish. A significant deficiency in genetic diversity was found in the introduced population, which was in part probably due also to a bottleneck following the introduction. This study shows that a loss of genetic variability may occur even at census population sizes much larger than the N_e commonly used in models.

One of the few direct studies which examined the genetic impacts of fish introductions in African lakes was undertaken on the clupeid *Limnothrissa miodon* (Hauser *et al.*, 1995). After a transfer of 57 400 clupeid fry from Lake

Tanganyika to Lake Kivu in 1959 (Collart, 1960), one of the two clupeid species occurring in Lake Tanganyika, *L. miodon*, became established. As the transferred fish were not identified to species, the exact size of the founder population is uncertain. In 1993, samples were collected from founder and transplant populations to compare genetic diversity using allozymes and mitochondrial DNA (mtDNA) analysis.

Evidence from allozyme data showed clearly that there was no significant reduction in genetic diversity following the introduction. In contrast, there was a significant reduction in mtDNA haplotype diversity in three of the four samples from Lake Kivu. The apparent disparity between allozyme and mtDNA data is not unusual, and follows conventional theory on the consequences of a smaller (one-quarter) effective population size of the mtDNA genome, relative to nuclear DNA (Billington and Hebert, 1991b).

Summing up, the introduction of small numbers of fish may result in a significant loss of rare alleles, whereas the heterozygosity is initially less affected. There may, however, be considerable loss of heterozygosity in the period following the introduction. If the rate of population increase is slow, either because the species has a low intrinsic rate of population growth, or because of unfavourable environmental conditions, heterozygosity may be greatly reduced, with poor prospects for recovery on a human time scale.

22.5 METHODS AND THEIR APPLICATIONS

Several methods have been employed to investigate the structure of fish populations in relation to introductions, ranging from the use of behavioural and morphological traits, to molecular markers. Here we will concentrate on the most common techniques and discuss how they may be used to monitor and examine fish introductions. Detailed accounts on the methodology, limitations and applications of molecular techniques, including allozyme electrophoresis, mitochondrial DNA analysis and DNA restriction enzyme and sequence analysis can be found elsewhere (Richardson *et al.*, 1986; Utter *et al.*, 1987; Hillis and Moritz, 1990; Kornfield, 1991; Hoelzel, 1992; Carvalho and Pitcher, 1994a).

Morphometrics and meristics

Anatomical characters have traditionally been used in ichthyology and fisheries biology to describe various categories of fish taxonomy (Ihssen *et al.*, 1981). Their application, especially at the intraspecific level, however, is complicated by the fact that the phenotypic variation expressed is not directly under genetic control but is subjected to environmental modification. Nevertheless, their usefulness compared with other techniques such as allozyme electrophoresis has been shown in a number of studies (e.g. Roby *et al.*, 1991; Melvin *et al.*, 1992); thus the use of anatomical features is still warranted, especially in circumstances where simplicity and cost are important considerations.

The aim of using morphometrics is to reduce fish shape to a series of one-dimensional measurements that can detect subtle differences and yet exclude the effect of allometric relationships (Ihssen *et al.*, 1981). The effect of fish size often obscures interpopulation differences and has to be removed by statistical methods such as size ratios, regressions or multivariate analysis (Kuhry and Marcus, 1977; Winans, 1987; Edge *et al.*, 1991; Schaefer, 1991), including principal component, discriminant function or factor analysis (Ihssen *et al.*, 1981; Winans, 1987).

Meristic characters are serially repeated elements such as fin rays, vertebrae and gill rakers. They are thus represented on a discrete scale, which means that their discriminative abilities are likely to be weaker than those of morphometric features. Accordingly it has been suggested that morphometric and meristic characters should be analysed separately (Ihssen *et al.*, 1981).

Morphometrics and meristic characters are mainly used to distinguish species, races and populations, and to identify hybrids. For studies on fish introductions, the most likely use is in the detection of hybridization (Campton, 1987). Meristic and morphometric characters are counted and measured to characterize particular groups, and in the case of hybridization, comparisons are made between putative hybrids and the two hypothesized parental species. Morphological characters are, however, not uniformly intermediate in hybrids (e.g. Ross an Cavender, 1981; Leary *et al.*, 1983), but often individually resemble those of either parental species. It may therefore be necessary to consider all characters simultaneously. A well-documented example (Elder and Garrod, 1961) was the detection of natural hybrids between *Oreochromis spirulus nigra* and *O. leucosticta* in Lake Naivasha, Kenya using a combination of characters including coloration, fin ray counts, scale counts, number of gill rakers, teeth, jaw-, paired fin-, and spine-lengths together with body depth.

Although sensitive statistical methods may discriminate among morphological and meristic characters with high variance, it may often be necessary to use some other technique in combination, such as allozyme electrophoresis, to provide a less equivocal characterization of hybrids (Campton, 1987).

Allozyme electrophoresis

Allozyme electrophoresis is the movement of charged enzymes in an electric field (Harris and Hopkinson, 1976). The net electric charge, molecular size and shape determine the distance a protein moves in an electric field across a gel, that is, its 'electrophoretic mobility'. Some changes at the nucleotide level may bring about corresponding changes in net charge, size or shape at the protein level through amino acid substitutions, producing enzymes with different electrophoretic mobilities. Such genetically controlled multiple enzymes or allozymes (Markert and Møller, 1959) provide the genetic markers for examining population structure. Genotype and gene frequencies are calculated from an array of enzyme phenotypes visualized on a gel (Richardson *et al.*, 1986).

Although allozyme electrophoresis is generally less sensitive in detecting genetic variation than nucleic acid analysis, it remains the most tractable and useful molecular tool for many studies of population structure (Ward and Grewe, 1994). The value of data obtained does, however, depend on the careful interpretation of gel banding patterns (Richardson *et al.*, 1986; Utter, *et al.*, 1987). Allozymes not only describe population differentiation and temporal genetic changes, but they also provide valuable estimates of genetic diversity and indicators of hybridization and introgression. There are four major ways that allozymes may provide data relevant to fish introductions: in the description of population differentiation, in estimating levels of diversity in founder populations, in examining the genetic basis of phenotypic changes since introduction, and in the detection of hybridization.

First, it is important to determine whether most genetic diversity is contained within or between populations (Nelson and Soulé, 1987; Ferguson, 1994) before introducing a species into a new environment. Such information is important when designing a sampling programme for obtaining founder fish with the aim of providing colonists that are representative of a species' gene pool. If, for example, allozymes detect significant differentiation between populations, founder individuals for introduction should be obtained from several such populations. If, however, genetic evidence points to interpopulation homogeneity, then individuals collected from one population should suffice. Such considerations are important not only to maximize levels of genetic diversity in the introduced population (Nelson and Soulé, 1987), but also to assess the potential for local adaptation. Locally adapted populations may show a marked reduction in fitness in a new environment, especially if selection pressures are different (Nelson and Soulé, 1987; Taylor, 1991; Carvalho, 1993).

Second, allozymes can be used to estimate the level of genetic diversity in an introduced population to determine whether founder effects or genetic drift during subsequent population bottlenecks have caused a reduction in diversity (Gall, 1987; Leberg, 1992; Carvalho and Pitcher, 1994b; Hauser *et al.*, 1995). Allozymes can be used to compare genetic diversity in native and introduced populations, which if found to be significantly lower in the latter, may indicate the need for supplementary stocking. Such comparisons of genetic diversity can also indicate whether there is a need to modify the number of founders in future introductions.

Third, allozymes can indicate whether phenotypic differences detected between native and introduced populations are associated with genetic divergence. For example, the introduced population of kapenta, *Limnothrissa miodon*, in Lake Kariba shows significant morphological and life history differences from the source population in Lake Tanganyika (Marshall, 1993; Chapter 24). Allozyme studies have indicated, however, no evidence for genetic divergence of the introduced kapenta after 25 years (Carvalho and Pitcher, 1994b), suggesting that the previously described differences may be largely phenotypic in origin. In contrast, allozymes have detected marked genetic divergence in the vendace, *Coregonus albula*, in SE Norway (Vuorinen *et al.*, 1991) 92 years after introduction. Genetic divergence is not only determined by time of separation, but also by differences in selection pressures between environments, together with the level of genetic polymorphism. It is important, for example, to determine the genetic component of such traits as age of maturity and scope for growth, because these will, in part, determine the levels of exploitation that can be sustained and are therefore critical to management.

Finally, allozymes may be of value in detecting hybridization and introgression (Campton, 1987). Disclosure of natural hybridization using allozymes may be relatively straightforward if the parental species are completely fixed for different alleles at two or more loci. Individuals will be homozygous for different alleles in each of the parental species, whereas the F_1 hybrids will be heterozygous for these alleles at diagnostic loci (Brassington and Ferguson, 1976; Campton, 1987). It is also possible to determine hybrid fertility and second-generation hybridization by examining the occurrence of recombinant genotypes (Campton, 1987), information of vital importance when considering the long-term impact of a species introduction. The predicted reduction in the fitness of hybrids between distinct gene pools has been documented using allozymes (Philipp, 1991).

Where two species are fixed for different alleles at only one locus, it is not possible to distinguish F_1 hybrids from backcross or F_2 hybrids. Here, the presence of a heterozygous individual at the diagnostic locus can only provide evidence that the two species have interbred; it says nothing about the extent of hybridization. All individuals with one diagnostic locus will express either the F_1 heterozygous phenotype, or one of the two parental species regardless of the level of hybridization or backcrossing.

Hybridization is more difficult to detect where no fixed differences at the allozymic loci occur. Statistical methods can, however, use gene frequencies based on a comparison of genotypic distributions in the sampled population, with those predicted from conditions of random mating within one or both species (Campton, 1987).

The utility of the above applications, speed of processing large samples, and the relatively low cost and degree of molecular expertise required, still renders allozyme electrophoresis the first choice for many fishery biologists (Ryman and Utter, 1987). There are, however, some limitations (Richardson *et al.*, 1986; Hillis and Moritz, 1990), the most relevant of which relates to the ability of allozymes to detect genetic differences. Allozyme electrophoresis underestimates the amount of genetic variation within a taxon, perhaps by 30% or more (Ramshaw *et al.*, 1979), and in some animal groups, including haplochromines (Kornfield, 1991), has largely failed to detect enzyme polymorphism. Carefully interpreted mobility differences in allozymes indicate genetic differences, but the converse is not true; the presence of co-migrating bands on a gel does not necessarily indicate genetic homogeneity. Due to the evolutionary time scale for new alleles to accumulate in the nuclear genome (Nei, 1972; Thorpe, 1983), there has been a search to identify both highly sensitive methods for detecting genetic differences, as well as a source of more rapidly evolving genetic markers (Park and Moran, 1994). The remainder of our review considers developments in nucleic acid analysis, and their application to fish introductions.

Mitochondrial DNA analysis

Mitochondrial DNA (mtDNA) restriction analysis has become increasingly adopted for use in fisheries management (Ferris and Berg, 1987; Ovenden, 1990; Billington and Hebert, 1991b; Carvalho and Pitcher, 1994a), especially in stock discrimination (Carvalho and Hauser, 1994) and studies of hybridization. mtDNA is extracted, often from liver or gonadal tissues, digested using restriction endonucleases which cut at specific nucleotide sequences, and the resulting fragments are separated by electrophoresis. After electrophoresis, the fragments are visualized either directly using UV stains, or indirectly using radiolabelled probes (Dowling *et al.*, 1990). A particularly valuable development is the examination of mtDNA variation using the polymerase chain reaction (PCR) products (Hoelzel and Green, 1992; Cronin *et al.*, 1993; Park and Moran, 1994). With PCR, specific nucleotide sequences are selectively amplified using sequence-specific oligonucleotide primers, which can often be used across divergent taxa. Variation in PCR products can be analysed by either direct sequencing or restriction enzyme analysis. The technique is much faster than conventional restriction analysis, removing the need for Southern blotting and hybridization, and can be used on somewhat degraded tissues.

mtDNA, a small circular molecule (14 000–26 000 base pairs, bp) is homologous among widely divergent taxa, and has several characteristics that render it particularly useful for population studies. The rate of nucleotide substitution is approximately 5–10 times that of nuclear DNA, so increasing the rate at which mutations accumulate, thus enhancing the sensitivity for detecting differences between recently separated populations. The smaller N_e for mtDNA, which is approximately one-quarter that for nuclear DNA, means that there is an increased chance of population bottlenecks with a consequent reduction in mtDNA variability and greater divergence between mtDNA genotypes. In fish species where females are philopatric (Shields, 1982), but where spawning males disperse, genetic divergence at the mtDNA level may still be detected, in contrast to allozymic data (Billington and Hebert, 1991b). Moreover, the maternal transmission of mtDNA genotypes allows relatively easy reconstruction of historical events without the complications arising from recombination. The net result is that mtDNA analysis may reveal differentiation where allozymes display spatial homogeneity.

As in allozyme studies, the effectiveness of mtDNA as a molecular tool depends on both the number of fish examined, and on the intensity of sampling of the genome (Billington and Hebert, 1991b; Zwangenburg *et al.*, 1992). Genotype diversity often increases both with sample size, and with the number of restriction enzymes utilized. It is therefore important when designing a sampling programme to correct for variation in sample size (Nei and Tajima, 1981) and to determine the optimum number of restriction enzymes to detect the genotypic diversity present (Bernatchez *et al.*, 1989). mtDNA genetic markers can be used in similar ways to allozymes, but may provide greater sensitivity for detecting genetic divergence and reduced genetic variability. mtDNA can also provide extra information on both the directionality of hybridization and the incidence of introgression (Avise and Saunders, 1984; Avise *et al.*, 1984; Billington *et al.*, 1988). Due to the effectively clonal inheritance of mtDNA, it is often easier than searching for fixed allozymic differences to monitor reproductive success in strains 'tagged' with distinct mtDNA genotypes introduced into a new environment (Billington and Hebert, 1990).

Nuclear DNA analysis

Direct analysis of the nuclear genome may generate useful genetic markers by exploiting diversity in nucleotide sequences directly through substitutions in restriction sites (Quinn and White, 1987; Dowling *et al.*, 1990). So-called restriction fragment length polymorphisms (RFLPs) of single-copy nuclear DNA are obtained using similar techniques as for mtDNA restriction analysis, though often are less sensitive for discriminating populations (Dowling *et al.*, 1990; Franck *et al.*, 1992; Bentzen *et al.*, 1993; Wright, 1993). Considerably greater polymorphism is detectable using repeated nuclear genes such as ribosomal DNA or minisatellite regions (Dowling *et al.*, 1990). Highly polymorphic minisatellite and microsatellite markers (Tautz, 1989; Park and Moran, 1994) have provided a wealth of genetic variants to examine fish populations (Taggart and Ferguson, 1990a,b; Bentzen *et al.*, 1991; Prodöhl, 1992; Prodöhl *et al.*, 1994; Wright, 1993; Wright and Bentzen, 1994), usually enhancing our ability to discriminate individuals and populations. Moreover, amplification of such a variable number of tandem repeats (VNTRs) using PCR (Saiki *et al.*, 1988; Bentzen *et al.*, 1991, 1993; Wright, 1993) removes the technical difficulties of repeated sequence isolation and cloning (Amos, 1992).

Several features of VNTRs render them valuable for examining fish population structure. First, they are usually non-coding, and therefore the variation should be largely independent of natural selection, except where they are closely linked to adaptively significant coding sequences. Second, allozyme studies require fresh or frozen tissue, which may constrain sampling logistics, whereas small quantities of blood or other tissues preserved in alcohol are usually adequate for analysing repetitive DNA. Often such tissues can be obtained without killing the animal. Finally, high levels of heterozygosity (Wright, 1993; Wright and Bentzen, 1994) usually ensure the provision of abundant variants to characterize populations.

22.6 SUMMARY

The conflict between the demand for short-term benefits in fisheries yield and the long-term conservation of biodiversity is nowhere more apparent than in African fresh waters. Demand for fish is high, representing a major source of affordable protein for many rural and urban communities. Artificial changes in African fresh waters through habitat destruction or exploitation have led to a marked reduction in the diversity and abundance of many fish species, some of which have become extinct before being fully described. It is through fish introductions, however, that some of the most alarming impacts on biodiversity have occurred (see Chapters 2–11 for further discussion). The rapid and direct effects of such perturbations on the population genetic structure of alien or native species through microevolutionary forces has led to extensive species changes, although the genetic impacts of many African fish introductions have yet to be evaluated. The great concern is that once initiated, a reduction in genetic variability, or introgression between species will persist over an evolutionary time scale and may therefore be irreversible in the short term. Conservation of genetic resources must include not only practices that minimize loss of genetic diversity, but also measures to preserve the associated genotypic variance in ecologically significant traits. It is the ultimate phenotypic expression of genotypic variability in viability and fecundity that will determine the adaptability and persistence of local populations and thus the evolutionary potential of a species.

Advances in predicting and evaluating the genetic impacts of species introductions depend heavily on incorporating data on factors affecting intraspecific genetic variability, together with the application of appropriate management tools. Such considerations as the source and size of the founder population, or the effect of an alien predator or competitor on effective population size require ecological data on species interactions, as well as the application of molecular techniques to describe population structure. It is only when such information has been integrated that an optimal strategy for minimizing detrimental species changes can be attained. Future research priorities should include genetic studies on the extent and consequences of hybridization, as well as a description of temporal and spatial patterns in genetic variability. Only when such data are available can the long-term benefits of genetic conservation be fully appreciated and thus incorporated into management decisions.

ACKNOWLEDGEMENTS

We are grateful to Dr D.O.F. Skibinski and an anonymous referee for commenting on the manuscript.

REFERENCES

Acere, T.O. (1985) Observations on the biology, age, growth, maturity and sexuality of Nile perch, *Lates niloticus*, and the growth of its fishery in the northern waters of Lake Victoria. *FAO Fish. Rep.*, No. 335, pp. 42–52.

Achieng, A.P. (1990) The impact of the introduction of Nile perch, *Lates niloticus* (L.) on the fisheries in Lake Victoria. *J. Fish Biol.*, **37**, (Suppl. A), 17–23.

Allendorf, F.W. and Ryman, N. (1987) Genetic management of hatchery stocks, in *Population Genetics and Fisheries Management* (eds N. Ryman and F. Utter), Univ. Washington Press, Seattle and London, pp. 141–60.

Amos, B. (1992) Analysis of polygamous systems using DNA fingerprinting, in *Biotechnology and the Conservation of Genetic Diversity* (eds H.D.M. Moore, W.V. Holt and G.M. Mace), Clarendon Press, Oxford, pp. 151–65.

Arthington, A.H. (1991) Ecological and genetic impacts of introduced and translocated freshwater fishes in Australia. *Can. J. Fish. aquat. Sci.*, **48** (Suppl. 1), 33–43.

Avise, J.C. and Saunders, N.C. (1984) Hybridization and introgression among species of sunfish (*Lepomis*): analysis by mitochondrial DNA and allozyme markers. *Genetics*, **108**, 237–55.

Avise, J.C., Bermingham, E., Kessler, L.G. and Saunders, N.C. (1984) Characterization of mitochondrial DNA in a hybrid swarm between subspecies of bluegill sunfish (*Lepomis macrochirus*). *Evolution*, **38**, 931–41.

Bagenal, T.B. (1973) Fish fecundity and its relation with stock and recruitment. *Rapp. P.-v. Réun. Cons. int. Explor. Mer*, **148**, 186–98.

Baker, A.J. and Moeed, A. (1987) Rapid genetic differentiation and founder effect in colonizing populations of common mynas (*Acridotheres tristis*). *Evolution*, **41**, 525–8.

Balon, E.K. (1975) Reproductive guilds of fishes: a proposal and definition. *J. Fish. Res. Bd Can.*, **32**, 821–64.

Balon, E.K. and Bruton, M.N. (1986) Introduction of alien species or why scientific advice is not heeded. *Env. Biol. Fishes*, **16**, 225–30.

Barel, C.D.N., Dorit, R., Greenwood, P.H., Fryer, G., Hughes, N., Jackson, P.B.N., Kawanabe, H., Lowe-McConnell, R., Nagoshi, M., Ribbink, A.J., Trewavas, E., Witte, F. and Yamaoka, K. (1985) Destruction of fisheries in Africa's lakes. *Nature, Lond.*, **315**, 29–30.

Barlow, C.G. and Lisle, A. (1987) Biology of the Nile perch *Lates niloticus* (Pisces: Centropomidae) with reference to its proposed role as a sport fish in Australia. *Biol. Cons.*, **39**, 269–89.

Bartell, S.M. (1981) Potential impact of size selective planktivory on P release by zooplankton. *Hydrobiologia*, **80**, 139–46.

Barton, N.H. (1989) Founder effect speciation, in *Speciation and Its Consequences* (eds D. Otte and J.A. Endler), Sinauer Assoc., Sunderland, MA, pp. 229–56.

Barton, N.H. and Charlesworth, B. (1984) Genetic revolutions, founder effects and speciation. *Annu. Rev. Ecol. System.*, **15**, 133–64.

Bentzen, P., Harris, A.S. and Wright, J.M. (1991) Cloning of hypervariable minisatellite and simple sequence microsatellite repeats for DNA fingerprinting of important aquacultural species of salmonids and Tilapia. In *DNA Fingerprinting: Approaches and Applications* (eds T. Burke, G. Dolf, A.J. Jeffreys and R. Wolff), Birkhäuser Verlag, Basel, pp. 243–62.

Bentzen, P., Taylor, E.B. and Wright, J.M. (1993) A novel synthetic probe for DNA fingerprinting salmonid fishes. *J. Fish Biol.*, **43**, 313–16.

Bernatchez, L., Dodson, J.J. and Boivin, S. (1989) Population bottlenecks: influence on mitochondrial DNA diversity and its effect in coregonine stock discrimination. *J. Fish Biol.*, **35** (Suppl. A), 233–44.

Berst, A.H. and Simon, R.C. (eds) (1981) *Proceedings of the Stock Concept Symposium. Can. J. Fish. aquat. Sci.*, **38**, 1457–1921.

Billington, N. and Hebert, P.D.N. (1990) Technique for determining mitochondrial DNA markers in blood samples from walleyes. *Am. Fish. Soc. Symp.*, **7**, 492–8.

Billington, N. and Hebert, P.D.N. (eds) (1991a) *The Ecological and Genetic Implications of Fish Introductions. Can. J. Fish. aquat. Sci.*, **48** (Suppl. 1), 1–181.

Billington, N. and Hebert, P.D.N. (1991b) Mitochondrial DNA diversity in fishes and its implications for introductions. *Can. J. Fish. aquat. Sci.*, **48** (Suppl. 1), 80–94.

Billington, N., Hebert, P.D.N. and Ward, R.D. (1988) Evidence of introgressive hybridization in the genus *Stizostedion*: interspecific transfer of mitochondrial DNA between sauger and walleye. *Can. J. Fish. aquat. Sci.*, **45**, 2035–41.

Bluhdorn, D.R., Arthington, A.H. and Mather, P.B. (1990) The introduced cichlid *Oreochromis mossambicus* in Australia: a review of distribution, population genetics, ecology, management issues and research priorities, in *Proc. Australian Society for Fish Biology's Workshop on Introduced and Translocated Fishes and Their Ecological Effects*, Magnetic Island, Townsville, August 1989 (ed. D.A. Pollard), Australian Govt Publ. Serv., Canberra, *Bur. Rural Resources Proc.*, no. 8, pp. 83–92.

Bonnell, M.L. and Selander, R.K. (1974) Elephant seals: genetic variation and near extinction. *Science*, **184**, 908–9.

Brassington, R.A. and Ferguson, A. (1976) Electrophoretic identification of roach (*Rutilus rutilus*), rudd (*Scardinius erythrophthalmus*), bream (*Abramis brama*) and their natural hybrids. *J. Fish Biol.*, **9**, 471–7.

Brookfield, J.F.Y. and Parkin, D.T. (1993) Use of single-locus probes in the establishment of relatedness in wild populations. *Heredity*, **70**, 660–63.

Bruford, M.W., Hanotte, O., Brookfield, J.F.Y. and Burke, T. (1992) Single-locus and multilocus DNA fingerprinting, in *Molecular Genetic Analysis of Populations* (ed. A.R.

Hoelzel), IRL Press, Oxford, pp. 225–69.

Bruton, M.N. (1990) The conservation of the fishes of Lake Victoria, Africa: an ecological perspective. *Env. Biol. Fishes*, **27**, 161–75.

Burke, T., Hanotte, O., Bruford, M.W. and Cairns, E. (1991) Multilocus and single locus minisatellite analysis in population biological studies, in *DNA Fingerprinting: Approaches and Applications* (eds T. Burke, G. Dolf, A.J. Jeffreys and R. Wolff), Birkhäuser Verlag, Berlin, pp. 154–68.

Campton, D.E. (1987) Natural hybridization and introgression in fishes: methods of detection and genetic interpretations, in *Population Genetics and Fishery Management* (eds N. Ryman and F. Utter). Univ. Washington Press, Seattle and London, pp. 161–92.

Carvalho, G.R. (1993) Evolutionary aspects of fish distribution: genetic variability and adaptation. *J. Fish Biol.*, **43** (Suppl. A), 53–73.

Carvalho, G.R. and Hauser, L. (1994) Molecular genetics and the stock concept in fisheries. *Rev. Fish Biol. Fish.*, **4**, 326–50.

Carvalho, G.R. and Pitcher, T.J. (eds) (1994a) *Molecular Genetics in Fisheries*. Special Issue of *Rev. Fish Biol. Fish.*, **43**(3), 269–399.

Carvalho, G.R. and Pitcher, T.J. (1994b) Genetic diversity and stock structure of Lake Tanganyika kapenta (*Limnothrissa miodon*) introduced to African lakes. Unpublished final report to the Overseas Development Administration (ODA), U.K. Fish Genetics Programme, 103 pp.

Collart, A. (1960) L'introduction du *Stolothrissa tanganicae* (ndagala) au Lac Kivu. *Bull. Agricole du Congo Belge*, **51**, 975–85.

Coulter, G.W. (1970) Population changes within a group of fish species in Lake Tanganyika following their exploitation. *J. Fish Biol.*, **2**, 329–53.

Coulter, G.W. (1976) The biology of *Lates* species (Nile perch) in Lake Tanganyika and the status of the pelagic fishery for *Lates* species and *Luciolates stappersi*. *J. Fish Biol.*, **9**, 235–59.

Coulter, G.W., Allanson, B.R., Bruton, M.N., Greenwood, P.H., Hart, R.C., Jackson, P.B.N. and Ribbink, A.J. (1986) Unique qualities and special problems of the African Great Lakes. *Env. Biol. Fishes*, **17**, 161–83.

Cronin, M.A., Spearman, W.J., Wilmot, R.L., Patton, J.C. and Bickham, J.W. (1993) Mitochondrial DNA variation in chinook (*Oncorhynchus tshawytscha*) and chum salmon (*O. keta*) detected by restriction enzyme analysis of polymerase chain reaction (PCR) products. *Can. J. Fish. aquat. Sci.*, **50**, 708–15.

Crow, J.F. and Kimura, M. (1970) *An Introduction to Population Genetics Theory*, Harper and Row, New York, 591 pp.

Cushing, D.H. (1973) *Recruitment and Parent Stocks in Fishes*, Univ. Washington Press, Seattle, Washington Sea Grant Publ. WSG 73-1, 295 pp.

Dowling, T.E., Moritz, C. and Palmer, J.D. (1990) Nucleic acids II: restriction site analysis, in *Molecular Systematics* (eds D.M. Hollis and C. Moritz), Sinauer Assoc., Sunderland, MA, pp. 250–317.

Edge, T.A., McAllister, D.E. and Qadri, S.U. (1991) Meristic and morphometrics variation between the endangered Acadian whitefish, *Coregonus huntsmani*, and the lake whitefish, *Coregonus clupeaformis*, in the Canadian maritime provinces and the State of Maine, USA. *Can. J. Fish. aquat. Sci.*, **48**, 2140–51.

Elder, H.Y. and Garrod, D.J. (1961) A natural hybrid of *Tilapia nigra* and *Tilapia leucosticta* from Lake Naivasha, Kenya Colony. *Nature, Lond.*, **191**, 722–4.

Elton, C.S. (1958) *The Ecology of Invasions by Animals and Plants*, Methuen, London.

Elton, C.S. and Miller, R.S. (1954) The ecological survey of animal communities: with a practical system of classifying habitats by structural characteristics. *J. Ecol.*, **42**, 460–96.

Emlen, J.M. (1991) Heterosis and outbreeding depression: a multi-locus model and application to salmon production. *Fish. Res.*, **12**, 187–212.

Falconer, D.S. (1989) *Introduction to Quantitative Genetics*, 3rd edn, Longman Scientific and Technical, Harlow, UK, 438 pp.

FAO/CIFA (1985) *Introduction of Species and Conservation of Genetic Resources.* Committee for Inland Fisheries of Africa, Sixth Session, Lusaka, Zambia, 7–11 Oct. 1985. CIFA/85/13, 18 pp.

FAO/UNEP (1981) Conservation of the Genetic Resources of Fish: Problems and Recommendations. *Report of the Expert Consultation on the Genetic Resources of Fish.* Rome, 9–13 June 1980. *FAO Fish. tech. Pap.*, no. 217, 43 pp.

Ferguson, A. and Thorpe, J.E. (eds) (1991) *Biochemical Genetics and Taxonomy of Fish. J. Fish. Biol.*, **39** (Suppl. A), pp. 1–357.

Ferguson, M.M. (1994) The role of molecular genetic markers in the management of cultured fishes. *Rev. Fish Biol. Fish.*, **4**, 351–73.

Fernando, C.H. (1991) Impacts of fish introductions in tropical Asia and America. *Can. J. Fish aquat. Sci.*, **48** (Suppl. 1), 24–32.

Ferris, S.D. and Berg, W.J. (1987) The utility of mitochondrial DNA in fish genetics and fishery management, in *Population Genetics and Fishery Management* (eds N. Ryman and F. Utter), Univ. Washington Press, Seattle and London, pp. 277–301.

Franck, J.C.B., Wright, J.M. and McAndrew, B.J. (1992) Genetic variability in a family of satellite DNAs from tilapia (Pisces: Cichlidae). *Genome*, **35**, 719–25.

Frankel, O.H. and Soulé, M.E. (1981) *Conservation and Evolution*, Cambridge Univ. Press, Cambridge, 327 pp.

Franklin, I.R. (1980) Evolutionary change in small populations, in *Conservation Biology: An Evolutionary-Ecological Perspective* (eds M.E. Soulé and B.A. Wilcox), Sinauer Assoc., Sunderland, MA, pp. 135–50.

Fryer, G. (1960) Concerning the proposed introduction of Nile perch into Lake Victoria. *East Afr. Agric. J.*, **25**, 267–70.

Fryer, G. (1968) The parasitic Crustacea of African freshwater fishes: their biology and distribution. *J. Zool.*, **156**, 45–95.

Fryer, G. (1972) Conservation of the African Great Lakes: a lesson and a warning. *Biol. Cons.*, **4**, 256–62.

Fryer, G. (1984) The conservation and rational exploitation of the biota of Africa's Great Lakes, in *Conservation of Threatened Natural Habitats* (ed. A.V. Hall), CSIR, Pretoria, *S. Afr. Natn. Scient. Prog. Rep.*, no. 92, pp. 135–54.

Fryer, G. and Iles, T.D. (1972) *The Cichlid Fishes of the Great Lakes of Africa: Their Biology and Evolution*, Oliver and Boyd, Edinburgh.

Gall, G.A.E. (1987) Inbreedng, in *Population Genetics and Fishery Management* (eds N. Ryman and F. Utter), Univ. Washington Press, Seattle and London, pp. 47–88.

Garcia-Marin, J.L., Jorde, P.E., Ryman, N., Utter, F. and Pla, C. (1991) Management implications of genetic differentiation between native and hatchery populations of brown trout (*Salmo trutta*) in Spain. *Aquaculture*, **95**, 235–49.

Gharret, A.J. and Thomason, M.A. (1987) Genetic changes in pink salmon (*Oncorhynchus gorbuscha*) following their introduction into the Great Lakes. *Can. J. Fish. aquat. Sci.*, **43**, 787–92.

Gregorius, H.-R. (1991) Gene conservation and the preservation of adaptability, in *Species Conservation: A Population-Biological Approach* (eds A. Seitz and V. Loeschcke), Birkhäuser Verlag, Basel, pp. 31–47.

Harris, H. and Hopkinson, D. (1976) *Handbook of Enzyme Electrophoresis in Human Genetics*, American Elsevier, New York.

Harrison, K. (1991) The taxonomy of East African Nile perch, *Lates* spp. (Perciformes,

Centropomidae). *J. Fish Biol.*, **38**, 175–86.

Hauser, L., Carvalho, G.R. and Pitcher, T.J. (1995) Morphological and genetic differentiation of the African clupeid. *Limnothrissa miodon* 34 years after its introduction to Lake Kivu. *J. Fish. Biol.*, **47** (Suppl. A), in press.

Hillis, D.M. and Moritz, C. (eds) (1990) *Molecular Systematics*, Sinauer Assoc., Sunderland, MA, 588 pp.

Hindar, K., Ryman, N. and Utter, F. (1991) Genetic effects of cultured fish on natural fish populations. *Can. J. Fish. aquat. Sci.*, **48**, 945–57.

Hoelzel, A.R. (ed.) (1992) *Molecular Genetic Analysis of Populations*, IRL Press, Oxford, 315 pp.

Hoelzel, A.R. and Green, A. (1992) Analysis of population-level variation by sequencing PCR-amplified DNA, in *Molecular Genetic Analysis of Populations* (ed. A.R. Hoelzel), IRL Press, Oxford, pp. 159–87.

Hoffman, G.L. (1970) Intercontinental and transcontinental dissemination and transformation of fish parasites with emphasis on whirling disease (*Myxosoma cerebralis*). *Spec. Publ. Am. Fish. Soc.*, **5**, 69–81.

Hughes, N.F. (1986) Changes in the feeding biology of the Nile perch, *Lates niloticus* (Pisces: Centropomidae), in Lake Victoria, East Africa since its introduction in 1960, and its impact on the native fish community of the Nyanza Gulf. *J. Fish Biol.*, **29**, 541–8.

Ihssen, P.E., Booke, H.E., Casselman, J.M., McGlade, J.M., Payne, N.R. and Utter, F. (1981) Stock identification: materials and methods. *Can. J. Fish. aquat. Sci.*, **38**, 1838–55.

Jackson, P.B.N. (1960) On the desirability or otherwise of introducing fishes to waters that are foreign to them. *Pub. Cons. Sci. Afr. S. Sahara*, **63**, 157–64.

Jeffreys, A.J., Wilson, V. and Thein, S.L. (1985) Hypervariable "minisatellite" regions in human DNA. *Nature, Lond.*, **314**, 67–73.

Kapuscinscy, A.R. and Hallerman, E.M. (1991) Implications of introduction of transgenic fish into natural ecosystems. *Can. J. Fish. aquat. Sci.*, **48** (Suppl. 1), 99–107.

Karl, S.A. and Avise, J.C. (1993) PCR-based assays of mendelian polymorphisms from anonymous single-copy nuclear DNA – techniques and applications for population genetics. *Mol. Biol. Evol.*, **10**, 342–61.

Kirpichnikov, V.S. (1992) Adaptive nature of intrapopulational polymorphism in fish. *J. Fish Biol.*, **40**, 1–16.

Knight, A.J., Hughes, R.N. and Ward, R.D. (1987) A striking example of the founder effect in the mollusc *Littorina saxatilis*. *Biol. J. Linn. Soc.*, **32**, 417–26.

Kornfield, I. (1991) Genetics, in *Cichlid Fishes: Behaviour, Ecology and Evolution* (ed. M.H.A. Keenleyside), Chapman and Hall, London, pp. 103–28.

Kricka, L.J. (1992) *Non-isotope DNA Probe Techniques*, Academic Press, London, 358 pp.

Krueger, C.C. and May, B. (1991) Ecological and genetic effects of salmonid introductions in North America. *Can. J. Fish. aquat. Sci.*, **48** (Suppl. 1), 66–77.

Krueger, C.C., Gharret, A.J., Dehring, T.R. and Allendorf, F.W. (1981) Genetic aspects of fisheries rehabilitation programmes. *Can. J. Fish. aquat. Sci.*, **38**, 1877–81.

Kuhry, B. and Marcus, L.F. (1977) Bivariate linear models in biometry. *Syst. Zool.*, **26**, 201–9.

Lande, R. and Barrowsclough, G.F. (1987) Effective population size, genetic variation, and their use in population management, in *Viable Populations for Conservation* (ed. M.E. Soulé), Cambridge Univ. Press, Cambridge, pp. 87–123.

Leary, R.F., Allendorf, F.W. and Knudsen, K.L. (1983) Consistently high meristic counts in natural hybrids between brook trout and bull trout. *Syst. Zool.*, **32**, 369–76.

Leberg, P.L. (1992) Effects of population bottlenecks on genetic diversity as measured by allozyme electrophoresis. *Evolution*, **46**, 477–94.

Loeschke, V., Tomiuk, J. and Jain, S.K. (1994) *Conservation Genetics*, Birkhäuser Verlag, Basel, 492 pp.

Margalef, R. (1968) *Perspectives in Ecological Theory*, Univ. Chicago Press, Chicago, 320 pp.

Markert, C.L. and Møller, F. (1959) Multiple forms of enzymes: tissue, ontogenetic, and species-specific patterns. *Proc. natl. Acad. Sci. USA*, **45**, 753–63.

Marshall, B.E. (1993) Biology of the African clupeid *Limnothrissa miodon* with reference to its small size in artificial lakes. *Rev. Fish Biol. Fish.*, **3**, 17–38.

Martin, A.P., Humphreys, R. and Palumbi, S.R. (1992) Population genetic structure of the armorhead, *Pseudopentaceros wheeleri*, in the North Pacific Ocean: application of the polymerase chain reaction to fisheries problems. *Can. J. Fish. aquat. Sci.*, **49**, 2386–91.

Mather, P.B. and Arthington, A.H. (1991) An assessment of genetic differentiation among feral Australian *Tilapia* populations. *Aust. J. mar. Freshwat. Res.*, **42**, 721–8.

May, R.M. (1975) Patterns of species abundance and diversity, in *Ecology and Evolution of Communities* (eds M.L. Cody and J.M. Diamond), Harvard Univ. Press, Cambridge, MA, pp. 81–120.

Melvin, G.D., Dadswell, M.J. and McKenzie, A. (1992) Usefulness of meristic and morphometric characters in discriminating populations of American shad (*Alosa sapidissima*) (Osteichthyes: Clupeidae) inhabiting a marine environment. *Can. J. Fish. aquat. Sci.*, **49**, 266–80.

Miller, D.J. (1989) Introductions and extinctions of fish in the African Great Lakes. *Trends Ecol. Evol.*, **4**, 56–9.

Moyle, P.B., Li, H.W. and Barton, B.A. (1986) The Frankenstein effect: impact of introduced fishes on native fishes in North America, in *Fish Culture in Fisheries Management* (ed. R.H. Stroud), Am. Fish. Soc., Bethesda, MD, pp. 415–26.

Nei, M. (1972) Genetic distance between populations. *Am. Nat.*, **106**, 283–92.

Nei, M. (1987) *Molecular Evolutionary Genetics*, Columbia Univ. Press, New York, 512 pp.

Nei, M. and Tajima, F. (1981) DNA polymorphism detectable by restriction endonucleases. *Genetics*, **97**, 145–63.

Nei, M., Maruyama, T. and Chakraborty, R. (1975) The bottleneck effect and genetic variability in populations. *Evolution*, **29**, 1–10.

Nelson, K. and Soulé, M. (1987) Genetical conservation of exploited fishes, in *Population Genetics and Fisheries Management* (eds N. Ryman and F. Utter), Univ. Washington Press, Seattle and London, pp. 345–68.

ODA (1991) *Biological Diversity and Developing Countries: Issues and Options*. Natural Resources and Environment Dept, Overseas Development Admin., 94 Victoria Street, London SW1 5JL, 50 pp.

Ogutu-Ohwayo, R. (1990a) The decline of the native fishes of Lakes Victoria and Kyoga (East Africa) and the impact of introduced species, especially the Nile perch, *Lates niloticus*, and the Nile tilapia, *Oreochromis niloticus*. *Env. Biol. Fishes*, **27**, 81–96.

Ogutu-Ohwayo, R. (1990b) The reduction in fish species diversity in Lakes Victoria and Kyoga (East Africa) following human exploitation and the introduction of non-native fish. *J. Fish Biol.*, **37** (Suppl. A), 207–8.

Ogutu-Ohwayo, R. (1992) The purpose, costs and benefits of fish introductions: with specific reference to the Great Lakes of Africa. *Mitt. int. Verein. Limnol.*, **23**, 37–44.

Ogutu-Ohwayo, R. and Hecky, R.E. (1991) Fish introductions in Africa and some of their implications. *Can. J. Fish. aquat. Sci.*, **48** (Suppl. 1), 8–12.

Oldfield, M.L. (1989) *The Value of Conserving Genetic Resources*, Sinauer Assoc., Sunderland, MA, 379 pp.

Ovenden, J.R. (1990) Mitochondrial DNA and marine stock assessment: a review. *Aust. J. mar. Freshwat. Res.*, **41**, 835–53.

Park, L.K. and Moran, P. (1994) Developments of molecular genetic techniques in fisheries. *Rev. Fish Biol. Fish.*, **4**, 272–99.

Philipp, D.P. (1991) Genetic implications of introducing largemouth bass, *Micropterus salmonides floridanus*. *Can. J. Fish. aquat. Sci.*, **48** (Suppl. 1), 58–65.

Prodöhl, P.A. (1992) Genetic variability within and between sympatric brown trout. *Hereditas*, **117**, 45–50.

Prodöhl, P.A., Taggart, J.B. and Ferguson, A. (1994) Single locus minisatellite variation in brown trout, *Salmo trutta* L., populations, in *Genetics and Evolution of Aquatic Organisms* (ed. A.R. Beaumont), Chapman and Hall, London, pp. 263–70.

Quinn, T.W. and White, B.N. (1987) Identification of restriction fragment length polymorphisms in genomic DNA of the lesser snow goose. *Mol. Biol. Evol.*, **4**, 126–43.

Ramshaw, J.A.M., Coyne, J.A. and Lewontin, R.C. (1979) The sensitivity of gel electrophoresis as a detector of genetic variation. *Genetics*, **93**, 1019–37.

Reisenbichler, R.R. and McIntyre, J.D. (1977) Genetic differences in growth and survival of juvenile hatchery and wild steelhead trout, *Salmo gairdneri*. *J. Fish. Res. Bd Can.*, **34**, 123–8.

Ribbink, A.J. (1987) African lakes and their fishes: conservation scenarios and suggestions. *Env. Biol. Fishes*, **19**, 3–26.

Ribbink, A.J., Marsh, B.A., Marsh, A.C., Ribbink, A.C. and Sharp, B.J. (1983) A preliminary study of the cichlid fishes of rocky habitats of Lake Malawi. *S. Afr. J. Zool.*, **18**, 149–310.

Richardson, B.J., Baverstock, P.R. and Adams, M. (1986) *Allozyme Electrophoresis: A Handbook for Animal Systematics and Population Studies*, Acad. Press, Sydney and London, 410 pp.

Ricker, W.E. (1972) Hereditary and environmental factors affecting certain salmonid populations, in *The Stock Concept in Pacific Salmon* (eds R.C. Simon and P.A. Larkin), MacMillan Lectures in Fisheries, Univ. British Columbia, Vancouver, pp. 19–160.

Roby, D., Lambert, J.D. and Sévigny, J.M. (1991) Morphometric and electrophoretic approaches to discrimination of capelin (*Mallotus villosus*) populations in the estuary and Gulf of St. Lawrence. *Can. J. Fish. aquat. Sci.*, **48**, 2040–50.

Ross, M.R. and Cavender, T.M. (1981) Morphological analysis of four experimental intergeneric cyprinid hybrid crosses. *Copeia*, **1981**, 377–87.

Ryman, N. and Utter, F. (eds) (1987) *Population Genetics and Fishery Management*, Univ. Washington Press, Seattle and London, 420 pp.

Ryman, N., Baccus, R., Reuterwall, C. and Smith, M.H. (1981) Effective population size, generation interval, and potential loss of genetic variability in game species under different hunting regimes. *Oikos*, **36**, 257–66.

Saiki, R.K., Gelfand, D.H., Stoffel, S., Scharf, S.J., Higuchi, R., Horn, G.T., Mullis, K. and Erlich, H.A. (1988) Primer-directed enzymatic amplification of DNA with a thermostable DNA polymerase. *Science*, **239**, 487–91.

Schaefer, S.A. (1991) Morphometric investigations in cyprinid biology, in *Cyprinid Fishes: Systematics, Biology and Exploitation* (eds I.J. Winfield and J.S. Nelson), Chapman and Hall, London, pp. 55–82.

Seitz, A. (1991) Introductory remarks: population biology, the scientific interface to species conservation, in *Species Conservation: A Population-Biological Approach* (eds A. Seitz and V. Loeschke), Birkhäuser Verlag, Basel etc., pp. 1–13.

Seitz, A. and Loeschcke, V. (eds) (1991) *Species Conservation: A Population-Biological Approach*, Birkhäuser Verlag, Basel etc., 315 pp.

Shafland, P.L. and Lewis, W.M. (1984) Terminology associated with introduced organisms. *Fisheries*, **9**(4), 17–18.

Shaw, P.W., Carvalho, G.R., Seghers, B.H. and Magurran, A.E. (1992) Genetic consequences of an artificial introduction of guppies (*Poecilia reticulata*) in N. Trinidad. *Proc. R. Soc. Lond.*, B, **248**, 111–16.

Shields, W.M. (1982) *Philopatry, Inbreeding and the Evolution of Sex*, State Univ. New York Press, New York, 237 pp.

Siddiqui, A.Q. (1977) Lake Naivasha (Kenya, East Africa) fishery and its management together with a note on the food habits of fishes. *Biol. Cons.*, **12**, 217–27.

Simberloff, D. (1988) The contribution of population and community biology to conservation science. *Annu. Rev. Ecol. Syst.*, **19**, 473–512.

Solbrig, O.T. (ed.) (1991a) *From Genes to Ecosystems: A Research Agenda for Biodiversity*, Int. Union Biol. Sci., Sunderland, MA, 55 pp.

Solbrig, O.T. (1991b) *Biodiversity. Scientific Issues and Collaborative Research Proposals*, UNESCO, Paris, *Mab Digest*, **9**, 87 pp.

Soulé, M.E. (1976) Allozyme variation: its determinants in space and time, in *Molecular Evolution* (ed. F.J. Ayala), Sinauer Assoc., Sunderland, MA, pp. 60–77.

Soulé, M.E. (1986) *Conservation Biology: The Science of Scarcity and Diversity*, Sinauer Assoc., Sunderland, MA, 584 pp.

Stewart, J.E. (1991) Introductions as factors in diseases for fish and aquatic invertebrates. *Can. J. Fish aquat. Sci.*, **48** (Suppl. 1), 110–17.

St. Louis, V.L. and Barlow, J.C. (1988) Genetic differentiation among ancestral and introduced populations of the Eurasian tree sparrow (*Passer montanus*). *Evolution*, **42**, 266–76.

Taggart, J.B. and Ferguson, A. (1990a) Hypervariable minisatellite DNA single locus probes for the Atlantic salmon, *Salmo salar* L. *J. Fish Biol.*, **37**, 991–3.

Taggart, J.B. and Ferguson, A. (1990b) Minisatellite DNA fingerprints of salmonid fishes. *Anim. Genet.*, **21**, 377–89.

Tautz, D. (1989) Hypervariability of simple sequences as a general source for polymorphic DNA markers. *Nucl. Acids Res.*, **17**, 6463–71.

Taylor, E.B. (1991) A review of local adaptation in Salmonidae, with particular reference to Pacific and Atlantic salmon. *Aquaculture*, **98**, 185–207.

Thorpe, J.P. (1983) Enzyme variation, genetic distance and evolutionary divergence in relation to levels of taxonomic separation, in *Protein Polymorphism: Adaptive and Taxonomic Significance* (eds G.S. Oxford and D. Rollinson), Acad. Press, London, pp. 131–52.

Tiedje, J.M., Colwell, R.K., Grossman, Y.L., Hodson, R.E., Lenski, R.E., Mack, R.N. and Regal, P. (1989) The planned introduction of genetically engineered organisms: ecological considerations and recommendations. *Ecology*, **70**(2), 298–315.

Trewavas, E. (1983) *Tilapiine Fishes of the Genera* Sarotherodon, Oreochromis *and* Danakilia, British Museum, London.

Turner, J.L. (1977) Changes in the size structure of cichlid populations of Lake Malawi resulting from bottom trawling. *J. Fish. Res. Bd Can.*, **34**, 232–8.

Utter, F., Aebersold, P. and Winans, G. (1987) Interpreting genetic variation detected by electrophoresis, in *Population Genetics and Fishery Management* (eds N. Ryman and F. Utter), Univ. Washington Press, Seattle and London, pp. 21–45.

Vanni, M.J. (1986) Fish predation and zooplankton demography: indirect effects. *Ecology*, **67**, 337–54.

Vuorinen, J., Næsje, T.F. and Sandlund, O.T. (1991) Genetic changes in a vendace *Coregonus albula* (L.) population, 92 years after introduction. *J. Fish Biol.*, **39** (Suppl. A), 193–202.

Waples, R.S. (1991) Genetic interactions between hatchery and wild salmonids: lessons from the Pacific Northwest. *Can. J. Fish. aquat. Sci.*, **48** (Suppl. 1), 124–33.

Ward, R.D. and Grewe, P.M. (1994) Appraisal of molecular genetic techniques in fisheries. *Rev. Fish Biol. Fish.*, **4**, 300–25.

Welcomme, R.L. (1984) International transfer of inland fish species, in *Distribution, Biology and Management of Exotic Fishes* (eds W.R. Courtenay and J.R. Stauffer), Johns Hopkins Univ. Press, Baltimore and London, pp. 22–40.

Welcomme, R.L. (1988) *International Introductions of Inland Fish Species. FAO Fish. tech. Pap.*, no. 294, 120 pp.

Winans, G.A. (1987) Using morphometric and meristic counts for identifying stocks of fish. *Proc. Stock Identification Workshop*, Panama City Beach, Florida, *NOAA tech. Memo.*, NMFS-SEFC-199, pp. 25–62.

Witte, F., Goldschmidt, T., Wanink, J., van Oijen, M., Goudswaard, K., Witte-Maas, E. and Bouton, N. (1992) The destruction of an endemic species flock: quantitative data on the decline of the haplochromine cichlids of Lake Victoria. *Env. Biol. Fishes*, **34**, 1–28.

Worthington, E.B. (1973) The ecology of introductions – a case study from the African lakes. *Biol. Cons.*, **5**, 221–2.

Wright, J.M. (1993) DNA fingerprinting of fishes, in *Biochemistry and Molecular Biology of Fishes*, Vol. 2 (eds P. Hochachka and T. Mommsen), Elsevier, New York, pp. 57–91.

Wright, J.M. and Bentzen, P. (1994) Microsatellites: genetic markers for the future. *Rev. Fish Biol. Fish.*, **4**, 384–8.

Zaret, T.M. (1979) Predation in freshwater fish communities, in *Predator–Prey Systems in Fisheries Management* (ed. H. Clepper), Sport Fishing Inst., Washington, DC, pp. 135–43.

Zaret, T.M. and Paine, R.T. (1973) Species introduction in a tropical lake. *Science*, **182**, 449–55.

Zwangenburg, K.C.T., Bentzen, P. and Wright, J.M. (1992) Mitochondrial DNA differentiation in western North Atlantic populations of haddock (*Melanogrammus aeglefinus*), *Can. J. Fish. aquat. Sci.*, **49**, 2527–37.

Chapter twenty-three

Thinking the unthinkable: a candidate model for predicting sustainable yields of introduced fish species in African lakes

Tony J. Pitcher

23.1 INTRODUCTION

In fresh waters world-wide, Welcomme (1988, 1992) analyses over 1.5 million introductions of 291 species into 148 countries since 1900. Many of these introductions have been designed to enhance fisheries, although the majority have been carried out for aquaculture. Outcomes have been mixed, and so it would be helpful to try to ensure that future introductions are beneficial. Adverse effects include predation, competition, hybridization, introduction of disease and parasites, crowding, stunting and degradation of water quality (Welcomme, 1984), and there is evidently a need to better predict the ecological role and impact of an introduced species, and to obtain international agreement on its introduction. Guidelines to this effect have been published by FAO (Tiews, 1985). As Welcomme (1986) points out, new introductions cannot be banned, and so this Chapter aims to assess the impact of potential new introductions of the freshwater sardine or Nile perch to

The Impact of Species Changes in African Lakes. Edited by Tony J. Pitcher and Paul J.B. Hart. Published in 1995 by Chapman & Hall, London. ISBN 0 412 55050 4.

African lakes by evaluating the balance of costs and benefits. A new model based on primary productivity is introduced to estimate potential sustainable yields.

With respect to putative fish introductions, it is worthwhile to distinguish human-made lakes, or impoundments, from natural lakes. Human-made lakes rarely have fish fauna adapted to lacustrine conditions (Chapter 24), and so many of the ethical and genetic arguments against introducing exotic species do not apply. In human-made lakes, ethical and genetic considerations become important only if escape to natural lakes in the same watershed is likely. A second factor is that there should be little risk that successful existing fisheries are compromised. Cowx and Kapasa (Chapter 15) point out that there are seven artificial African lakes with areas greater than 1000 km^2, another 32 larger than 100 km^2, and 74 larger than 10 km^2. The total area of the large lakes is around 29 000 km^2, and the smaller ones from the list above amount to approximately 15 000 km^2, giving a grand total of around 35 000 km^2. This adds up to a very large area, equivalent to a human-made lake the size of Lake Tanganyika, or half that of Lake Victoria. Many of these human-made lakes may be candidates for establishment of fisheries for introduced sardines and/or Nile perch.

The 13 largest natural lakes in Africa amount to a surface area of around 176 000 km^2, a huge area equivalent to the Gulf of St Lawrence, the Persian Gulf, or one-third that of the North Sea. In natural lakes, consideration of the value of endemic biodiversity may inhibit new fish introductions. But in reality, unthinkable events may occur under economic and political pressure. Reported adverse affects on biodiversity have led many to argue that such introductions should not be contemplated (Ribbink, 1987) and indeed, at present, fresh introductions to natural lakes in Africa are unlikely because most of the region's governments have a brief to support conservation and would be loath to attract the opprobrium attached to the Nile perch introduction that is the legacy of conservation-orientated publicity in the mid 1980s. Such perceptions have a tangible economic worth in affecting the cash value of tourism, or the earnings of the aquarium trade. Furthermore, we should also not forget the impact of economic sanctions and pressures imposed on Third World governments by First World governments, organizations and individuals. These pressures may be both official (e.g. threats to economic relations with whaling, tuna or ivory-exporting countries by the US), and unofficial in the form of boycotts organized by Greenpeace and similar groups (e.g. those campaigning against the fur trade, ivory or whaling).

It is, nevertheless, possible to envisage scenarios of such desperate need for protein, or such high economic and political pressure to earn foreign exchange, that new fish introductions to African lakes may be contemplated. This may be especially likely to occur following dramatic change in the political complexion of a regime or a *coup d'état* and, moreover, as the human population grows in the region, the likelihood of such a scenario increases.

This Chapter first considers the costs and benefits that may affect decisions about new introductions; second, describes the new model; goes on to assess yields that could be achieved; and finally, surveys the likelihood of new introductions enhancing fishery yields of the major human-made and natural lakes in Africa.

23.2 WHAT FACTORS MIGHT AFFECT DECISIONS TO MAKE NEW FISH INTRODUCTIONS?

How can we assess the arguments affecting the decision about introducing Nile perch or sardines to a new lake? A balance sheet of costs deriving from loss of the endemic fauna needs to be set against the estimated benefits of a fishery deriving from the introduction.

The relatively minor costs of implementing an introduction are not likely to affect the decision. Moreover, it is evident that establishment after introduction is virtually unpredictable using existing ecological knowledge (Chapter 1), so it could be argued that introduction should therefore not be undertaken. But if the alternative outcomes are having the benefit of the introduced species, or the *status quo* with no benefits, ecological success or failure of establishment is irrelevant to the decision to go ahead.

On the debit side, a large number of questions may be asked, and it will be difficult to attach precise values to many of the answers. Nevertheless, anything not given some measure of quantification will be unlikely to count for much in making the decision. In the first intance, we need to estimate how much of the endemic fish fauna will survive: we may as well assume a complete loss if a piscivore is introduced. In each case we need estimates, first of the economic value of the present situation, and secondly, of the potential maximum sustainable value for the new resource. The following list of costs is not exhaustive, but is intended to illustrate the kind of evaluation that should take place.

1. *Loss of economic value of existing fishery.* What is the current economic value of the fishery for endemic species? What is the potential economic value if the existing fishery is managed optimally?
2. *Loss of socio-economic value.* How many jobs will be lost in the traditional fisheries? What changes may be envisaged to the equity of the present fishery benefits? Which sectors of the population will be disadvantaged by the loss of the fishery for endemic species?
3. *Costs of satisfying existing fishery sectors.* What part of the future benefits

of the introduction would have to be given to the existing artisanal and commercial fishers for them to accept the introduction politically?

4. *Instability and non-sustainability of gains.* What is the cost of potential ecological instability that would compromise the putative gains from a fishery for an introduced species? What is the likelihood of failure to maintain the early yields from an introduced species?
5. *Loss of genetic diversity.* What would be the consequences if loss in genetic diversity provided less buffering for future environmental change? For example, endemic stocks may be more resilient to the effects of global warming. Increased variance and more frequent occurrence of climatic extremes would probably be a more significant factor in the tropics than higher average temperatures. This is likely to be seen in lakes as greater variation of water levels, nutrient inputs, wind-driven mixing and the kind of tectonic events that are known to have shaped the recent history of fish faunas in the Rift Valley.
6. *Loss of aquaculture potential of endemic species.* What is the actual and potential future use for aquaculture of species endemic to the region? Are any of the species under threat being farmed already and what is the actual and potential value of this enterprise?
7. *Loss of cultural benefits from endemic fauna.* Do parts of the endemic fish fauna at risk from the introduction have a cultural and ethnic value for aboriginal peoples, or as a national heritage? How can the cost of the loss of this resource be quantified? Does this factor effectively veto any estimated gain from the introduction, or are there such high levels of economic gain that would override this veto? Could wealth generated from the introduction be injected into critical sectors of the community in compensation for the loss of culture, and if so, how much would buy off the veto? Would the preservation of stocks by farming be an acceptable compromise in mitigation of the loss of culture?
8. *Loss of tourism benefits of endemic fauna.* What is the value of tourism for sport fishing, or to view fish by scuba diving? How would this be affected if the natural fauna were to be lost?
9. *Loss of aquarium-trade benefits of endemic fauna.* What is the current and potential value of the aquarium trade in fishes? For the introduction of a predator, the balance sheet would assume this would all be lost if the introduction took place.
10. *Loss of medicinal benefits of endemic fauna.* Do some fish have a pharmacological or medicinal value? The present and potential value of rare compounds, pigments or poisons and the medicinal properties of livers, gonads, scales, swim bladders can be taken into consideration. Of course it is sometimes difficult to know if such properties exist, for example the chance of curing cancer with dodo livers has been lost for all time, so the calculation must be made with extant knowledge only.

11. *International retribution for loss of endemic biodiversity.* What is the potential cost of sanctions against the destruction of endemic biodiversity? These may arrive as threats to aid and economic support by First World governments, or as campaigns to boycott goods and tourism run by conservationist lobbies, or in extreme cases the actions of ecoterrorist groups.

The positive side of the balance sheet affecting the decision is less complex.

1. *Economic gain.* What is the direct economic gain to be had from the introduction? This must include estimates of the net value of the fish sold as catch after operating costs have been deducted, using a simple bioeconomic model of the new fishery.
2. *Socio-economic gain in the fishery sector.* What benefits in employment will result? New jobs may be created not only in the fishery but also in onshore processing, transport and service industries. What sectors of the population will gain from the new fishery and what is the likely extension of equity?
3. *Nutritional gain.* What is the value of the likely nutritional gain to be had from the introduction? This can be costed by looking at the cost of current and projected food imports, the costs of remedies to malnutrition and associated community health service costs.
4. *Sport fishery income.* Are there likely ancillary gains in the creation of a sport fishery? The indirect benefits of increased tourism should be included with the revenue for the sale of licences and fishing gear.
5. *Aquaculture opportunity.* Are there likely ancillary gains to be made in enhancing the opportunities for aquaculture?

If the probable benefits exceed the likely assessed costs by a margin estimated to be a significant improvement on the *status quo*, if the risks are thought to be small in relation to those benefits, and if the current situation is perceived to be sufficiently bad, the pressure to carry out a new introduction may be irresistible irrespective of the existence of international codes of practice.

23.3 ESTIMATING SUSTAINABLE YIELDS FROM NEW INTRODUCTIONS

Before we can analyse possible further introductions in the African lakes, we must have a way of estimating the likely sustainable yields of the new fisheries. This Chapter considers only two species, Nile perch, *Lates niloticus*, a centropomid piscivore, and the freshwater (inshore) sardine *Limnothrissa miodon*, a zooplanktivorous clupeid, because there is good information and assessments about successful fisheries for these introduced species. Nile tilapia, *Oreochromis niloticus*, a herbivore, is a third candidate species for further introduction, and has already been frequently introduced, often as an aquaculture subject, around the world (Welcomme, 1988). However, fisheries information from successful Nile tilapia fisheries in Africa is not yet good enough to support an analysis.

Relating lake fish yield to primary production

Fishery managers would find it convenient if fish yields could be related to primary production in some easy fashion. This concept has been attacked as unhelpful and simplistic from two directions.

First, limnologists do not support the idea because the type and rate of primary production bears a complex relationship with the nutrient profile, hydrology and morphology of lakes.

Secondly, fish ecologists attack the concept because the use of total fish production ignores the details of trophic community structure and large differences among and between the ichthyofauna of tropical and temperate lakes. For example, Briand (1985) emphasized the differences among freshwater communities resulting from depth, primary production, transfer efficiencies and variance in physical parameters. Evans *et al.* (1987) summarize ecological complexity in the North American Great Lakes, concluding that production of fish species is strongly influenced by the structure of trophic interactions.

Despite these misgivings, the practical fishery manager can hold a different perspective. Estimating the total yield from a fishery with a known level of confidence is a legitimate goal and if predictions can be made with moderate precision, there may be no need to be overly concerned with details of ecological relationships.

The morphoedaphic index (MEI) which relates various measures of primary production and mean lake depth to fish yield (Ryder, 1965, 1978, 1982; Ryder *et al.*, 1974), has undergone a revival recently with up-to-date statistical analysis (Schnieder and Headrich, 1989; Badjik and Schnieder, 1991). But it is unlikely that more rigorous statistics can overcome the problem that the MEI is not very successful for African lakes (Henderson *et al.*, 1973; Henderson and Welcomme, 1974). However, although predictive precision is not high, the increase in fish yields consequent on introduction to Lakes Kariba and Victoria are especially clear from an MEI-style plot adapted from Fryer and Iles' (1972) classic book on the African Great Lakes (Fig. 23.1).

There is a contrast between these two lakes here: in the case of Lake Kariba, increased fish yield is a consequence of the introduced sardines filling an empty niche. In Lake Victoria, catch increased as the fishery expanded, although on the basis of ECOPATH modelling, Moreau (Chapter 8) suggested that there has in addition been an increase in overall fish production rate.

In contrast to the lack of predictive success with the MEI, there has been some success in approximate prediction of total fish yield from average primary production rates in both temperate and tropical lakes. There has been a

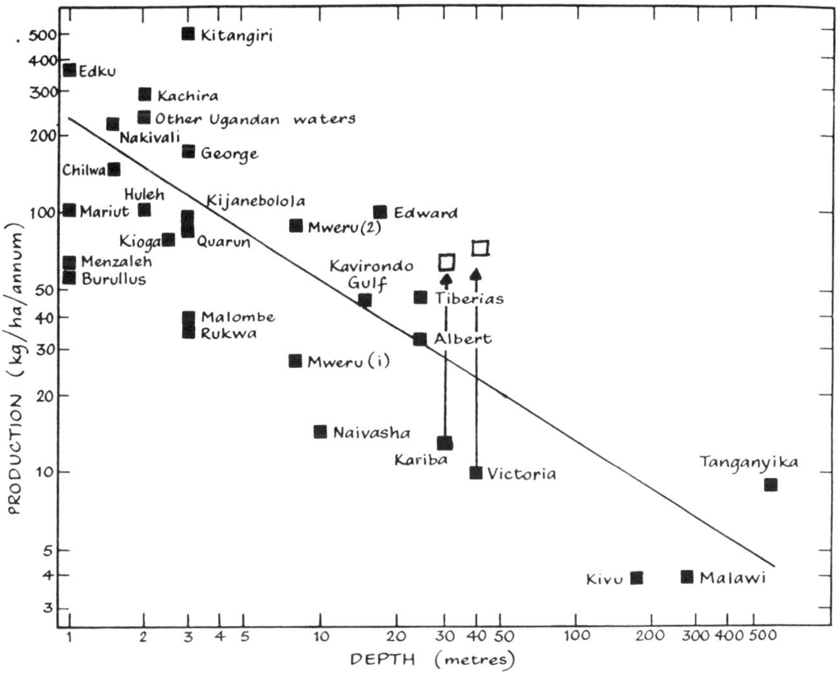

Fig. 23.1 Plot of fish yield per unit area (kg ha^{-1} year^{-1}; divide by 10 for tonnes km^{-2} year^{-1} as used in text) against mean depth for a group of African lakes, showing the enhanced fish yield (□) achieved by the introduction of Nile perch in Lake Victoria and sardines in Lake Kariba (adapted from Fryer and Iles, 1972). Closed and open symbols represent estimated sustainable fish yield before and after introduction. Victoria 'before' figures from Fryer and Iles ignore the possibility of a large trawl fishery for haplochromines as suggested by George Turner. Logarithmic scales.

considerable amount of recent work on this topic since Melack (1976) published a significant relationship from eight tropical lakes. For example, Oglesby (1977) demonstrated that logarithmic regressions of total fish yield on primary production rate in 34 lakes differing widely in provenance, climate and size, provided better predictors than the MEI. Applying more modern statistical treatment, Downing *et al.* (1990) confirmed this finding using data on primary production, phosphorus and the MEI from 20 contrasting lakes worldwide.

Most models have shown a logarithmic relationship of fish yield with primary production because at higher production levels, rate of increase of fish yield falls off. At very high production levels, fish yields seem to decline further; for example, Laing *et al.* (1982) fitted a logistic curve to fish yield and primary production in Chinese lakes and fish ponds. The most thorough published investigation appears to be that of Oglesby *et al.* (1987), where a dome-shaped model best fitted the data from 21 North American lakes that differed by an order of magnitude in levels of primary production. However, Oglesby *et al.* found that *Stizostedion* (walleye), the main target species in the investigation, was replaced in lakes with primary production lower than about $2\,\mathrm{mg\,m^{-3}}$ chlorophyll and higher than $11\,\mathrm{mg\,m^{-3}}$ by other predatory fish taxa, namely coregonid/salmonid and centrarchid/cyprinid communities respectively. Hence, Oglesby *et al.* thought that a dome-shaped relationship would apply to any one taxon, but that a second relationship summed over all taxa gave total fish yield that increased with the log of primary production. The overall relationship was achieved as one taxon supplanted another in the same niche. Leach *et al.* (1987) reviewed MEI, primary productivity, and a range of other indices involving algae, benthos, climate and morphometry as ways of predicting potential fish yields in the North American Great Lakes and concluded that primary production is one of the most useful approximate indicators. Unfortunately, for the African lakes, Crul (1992) concluded that the basic limnological information needed to use even the less-sophisticated published fish yield prediction models was lacking in most cases.

The sustainable yield prediction model

To get around these problems, this Chapter suggests the use of fishery information from a successful fish introduction to enable us to make a short cut. The candidate model put forward here uses estimated sustainable yield per $\mathrm{km^2}$ from a baseline lake where the species forms a successful fishery as a 'predictor of ecology'. Here we are concerned with a thought experiment (*gedankenexperiment*) of predicting, not the sustained yields of endemic fish, but possible yield of an introduced species should successful new establishment occur in a target lake. The model assumes that, if successful colonization occurs, then the ecological similarities are sufficient to make a prediction based simply on the ratio of the log primary production levels in the target and baseline lakes. In the model, all of the minutiae of ecological interactions are subsumed within the production parameters and the fact that the species is fully established in the baseline lake. Hence, the only necessary information for this model is the sustainable yield in the baseline lake and the primary production levels in the two lakes.

The sustainable level of the fishery in the baseline lake is assumed to be assessed with known confidence limits, and the technology and economic infrastructure of the fishing industry in the target lake is assumed equivalent to that in the baseline lake.

One simple method might be to use the yield per unit area in proportion to the ratio of productivity in the baseline lake and the target lake. This makes the assumption that fish production is linearly related to primary production:

$$Y = aP \quad (23.1a)$$

where Y is potential fish yield, P is primary production, and a is the slope.

Using this relationship to predict fish yield in a target lake, given production in the baseline lake and the two primary production figures, we get:

$$Y_t = (Y_b/A_b) \times A_t \times (P_t/P_b) \quad (23.1b)$$

where Y_t is the estimated annual sustainable yield in the target lake (tonnes $year^{-1}$), Y_b is the sustainable annual yield in the baseline lake (tonnes $year^{-1}$), A_b is the area of the baseline lake (km^2), A_t is the area of the target lake (km^2), P_t is the average primary production of the target lake ($gC\,m^{-2}\,day^{-1}$), and P_b is the average primary production of the baseline lake ($gC\,m^{-2}\,day^{-1}$).

The calculation takes the annual yield of fish per km^2 in the baseline lake and multiplies this by the area of the target lake. The result is then scaled by the ratio of primary productivity in the two lakes. However, as we have seen from the review of the literature above, fish yields are better modelled using a logarithmic relationship with increasing primary productivity. The relationship of primary production to fish production is therefore:

$$Y = a \ln P \quad (23.2a)$$

and with reference to the baseline lake, the model now becomes:

$$Y_t = (Y_b/A_b) \times A_t \times [\ln(P_t + 1)/\ln(P_b + 1)] \quad (23.2b)$$

where Y_t is the estimated annual yield in the target lake (tonnes $year^{-1}$), Y_b is the sustainable annual yield in the baseline lake (tonnes $year^{-1}$), A_b is the area of the baseline lake (km^2), A_t is the area of the target lake (km^2), P_t is the average primary production of the target lake ($gC\,m^{-2}\,day^{-1}$), and P_b is the average primary production of the baseline lake ($gC\,m^{-2}\,day^{-1}$).

The addition of 1 to the primary production values in Equation 23.2b ensures that the logs do not become negative for lakes where primary production is less than 1 $gC\,m^{-2}\,day^{-1}$. In addition, the curve intersects the linear model at the values for the baseline lake.

Sources of data

Primary production values used in the model have been taken from the literature and evaluated using Lowe-McConnell (1987). Sustainable yields of African lake fisheries are taken as 90% of published MSY values where available, or are 90% of approximations of MSY made using available catch figures. Sources of data on primary production rates are Craig (1992) and Vanden Bossche and Bernacsek (1990, 1991). Sustainable fish yields for Lakes Victoria and Kariba are taken from Pitcher and Bundy (1994, Chapter

9), Lake Tanganyika from Hecky (1992) and Coulter (1992a), Lake Turkana estimated from Kolding (Chapter 16), and Lakes Albert and Chad estimated from figures in Craig (1992).

This model might be refined by including more differences among the lakes with regard to nutrients, morphology etc., or perhaps by incorporating some form of the morphoedaphic index. But to do this we would need more data on sustainable catches from fisheries resulting from successful introductions. Such data may be available for sardines in about 10 years' time as fisheries become established in Cahora Bassa and Lake Itezhi-tezhi.

Assessment of uncertainties in the estimates made using the model

Each estimation of potential yield was subjected to 2000 bootstrapped estimations using the error distributions of the principal model parameters, P_t, P_b and Y_b (Crystal Ball Software, Decisioneering Inc., Denver, CO, USA). In this way 95% confidence limits deriving from estimation error were attached to each yield estimate.

Sustainable yield values

Estimates of sustainable yield are notoriously subject to a great deal of error. Published error values have been used in a Normal distribution where estimation errors have been assessed, as with Nile perch in Lake Victoria (Pitcher and Bundy, 1994), but in other cases uniform distributions around the calculated figure have been taken. Values are shown in Table 23.1.

Primary production rates

Averaging primary productivity over a year ignores a large seasonal increase of the order of 50% as pelagic production is enhanced by nutrients released during seasonal upwelling and lake overturn. But the average annual value is useful for this exercise because these seasonal fluctuations, although reflected in fish growth and mortality, should be effectively integrated in fish production over the year. I have taken a value of 20% for the likely error in estimating average annual primary production. For example, this would mean that in Lake Victoria, for the published mean value of 2.64 $gC\,m^{-2}\,day^{-1}$, the confidence limits used in the model are 2.25 to 3.03 $gC\,m^{-2}\,day^{-1}$. Although there are few time series of mean lake production, this may not be wide enough to cover interannual variation in some lakes. For example, in Lake Malaŵi (Nkhotakota Bay), primary production averaged 0.9 $gC\,m^{-2}\,day^{-1}$ in 1992 and 1.4 $gC\,m^{-2}\,day^{-1}$ in 1993 (A. Menz, pers. comm.).

Table 23.1 Uncertainties investigated for the yield prediction model. Symbols as in text; RTF is the relative trophic transfer efficiency assumed in the lake (see text)

Baseline lake	Species	Parameter							
		Y_b		P_b		P_t		RTF	
		95% iles*	Dist.†	95% iles	Dist.	95% iles	Dist.	95% iles	Dist.
Victoria	Nile perch	28%	N	20%	N	20%	N	1	F
Chad	Nile perch	50%	U	20%	N	20%	N	1	F
Turkana	Nile perch	50%	U	20%	N	20%	N	1	F
Albert	Nile perch	50%	U	20%	N	20%	N	1	F
Kariba	Sardine	30%	U	20%	N	20%	N	1	F
Tanganyika	Sardine	30%	U	20%	N	20%	N	1.5–2.2	U

* The 95-percentiles of the distribution around the mean.
† Type of probability distribution employed: F, fixed; N, normal; U, uniform.

Lake Tanganyika energy anomaly

In Lake Tanganyika, most estimators of fish yield provide figures lower than the present and historical catches per km^2 that are known to have been sustained for many years (Bayona, 1992). There appear to be two reasons for this anomaly. First, the great antiquity of the lake appears to have allowed the evolution of highly efficient pelagic species so that carbon accumulates in planktivorous fish rather than as algae. Secondly, there is an additional source of energy fixation from heterotrophic bacteria in the immense deep anoxic zone (Hecky and Fee, 1981; Hecky *et al.*, 1981). Assessing the evidence that fish production levels imply twice the normal trophic transfer efficiency, Hecky (1992) showed that Lake Tanganyika has more affinity to marine and estuarine systems than the other African Great Lakes. For Lake Tanganyika calculations in the model I have therefore included uncertainty in the form of a trophic transfer efficiency multiplier (uniform probability distribution from 1.5 to 2.2 with a median of 2 – Table 23.1).

Choice of baseline lake

Results of this model clearly depend upon the choice of baseline lake. In general, I have attempted to compare results using as many baseline lakes as are feasible for each species. There are fisheries for introduced Nile perch in Lakes Victoria and Kyoga, with fisheries for similar natural *Lates* species in Lakes Albert, Turkana and Chad. Of these, data from Lake Kyoga are very uncertain given the difficulty of access (Chapter 4). So Victoria, Chad, Turkana and Albert are used in turn as baseline lakes for Nile perch introductions. Sardine fisheries in Lake Tanganyika (natural) and Lake Kariba (introduced) have been reasonably well assessed and are used as baseline lakes for sardine introductions; fishery data from Lake Kivu have not to date been good enough to use for MSY and therefore cannot be used as a baseline (Chapter 13).

23.4 RESULTS

Results pertaining to possible Nile perch introductions are shown on a logarithmic scale in Fig. 23.2(a). The 20 lakes are listed in order of area. For each lake there are four fish yield estimates from left to right, resulting from use of Lakes Victoria, Chad, Albert and Turkana as the baseline lake respectively. Arrows indicate the sustainable yields (0.9 MSY) used for each baseline lake. Error bars show the 95-percentiles from bootstrapping. Following a similar convention, results for possible introductions of sardines are shown in Fig. 23.2(b) using Lakes Tanganyika and Kariba as baseline lakes for the model.

(a) Nile perch introductions

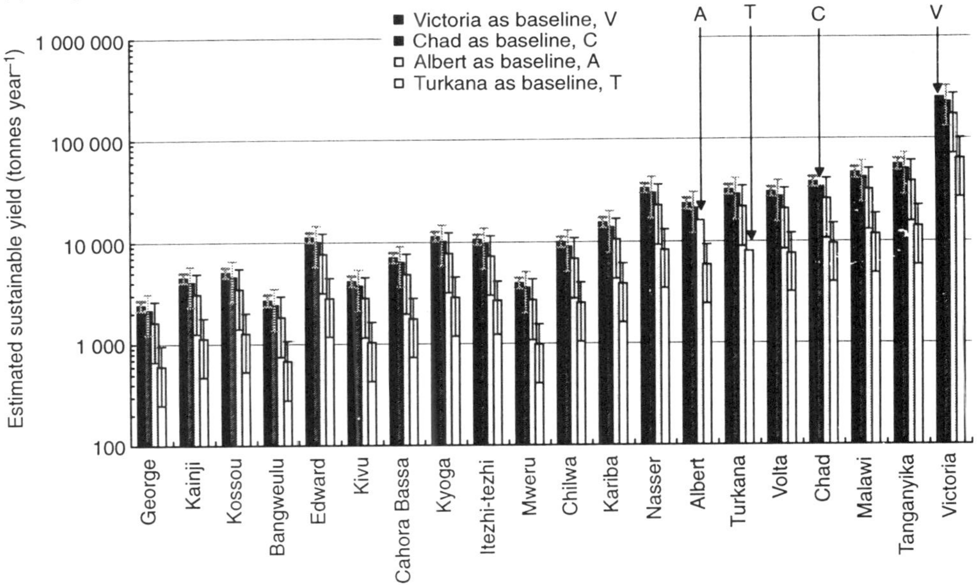

Fig. 23.2 Sustainable yields for fish introductions from 20 African lakes as estimated by the relative log primary production model described in the text. Columns show estimates for each lake, bars show bootstrapped 95% confidence limits. Baseline lakes for each run of the model are indicated by arrows. (a) Nile perch introductions, with four alternative baseline lakes. (b) Sardine introductions, with two alternative baseline lakes.

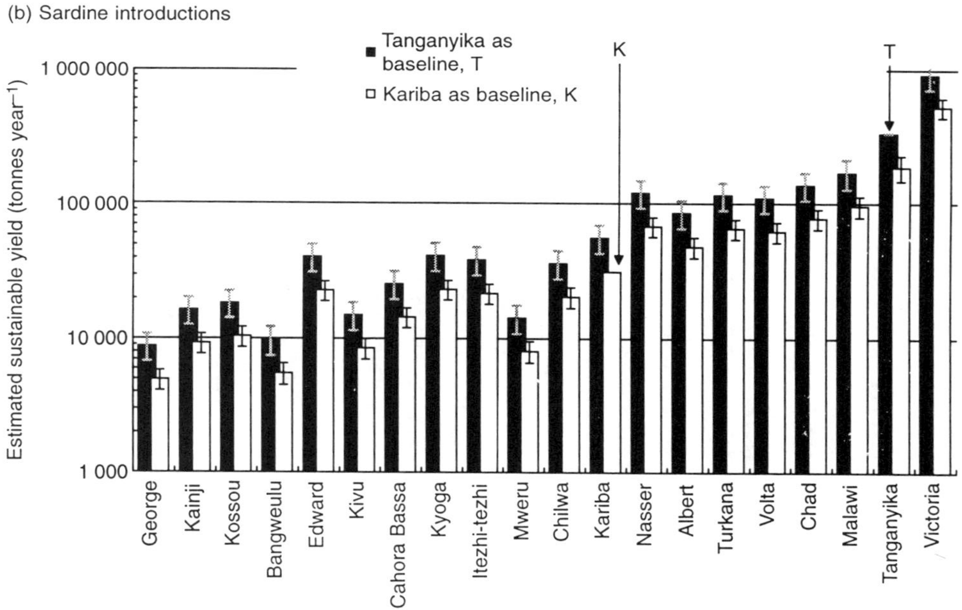

Figure 23.2 (continued).

23.5 DISCUSSION

First, it should be emphasized that use of the model is not intended to imply that an introduced fish would become established, or that the introduction is feasible on ecological grounds. For example, Lake Chilwa is too shallow and seasonal for either sardine or Nile perch, while Lake Mweru has its own three endemic sardine species.

Secondly, the consideration of estimation errors through the bootstrapping technique is not intended to subsume process errors. If the assumptions of the model are not met, its predictions will be invalid. The key assumption of the model is that successful establishment in the target lake means that there is sufficient ecological similarity to predict MSY using relative log primary production levels, within the limits of the estimation errors.

Before surveying the implications for selected major lakes one by one, some general features of the results are discussed.

Nile perch introductions

Estimates using Lake Victoria as the baseline lake for the model have the lowest estimation errors (Fig. 23.2(a)), which is not surprising because a quantitative assessment of the fishery is available. Using Lake Turkana as baseline produces figures considerably lower. Of all the lakes, sustainable yields of Nile perch in Lake Turkana are the most uncertain as they are based on a small artisanal fishery serving fluctuating and small local demand for fish. Kolding (Chapter 16) considers that environmental factors and lake level have been the most important factors in determining historical catch levels. It is encouraging, however, that using Lakes Victoria, Chad or Albert as baseline lakes for the model produces estimates that are not significantly different. Moreover, the Chad and Albert models provide estimates for Lake Victoria that are not significantly different from the actual value. Using the highest and lowest confidence limits from these three yield estimates, we obtain the total yields and rates per km^2 listed in Table 23.2.

Approximate world market prices for Nile perch sold as frozen fillets produce the range of possible economic returns listed in Table 23.2. Clearly, not all catch from a fishery will be sold as frozen blocks, but prices for other Nile perch products will be geared to this price. Totals are given for natural and human-made lakes.

In natural lakes, Table 23.2 indicates that the total annual economic return could be US$ 175–768 million from 144 000 to 632 000 t (tonnes) per annum sustainable yield. From artificial lakes the total yield might be worth US$ 36–166 million from 30 000 to 137 000 t. These figures are at the same time both pessimistic and optimistic. Assuming no additional catch from any

Table 23.2 Estimated yields of introduced Nile perch: high and low values were obtained from the bootstrapped model described in the text. Nile perch world market price of US$ 2700 per tonne frozen blocks from *Infofish*, March 1993, assuming 45% wet weight conversion to blocks

Lake*	Total yields (t year^{-1})		t km^{-2} year^{-1}		Value (million US$)	
	Low	High	Low	High	Low	High
George	672	3 111	0.90	4.15	0.82	3.78
Kainji (a)	1 261	5 829	0.99	4.55	1.53	7.08
Kossou (a)	1 412	6 548	0.88	4.09	1.72	7.96
Bangweulu	746	3 487	0.36	1.68	0.91	4.24
Edward	3 143	14 426	1.35	6.20	3.82	17.53
Kivu	1 157	5 336	0.49	2.25	1.41	6.48
Cahora Bassa (a)	1 980	9 094	0.74	3.41	2.41	11.05
Kyoga	3 165	14 658	1.17	5.43	3.85	17.81
Itezhi-tezhi (a)	2 975	13 700	0.80	3.70	3.61	16.65
Mweru	1 086	5 073	0.23	1.09	1.32	6.16
Chilwa	2 768	12 937	0.55	2.59	3.36	15.72
Kariba (a)	4 318	19 836	0.80	3.70	5.25	24.10
Nasser (a)	9 265	42 590	1.49	6.85	11.26	51.75
Albert	11 970	30 226	1.76	4.45	14.54	36.72
Turkana	8 943	41 076	1.18	5.43	10.87	49.91
Volta (a)	8 397	38 959	1.02	4.71	10.20	47.33
Chad	10 626	41 890	1.06	4.19	12.91	50.90
Malaŵi	12 997	60 232	0.42	1.96	15.79	73.18
Tanganyika	15 575	72 355	0.47	2.19	18.92	87.91
Victoria	70 999	326 935	1.04	4.81	86.26	397.23
Artificial (7)	29 608	136 555			35.97	165.91
Natural (13)	143 848	631 743			174.78	767.57
Total (20)	173 456	768 297			210.75	933.48

*Lakes marked (a) are human-made. Lakes are listed in order of increasing area.

other remaining fisheries in the lake is pessimistic. Assuming that all these introductions were both approved and established is optimistic.

Sardine introductions

With Lake Tanganyika as baseline, the yield predictions are about twice as high as using Lake Kariba. This may simply be because both inshore (*Limnothrissa*) and off-shore (*Stolothrissa*) species of sardines are included in the Lake Tanganyika figures. If reduced by half they are similar to the Lake Kariba baseline for *L. miodon*. However, I have no way of knowing what the appropriate reduction factor should be, and, following the procedure outlined for the Nile perch model above, have therefore used the Kariba baseline alone

Table 23.3 Estimated yields of freshwater sardine: high and low values are obtained from the bootstrapped model described in the text. Ranges for Lake Kariba were from assessment. Sardine values are taken as $1000 per tonne (see text)

Lake*	Total yields (t year^{-1})		t km^{-2} year^{-1}		Value (million US$)	
	Low	High	Low	High	Low	High
George	4 098	5 792	5.46	7.72	4.10	5.79
Kainji (a)	7 693	10 830	6.01	8.46	7.69	10.83
Kossou (a)	8 613	12 172	5.38	7.61	8.61	12.17
Bangweulu	4 515	6 537	2.18	3.15	4.51	6.54
Edward	19 098	26 752	8.21	11.51	19.10	26.75
Kivu	6 986	9 950	2.95	4.20	6.99	9.95
Cahora Bassa (a)	12 013	16862	4.51	6.33	12.01	16.86
Kyoga	19 405	27 164	7.19	10.06	19.41	27.16
Itezhi-tezhi (a)	18 041	25 416	4.88	6.87	18.04	25.42
Mweru	6 600	9 492	1.42	2.04	6.60	9.49
Chilwa	16 896	24 086	3.38	4.82	16.90	24.09
Kariba (a)	26 775	36 225	4.99	6.75	26.78	36.23
Nasser (a)	56 568	78 718	9.10	12.66	56.57	78.72
Albert	39 820	56 182	5.86	8.26	39.82	56.18
Turkana	54 478	76 088	7.20	10.05	54 48	76.09
Volta (a)	51 414	72 482	6.22	8.76	51.41	72.48
Chad	64 402	90 305	6.44	9.03	64.40	90.31
Malaŵi	78 639	112 468	2.55	3.65	78.64	112.47
Tanganyika	147 256	225 600	4.46	6.84	147.26	225.60
Victoria	433 216	605 647	6.37	8.91	433.22	605.65
Artificial (7)	188 102	262 654			188.10	262.65
Natural (13)	888 423	1 266 113			888.42	1 266.11
Total (20)	1 076 525	1 528 767			1 076.52	1 528.77

*Lakes marked (a) are human-made. Lakes are listed in order of increasing area.

to generate the likely range of economic returns from sardine introductions in Table 23.3.

The March 1993 price of Lake Kariba sardines was about ZIM$ 6 per kilogram, wet weight (or US$ 12 per kilogram dried). This is approximately US$ 1000 per tonne wet weight caught (1994 prices have been about 25% higher – *Eds*).

In natural lakes, Table 23.3 indicates that the total annual economic return could be US$ 0.9–1.3 billion from 888 000 to 1 266 000 t per annum sustainable yield. From human-made lakes the total sardine yield might be US$ 188–263 million from 188 000 to 263 000 t per annum. These figures are subject to the same qualifications as for Nile perch above.

The major West African artificial lakes, Kainji, Kossou and Volta, contain endemic clupeids from the original riverine fauna, comprising *Pellonula*, *Sierrathrissa* and *Cynothrissa* species. It is encouraging that the 'sardine' model predictions, of 51 000 to 72 000 t in Lake Volta and 7600 to 10 800 t in Lake Kainji, are in close accord with previous estimates of the potential clupeid fishery yields from these lakes (60 000 and 10 000 t respectively; Marshall, 1984).

Human-made lakes

The three human-made lakes in the Zambezi Basin seem ideal candidates for introductions because high falls prevent colonization up the Shire River to Lake Malaŵi, there are no natural lakes in the rest of the basin, and existing river fisheries are very minor. *Limnothrissa miodon* has been introduced to Lakes Kariba and Itezhi-tezhi, and colonized Cahora Bassa down the Zambezi from Lake Kariba. Model predictions for Lakes Nasser, in the lower Nile Basin, and Kainji, Nigeria, are also considered here.

Lake Kainji

Yield from the lake exceeded 30 000 t year^{-1} in the early 1970s, of which more than half was Nile perch. By 1977 (the last year for which good records are available), total catch was down to around 10 000 t of which only about 3000 t was Nile perch. This has been represented as the 'collapse' of the Nile perch fishery (Turner, 1995), but comparison with the model suggests that this may be more a fishing down of virgin biomass rather than a collapse from overexploitation. The model predicts a Nile perch sustainable yield of between 1300 and 5000 t year^{-1} and a sardine yield from 7700 to 10 800 t. Total sustainable yield from Nile perch and sardines combined would be approximately 9000–15 000 t year^{-1}. The sardine fishery appears to have expanded (Turner, pers. comm.). New data on the fishery should be available within the next couple of years and will provide a test of the model.

Lake Kariba

Introduced sardines currently provide a catch of around 5.5 t km^{-2}, which is approximately the maximum sustainable (Pitcher and Bundy, 1994), constitutes a respectable yield for any fishery, and is worth around $31 million annually to Zambia and Zimbabwe. The model estimates a range of 5 to 6.75 t km^{-2}, representing annual catches between 27 000 and 36 000 t (Kariba baseline, so confidence limits obtained from bootstrapped errors on primary production and yield figures). If the estimated 10 t km^{-2} sustainable yield from Lake Tanganyika could also be achieved in Lake Kariba, we would see a larger harvest of around 45 000 to 70 000 t from the whole lake, but this is unlikely as the Lake Tanganyika figure includes both species of sardine. The offshore sardine, *Stolothrissa*, failed to establish in Lake Kariba (Marshall,

1992). The relatively shallow depth of Lake Kariba (about 30 m) makes it more akin to the inshore regions of Lake Tanganyika, where *Limnothrissa* rather than *Stolothrissa* is predominant.

One option in this human-made lake would be to introduce Nile perch. The sardine population is evidently able to withstand a lot of predation, including that from tigerfish (*Hydrocynus forskhalii*, newly turned pelagic), birds, crocodiles and humans. So *Lates* probably would not wipe out the sardine fishery. The choice of gear for the perch fishery would be tricky because of drowned trees, but some sort of longline that would float at sardine depth above tree level might be devised. The effect on the tigerfish sport fishery on Lake Kariba would be unpredictable, but a recreational fishery for large *Lates* would also be popular, so sport fish income might be maintained in the lake even if tigerfish were reduced by competition from Nile perch. The Nile perch would presumably also colonize the Zambezi and Cahora Bassa, but at present would not be able to get up the Shire River into Lake Malaŵi. (However, there are rumours of a new falls by-pass canal under construction that would alter this situation.) An alternative, not covered by the existing model as we have no baseline, is to introduce one of the pelagic *Lates* species from Lake Tanganyika that coexist with sardines, such as *Lates stappersi*.

According to Table 23.2, the model suggests that we might expect about 12 000 t (confidence limits 4500–20 000) of Nile perch. The Nile perch fishery could create a lakeshore processing industry and frozen block exports worth between US$ 5–24 million per year. If the sardine fishery were completely lost, the conclusion is that there is obviously no economic gain to be made from introducing Nile perch. If we assume optimistically that a reduced sardine fishery of half the current amount can coexist with Nile perch, then we get a total 95% likelihood band of US$ 18–42 million in revenue. This means that there is a 55% chance that post-Nile-perch-introduction lake income would be less than that from the current fishery, and clearly, these are not very encouraging odds.

But this conclusion is sensitive to the sardine price: at half the assumed sardine price, Nile perch begins to look more attractive (US$ 16 million from sardines alone versus US$ 13–30 million from Nile perch plus a reduced sardine fishery, with an 80% chance of increasing revenue). This calculation becomes much less convincing, however, when we realize how sensitive it is to the size of the remaining sardine fishery, a very uncertain figure.

In summary, none of these considerations favours introduction of Nile perch to Lake Kariba.

Cahora Bassa

After the cessation of the recent war in Mozambique, an embryonic fishery has commenced in Cahora Bassa. According to Marshall (1992) there are many uncertainties in setting up a sardine fishery. In particular, volatility in sardine production levels may be exacerbated by large amounts of suspended clay, especially in the shallow upstream end of the lake. It was thought that the same problem might cause difficulties in light attraction to the lift net gear but this has not been borne out in practice (Mtsambiwa, pers. comm.).

Potential annual sardine yield predicted by the model is 12 000 to 17 000 t, rather more than the 8000 t predicted by Bernacsek (1984), but in line with Marshall's views (1992). Nile perch might produce around 2000–9000 t, worth in the order of US$ 2–11 million, but indications against its introduction are the same as in Lake Kariba.

Lake Itezhi-tezhi

There is little published information on Lake Itezhi-tezhi, apart from that in Chapter 15. Assuming primary production equal to that in Lake Kariba, the model estimates 18 000 to 25 000 t of sardines worth US$ 18–25 million annually to Zambia. This estimate should be modified once primary production figures for the lake become available. Actual sustainable yield once the fishery is operating will provide a good test of this model. At present there is some doubt concerning the establishment of *Limnothrissa* in the lake (Mtsambiwa, pers. comm.).

Lake Nasser–Nubia

In 1981 the reported catch from Lake Nasser–Nubia was about 35 000 t, 90% of which was tilapias and the rest mainly cyprinids (Abdel-Latif, 1984; Ali, 1984). Recent figures appear not to be available in the published literature. The model suggests that the productive Lake Nasser (including Lake Nubia) might produce over 56 000 to 78 000 t of sardines, but this ignores the reported low utilization of phytoplankton. The zooplankton-feeding fish community appears to be very small (Abdel-Latif, 1984) and so the future for a sardine introduction, suggested by Marshall (1984 and see Chapter 24), seems uncertain until the fisheries ecology of Lake Nasser is investigated in more detail.

Nile perch, naturally present in the lake since impoundment of the Nile by the Aswan High Dam, constituted only about 500 t of the catch in 1984. Therefore, with the limited information available, the model applied to Lake Nasser appears misleading.

Possibly a Nile tilapia version of the model would be more useful for Lake Nasser as, overall, the performance of the sardine and Nile perch models are less than satisfactory in this lake. Another option for Lake Nasser might be the introduction of Malaŵi chambo, *Oreochromis lidole*, which could feed on the high production of bluegreen phytoplankton (Pitcher *et al.*, 1989). We need up-to-date and reliable information about current fisheries in Lake Nasser to improve the evaluation.

Potential total yield from African human-made lakes

It is worth considering the potential fish yield for all the human-made African lakes. Unfortunately, primary production and sustainable fishery yields are available for only a few of the 104 artificial lakes referred to by Cowx and Kapasa (Chapter 15). However, we have sufficient information on the seven largest human-made lakes in Tables 23.2 and 23.3 with which to draw some tentative conclusions.

The model predicts a total potential sardine yield in these seven lakes of approximately 188 000 to 262 000 t year^{-1} (without Nasser 161 000 to 183 000), equivalent to 7.9–11.0 t km^{-2} overall. At a value of US$ 1000 per tonne this is worth in the region of US$ 200 million per year. Adding the smaller artificial lakes at a conservative average of 4 t sardines per km^2 produces another 30 000 to 65 000 t. For Nile perch, the model predicts a total of 29 000 to 136 000 t, worth about US$ 36–166 million per year under the assumptions set out above. Evidently, the sardines are a more valuable harvest. Alongside the model predictions, evaluations of the costs and benefits of introductions are generally favourable for sardines in most of these human-made lakes. Conversely, evaluation does not support Nile perch introductions, but this conclusion may be sensitive to sardine price.

Natural lakes

I consider here the model results and evaluation of sardine and/or Nile perch introductions for five major African lakes: Kyoga, Victoria, Tanganyika, Malaŵi and Kivu.

Lake Kyoga

In the mid 1970s, 70 000 t of Nile perch was harvested annually from Lake Kyoga, a yield of 26 t km^{-2} (Twongo, 1988). By the late 1980s, catches had declined to around 15 000 t. This history is generally represented as a 'collapse' of an unstable Nile perch fishery (e.g. Witte *et al.*, 1992; Chapters 4, 6). Yet the model estimates 3000 to 15 000 t as the likely sustainable yield of Nile perch in Lake Kyoga. So although the Lake Kyoga catch history may indeed represent a collapse in the sense of a fishery-driven stock reduction, the early high yields are only what would be expected in fishing down a virgin stock. Reduction of effort during the years in which access to the lake was problematical because of security problems in Uganda may have been a blessing in disguise, as the current Nile perch catch levels appear to be about right. Most of the catch (about 80%) now consists of the introduced Nile tilapia (Oguto-Ohwayo, 1990), and so, with reliable catch data, Kyoga might make a suitable baseline lake for estimating sustainable yields of that species.

Lake Victoria

Model predictions of Nile perch yields (Table 23.2) are 71 000 to 327 000 t. Although this band encompasses the MSY assessment, the lower range of values within the confidence intervals reflects the use of the uncertain Lake Chad and Lake Albert baselines. Clearly, in this case the more precisely estimated MSY value from the full assessment of Nile perch in Lake Victoria (Chapter 9) is more credible and useful. Nevertheless, the results from the model reinforce alarm that the present Nile perch fishery exceeds even the upper limit of the model's predictions for sustainable yield.

Potential yields of introduced sardine may be considered in relation to the endemic cyprinid *Rastrineobola* (dagaa, omena) (Table 23.3). The model estimates pelagic sardine yield (Lake Kariba baseline) at 433 000 to 606 000 t year^{-1} (6.4–8.9 t km^{-2}), which could be worth about 2–6 times as much as the present Nile perch fishery. Unfortunately, there is almost no information available from the emerging *Rastrineobola* fishery. Assessments of potential dagaa yield should be compared with this sardine figure when they become available. If the endemic *Rastrineobola* is unable to match this figure, sardine introduction might be contemplated. If it failed, the *status quo* would prevail. If it succeeded, pelagic yield would be enhanced. The presence of the lake prawn, *Caridina*, would ensure sardine nutrition, but it is likely that competitive exclusion between the sardines and dagaa would mean that only one of the two species would survive.

There seems to be no ethical case against new introductions in a lake that has undergone such drastic human-caused changes, many of which predate the Nile perch upsurge (Hecky, 1993), and which is continuing to change as the water hyacinth invades and as nutrient enrichment from the industries along the northern lakeshore accelerates. It is important to remember that the traditional Lake Victoria fisheries were severely depleted by overexploitation long before the Nile perch upsurge (Bruton, 1990; Chapter 7). The rationale behind the Nile perch, tilapia and the failed carp introductions was to mitigate this overexploitation. Rather than considerations of biodiversity, a paramount consideration for any new introduction to Lake Victoria is that it should not disrupt existing successful fisheries for Nile perch, tilapias and the emerging harvest of small pelagic species.

In this respect, other introductions and restorative management programmes might be worth considering. Introduction of fish such as air-breathers that could withstand low oxygen in the new lake regime might be considered. Carp failed to establish, but other benthic grazers and small predators from SE Asia or South America might be usefully canvassed (cyprinids, characins, catfish, anabantids or snakeheads). Some consideration could be given to encouraging species of high value in the traditional fishery. This has already been suggested for lost endemic haplochromine species (Ribbink, 1987). Haplochromine species proven to survive Nile perch predation may be helped by establishing suitable habitat refuges and reserves. Nevertheless, it would be worthwhile putting some effort into the restoration of some high-value traditional fishery species such as *Oreochromis esculentus*, the air-breathing *Protopterus* and mormyrids. This might be through small-scale ranching and aquaculture enterprises associated with habitat reconstruction or lakeshore pond culture. Feasibility studies would be required at the start, and one problem to be solved is the hybridization that may have occurred between *O. esculentus* and *O. niloticus*. Experimental 'mesocosm' experiments might help in this venture.

Lake Tanganyika

A potential sardine yield of 147 000 to 226 000 t of sardines is predicted by the Lake Kariba baseline model, and this is more than the present reported catch of around 70 000 t. But these figures represent only 4.5–6.8 t km^{-2}, while a range of independent sustainable yield estimates based on primary production and on sardine biomass suggest a lakewide average of 10 t km^{-2} (Coulter, 1992a; Hecky, 1992; Marshall, 1993), as used in the Lake Tanganyika baseline model described above. In fact, total pelagic yields of up to 19–30 t km^{-2} may be achievable in the productive Burundi and Zambian regions at the north and south ends of the lake respectively. The problem with the Lake Kariba-based model is first, that it is based only on *Limnothrissa*, whereas pelagic catches in Lake Tanganyika include *Stolothrissa* and small pelagic *Lates stappersi*. Secondly, the higher energy transfer efficiency in the Lake Tanganyika ecosystem makes the model's predictions more uncertain. A considerable potential expansion of sardine catches is evident from both the model and the published assessments of sardines in this lake, although care about underreporting of catch statistics and local depletion of stocks has to be borne in mind.

The model predicts a 'Nile perch' catch of between 16 000 and 72 000 t, but application of the model is complicated by the presence of four endemic *Lates* species. The only one of these currently sustaining large catches is *Lates stappersi*, a specialized pelagic predator that seems to be resilient to overfishing because its fry out-compete sardines and it can switch its diet from sardine to shrimps (Coulter, 1992a). The larger pelagic species, *L. microlepis*, declines rapidly when fished. The model is not relevant to these species.

Catches of the large benthic species, *L. angustifrons* and *L. mariae*, declined rapidly in areas where they were exploited in Zambia and in Burundi (Coulter, 1992b). Data show that the species closest to Nile perch, *L. mariae*, considered the keystone of the unexploited fish community (Paine, 1969), declined in abundance in the SW Arm of the Zambian sector of the lake during expansion of the artisanal gill net fishery in the 1960s. The Nile perch model sheds some light on this decline, suggesting a sustainable yield of 0.5–2.2 t km^{-2}. Catch data for 1972 given in Coulter (1992b) for the 2000 gill nets in the SW Arm imply an *L. mariae* catch of 2.2–4.1 t km^{-2}, if we assume fishing 250 nights per year (or 3.2–6.0 t km^{-2} for 365 nights' fishing). The model suggests that catches may have been at least twice the estimated sustainable level given above, which calculation can show is roughly equivalent to a nightly catch of 1–4 kg of *Lates* per net rather than the 6–10 kg per net recorded catch at that time.

A much more rapid decline in *L. mariae* catches (and mean size) in the SE Zambian arm was likely due to the expansion of a nocturnal pelagic purse seine fishery operating with light attraction ('chiromila' net). Although this gear targets pelagic sardines and *L. stappersi*, Coulter (1992b) points out that nocturnal vertical migration of *L. mariae* renders the species vulnerable to this gear.

If these analyses are correct, the recorded decline of *L. mariae* in the Zambian sector from overfishing is not surprising, but is no reason to extend the tendency to decline to all *Lates* fisheries irrespective of fishing effort. If the model is correct, reasonable sustainable catches of large endemic *Lates* are possible in Lake Tanganyika.

In Lake Tanganyika not only is there a strong ethical case against any introductions by virtue of the unique cichlid fauna, but also there is no question of empty niches because the pelagic zone is fully occupied. The main management thrust should be to increase its endemic fish production towards its potential, whilst regulating all fisheries more carefully in the areas exhibiting symptoms of overexploitation. Coulter's (1992a) estimate of potential sustainable yield of 10 t km^{-2} is at the lower end of the range of possible fish yields: even at this level, Lake Tanganyika is potentially one of the richest fish-producing lakes in the world.

Lake Malaŵi

In Lake Malaŵi, the ethical case against fish introductions based on the intrinsic value of species flocks of endemics may be tempered by the need to feed a large undernourished population growing at a rate of 3% per year, a problem exacerbated by refugees from the war in Mozambique. Despite a highly fertile food-production base, these factors have placed a severe strain on the agricultural and fisheries production of the country. One politician has been reported as saying "... of course we do not want to spoil the lake, but we cannot afford to keep it simply as a living museum".

The existing lake fisheries, which currently supply 50% of the animal protein in the country, cannot be significantly increased (Chapter 17). At one stage it was thought that pelagic planktivores might be underexploited, but recent work by ODA (A. Menz, unpublished data and pers. comm.) has shown that this is not the case. There is some scope for increasing aquaculture in the country, but this is unlikely to contribute more than 10% of requirements. Accordingly, at some stage it may be expedient to consider if the current fish yield from Lake Malaŵi could be increased by species introductions.

As in Lake Victoria, the potential yields of introduced sardines have to be considered in relation to an endemic cyprinid *Rastrineobola (Engraulicypris) sardella* (usipa). The model suggests that sardines established in Lake Malaŵi could provide 79 000 to 112 000 t, or 2.6–3.6 t km^{-2}. But it seems likely that existing pelagic planktivores, including usipa, two groups of cichlids and a catfish currently produce more than this. *Diplotaxodon limnothrissa* is a previously unreported mesopelagic species that constitutes 65% of the demersal trawl fishery (Turner, 1994; Chapter 18). Pelagic surveys (A. Menz, pers. comm.) have revealed unexpectedly high numbers of *Rhamphochromis* (two species), *Synodontis* and two species of *Diplotaxodon*. Together these species could likely provide an annual catch 1.5–2 times the amount estimated by the model for sardines. In fact, usipa alone may produce a catch of this magnitude in some years (Lewis and Tweddle, 1990). Usipa year class strengths vary quite widely (Tweddle and Lewis, 1990) and food limitation on recruitment was confirmed by Menz (pers. comm.), who carried out twelve lakewide surveys in 1992–1993, finding many usipa larvae but very few adults. Recent invertebrate surveys (Menz, pers. comm.) analysed by ECOPATH modelling suggest that there is no excess of secondary production in the form of lake fly larvae, *Chaarborus edulis*, available to be cropped by a sardine, an introduction suggested by J. Turner (1982; Chapter 17). Trophic transfer efficiency is estimated at 8.2%, about what one would expect for a system with four to five trophic levels. So the prediction of the new model here supports both the ECOPATH finding and Iles' (1960) speculation that usipa is limited by zooplankton production and that there is no reason to suppose that a clupeid could do any better.

The model suggests that a sustainable Nile perch catch in Lake Malaŵi would be between 13 000 and 60 000 t, worth US$ 16–73 million if all was converted to frozen fillets. There is no large fast-moving piscivore in the lake, so the perch would find an empty niche and it is unlikely that there would be any difficultly in establishment, as the lake teems with suitable pelagic and benthic prey. Successful establishment of Nile perch in Lake Malaŵi would drastically alter the ecosystem and quite likely, as a worst-case scenario, bring about the collapse of most existing fisheries. Exceptions might be usipa and other pelagic zooplanktivores, because the Lake Victoria *Lates* coexists with a pelagic cyprinid. A number of the inshore tilapias would probably survive, as *Oreochromis* spp. do in Lake Victoria. Some rocky shore species of haplochromines would probably find refuges, but other fish of commercial significance could be assumed to be lost. Almost certainly the balance sheet for the introduction should assume that the offshore chambo (*Oreochromis*) species would be eaten to extinction. Most would regard these drastic changes to the Lake Malaŵi ecosystem as totally unacceptable on ethical grounds.

But two further questions are important for the balance sheet of such an introduction in Lake Malaŵi. Would more money be earned? Would more fish be produced?

The current value of the catch in Lake Malaŵi amounts to 75 000 to 100 000 t at 0.5 Kwacha per kilogram, giving a total value of 37–50 million Kwacha, or about US$ 13–17.5 million (Mkoko, 1992). Virtually all of this income is made internal to the country on a small local scale. The model suggests that the Nile perch catch would be worth 1.5 to 4 times as much as this and moreover, would be in foreign currency. The Nile perch fishery would also create a number of jobs in lakeshore processing factories, and, as in Lake Victoria (Chapter 11) there would be large socio-economic changes. The answer to the first question is therefore that such an introduction of Nile perch might increase sustainable revenue. Short-term economic gains as Nile perch hoovered up the most vulnerable prey might be large, and the temptation to expand the fishery infrastructure at this stage should be resisted.

On the other hand, given the likely destruction of most of the existing fisheries, comprising a sustainable 75 000 to 120 000 t, long-term sustainable yield is not going to be increased by introducing Nile perch. Because the immediate and primary need is food protein in Malaŵi, the introduction could not be justified on these grounds. Furthermore, in calculating the estimated value of the Nile perch above it is assumed that all of the catch would be sent for export, so the Nile perch fishery would not alleviate the present food and protein shortage. Some of the money earned could be used to buy grain or other vital foodstuffs, and some of the revenue might filter back into the local economy through job creation, but the Lake Victoria situation suggests that such socio-economic benefits would be patchy (Chapter 11). Of course, a portion of the catch could be exported and the rest eaten at home, and a whole range of trade-offs of this kind might be envisaged. As in Lake Victoria, the high price of Nile perch for the export market would likely mean that the total amount of fish consumed at home is likely to be less than that prior to the introduction, except perhaps during the initial years after establishment.

In summary, the balance of costs and benefits favours neither a sardine nor a Nile perch introduction into Lake Malaŵi.

Lake Kivu

The model predicts 7000–10 000 t of sardines (3–4.2 t km^{-2}), which have, of course, already been introduced. Potential catch has been estimated widely at between 3000 and 67 000 t, but the model is in accord with Marshall's (1992) calculation that the most likely value is in excess of 6000 t. The actual (recorded) Kivu catch of 2500 t from Rwanda and Zaïre seems so much less than the potential catch of this lake that management should be directed at this shortfall in the first instance. Sardine introduction was followed by major changes to the plankton community (Chapter 13). There is a small species flock of haplochromines in the lake but the effects of the sardine introduction on them are unknown (Marshall, 1992).

Nile perch might yield only 1100–5300 t, worth US$ 1–6 million. As in Lake Malaŵi, this would only be worth doing if there was some indication that the sardine fishery would survive. At present so little is known of the ecology and of the existing fishery in the lake that few of the questions germane to the decision could be answered with any confidence.

Total fish yields of large natural lakes in Africa

The thirteen largest natural lakes in Africa, listed in Tables 23.2 and 23.3, amount to a surface area of around 176 000 km^2, an area equivalent to the Gulf of St Lawrence, the Persian Gulf, or one-third that of the North Sea. The estimated total sardine yield from the model is 0.9–1.3 million tonnes, and Nile perch yield about 132 000–631 000 t. The approximate total value of this catch would be US$ 1–2 billion. However, in nearly all cases the balance of costs and benefits considered here does not favour fresh introductions of either Nile perch or sardines.

A test of the model: estimating yields from haplochromine fisheries

It has been argued (G.F. Turner, pers. comm.) that the upsurge of Nile perch has forced the Lake Victoria fishery to forego yields of haplochromines. Turner has made the interesting suggestion that the present model may be put to the test in estimating what the sustainable yield of the Lake Victoria haplochromine fishery could have been had the introduced Nile perch not taken over. Lake Malaŵi is a suitable baseline lake for the model because there is a long-established and well-assessed fishery for haplochromines (Chapters 17, 18). Alternatively, the SE Arm of Lake Malaŵi, the shallow, most productive region, might be used as a baseline.

The model estimates that Lake Victoria could produce haplochromine yields of around 5–8 t km^{-2}, equivalent to 260 000 to 770 000 t of fish. Lake Malaŵi itself might produce between 40 000 and 140 000 t of haplochromines at 2–4 t km^{-2}. The upper range of these figures appears to be unrealistic as they are based on the SE Arm of Lake Malaŵi and it is probably not wise to use small portions of lakes in the model. Using the whole of Lake Malaŵi as baseline produces a more believable range of estimates for Lake Victoria of 266 000 to 550 000 t.

Thus, the Nile perch, the sustainable yield of which is in the order of 280 000 t (Chapter 9), has meant that up to 270 000 t of sustainable yield from Victoria's haplochromines may have been foregone, although it is perhaps relevant to point out that the confidence limits of this estimate include zero.

23.6 SUMMARY

This Chapter sets out a detailed balance sheet of biotic, economic and social costs and benefits for the evaluation of fish introductions in African lakes. A new model for estimating sustainable fishery yields for introduced Nile perch or freshwater sardine is based on using relative log primary productivity to scale the sustainable fishery yield per unit area from a reasonably well-studied baseline lake. Implementation of the model employs bootstrapped confidence limits and comparisons of results from several alternative baseline lakes. The better the information from the baseline lakes, the more accurate the model predictions. The results of the candidate model are explored through application to 20 human-made and natural lakes.

Results are convincing and helpful for most of the lakes considered, with the exception of Lake Nasser. In general, the model suggests that new introductions of Nile perch will not result in an increase in fish yield, nor in higher economic returns. Results illuminate the collapses of overexploited *Lates* fisheries in Lakes Kainji, Kyoga and Tanganyika. The model predicts that introduction of sardines is unlikely to increase pelagic yield in Lake Malaŵi over indigenous cyprinid and cichlid fish. The situation is unclear in Lake Victoria, where an existing fishery for a pelagic cyprinid remains unassessed. In Lake Tanganyika, the model suggests that a considerable increase in sardine yield may be possible. Future sardine fishery yields from human-made Lakes Itezhi-tezhi, Kainji and Cahora Bassa will prove good tests of the model. In conjunction with detailed consideration of the costs and benefits, the model can be used to evaluate new fish introductions to lakes in Africa.

ACKNOWLEDGEMENTS

I am grateful to Ray Beverton, Alida Bundy, Daniel Pauly and George Turner for comments on earlier versions of this Chapter. This work was prepared with support from the Overseas Development Administration of the United Kingdom government.

REFERENCES

Abdel-Latif, A-F. (1984) Lake Nasser, in *Status of African Reservoir Fisheries* (eds J.M. Kepetsky and T. Petr), *FAO/CIFA tech. Pap.*, no. 10, pp. 193–246.

Ali, M.el-T. (1984) Lake Nubia, in *Status of African Reservoir Fisheries* (eds J.M. Kepetsky and T. Petr), *FAO/CIFA tech. Pap.*, no. 10, pp. 247–60.

Badjik, C.D. and Schnieder, D.C. (1991) Models of the fish yield from lakes: does the random component matter? *Can. J. Fish. aquat. Sci.*, **48**, 619–22.

Bayona, J.D.R. (1992) A review of the biological productivity and fish production in Lake Tanganyika. *FAO/CIFA occ. Pap.*, no. 15, pp. 1–17.

Bernacsek, G.M. and Lopes, S. (1984) Cahora Bassa, in *Status of African Rservoir Fisheries* (eds J.M. Kepetsky and T. Petr), FAO, Rome, *CIFA tech. Pap.*, no. 10, pp. 21–42.

Briand, F. (1985) Structural singularities of freshwater food webs. *Verh. int. Verein. Theor. Angew. Limnol.*, **22**, 3356–64.

Bruton, M.N. (1990) The conservation of the fishes of Lake Victoria, Africa: an ecological perspective. *Env. Biol. Fishes*, **27**, 161–75.

Coulter, G.W. (1992a) Fisheries, in *Lake Tanganyika and its Life* (ed. G.W. Coulter), Oxford, Oxford Univ. Press, pp. 139–50.

Coulter, G.W. (1992b) Benthic fish community, in *Lake Tanganyika and Its Life* (ed. G.W. Coulter), Oxford, Oxford Univ. Press, pp. 151–99.

Craig, J.F. (1992) Human-induced changes in the composition of fish communities in the African Great Lakes. *Rev. Fish Biol. Fish.*, **2**, 125–43.

Crul, R.C.M. (1992) Models for estimating potential fish yields of African inland waters. *FAO/CIFA Occ. Pap.*, no 16, 22 pp.

Downing, J.A., Plante, C. and Lalonde, S. (1990) Fish production correlated with primary productivity not the morphoedaphic index. *Can. J. Fish. aquat. Sci.*, **47**, 1929–36.

Evans, D.O., Henderson, B.A., Bax, N.J., Marshall, T.R., Oglesby, R.J. and Christie, W.J. (1987) Concepts and methods of community ecology applied to freshwater fisheries management. *Can. J. Fish. aquat. Sci.*, **44** (Suppl. 2), 448–70.

Fryer, G. and Iles, T.D. (1972) *The Cichlid Fishes of the Great Lakes of Africa*, Edinburgh, Oliver & Boyd, 641 pp.

Hecky, R.E. (1992) The pelagic community, in *Lake Tanganyika and Its Life* (ed. G.W. Coulter), Oxford, Oxford Univ. Press, pp. 90–110.

Hecky, R.E. (1993) The eutrophication of Lake Victoria. *Verh. int. Verin. Limnol.*, **25**, 39–48.

Hecky, R.E. and Fee, E.J. (1981) Primary production and rates of algal growth in Lake Tanganyika. *Limnol. Oceanogr.*, **26**, 532–47.

Hecky, RE., Fee, E.J., Kling, H.J. and Rudd, J.W.M. (1981) Relationship between primary production and fish production in Lake Tanganyika. *Trans. Am. Fish. Soc.*, **110**, 336–45.

Henderson, H.F. and Welcomme, R.L. (1974) The relationship of yield to MEI and numbers of fishermen in African Inland waters. *FAO/CIFA Occ. Pap.*, no. 1, 19 pp.

Henderson, H.F., Ryder, R.A. and Kudhongania, A.W. (1973) Assessing fishery potentials of lakes and reservoirs. *J. Fish Res. Bd Can.*, **30**, 2000–9.

Iles, T.D. (1960) A group of zooplankton feeders of the genus *Haplochromis* in Lake Nyasa. *Ann. Mag. nat. Hist.*, **2**, 257–80.

Laing, Y., Melack, J.M. and Wang, J. (1982) Primary production and fish yields in Chinese ponds and lakes. *Trans. Am. Fish Soc.*, **110**, 346–50.

Leach, J.H., Dickie, L.M., Shuter, B.J., Borgmann, U., Hyman, J. and Lysack, W. (1987) Review of methods for production of potential fish production with application to the Great Lakes and Lake Winnipeg. *Can. J. Fish aquat. Sci.*, **44** (Suppl. 2), 471–85.

Lewis, D.S.C. and Tweddle, D. (1990) The yields of usipa (*Engraulicypris sardella*) from the Nankumba Peninsula, Lake Malawi (1985–1986). *Coll. Rep. Fish. Res. Malawi, Occ. Pap.*, **1**, 57–66.

Lowe-McConnell, R.H. (1987) *Ecological Studies in Tropical Fish Communities*, Cambridge, Cambridge Univ. Press, 382 pp.

Marshall, B.E. (1984) Small pelagic fishes and fisheries in African inland waters. FAO, Rome, *CIFA tech. Pap.*, no. 14, 25 pp.

Marshall, B.E. (1992) *Limnothrissa miodon* and its fisheries in the African lakes. Consultancy report to Renewable Resources Assessment Group, London, 63 pp.

Marshall, B.E. (1993) Biology of the African clupeid *Limnothrissa miodon* with reference to its small size in artifical lakes. *Rev. Fish Biol. Fish.*, **3**, 17–38.

Melack, J.M. (1976) Primary productivity and fish yields in tropical lakes. *Trans. Am. Fish. Soc.*, **105**, 575–80.

Mkoko, B.J. (1992) Opening Address, page 100 in *Fisheries Management of the South East Arm of Lake Malawi* (eds FAO), FI: DP/MLW/86/013, *tech. Pap.*, no. 1, 106 pp.

Oglesby, R.T. (1977) Relationships of fish yield to lake phytoplankton standing crop, production and morphoedaphic factors. *J. Fish. Res. Bd Can.*, **34**, 2271–9.

Oglesby, R.T., Leach, J.H. and Forney, J. (1987) Potential *Stizostedion* yields as a function of chlorophyll concentration with special reference to Lake Erie. *Can. J. Fish. aquat. Sci.*, **44**, 166–70.

Oguto-Ohwayo, R. (1990) The decline of the native fishes of Lakes Victoria and Kyoga (East Africa) and the impact of introduced species, especially the Nile perch, *Lates niloticus* and the Nile tilapia, *Oreochromis niloticus. Env. Biol. Fishes*, **27**, 81–96.

Paine, R.T. (1969) A note on trophic complexity and community stability. *Am. Nat.*, **103**, 91–3.

Pitcher, T.J. and Bundy, A. (1994) Successful species introductions in the African lakes: assessment, uncertainties and strategies for fishery management, in *Management Strategies for Exploited Fish Populations* (eds G. Kruse, D.M. Eggers, R.J. Marasco, C. Pantzko and T.J. Quinn), Alaska Sea Grant College Program 98-02, Fairbanks, AL, pp. 545–70.

Pitcher, T.J., Turner, G.F. and Grimm, A.S. (1989) The cichlid fish *Oreochromis lidole* as a candidate for introduction to tropical impoundments. Report NRD/R4334 to Overseas Development Administration, London, UK, 103 pp.

Ribbink, A.J. (1987) African lakes and their fishes: conservation scenarios and suggestions. *Env. Biol. Fishes*, **19**, 3–26.

Ryder, R.A. (1965) A method for estimating the potential fish production of North Temperate lakes. *Trans. Am. Fish. Soc.*, **94**, 214–18.

Ryder, R.A. (1978) Fish yield assessment of large lakes and reservoirs – a prelude to management, in *Ecology of Freshwater Fish Production* (ed. S.D. Gerking), Oxford, Blackwell, pp. 403–23.

Ryder, R.A. (1982) The MEI – use, abuse and fundamental concepts. *Trans. Am. Fish. Soc.*, **111**, 154–64.

Ryder, R.A., Kerr, S.R., Loftus, K.H. and Regier, H.A. (1974) The MEI, a fish yield estimator – review and evaluation. *J. Fish Res. Bd Can.*, **31**, 663–88.

Schnieder, D.C. and Headrich, R.L. (1989) Prediction limits of allometric equations: a reanalysis of Ryder's morphoedaphic index. *Can. J. Fish. aquat. Sci.*, **46**, 503–8.

Tiews, K. (1985) EIFAC-Richtlinie ueber Einfuehrung nicht einhemimischer aquatischer organismen. *Arb. Dtsch. Fisch. Verb.*, no. 37a, 88–103.

Turner, G.F. (1994) A description of a commercially important new pelagic species of the genus *Diplotaxodon* (Pisces: Cichlidae) from Lake Malawi. *J. Fish Biol.*, **44**, 799–807.

Turner, G.F. (1995) Maximisation of yields from African Lakes, in *Stock Assessment in Inland Fisheries* (ed. I.G. Cowx), Oxford, Fishing News Books. (In press.)

Turner, J.L. (1982) Lake flies, water fleas and sardines. *Biological studies on the pelagic ecosystem of Lake Malawi*. Rome, FAO, FI: DP/MLW/75/019 *tech. Rep.* **1**, 165–73.

Tweddle, D. and Lewis, D.S.C. (1990) The biology of usipa (*Engraulicypris sardella*) in relation to fluctuations in productivity in Lake Malawi and species introductions. *Coll. Rep. Fish. Res. Malawi. Occ. Pap.*, **1**, 57–66.

Twongo, T (1988) Recent trends in the fisheries of Lake Kioga, Uganda, in Predator-prey relationships, population dynamics and fisheries productivities of large African lakes (ed. D. Lewis), *FAO/CIFA occ. Pap.*, no. 15, pp. 140–51.

Vanden Bossche, J.P. and Bernacsek, G.M. (1990) Source book for the inland fishery resources of Africa, Vol. 1. FAO, Rome, *CIFA tech. Pap.*, **18**(1), 411 pp.

Vanden Bossche, J.P. and Bernacsek, G.M. (1991) Source book for the inland fishery resources of Africa, Vol. 2. FAO, Rome, *CIFA tech. Pap.*, **18**(2), 240 pp.

Welcomme, R.L. (1984) International transfers of inland fish species, in *Distribution, Biology and Management of Exotic Fishes* (eds W.R. Courtenay, jr and J.R. Stauffer), FAO, Rome, pp. 22–40.

Welcomme, R.L. (1986) International measures for the control of introductions of aquatic organisms. *Fisheries*, **11**(2), 4–9.

Welcomme, R.L. (1988) International introductions of inland aquatic species. *FAO Fish. tech. Pap.*, no. 294, 318 pp.

Welcomme, R.L. (1992) A history of international introductions of inland aquatic species. *ICES mar. Sci. Symp.*, **194**, 3–14.

Witte, F., Goldschmidt, T., Wanink, J., Van Oijen, M., Goudswaard, K., Witte-Mas, E. and Bouton, N. (1992) The destruction of an endemic species flock: quantitative data on the decline of the haplochromine cichlids of Lake Victoria. *Env. Biol. Fishes*, **34**, 1–28.

Chapter twenty-four

Why is *Limnothrissa miodon* such a successful introduced species and is there anywhere else we should put it?

Brian E. Marshall

24.1 INTRODUCTION

The food crisis facing sub-Saharan Africa is a major international problem which shows few signs of abating (Anon., 1990). Per capita food production is decreasing and the region relies more heavily than any other on imported food. It has the greatest proportion of land with declining fertility and the highest percentage of population suffering from malnutrition. Moreover, Africa has the lowest per capita production of meat, milk and fish (Anon., 1990) and an increase in the availability of protein is urgently needed in most countries.

The continent's fisheries are expected to contribute to these requirements and many water bodies are intensively fished. Some are perceived as being unproductive, with fish yields that do not reflect their potential productivity. A case in point was Lake Malaŵi where, it was claimed, the potential productivity of the pelagic waters was not being fully realized (Turner, 1982) because the native planktivore *Rastrineobola* (= *Engraulicypris*) *sardella* was apparently unable to utilize pelagic resources adequately (Chapters 17, 23).

The Impact of Species Changes in African Lakes. Edited by Tony J. Pitcher and Paul J.B. Hart. Published in 1995 by Chapman & Hall, London. ISBN 0 412 55050 4.

It was suggested that the problem could be solved by introducing the Lake Tanganyika clupeid *Limnothrissa miodon*, a freshwater 'sardine'.

The introduction of alien fish species into lakes and rivers is a management practice with a long history and many have been translocated from one continent to another or from different parts of the same continent (Welcomme, 1981; Chapter 23). If successful, introductions can transform a water body's fish production, sometimes quite dramatically. The Nile perch, *Lates niloticus*, introduced to Lake Victoria about 30 years ago, and *Limnothrissa miodon*, introduced to Lakes Kariba and Kivu, both illustrate this point (Chapter 23, Fig. 23.1).

In Lake Kariba the pelagic fishery, consisting almost entirely of *Limnothrissa*, now produces nearly ten times as much as the inshore fishery, which utilizes the indigenous fish species (Table 24.1). There is no evidence of overfishing (Machena and Mabaye, 1987) and a considerable increase in the yield might be possible (Marshall, 1992a; Chapter 23). Much the same applies in Lake Kivu where the catch has risen from a few hundred tonnes before the sardines were stocked to about 4300 t per annum (M. Lamboeuf, pers. comm.) (Chapter 13).

The introduction of *Limnothrissa* into Lake Kariba was one of the most cost-effective fisheries development projects undertaken. The total cost of the introduction, including research, was around US$ 500 000 (Jackson, cited in Eccles, 1985) whilst the cumulative value of the catch, since the fishery began in 1974, now exceeds US$ 200 million. This is in marked contrast with other fisheries developments in Africa, such as aquaculture, where the returns have seldom matched the investment (Cross, 1991).

However, the benefits of alien species come at the price of irreversible changes to the ecosystem. For example, the piscivorous Nile perch appears to have helped to eliminate some of the endemic cichlid fauna of the lake (Ogutu-Ohwayo, 1990; Chapter 7) and its introduction has been almost universally condemned by biologists (Barel *et al.*, 1985) in spite of its dramatic

Table 24.1 Fish production (tonnes) and productivity of Lake Kariba in 1989. Based on unpublished data from the Lake Kariba Fisheries Research Institute (Zimbabwe) and the Department of Fisheries (Zambia); an average lake area of 4900 km^2 is assumed for calculating productivity

	Production (t)			Productivity ($t\,km^{-2}$)
	Zimbabwe	Zambia	Total	
Inshore catch	1 530	2 060	3 590	0.73
Pelagic catch	21 426	8 349	29 775	6.08
Total	22 956	10 409	33 365	6.81

economic impact (considered in Chapters 10 and 11 in this volume). *Limnothrissa* is not, of course, a major fish predator and its ecological effects have been less dramatic (Marshall, 1991). In spite of this, it has been described as an "ecodisaster" in Lake Kivu (Dumont, 1986) and Turner's (1982) proposal to introduce it into Lake Malaŵi has been severely criticized (Eccles, 1985; McKaye *et al.*, 1985; Balon and Bruton, 1986; Chapter 23).

This chapter addresses three questions about *Limnothrissa* which need to be answered if this species is to be introduced elsewhere. Why is it so successful? What are the ecological consequences of introduction? Where should it be introduced?

24.2 WHY IS *L. MIODON* SUCH A SUCCESSFUL SPECIES?

Presence of a clearly identifiable ecological niche

The lack of an obvious ecological niche has led to the failure of many introductions, such as the various European fish species that were brought to Zimbabwe in the 1930s and 1940s (Toots, 1970). A vacant niche should clearly be available before any introduction can be justified. This requirement was met in both Lake Kariba and Lake Kivu because they each lacked pelagic planktivorous fish. This is obviously the main reason why *L. miodon* was able to establish itself and spread so rapidly (Fig. 24.1).

Balon (1974) suggested that the characid *Brycinus lateralis* would eventually have colonized the open waters of Lake Kariba. This view is debatable because although *Brycinus* does eat zooplankton, it is relatively unimportant in the diet (Mitchell, 1976). Furthermore, *Brycinus* had not colonized the open water of the lake in the 10 years following impoundment whilst *Limnothrissa* did so competely in less than 2 years. This suggests that *Brycinus* is poorly adapted to feeding on zooplankton and that there was very little competition between the two species.

Biological factors that favour *Limnothrissa* as an introduced species

Limnothrissa has not yet been introduced into a lake with an already established planktivore so the outcome of competition between it and another species cannot be predicted. Nevertheless, it is a versatile species and possesses a number of attributes which suggest that it could compete very effectively.

Its ability to survive passage down the Zambezi and colonize Cahora Bassa is a good example of its adaptability. The Zambezi below Kariba is a harsh environment for small fish because its water is clear and flows strongly through unvegetated sandbanks. This means that they cannot easily escape predation by tigerfish, *Hydrocynus vittatus*. These fish are such effective

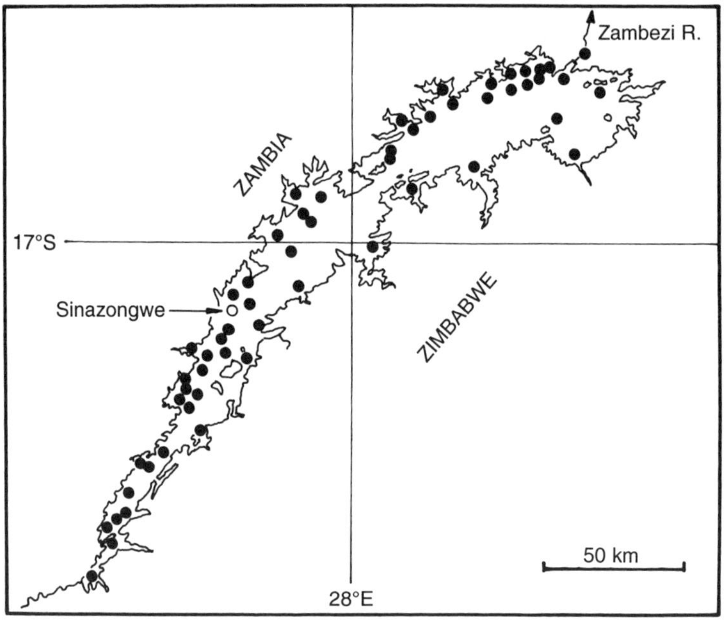

Fig. 24.1 Spread of *Limnothrissa* in Lake Kariba. The fish were introduced off Sinazongwe in Zambia in 1967 and 1968 (open circle) and were found throughout the lake and in the Zambezi in 1970 (closed circles). Based on data in Junor and Begg (1971) and Balon (1974).

predators that prior to Kariba's construction only eight of the 28 species living in the main stream had adults that were less than 18 cm in length (Jackson, 1961). In spite of these hazards, the sardines survived a passage of at least 270 km down the river before reaching Cahora Bassa.

It has already been suggested that it would utilize the pelagic resources of Lake Malaŵi more effectively than the native species *Engraulicypris sardella* (Turner, 1982). The biological characteristics that might give *Limnothrissa* an advantage in competition with other species should therefore be considered. They include flexible feeding behaviour, appropriate reproductive strategy, commitment to reproduction, rapid growth and flexible life history.

Flexible feeding behaviour

Although *Limnothrissa* is a specialized zooplanktivore it readily feeds on other items, especially when it is in shallow water. Its diet includes phytoplankton, attached algae, benthic insects and brewery wastes (Begg, 1974; de Iongh *et al.*, 1983). In Lake Tanganyika it feeds on *Stolothrissa*, whilst fish scales and the small cichlid *Pseudocrenilabrus philander* have been found in the stomachs of specimens from Lake Kariba (Mitchell, 1976). Cannibalism is known to be widespread in the Lake Kivu population (Reusens, 1987) and is now believed to be a significant cause of fry mortality in Lake Kariba (H. Paulson, pers. comm.). By contrast, *Stolothrissa* seems to be a more specialized planktivore (Coulter, 1970) which may account for its failure in Lake Kivu, even though it had been the preferred species on account of its importance in the Lake Tanganyikan fishery (Collart, 1960).

Appropriate reproductive strategy

An introduction may fail if the species concerned has an inappropriate reproductive strategy. For example, the common carp, *Cyprinus carpio*, rarely becomes abundant in tropical African waters, possibly because the eggs stick to vegetation (Vostradovsky, 1973) where they are liable to be eaten by tilapias which scrape attached algae and other organisms from plants. The effects from this are illustrated by data from Savory Dam, Zimbabwe (12 ha), into which 60 carp were stocked in 1958 and in which only 110 (none <3 kg) were present when the dam was drained in 1980 (Evans, 1982).

An appropriate reproductive strategy might also explain why *Limnothrissa* succeeded in Kivu and Kariba whilst *Stolothrissa* did not (assuming that some were, in fact, introduced). According to Matthes (1965–66), *Limnothrissa* eggs are laid in shallow water and sink to the bottom whilst those of *Stolothrissa* are laid in deep water, sinking slowly until the larvae are fully developed. This might be an appropriate strategy in a deep lake like Lake Tanganyika, where the thermocline is at about 150 m depth (Beale, 1981), but not for shallower ones like Lake Kariba where the thermocline lies at about 20 m depth (Begg, 1970). Even Lake Kivu might be effectively too shallow because it is deoxygenated from a depth of 70 m (Verbeke, 1957) and any eggs that fell into this zone would fail to hatch.

A commitment to reproduction

To be successful, an introduced species should be capable of rapid reproduction, and it is obvious from the rate at which *Limnothrissa* colonized Kariba that it is. Its commitment to reproduction is demonstrated by its relative fecundity, which is about 5–10 times greater than that of any other species in the lake, including some species of about the same size (Table 24.2). This investment in reproduction is bound to be an advantage in the face of harsh environmental factors, such as intense predation or competition.

Little is known about the fecundity of other pelagic species in the African lakes, but there is some evidence that *Limnothrissa* is relatively more fecund than the cyprinid *Rastrineobola argentea* from Lake Victoria. Its relative fecundity ($RF = 10 \times AF \times SL^{-3}$, where AF is absolute fecundity and SL is

Table 24.2 Relative fecundity (eggs g^{-1}) in relation to length of *Limnothrissa miodon* and several other fish species from Lake Kariba. Calculated from fecundity and length–weight data in Bowmaker (1973), Begg (1974), Marshall (1987) and Torres (1992)

Species	Length (cm)								
	5	10	15	20	30	40	50	60	70
Limnothrissa miodon	1290	1064							
Brycinus lateralis	175	182							
Brycinus imberi		155	145	99					
Hippopotamyrus discorhynchus		53	35						
Synodontis zambezensis			95	76	42				
Oreochromis mortimeri			1	3	3	2			
Schilbe intermedius				82	90				
Hydrocynus vittatus					98	207	183	145	
Labeo altivelis					187	248	210		
Clarias gariepinus						59	47	38	30
Mormyrus longirostris						19	21	19	15

standard length in cm) was 87.1 ± 31.8 and 86.6 ± 26.9 before and after the Nile perch introduction (Wanink, 1991). By contrast, the relative fecundity of *Limnothrissa* in Lake Kariba was 137.8 ± 99.3 (based on data in Begg, 1974).

Rapid growth

A rapid growth rate will be an advantage to any species colonizing a new habitat because it will reach maturity before any possible competitors. The scanty data on the growth rates of pelagic species in African lakes suggest that the clupeids grow much faster than the cyprinids (Table 24.3). Obviously, much more work needs to be done on the comparative growth of small pelagic species in African lakes, but with growth performance indices that may be twice as high as those of comparable small pelagic cyprinids such as dagaa or usipa it seems likely that the clupeids could outperform them in most situations.

A flexible life history

One of the most striking features of *Limnothrissa* is its small size in Lake Kariba and Cahora Bassa compared with that in Lakes Kivu and Tanganyika. It was first suggetsed that this was caused by changes in the growth rate

Table 24.3 A comparison of growth in some pelagic clupeid and cyprinid species. L_∞ (cm) and k (annual rate) are parameters of the von Bertalanffy growth equation, $\Phi' = \log_{10} k + 2 \log_{10} L_\infty$ and is an index of growth performance (Pauly and Munro, 1984). Values for Lakes Tanganyika and Kivu are means

Species	L_∞	k	Φ'	Source*
Clupeids				
Stolothrissa tanganicae, Lake Tanganyika	10.08	2.53	2.41	1.2
Limnothrissa miodon, Lake Kariba	7.42	3.05	2.33	3
Limnothrissa miodon, Lake Kariba	8.10	1.74	2.06	4
Limnothrissa miodon, Lake Tanganyika	16.63	1.13	2.33	5
Limnothrissa miodon, Lake Kivu	15.00	1.45	2.51	6
Cyprinids				
Engraulicypris sardella, Lake Malaŵi	13.75	0.09	1.23	7

* Sources: 1, Roest (1978); 2, Chapman and van Well (1978); 3, Marshall (1987); 4, Cochrane (1984); 5, Moreau *et al.* (1991); 6, Mannini (1990); 7, Rufli and Van Lissa (1982).

(Marshall, 1987) but this now seems to be incorrect (Marshall, 1993). Their small size is most likely to be caused by a greatly reduced life span as there seems to be only one age class in Lake Kariba compared with two in Lakes Kivu and Tanganyika (Fig. 24.2). This is reflected by a much higher mortality rate in the Lake Kariba fish (Mannini, 1990) and very few of them survive to the age of 12 months (Marshall, 1987). Moreover, the Lake Kariba fish reach reproductive maturity much sooner, at 35 mm in length ($\approx$ 3 months old) compared with 75 mm ($\approx$ 7 months) (Ellis, 1971; Begg, 1974; Splicthoff *et al.*, 1983).

The reasons for this phenemenon are still unclear. It may be related to food abundance, to the relative instability of the human-made lakes or to predation (Marshall, 1993). The importance of predation is demonstrated by the reduction in the mean length of *Rastrineobola* that occurred after the establishment of Nile perch in Lake Victoria (Wanink, 1991). Lake Kivu lacks predatory fish species and there is no need for its sardines to reproduce as soon as they do in Lake Kariba. Thus, they can afford a longer life cycle which enables them to grow to a much larger size. The widespread cannibalism that occurs in the Lake Kivu fish (Reusens, 1987) suggests that the larger individuals are behaving like the fish predators that occur in the other lakes.

A general consequence of stunting is that the same proportion of mature fish is maintained in a population (Iles, 1973). This can be seen in *Limnothrissa* in Lakes Kariba and Cahora Bassa, where the pelagic biomass is effectively the same as it is in Lake Tanganyika (Table 24.4). The biomass in Lake Kivu seems to be relatively low, however, even though the lake is apparently as productive as the others (Marshall, 1993). It is not clear why the biomass should be so much lower.

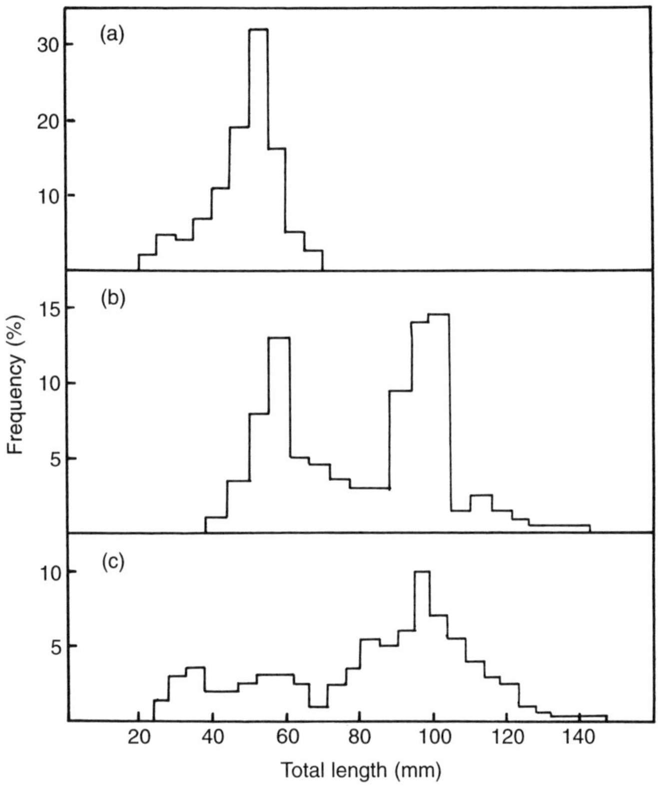

Fig. 24.2 Mean length-frequency distribution of *Limnothrissa* in commercial catches in (a) Lake Kariba (n = 32 776), (b) Lake Kivu and (c) Lake Tanganyika (n = 22 236). Based on data in Marshall (1992b), Reusens (1987) and N'Sibula and Mannini (1992), respectively. The number of fish in the Lake Kivu sample was not given in the source; their lengths were originally given as fork length and were multiplied by 1.10 to convert to total length (based on unpublished data). Lake Tanganyika fish were measured to standard length and their lengths were multiplied by 1.182 to convert them to total length (Torres, 1992). Note differing vertical scales.

Table 24.4 Biomass of the pelagic communities in relation to surface area and volume in Lakes Tanganyika, Kivu, Kariba and Cahora Bassa. The habitable volume is here considered to be the upper 125 m in Lake Tanganyika and the upper 60 m in Lake Kivu; the entire volume of Lakes Kariba and Cahora Bassa is habitable, at least for part of the year

Lake	Area (km^2)	Habitable volume (km^3)	Biomass ($t\,km^{-2}$)	Biomass ($t\,km^{-3}$)	Sources*
Tanganyika	32 900	4 112	30.0	240	1
Kivu	2 370	142	16.0	43	2, 3
Kariba	5 400	160	5.9	199	4
Cahora Bassa	2 303	43	4.2	225	5

* Sources: 1, Moreau and Nyakageni (1989); 2, Reusens (1987); 3, Lamboeuf *et al.*, (1990); 4, Marshall (1988); 5, Vostradovsky (1986).

24.3 WHAT ARE THE ECOLOGICAL CONSEQUENCES OF *LIMNOTHRISSA* INTRODUCTIONS?

No introduced species could be as successful and abundant as *Limnothrissa* has been without having some ecological impact. These impacts should be evaluated before its introduction to any other lake is contemplated. In Lake Kariba, its ecological effects were most pronounced in these areas (Marshall, 1991): on competitors and predators, on the plankton, and on nutrient cycling.

Competitors and predators

So far, *Limnothrissa* has never been put in a lake with any other pelagic planktivore. There is no clear evidence that *Brycinus lateralis* would have become one in Lake Kariba, as suggested by Balon (1974), because they were only ever abundant in shallow water to which they are now restricted. Predator species, by contrast, increased considerably. The abundance of *Hydrocynus vittatus*, in particular, rose from 5% to 10% of the gill net catch after the arrival of the sardines. Its population seems now to be partly regulated by the abundance of sardines, and both species decreased in the 1970s because of environmental conditions in the lake (Marshall, 1992a).

Liitle is known about the population dynamics and biology of the other fish species in the lake, but many of them eat fish to some extent (Table 24.5). They can probably feed on juvenile sardines in shallow water and could therefore have benefited from the introduction of *Limnothrissa*. It has been argued that the survival of their fry could be enhanced because *Hydrocynus* would prey on the sardines instead (Kenmuir, 1971) but there is no evidence to support or refute this idea.

Table 24.5 Incidence of piscivory amongst the indigenous fish species of Lake Kariba. Values indicate the percentage of stomachs in which fish were recorded; + denotes the presence of fish in some stomachs. Data from Bowmaker (1973) and Mitchell (1976)

Species*	1967–1968	1973–1974
Mormyrops deliciosus	22	8.1
Hydrocynus vittatus	98	99.9
Heterobranchus longifilis		66.5
Clarias gariepinus	+	29.7
Eutropius depressirostris	18	22.1
Synodontis zambezensis	+	6.2
Tilapia rendalli		6.6
Serranochromis codringtonii		6.2
Pseudocrenilabrus philander		9.3
Pharyngochromis darlingi		20.8

* This list should also include the predatory cichlid *Serranochromis macrocephalus* which was rare in the 1960s and 1970s. It has since spread to all areas of the lake and its population is probably still increasing. It almost certainly feeds on sardines because it takes other fish species (Bell-Cross and Minshull, 1988).

The population of piscivorous birds on Lake Kariba is relatively low and only the white-winged black-tern, *Chlidonias leucoptera*, can feed on sardines and has recently learnt to hunt at night around the fishing boats. The grey-headed gull, *Larus cirrocephalus*, is another species that has taken advantage of the sardines by scavenging fish from the racks on which they are sun-dried. This gull was once a vagrant to the lake but there is now a considerable resident population which has begun to breed on small islands.

The situation in Cahora Bassa is probably very similar and it is known that predatory species like the tigerfish and the butter catfish, *Schilbe intermedius*, feed on *Limnothrissa* there (Vostradovsky, 1986). Their numbers are possibly greater than they would have been if the sardines had not colonized the lake.

Little is known about the small cichlid species-flock in Lake Kivu and some of the species have not yet been formally described (Snoeks, 1986). Their ecology is largely unknown and it is impossible to determine the impact of *Limnothrissa* on them. This is unfortunate because such knowledge might have contributed to the debate over the introduction of this fish into Lake Malaŵi or any of the other African Great Lakes.

Effects on the plankton

The effects of zooplanktivorous fish on zooplankton populations are well known and have been confirmed experimentally (Lazzaro, 1987). The most obvious consequence is that larger species are preyed upon and decrease, or

Table 24.6 Changes in the composition (%) of the zooplankton in Lake Kariba. From Marshall (1991). Values in column *W* are based on dry weight per unit volume ($mg\,m^{-3}$) whilst those in column *n* are based on numbers per unit volume (no. m^{-3}); + denotes a value of less than 0.01%. Data from Bowmaker (1973) and Green (1985)

Component	1967–1968		1970	1972	1975–1976	1983
	W	*n**	*n*	*n*	*n*	*n*
Diaptomids	18.7	14.2	0.2	0	+	0
Cyclopoids	17.6	4.6	12.5	10.1	13.8	1.1
Nauplii	9.3	1.1	16.4	17.1	3.1	14.8
Bosmina	4.8	7.3	15.9	17.1	10.5	2.8
Ceriodaphnia	43.0	69.1	14.3	0	0.3	0
Diaphanosoma	10.5	2.1	0.2	0	+	0
Daphniids	0.3	0.8	0.2	0	+	0
Rotifers	4.0	+	40.3	55.7	72.3	81.3
Cladoceran/ copepod ratio	1.7	4.2	2.4	1.7	0.8	0.2

*Based on one sample only.

even disappear, whilst smaller species become dominant. The total biomass also decreases but this is matched by an increase in the turnover rate of the population.

Limnothrissa has brought about changes of this nature in the lakes into which it has been introduced. In Lake Kariba, diaptomids and larger cladocerans decreased rapidly after the sardines were introduced (Table 24.6) and *Chaoborus* has apparently disappeared. The ratio of cladocerans to copepods decreased markedly and it might be possible to use this ratio as an index of zooplanktivory. Rotifers and nauplii larvae became increasingly important constituents of the zooplankton, whilst the mean biomass decreased from 2280 $mg\,m^{-3}$ (dry weight) in 1967–1968 to 28 $mg\,m^{-3}$ in 1978 (Bowmaker, 1973; Magadza, 1980).

The larger planktonic species also declined in Lake Kivu, leading to the almost complete disappearance of *Daphnia* and an increase of ciliates and other protozoa (Fig. 24.3). Larger species declined less rapidly in Cahora Bassa because its turbid water provided them with some protection. Their populations declined as the transparency of the water increased and the largest species were the first to be affected (Fig. 24.4).

Effects on nutrient cycling

Any fish species must have some effect on the nutrient dynamics of its ecosystem and these are likely to be pronounced when a species is abundant.

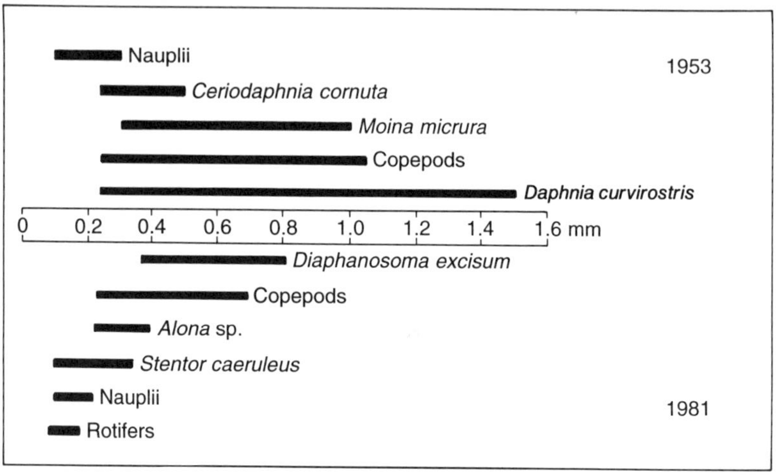

Fig. 24.3 Changes in the composition and size range of zooplankton in Lake Kivu following the introduction of *Limnothrissa* in 1960. Redrawn from Reyntjens (1982).

Unfortunately, little is known about the effect of *Limnothrissa* on nutrient cycling but Marshall and Junor (1981) have suggetsed that it contributed to the decline of the fern *Salvinia molesta* on Lake Kariba. The floating mats of this plant retained large quantities of nutrients (Mitchell, 1973) which were evidently released when they collapsed from 1970 to 1975. There was a considerable increase in the relative abundance of several fish species at the time (Fig. 24.5). If the sardines were responsible for the collapse of *Salvinia* then they clearly have the capacity to influence nutrient cycling in a lake. The effects of such changes are unpredictable.

24.4 WHERE SHOULD WE INTRODUCE *LIMNOTHRISSA*?

Limnothrissa has shown itself to be an adaptable species and could almost certainly become established in any African Lake. Furthermore, it could probably compete successfully against any of the pelagic cyprinids or characids that occur in some of them, because of its rapid growth, high fecundity and catholic feeding habits.

Events in Lake Victoria suggest that the fish communities of the African Great Lakes, especially those dominated by specialized haplochromines, are unable to compete against introduced species. It has been suggested (Turner, 1982) that planktivorous fish in Lake Malaŵi do not utilize zooplankton

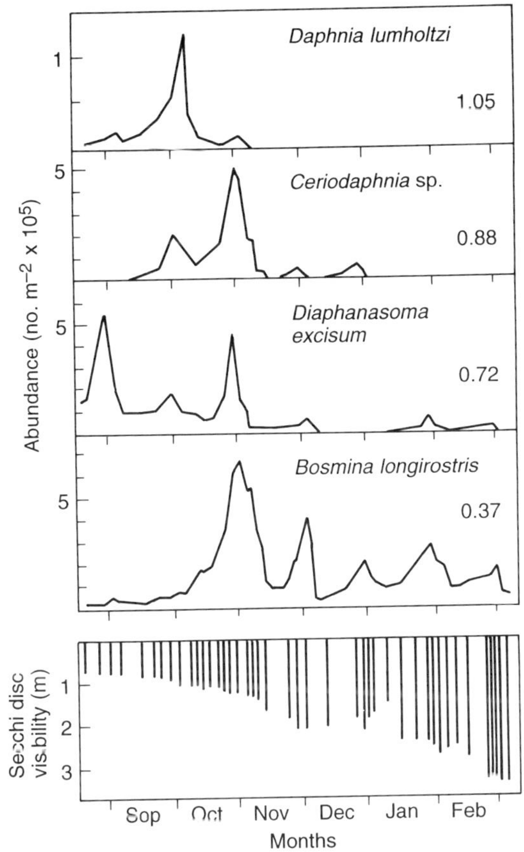

Fig. 24.4 Abudance of four cladoceran species in Lake Cahora Bassa in relation to water transparency, showing how the larger species tend to disappear earlier as the water transparency increases. Values in each panel denote the mean size of maturity (mm) of each species. Redrawn from Gliwicz (1985).

effectively, but this seems now to be incorrect (Menz, pers. comm.). In fact, the lake supports a complex and poorly understood pelagic community consisting of the cyprinid *Engraulicypris sardella*, several species of *Oreochromis* (chambo) and some hitherto undescribed haplochromines (Chapter 17). The rock-dwelling and inshore haplochromines (mbuna and utaka), which make up the bulk of Lake Malaŵi's endemic fish fauna, also feed extensively on zooplankton (McKaye *et al.*, 1985; Ribbink, 1991).

With so many species utilizing this resource, there is no clearly vacant

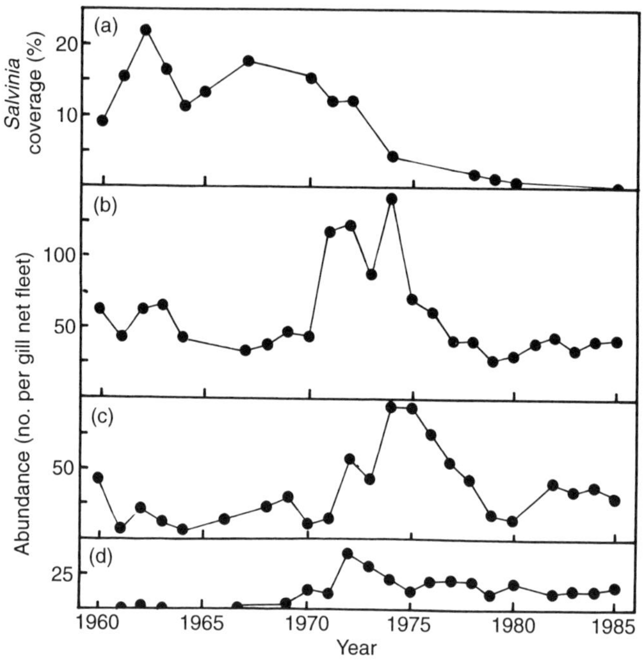

Fig. 24.5 Relative abundance of (b) *Hydrocynus vittatus*, (c) cichlids and (d) mormyrids in relation to (a) the extent of *Salvinia molesta* on Lake Kariba, 1960–1985. Redrawn from Marshall (1993).

niche in the lake, nor is it obvious that the zooplankton is underused. Although its density is greater in Lake Malaŵi than it is in Lakes Tanganyika, Kivu or Kariba, the cladoceran/copepod ratio is similar (Table 24.7). If this a reliable measure of a planktivore's impact, then it is reasonable to assume that the

Table 24.7 Mean abundance of zooplankton (no. m^{-3}) in Lakes Kariba, Cahora Bassa, Kivu and Malaŵi. Data from Cochrane (1978), Gliwicz (1984), Reusens (1987) and McKaye *et al.* (1985); n.d., not determined

Lake	Cladocerans	Copepods	Nauplii	Rotifers	Clad./cop. ratio
Kariba	1 280	1 630	3 600	5 293	0.8
Cahora Bassa	482	3 491	n.d.	5 525	0.1
Kivu	1 032	3 454	2 163	5 525	0.3
Malaŵi	1 754	10 786	40 630	n.d.	0.2

plankton in Lake Malaŵi is already being exploited by fish. If it were introduced, *Limnothrissa* would almost certainly affect the zooplankton and compete with many of the existing fish species. The result could be a series of unpredictable changes which might cause many of the endemic fish species to decrease or even become extinct. There is, therefore, nothing to justify the introduction of *Limnothrissa*, or any other clupeid, into Lake Malaŵi or any other African Great Lake (see discussion in Chapter 23).

The same cannot be said of human-made lakes, and the sardine could be introduced into almost any African reservoir. The fish communities in reservoirs are less diverse, relatively unspecialized and unlikely to compete with *Limnothrissa*. It should be remembered that *Limnothrissa* has demonstrated its capacity to invade systems by moving down the Zambezi to Cahora Bassa and it should not be stocked into reservoirs that might drain into a natural lake. Its recent introduction into Lake Itezhi-tezhi in Zambia (Chapter 15) is acceptable because the Kafue river drains into the Zambezi where the sardine already occurs.

24.5 SUMMARY

The food crisis in sub-Saharan Africa is such that there are compelling reasons for making further introductions of alien species into lakes and reservoirs. This is because they have, in some cases, dramatically improved their fisheries productivity. The clupeid *Limnothrissa miodon* is one such species, having been introduced into Lakes Kivu and Kariba where it now supports thriving fisheries. Its capacity to colonize new areas is epitomized by the fact that it had spread throughout Lake Kariba within 2 years of being stocked. It also established itself in the Zambezi below the Kariba Dam and invaded Cahora Bassa after its creation in 1975. The factors responsible for its success include the presence of a vacant niche, a high investment in reproduction, an appropriate reproductive strategy and the ability to change its life history. Its ecological effects include an increase in some species of predatory fish, major changes to the structure of the zooplankton and a possible alteration of the existing patterns of nutrient cycling. It is concluded that this species is probably capable of being successful in any lake. Further introductions should be limited to human-made lakes, where stable communities have not yet become established and ecological changes are proceeding rapidly (see also discussion in Chapter 23 – *Eds*). The possibility that it could cause extensive, and unpredictable changes in natural lakes with relatively stable, or at least slowly changing ecosystems, suggests that it would be unwise to introduce it into them.

ACKNOWLEDGEMENTS

My attendance at the symposium on Species Change in African Lakes was supported by sponsorship from the Renewable Resources Assessment Group, Imperial College, London. I am especially grateful to Tony Pitcher who made it possible.

REFERENCES

Anon. (1990) *World Resources, 1990–91*, World Resources Institute, New York, 383 pp.

Balon, E.K. (1974) Fish production of a tropical ecosystem, in *Lake Kariba: a Man-made Tropical Ecosystem in Central Africa* (eds E.K. Balon and A.G. Coche), Dr W. Junk, The Hague, pp. 249–573.

Balon, E.K. and Bruton, M.N. (1986) Introduction of alien species or why scientific advise is not heeded. *Env. Biol. Fishes*, **16**, 225–30.

Barel, C.D.N., Dorit, R., Greenwood, P.H., Fryer, G., Hughes, N., Jackson, P.B.N., Kawanabe, H., Lowe-McConnell, R.H., Nagoshi, M., Ribbink, A.J., Trewavas, E., Witte, F. and Yamaoka, K. (1985) Destruction of fisheries in Africa's lakes. *Nature, Lond.*, **315**, 19–20.

Beadle, L.C. (1981) *The Inland Waters of Tropical Africa*, 2nd edn, Longman, London, 474 pp.

Begg, G.W. (1970) Limnological observations on Lake Kariba during 1967, with emphasis on some special features. *Limnol. Oceanogr.*, **15**, 776–88.

Begg, G.W. (1974) Investigations into the biology and status of the Tanganyika sardine, *Limnothrissa miodon* (Boulenger), in Lake Kariba, Rhodesia. Lake Kariba Fish. Res. Inst., Project Rep., no. 17, 151 pp.

Bell-Cross, G. and Minshull, J.L. (1988) *The Fishes of Zimbabwe*, National Museums and Monuments of Zimbabwe, Harare, 294 pp.

Bowmaker, A.P. (1973) An hydrobiological study of the Mwenda River and its mouth, Lake Kariba. PhD thesis, Univ. Witwatersrand, Johannesburg, 576 pp.

Chapman, D.W. and van Well, P. (1978) Growth and mortality of *Stolothrissa tanganicae*. *Trans. Am. Fish. Soc.*, **107**, 26–35.

Cochrane, K.L. (1978) Seasonal fluctuations in the catches of *Limnothrissa miodon* (Boulenger) in Lake Kariba. Lake Kariba Fish. Res. Inst., Project Rep., no. 29, 163 pp.

Cochrane, K.L. (1984) The influence of food availability, breeding seasons and growth on commercial catches of *Limnothrissa miodon* (Boulenger) in Lake Kariba. *J. Fish Biol.*, **24**, 623–35.

Collart, A. (1960) L'Introduction du *Stolothrissa tanganicae* (Ndagala) au lac Kivu. *Bull. agric. Congo Belge*, **51**, 975–85.

Coulter, G.W. (1970) Population changes within a group of fish species in Lake Tanganyika following their exploitation. *J. Fish Biol.*, **2**, 329–53.

Cross, D. (1991) FAO and aquaculture: ponds and politics in Africa. *The Ecologist*, **21**, 73–6.

Dumont, H.J. (1986) The Tanganyika sardine in Lake Kivu: another ecodisaster for Africa? *Env. Conserv.*, **13**, 143–8.

Eccles, D.H. (1985) Lake flies and sardines – a cautionary note. *Biol. Conserv.*, **33**, 309–33.

Ellis, C.M.A. (1971) The size at maturity and breeding seasons of sardines in southern Lake Tanganyika. *Afr. J. trop. Hydrobiol. Fish.*, **1**, 59–66.

Evans, R.E. (1982) The rehabilitation of the Savory Dam. *Zimbabwe agric. J.*, **79**, 47–56.

Gliwicz, Z.M. (1984) Limnological study of Cahora Bassa reservoir with special regard to sardine fishery expansion. Rome, FAO, FAO/GCP/MOZ/006/SWE Field Doc., no. 8, 71 pp.

Gliwicz, Z.M. (1985) Predation or food limitation: an ultimate reason for extinction of planktonic cladoceran species. *Arch. Hydrobiol. Ergebn. Limnol.*, **21**, 419–30.

Green, J. (1985) Horizontal variations in associations of zooplankton in Lake Kariba. *J. Zool., Lond.*, **206A**, 225–39.

Iles, T.D. (1973) Dwarfing or stunting in the genus *Tilapia*: a possibly unique recruitment mechanism. *Rapp. P.-V. Réun. Cons. int. Explor. Mer*, **164**, 247–54.

de Iongh, H.H., Spliethoff, P,C. and Frank, V. (1983) Feeding habits of the clupeid *Limnothrissa miodon* (Boulenger) in Lake Kivu. *Hydrobiologia*, **102**, 113–22.

Jackson, P.B.N. (1961) Ichthyology, the fish of the Middle Zambezi. *Kariba Stud.*, no. 1, 1–36.

Junor, F.J.R. and Begg, G.W. (1971) A note on the successful introduction of *Limnothrissa miodon* (Boulenger) "the Lake Tanganyika sardine" to Lake Kariba. *Newsl. Limnol. Soc. Southern Afr.*, no. 16, 8–14.

Kenmuir, D.H.S. (1971) An analysis of data on the tigerfish *Hydrocynus vittatus* Castelnau obtained from Kariba International Tigerfish Tournaments. *Newsl. Limnol. Soc. Southern Afr.*, no. 17, 33–8.

Lamboeuf, M., Mulindabigwi, A. and Mutamba, A. (1990) Estimation de l'abondance du stock d'Isambaza (*Limnothrissa miodon*) resultats de la prospection acoustique de Mars 1990. Projet de Développement de la Pêche au lac Kivu, UNDP-FAO/RWA/87/012, Aide-Memoire, no. 26, 8 pp.

Lazzaro, X. (1987) A review of planktivorous fishes: their evolution, feeding behaviour, selectivities and impacts. *Hydrobiologia*, **146**, 97–167.

Machena, C. and Mabaye, A.B.E. (1987) Some management aspects and constraints in the Lake Kariba fishery. *NAGA, ICLARM Q.*, **10**(4), 10–12.

McKaye, K.R., Makwinja, R.D., Menyani, W.W. and Mhone, O.K. (1985) On the possible introduction of non-indigenous zooplankton-feeding fishes into Lake Malawi. *Biol. Conserv.*, **33**, 289–309.

Magadza, C.H.D. (1980) The distribution of zooplankton in the Sanyati Bay, Lake Kariba. *Hydrobiologia*, **70**, 57–67.

Mannini, P. (1990) Parametres de la population de *Limnothrissa miodon* du lac Kivu (1980–1989). Projet de Développement de la Pêche au lac Kivu, UNDP-FAO/RWA/87/012, Doc. de Travail, no. 32, 40 pp.

Marshall, B.E. (1987) Growth and mortality of the introduced Lake Tanganyika clupeid, *Limnothrissa miodon*, in Lake Kariba. *J. Fish Biol.*, **31**, 603–15.

Marshall, B.E. (1988) A preliminary assessment of the biomass of the pelagic sardine, *Limnothrissa miodon*, in Lake Kariba. *J. Fish Biol.*, **32**, 515–24.

Marshall, B.E. (1991) The impact of the introduced sardine *Limnothrissa miodon* on the ecology of Lake Kariba. *Biol. Conserv.* **55**, 151–65.

Marshall, B.E. (1992a) The relationship between catch and effort and its effect on predicted yields of the sardine fishery in Lake Kariba. *NAGA, ICLARM Q.*, **15**(4), 36–9.

Marshall, B.E. (1992b) *Limnothrissa* in man-made lakes: do we understand the implications of their small size? in *Report on the symposium of biology, stock assessment and exploitation of small pelagic fish species in the African great lakes region* (ed. E.J. Coenen), FAO/UNDP, Bujumbura (Burundi) [FAO/UNDP RAF/87/D99-TD/48/92], p. 11.

Marshall, B.E. (1993) The biology of the African clupeid *Limnothrissa miodon*, with special reference to its small size in artificial lakes. *Rev. Fish Biol. Fish.*, **3**, 17–38.

Marshall, B.E. and Junor, F.J.R. (1981) The decline of *Salvinia molesta* in Lake Kariba. *Hydrobiologia*, **83**, 477–84.

Matthes, H. (1965–66) Preliminary investigations into the biology of Lake Tanganyika Clupeidae. *Fish. Res. Bull. Zambia*, **4**, 39–45.

Mitchell, D.S. (1973) Supply of plant nutrient chemicals in Lake Kariba. *Geophys. Monogr.*, **17**, 165–9.

Mitchell, S.A. (1976) The marginal fish fauna of Lake Kariba. *Kariba Stud.*, no. 8, 109–62.

Moreau, J. and Nyakageni, B. (1989) Les relations trophiques dans la zone pelagique du lac Tanganyika (secteur Burundi). Essai d'evaluation. *Revue Hydrobiol. trop.*, **21**, 357–64.

Moreau, J., Munyandorero, J. and Nyakageni, B. (1991) Evaluation des paramètres démographiques chez *Stolothrissa tanganicae* et *Limnothrissa miodon* du lac Tanganyika. *Verh. int. Verein. Limnol.*, **24**, 2552–8.

N'Sibula, M. and Mannini, P. (1992) Demographic features of *Stolothrissa tanganicae, Limnothrissa miodon* and *Lates stappersi* in the northwestern end (Zaïre) of Lake Tanganyika, in *Report on the symposium on biology, stock assessment and exploitation of small pelagic fish species in the African great lakes region* (ed. E.J. Coenen), FAO/UNDP, Bujumbura, (Burundi) [FAO/UNDP RAF/87/D99-TD/48/92], p. 17.

Ogutu-Ohwayo, R. (1990) The decline of the native fishes of Lakes Victoria and Kyoga (East Africa) and the impact of introduced species, especially the Nile perch, *Lates niloticus*, and the Nile tilapia, *Oreochromis niloticus. Env. Biol. Fishes*, **27**, 81–96.

Pauly, D. and Munro, J.L. (1984) Once more on the comparison of growth in fish and invertebrates. *Fishbyte*, **2**, 21.

Popova, O.A. (1978) The role of predaceous fish in ecosystems, in *Ecology of Freshwater Fish Production* (ed. S.D. Gerking), Blackwell, Oxford, pp. 215–49.

Reusens, M. (1987) Programme de recherche halietique au lac Kivu en particulier sur le *Limnothrissa miodon*. Projet de Développement de la Pêche au lac Kivu, UNDP-FAO/RWA/87/012, Doc. de Travail, no. 3, 114 pp.

Reyntjens, D. (1982) *Bijdrage tot de Limnologie van het Kivu-meer. Rijksuniversiteit te Gent, Faculteit van de Landbouwetenschappen, Academiejaar 1981–82.* 95 pp.

Ribbink, A.J. (1991) Distribution and ecology of the cichlids of the African great lakes, in *Cichlid Fishes: Behaviour, Ecology and Evolution* (ed. M.H.A. Keenleyside), Chapman & Hall, London, pp. 36–59.

Roest, F.C. (1978) *Stolothrissa tanganicae*: population dynamics, biomass evolution and life-history. FAO, Rome, *CIFA Occ. Pap.*, no. 15, 104–29.

Rufli, H. and Van Lissa, J. (1982) Age and growth of *Engraulicypris sardella* in Lake Malawi. FAO, Rome, FI:DP/MLW/75/019, Studies on the Pelagic Ecosystem of Lake Malawi, tech. Rep., No. 1, pp. 85–97.

Snoeks, J. (1986) Some problems in taxonomic research of the haplochromine taxa from Lake Kivu, in *Proc. Third European Workshop on Cichlid Biology* (eds M. -D. Crapon de Caprona and B. Fritzsch), Musee Royale de l'Afrique Centrale, Tervuren, pp. 135–8.

Spliethoff, P.C., de Iongh, H.H. and Franck, V. (1983) Success of the introduction of the freshwater clupeid *Limnothrissa miodon* (Boulenger) in Lake Kivu. *Fish. Manage.*, **14**, 17–31.

Toots, H. (1970) Exotic fishes in Rhodesia. *Rhodesia agric. J.*, **67**, 83–8.

Torres, F.S.B. (1992) Length–weight relationships of Lake Kariba fishes. *NAGA, ICLARM Q.*, **15**(4), 42–3.

Turner, J.L. (1982) Lake flies, water fleas and sardines. FAO, Rome, FI:DP/MLW/75/019, Studies on the Pelagic Ecosystem of Lake Malawi, tech. Rep., No. 1, pp. 165–73.

Verbeke, J. (1957) Recherche écologique sur la faune des grands lacs de l'est du Congo Belge. *Explorations Hydrobiologiques du lacs Kivu, Edouard et Albert (1952–1954)*, **3**(1), Institute Royale des Sciences Naturelles de Belgique, Bruxelles, 177 pp.

Vostradovsky, J. (1973) *Freshwater Fishes*, Hamlyn, London, 252 pp.

Vostradovsky, J. (1986) On the ichthyofauna and possibilities of fishery utilisation of the Cahora Bassa Reservoir on the Zambezi River. *Prace VURH Vodnany (Czechoslovakia)*, **32**, 143–7.

Wanink, J.H. (1991) Survival in a perturbed environment: the effects of Nile Perch introduction on the zooplanktivorous fish community of Lake Victoria, in *Terrestrial and Aquatic Ecosystems: Perturbation and Recovery* (ed. O. Ravera), Ellis Horwood, New York, pp. 269–75.

Welcomme, R.L. (1981) Register of international transfers of inland fish species. *FAO Fish. tech. Pap.*, no. 213, 120 pp.

Appendix

Summary of characteristics of major African lakes

Nineteen lakes are included, giving details of location, depth and area, conductivity, pH, primary production, fish fauna, main commercial species status of introductions and, where available, approximate yield figures and the main ecological considerations. Values are taken from Vanden Bossche and Bernacsek (1990, 1991), Craig (1992) and relevant chapters of this book.

The Impact of Species Changes in African Lakes. Edited by Tony J. Pitcher and Paul J.B. Hart.
Published in 1995 by Chapman & Hall, London. ISBN 0 412 55050 4.

LAKE ALBERT (MOBUTU)

Natural lake, Western Rift | Zaire and Uganda, | 01°50′N, 30°40′E | (location map, Fig. 5.1)

Physical data

Area (km^2)	Volume (km^3)	Depth (m)			Drawdown (m)	Outflow/ vol. ratio
		Mean	Max.	Secchi		
6800	140	25	58	2.0–6.0		

pH	Conductivity (μmho cm^{-1} at 20 °C)	Primary productivity		Temp. (°C)	O_2 ($mg\,l^{-1}$)
		$gC\,m^{-2}\,day^{-1}$	$gC\,m^{-2}\,year^{-1}$		
8.9–9.5	730	0.8 – 3.8	970		

Other physical information Stratified.

Fish

Total no. families	Total no. species	Non-cichlids*		Cichlids*	
		Genera	Species	Genera	Species
14	46	26 (0)	37 (2)	2 (0)	9 (4)

Introductions None.
Main commercial species *Alestes (A. baremose), Hydrocynus, Lates (L. macrophthalmus* – lives offshore), *Bagrus, Oreochromis niloticus, O. leucostictus, Sarotherodon galilaeus, Tilapia zillii* and *Citharinus citharus.*
Fishing gears used Gill nets, beach seines, longlines, rod and line.
Total landings 12 548 $t\,year^{-1}$ in 1963, 12 532 $t\,year^{-1}$ in 1988.
Fish production 125 $kg\,ha^{-1}$ (1.3 $gC\,m^{-2}\,year^{-1}$).
Ecological considerations Fish species are similar to riverine Nile fish.

*Numbers in parentheses are endemic.

LAKE BANGWEULU

Natural lake, Zambia, 10°55′S, 30°10′E (location map, Fig. 1.1)
Natural shallow lakes, swamps and seasonally inundated floodplains

Physical data

Area (km^2)	Volume (km^3)	Depth (m)			Drawdown (m)	Outflow/ vol. ratio
		Mean	Max.	Secchi		
2072		2.5	10	0.8–1.05		

pH	Conductivity (μmho cm^{-1} at 20 °C)	Primary productivity		Temp. (°C)	O_2 ($mg\,l^{-1}$)
		$gC\,m^{-2}\,day^{-1}$	$gC\,m^{-2}\,year^{-1}$		
7.0–8.3	14–35		270		

Fish

Total no. families	Total no. species	Non-cichlids*		Cichlids*	
		Genera	Species	Genera	Species
	68	31 (0)	59 (3)	5 (0)	9 (1)

Introductions None.
Main commercial species Cichlids, mormyrids, clariids, bagrids, mochokids, characids.
Fishing gears used Gill nets.
Total landings 12 000–15 000 t $year^{-1}$.
Potential fish yields 10–35 kg ha^{-1} $year^{-1}$.

*Numbers in parentheses are endemic.

LAKE CAHORA BASSA

Human-made lake, Mozambique, 16°00′S, 31°00′E (location map, Fig. 5.1)
Middle Zambezi dam closed in 1974

Physical data

Area (km^2)	Volume (km^3)	Depth (m)			Drawdown (m)	Outflow/ vol. ratio
		Mean	Max.	Secchi		
2665	70	21	156		6	2:1

pH	Conductivity (μmho cm^{-1} at 20 °C)	Primary productivity		Temp. (°C)	O_2 ($mg\,l^{-1}$)
		$gC\,m^{-2}\,day^{-1}$	$gC\,m^{-2}\,year^{-1}$		
7.0–8.2	112–132				6.0–7.0

Other physical information High clay loadings.

Introductions *Limnothrissa miodon* by natural migration down the Zambezi.
Main commercial species *Hydrocynus vittatus*, *Labeo* spp., *Limnothrissa miodon*, *Clarias gariepinus*, *Oreochromis mortimeri*, *Distichodus schenga*.
Fishing gears used Gill nets.
Potential yield of *L. miodon*, 8000 t $year^{-1}$. No time series data available.
Fishing density 0.46 fishermen per km^2 in 1984; critical density, 1.5 per km^2.
Ecological considerations Fisheries development has been impaired by warfare, lack of government infrastructure inputs, insufficiencies of fishing gears and crafts, inappropriate processing practices, difficulty in marketing catch and conflicts between fishermen and some official bodies.

LAKE CHAD

Natural lake, Chad and Nigeria, 13°55′N, 13°40′E (location map, Fig. 1.1)

Physical data

Area (km^2)	Volume (km^3)	Depth (m)			Drawdown (m)	Outflow/ vol. ratio
		Mean	Max.	Secchi		
10 000–25 000			12	0.15–1.27		

pH	Conductivity (μmho cm^{-1} at 20 °C)	Primary productivity		Temp. (°C)	O_2 ($mg\,l^{-1}$)
		$gC\,m^{-2}\,day^{-1}$	$gC\,m^{-2}\,year^{-1}$		
7.1–8.3	40–88	0.7–2.7	370		

Fish

Total no. families	Total no. species	Non-cichlids*		Cichlids*	
		Genera	Species	Genera	Species
	93	41 (0)	84 (8)	3 (0)	9 (1)

Introductions None.
Main commercial species *Lates niloticus, Hydrocynus vittatus, Alestes baremose.*
Total landings 22 400 t $year^{-1}$.
Ecological considerations The lake has been subjected to serious droughts as it lies just south of the Sahara and has oscillated in size over the years. Fish biomass has varied as a result of lake size from 30 to 5620 kg ha^{-1}. Fish species are strongly influenced by inflowing rivers into which many fish enter to spawn. The lake is now almost dried up due to climatic conditions.

*Numbers in parentheses are endemic.

LAKE CHILWA

Natural lake, Malaŵi, 15°30′S, 35°30′E (location map, Fig. 1.1)

Physical data

Area (km^2)	Volume (km^3)	Depth (m)			Drawdown (m)	Outflow/ vol. ratio
		Mean	Max.	Secchi		
7500[a]			2.5[b]	0.2[b]		

pH	Conductivity (μmho cm^{-1} at 20 °C)	Primary productivity		Temp. (°C)	O_2 (mg l^{-1})
		gC m^{-2} day^{-1}	gC m^{-2} $year^{-1}$		
8.6[c]	1200[d]			21–31	5.2[e]

Generalized water balance, 1961–1971 Rainfall, 893 mm; inflow, 410 mm; evaporation, 1303 mm.

Fish

Total no. families	Total no. species	Non-cichlids*		Cichlids*	
		Genera	Species	Genera	Species
3	13	6 (0)	8 (0)	3 (0)	5 (0)

Introductions None.
Main commercial species *Clarias mossambicus, Barbus paludinosus, Oreochromis shiranus chilwae.*
Fishing gears used Gill nets, beach seines.
Total landings 100–9800 t $year^{-1}$. Depends on water level. In 1976, 159 kg ha^{-1} $year^{-1}$. One-third of national fish landings in Malaŵi.

Year	Mean depth (m)	Total biomass (t)	*Sarotherodon* (%)	*Barbus* (%)	*Clarias* (%)
1965	1.08	8 820	48	26	26
1968	0.00	97	4	7	89
1970	1.05	4 166	8	35	57
1973	0.02	1 903	?	?	?
1976	2.00	19 746	14	52	34

Ecological considerations In times of drought, fish take refuge in lagoons and streams. *Clarias mossambicus* can withstand desiccation and high conductivity.

*Numbers in parentheses are endemic.
[a] Chilwa Basin.
[b] Variable.
[c] In open water. Value in midswamp is 8.7; at land edge of swamp, 7.6.
[d] In open water. In midswamp, 1250; at land edge of swamp, 1000.
[e] In open water. In midswamp, 1.1; at land edge of swamp, 0.0.

LAKE EDWARD (IDI AMIN)

Natural lake, Western Rift | Zaïre, | 0°25′S, 29°40′E (location map, Fig. 5.1)

Physical data

Area (km^2)	Volume (km^3)	Depth (m)			Drawdown (m)	Outflow/vol. ratio
		Mean	Max.	Secchi		
2325	78	34	117	1.8–3.0		

pH	Conductivity (μmho cm^{-1} at 20 °C)	Primary productivity		Temp. (°C)	O_2 ($mg\,l^{-1}$)
		$gC\,m^{-2}\,day^{-1}$	$gC\,m^{-2}\,year^{-1}$		
8.7–9.3	840	4.3	1500		

Fish

Total no. families	Total no. species	Non-cichlids*		Cichlids*	
		Genera	Species	Genera	Species
8	77	10 (0)	17 (2)	5 (1)	60 (59)

See Lake George data.

*Numbers in parentheses are endemic.

LAKE GEORGE

Natural lake, Uganda, 0°02′N, 30°25′E (location map, Fig. 5.1)
West limb Rift Valley. Outflow Kazinga Channel to Lake Edward.

Physical data

Area (km^2)	Volume (km^3)	Depth (m)			Drawdown (m)	Outflow/ vol. ratio
		Mean	Max.	Secchi		
270	0.7	2.5	3.0	0.4		

pH	Conductivity (µmho cm^{-1} at 20 °C)	Primary productivity		Temp. (°C)	O_2 ($mg\,l^{-1}$)
		$gC\,m^{-2}\,day^{-1}$	$gC\,m^{-2}\,year^{-1}$		
8.5–9.5	200	5.4	1980	25–35	Can be supersaturated 259%

Other physical information Stratified by day, but this breaks down at night.

Fish

Total no. families	Total no. species	Non-cichlids*		Cichlids*	
		Genera	Species	Genera	Species
8	77	10 (0)	17 (2)	5 (1)	60 (59)

Introductions None.
Main commercial species *Oreochromis niloticus, Bagrus docmac, Clarias lazera, Protopterus aethiopicus.*
Fishing gears used Gill nets for cichlids. Minimum size limit 5 inches (12.5 cm). Use canoes for fishery. Baited longlines for predators.
Total landings 12 031 t $year^{-1}$ in 1963 and 5936 t $year^{-1}$ in 1988.
Biomass 45 000–55 000 fish ha^{-1} and $29 \pm 5\,g\,m^{-2}$. *Haplochromis nigripinnis* 30 000 ha^{-1} and 6.7 $g\,m^{-2}$, *H. angustifrons* 15 500 ha^{-1} and 2.1 $g\,m^{-2}$, *H. squamipinnis* 172 ha^{-1} and 0.4 $g\,m^{-2}$, *Oreochromis niloticus* 130 ha^{-1} and 3.7 $g\,m^{-2}$ and *Protopterus aethiopicus* 6 ha^{-1} and 1.5 $g\,m^{-2}$.
Ecological considerations A eutrophic lake where reduction in oxygen can lead to fish kills. Of the total biomass in open water, 95% is phytoplankton. Standing crops of other organisms are not high but fluctuate little throughout the year. Species composition of the open water is dominated by bluegreen algae and three herbivores, one cyclopoid copepod and two fish. Inshore areas carry a greater variety of animals which are preyed upon by a great multiplicity of fish, particularly haplochromids.

*Numbers in parentheses are endemic.

LAKE ITEZHI-TEZHI (KAFUE GORGE)

Human-made reservoir, Kafue River,	Zambia, dam closed in 1977	15°45′S, 26°30′E	(map, Fig. 15.1)

Physical data

Area (km^2)	Volume (km^3)	Depth (m)			Drawdown (m)	Outflow/ vol. ratio
		Mean	Max.	Secchi		
3700					7	

Fish

Total no. families	Total no. species	Non-cichlids		Cichlids	
		Genera	Species	Genera	Species
	67				

Introductions *L. miodon* in 1992, not yet fished.
Main commercial species *Oreochromis andersoni, O. macrochir, Clarias gariepinus, C. ngamensis, Serranochromis angusticeps, Tilapia rendalli, T. sparmani, Schilbe mystus, Hepstus odoe, Labeo molybdinus.*
Fishing gears used Gill nets, beach seines.
Total landings 8634 t $year^{-1}$ in 1978 (16% of Zambia's production), less than 1000 tonnes in 1988.

LAKE KAINJI

Human-made reservoir, River Niger	Nigeria, dam closed in 1968	10°00′N, 4°35′E	(location map, Fig. 1.1)

Physical data

Area (km^2)	Volume (km^3)	Depth (m)			Drawdown (m)	Outflow/ vol. ratio
		Mean	Max.	Secchi		
1280	15.8	12.3	50	0.1–3	10	4:1

pH	Conductivity (μmho cm^{-1} at 20 °C)	Primary productivity		Temp. (°C)	O_2 ($mg\,l^{-1}$)
		$gC\,m^{-2}\,day^{-1}$	$gC\,m^{-2}\,year^{-1}$		
6.0–8.0	73		890	23–31	

Generalized water budget Rainfall, 660–1352 mm; inflow, 80 000 × $10^6\,m^3$; evaporation 1500–2000 mm. Outflow to volume ratio is high compared with other reservoirs – cf. Volta, Kariba.
Other physical information Nutrient poor. Stratified: February to May in deep areas. Incomplete stratification May to November. Homothermy and oxygen well distributed December to January.

Fish

Total no. families	Total no. species	Non-cichlids		Cichlids	
		Genera	Species	Genera	Species
22	101[a]				

Introductions None?
Main commercial species *Sarotherodon galilaeus* (dominant in the catch), *Lates niloticus*, *Bagrus bayad*, *B. docmac*, *Oreochromis niloticus*, *Tilapia zillii*, *Hydrocynus forskahlii*, *H. brevis*.
Fishing gears used Cast nets from the shore and canoes. Until recently, lack of adequate fishing methods for clupeids.
Total landings 4500–6000 t $year^{-1}$ in 1970 and 1971, then declined and stabilized by 1974/75. 4500 t $year^{-1}$ in 1978.
Yield 35 kg ha^{-1} $year^{-1}$, similar to that of the floodplains.
Biomass by species/group/area Cichlids, 105 kg ha^{-1}; bagrids, 36 kg ha^{-1}. Average inshore, biomass 240 kg ha^{-1}. Pelagic fishery may now be exploited; two small clupeids (*Pellonula afzeliusi*, *Sierrathrissa leonensis*) abundant, total biomass of 18 kg ha^{-1}.
Ecological considerations *H. forskahlii* feeds on clupeids and *H. brevis* feeds on cichlids. Most mormyrids declined or disappeared after filling, although a few species increased after the initial decline.

[a] Similar species to Lake Volta.

LAKE KARIBA

Human-made reservoir, Zambezi River,	Zimbabwe and Zambia, dam closed in 1958	17°15′S, 27°55′E	(map, Fig. 12.1)

Physical data

Area (km^2)	Volume (km^3)	Depth (m)			Drawdown (m)	Outflow/ vol. ratio
		Mean	Max.	Secchi		
5364	156.5	29.2	120	1–12		1:3

pH	Conductivity (μmho cm^{-1} at 20 °C)	Primary productivity		Temp. (°C)	O_2 ($mg\,l^{-1}$)
		$gC\,m^{-2}\,day^{-1}$	$gC\,m^{-2}\,year^{-1}$		
8.0[a]	80	1.7		17 (June) to 30 (Jan.)	

Generalized water budget Rainfall, 5.0 km^3; inflows, 60.9 km^3; evaporation, 8.6 km^3; outflows, 46.6 km^3; outflow to volume ratio is low.
Other physical information Nutrient poor. Stratified October to June. Hypolimnion: anaerobic during stratification. Becomes aerobic at turnover time (July) when nutrients are released and zooplankton increases at this time.

Fish

Total no. families	Total no. species	Non-cichlids		Cichlids	
		Genera	Species	Genera	Species
16	40				

Introductions Clupeids *Limnothrissa miodon* and *Stolothrissa tanganicae* from Lake Tanganyika (latter was not successful), 1967 and 1968.
Main commercial species *Hydrocynus vittatus, Limnothrissa miodon, Clarias gariepinus, Tilapia rendalli, Serranochromis*. Brown squeaker now increasing.
Fishing gears used Artisanal gill net fishery and commercial/industrial pelagic fishery with lights and lift nets.
Total landings 1500 t in 1974, 26 000 t in 1985, 35 000 t in 1991.
MEI yield 23.2 $kg\,ha^{-1}$ $year^{-1}$. Artisanal fishery for other species 4.6 $kg\,ha^{-1}$ or 2000 $t\,year^{-1}$.
Biomass of inshore fish <10 000 t. Range 31 to 650 $kg\,h^{-1}$.
Ecological considerations There is the potential for an eel fishery and a squeaker fishery (Chifamba, pers. comm.). Zooplankton were thought to be 3 to 44 times lower, depending on taxon, approximately 10 years after the introduced clupeid became abundant.

[a] At the surface, hypolimnion, 6.8–7.0.

LAKE KIVU

Natural lake, Western Rift | Zaïre and Rwanda, | 1°45′S, | 28°55′E | (map, Fig. 13.1)

Physical data

Area (km^2)	Volume (km^3)	Depth (m)			Drawdown (m)	Outflow/ vol. ratio
		Mean	Max.	Secchi		
2700	583	240	489	2.5–9.35		

pH	Conductivity (μmho cm^{-1} at 20 °C)	Primary productivity		Temp. (°C)	O_2 ($mg\,l^{-1}$)
		$gC\,m^{-2}\,day^{-1}$	$gC\,m^{-2}\,year^{-1}$		
6.5–9.5	1240	0.66–1.0	240–540	24[a]	[b]

Other physical information Nutrient poor. Stratified permanently. Residence time of water, 110 years; refill time (run-off alone), 190 years.

Fish

Total no. families	Total no. species	Non-cichlids*		Cichlids*	
		Genera	Species	Genera	Species
16	40	3 (0)	9 (0)	2 (0)	8 (7)

Introductions *Limnothrissa miodon*, *Stolothrissa tanganicae* from Lake Tanganyika (latter was not successful) from 1958 to 1960.
Main commercial species *L. miodon.*
Fishing gears used For *L. miodon*, artisanal light attraction and lift nets.
Yield Welcomme (1972) estimated an MSY of 30 000 t $year^{-1}$ of clupeids. After one year of exploitation by the artisanal fishery, yield was 42 kg ha^{-1} $year^{-1}$.
Ecological considerations Success of introduction has been due to lack of predators and the abundance of pelagic plankton.

*Numbers in parentheses are endemic.
[a] At surface; minimum at 70 m of 22.8 °C.
[b] 6–7.5 to a depth of 20 m, 3 at 50 m, anaerobic at 70 m

LAKE KYOGA

Natural lake, Inter Rift | Uganda, | 1°30′N, | 32°45′E | (maps, Figs 4.1, 5.1)

Physical data

Area (km^2)	Volume (km^3)	Depth (m)			Drawdown (m)	Outflow/ vol. ratio
		Mean	Max.	Secchi		
2700	16	6	8	0.9		

pH	Conductivity (μmho cm^{-1} at 20 °C)	Primary productivity		Temp. (°C)	O_2 (mg l^{-1})
		gC m^{-2} day^{-1}	gC m^{-2} $year^{-1}$		
7.2–9.0	245–365		1204		

Other physical information Very high phytoplankton biomass, 20–40 μg Chl *a*.

Fish

Total no. families	Total no. species	Non-cichlids		Cichlids	
		Genera	Species	Genera	Species
12[a]	288[a]				

Introductions *Lates niloticus*. Tilapias: About four non-indigenous species introduced from 1951 onwards including: *Oreochromis leucostictus*, *O. niloticus*, *Tilapia rendalli* and *T. zillii*. *O. leucostictus* does not appear to have competed with indigenous fish of more open water. It occupies pools and shallow areas behind papyrus beds. *T. zillii* and *O. niloticus* displaced the endemic *O. variabilis* and *O. esculentus* by 1970. *O. niloticus* is a superior competitor as it has a wider food spectrum, is more fecund, has a faster growth rate, grows to a larger size and lives longer than endemic *Oreochromis* spp. Possible hybridization between *O. niloticus* and *O. variabilis*.

Main commercial species As for Lake Victoria.

Fishing gears used Gill nets, beach seines, longlines.

Total landings 4500 t in 1956; 160 000 t in 1977 (introduced species contributed about 80% of the latter), 68 000 t in 1988.

Ecological considerations In the last decade there has been a steady drop in the water level (1.5 m) which has caused a reduction in the breeding and nursery areas. Not all areas have been exploited to their full potential for political reasons.

[a] As for Lake Victoria, into which Lake Kyogo drains.

LAKE MALAWÎ (NYASA)

Natural lake, Western Rift,	Malawî, Mozambique, and Tanzania, Age about 1 to 2 million years	10°45′S, 34°30′E	(map, Fig. 17.1)

Physical data

Area (km^2)	Volume (km^3)	Depth (m)			Drawdown (m)	Outflow/ vol. ratio
		Mean	Max.	Secchi		
30 800	8400	426	756	17		

pH	Conductivity (μmho cm^{-1} at 20 °C)	Primary productivity		Temp. (°C)	O_2 ($mg\,l^{-1}$)
		$gC\,m^{-2}\,day^{-1}$	$gC\,m^{-2}\,year^{-1}$		
7.7–8.6	220	0.24–1.14	[a]	[b]	

Fish

Total no. families	Total no. species	Non-cichlids*		Cichlids*	
		Genera	Species	Genera	Species
11	545	19 (1)	45 (28)	23 (20)	500 (495)

Introductions None.
Main commercial species *Haplochromis* spp., *Oreochromis* spp., *Rhamphochromis* spp.
The fishery Estimated sustainable fish yield is 30 000–120 000 t. Total landings, 12 445 t $year^{-1}$ in 1968 and about 30 000 t $year^{-1}$ (Malawî waters) in recent years, of which 25% is caught by the commercial sector. Commercial demersal trawling began at the south of the lake in 1968. In all, 160 fish species occurred in the catch; 80% were small cichlids. Eleven species of large catfish, including *Bagrus meridionalis*, and minor species (mormyrids, cyprinids and large *Oreochromis* spp.) were caught. In 4 years, 1971–1974, 20% of the species disappeared from the catch. A bottom gill net fishery in the south catches *Labeo* spp., *Bagrus meridionalis* and *Oreochromis* spp. in decreasing order of importance. *Labeo* and *Bagrus* declined continuously from the 1950s through the 1970s. Catches are 5–40 kg per net in the SE Arm (this is the most productive area of the lake) and from 3.5 to 20 kg per net in the SW Arm. Those species that spawn in inflowing streams (*Labeo* spp., *Clarias gariepinus*) are particularly vulnerable when they congregate in the lake before entering the streams. The commercial fishery now consists of ring netting, midwater trawling and demersal trawling. There is an artisanal gill net fishery.

There are two groups of haplochromines in inshore areas: sand dwellers and rock dwellers. Purse seine catches decline with distance offshore – at 1 mile offshore they are about half, and 3 miles offshore about one quarter the nearshore catch. *Haplochromis* spp. make up 72% of the nearshore catch.

Pelagic fish are *Engraulicypris sardella* (usipa) and *Diplotaxodon limnothrissa*. *Chaoborus* (midge larvae) are found in the pelagic zone. No relationship has been found between spawning stock and year class strength (YCS) of *E. sardella*. The YCS is highly variable. Total pelagic fish biomass estimated from acoustic gear is 86–96 kg ha^{-1} and from purse seine catches, 75 kg ha^{-1}. Yield is 30–40 kg ha^{-1}. Acoustic surveys, purse seine catches and monofilament net catches indicated that the pelagic fish population was too small to support a commercial pelagic fishery. Lake Tanganyika has a pelagic fish biomass 1.6 times as great as that in Lake Malaŵi although secondary productivity is similar. A hypothesis has been put forward that secondary productivity is fully utilized by the clupeids in Lake Tanganyika but that much in Lake Malaŵi is 'lost' through its consumption by *Chaoborus* but a recent ODA/SADACC project has disproved this hypothesis.

Ecological considerations Lake Malaŵi National Park was formed in 1984 to protect endemic species and populations.

*Numbers in parentheses are endemic.

[a] Epilimnion 23.0–27.3; Hypolimnion 22.5.

[b] Hypolimnion 430 m, anoxic and rich in nutrients.

LAKE MWERU

Natural lake, Zaire and Zambia, 8°50′S, 28°50′E (location map, Fig. 1.1)

Physical data

Area (km^2)	Volume (km^3)	Depth (m)			Drawdown (m)	Outflow/ vol. ratio
		Mean	Max.	Secchi		
5200		12	37	0.76–1.22		

pH	Conductivity (μmho cm^{-1} at 20 °C)	Primary productivity		Temp. (°C)	O_2 ($mg\,l^{-1}$)
		$gC\,m^{-2}\,day^{-1}$	$gC\,m^{-2}\,year^{-1}$		
6.6–9.3	70–150				

Fish

Total no. families	Total no. species	Non-cichlids*		Cichlids*	
		Genera	Species	Genera	Species
16	94	39 (0)	83 (4)	5 (0)	11 (1)

Introductions None?
Main commercial species *Tilapia*, *Auchenoglanis*, Chrysichthys, *Gnathonemus*, *Mormyrus*.
Fishing gears used Gill nets.
Total landings 12 445 t $year^{-1}$ in 1968.

*Numbers in parentheses are endemic.

LAKE NASSER/NUBIA

Human-made reservoir, River Nile	Egypt and Sudan, Aswan Dam closed in 1964	23°50′N, 32°50′E	(location map, Fig. 1.1)

Physical data

Area (km^2)	Volume (km^3)	Depth (m)			Drawdown (m)	Outflow/ vol. ratio
		Mean	Max.	Secchi		
2585[a]	55.6[a]	21.5[a]	110[a]			
5238[b]	132.5[b]	25.2[b]	130[b]			
3057[c]	65.9[c]	21.6[c]	110[c]	[e]		1:2[f]
6216[d]	156.9[d]	25.2[d]	130[d]			

pH	Conductivity (μmho cm^{-1} at 20 °C)	Primary productivity		Temp. (°C)	O_2 ($mg\,l^{-1}$)
		$gC\,m^{-2}\,day^{-1}$	$gC\,m^{-2}\,year^{-1}$		
Nasser 7.8–8.9, Nubia 8.2–9.1	198.5–265.5			Nubia 15 (Feb.) to 30 (July)	Nubia 7.5–10.7

Fish

Total no. families	Total no. species	Non-cichlids		Cichlids	
		Genera	Species	Genera	Species
10[g]	26[g]				

Introductions None?
Main commercial species *Sarotherodon galilaeus, Labeo niloticus, Oreochromis niloticus, Hydrocynus forskahlli, Lates niloticus, Alestes baremose, Eutropius niloticus.* Commercially important at start but now negligible in occurrence: *Distichodus, Citharinus, Bagrus.*
Fishing gears used Gill nets.
Total landings Nasser and Nubia, 600 t $year^{-1}$ (both well below potential; Nasser open-water areas under-utilized). Nasser, 7000 t $year^{-1}$.
Estimated sustainable fish yield Nasser, 10 000 t $year^{-1}$; Nubia, 5000 t $year^{-1}$

[a] Values are for Lake Nasser, 160 m level.
[b] Values are for Lake Nasser, 180 m level.
[c] Values are for Lakes Nasser and Nubia, 160 m level.
[d] Values are for Lakes Nasser and Nubia, 180 m level.
[e] For Lake Nubia, 0.10 m (Aug.), 0.22 m (March).
[f] Value for Lakes Nasser and Nubia.
[g] In Lake Nubia.

LAKE TANGANYIKA

Natural lake,	Burundi, Tanzania,	5°15′S,	29°40′E	(maps, Figs 20.1, 21.1)
Western Rift,	Zambia and Zaire,			
	Age 2–4 million years			

Physical data

Area (km^2)	Volume (km^3)	Depth (m)			Drawdown (m)	Outflow/ vol. ratio
		Mean	Max.	Secchi		
33 000	18 800	570	1470	3.3–22		

pH	Conductivity (μmho cm^{-1} at 20 °C)	Primary productivity		Temp. (°C)	O_2 ($mg\,l^{-1}$)
		$gC\,m^{-2}\,day^{-1}$	$gC\,m^{-2}\,year^{-1}$		
8.6–9.2	566–620	0.40–3.10[a]	1292	[b]	[c]

Generalized water budget Rainfall on lake surface, 900 mm; inflows, 530 mm; evaporation, 1350 mm; outflow, 80 mm; residence time of water, 430 years; inflow to volume ratio, 1:1500; refill time (run-off alone), 1000 years.
Other physical information Stratified permanently. Thermocline may break down in certain years due to strong winds. Hypolimnion: 1200 m deep.

Fish

Total no. families	Total no. species	Non-cichlids*		Cichlids*	
		Genera	Species	Genera	Species
14	240	42 (8)	75 (52)	37 (33)	165 (164)

Introductions None.
Main commercial species *Limnothrissa miodon, Stolothrissa tanganicae, Lates mariae, L. microlepis, L. angustifrons, L. stappersi.*
Fishing gears used Commerical purse seines in the pelagic zone. Gill nets of mostly small mesh size for benthic fish. Artisanal inshore fishery uses beach seines, handlines, longlines, weirs and gill nets.
Total landings 73 000 t $year^{-1}$ (22.5 kg ha^{-1}). After 7 years in the 1960s, at the south of the lake, purse seine pelagic catches of *Lates* sp. were halved but clupeid catches approximately doubled. There was a similar change in the north end of the lake. Total biomass of *Lates* sp. = total biomass of clupeid prey (approximately). Pelagic clupeid abundance is now cyclic in periodicity of several years alternating with strong oscillations of *L. stappersi*, which has become the dominant predator. The top benthic predator, *L. mariae*, also caught in the pelagic zone by purse seines, has diminished considerably.

Estimated sustainable fish yield 300 000 t year^{-1}, 240 000 t year^{-1} clupeids and 30 000 t year^{-1} *L. stappersi*.

Biomass by species/group/area Clupeid 529 kg ha^{-1}.

Ecological considerations Benthic fish populations occur down to 200 m depth on the extensive shelf areas at the south and comprise about 60 species. Population numbers have not declined in the unexploited south west arm of the Sumbu National Wildlife Park. Primary production is not high compared with that in other lakes, but the lake fisheries yield (125 kg ha^{-1} year^{-1}) is very high. The great antiquity of the lake may have allowed selection of efficient pelagic species resulting in a highly efficient pelagic ecosystem where carbon accumulates at the top of the trophic pyramid and not in the algae. However, estimates of heterotrophic bacterial production equal or exceed algal primary production. The immense volume of anoxic deep water is the probable source of energy fixed in reduced substances and allowing high rates of bacterial production. *Chaoborus*, the lakefly, is not present in Lake Tanganyika as it is in Lake Malaŵi.

* Numbers in parentheses are endemic.

[a] Median 0.8.

[b] Epilimnion, 23.5–27; hypolimnion, 23.25–23.35.

[c] Hypolimnion anoxic.

LAKE TURKANA (RUDOLF)

Natural lake, Eastern Rift: semi-desert region	Kenya and Ethiopia, Influenced by River Ono in Ethiopia	3°30′N,	36°05′E	(map, Fig. 16.1)

Physical data

Area (km^2)	Volume (km^3)	Depth (m)			Drawdown (m)	Outflow/ vol. ratio
		Mean	Max.	Secchi		
7200	273	31	120			

pH	Conductivity (μmho cm^{-1} at 20 °C)	Primary productivity		Temp. (°C)	O_2 ($mg\,l^{-1}$)
		$gC\,m^{-2}\,day^{-1}$	$gC\,m^{-2}\,year^{-1}$		
9.4–10.0	3000–3500	0.25 to 6.2	[a]		

Fish

Total no. families	Total no. species	Non-cichlids*		Cichlids*	
		Genera	Species	Genera	Species
15	47	22 (0)	35 (6)	3 (0)	12 (11)

Introductions None.
Main commercial species *Lates niloticus, Oreochromis niloticus, Citharinus, Alestes.*
Fishing gears used Gill nets, beach seines and traditional methods. Nineteen species have featured in the developing fishery since 1963. Fishery is mostly inshore (< 15 m depth). From 1963 to 1965, *Sarotherodon* spp. and *L. niloticus* were exploited using beach seines.
Total landings 2045 t $year^{-1}$ in 1968, 3510 t $year^{-1}$ in 1974.
Biomass by species/group/area *Lates niloticus* made up 22% of the total gill net catch in the early 1970s, 348 t $year^{-1}$ in 1972/73 and 470 t $year^{-1}$ in 1973/74. Standing crop of *L. niloticus* is 1900 t.
Ecological considerations Similar fish fauna to Lake Albert, basically Nilotic. Twelve species riverine and confined to the region of the Ono Delta. Seasonality leaves growth checks on bony structures in many fish. *Lates longispinus*, a dwarf and offshore-living species, is comparable to the offshore form of *Lates* in Lake Albert.

* Numbers in parentheses are endemic.
[a] Northern sector, 1315–6220; central sector, 194–3936; southern sector, 259–293.

LAKE VICTORIA

Natural lake, Inter Rift	Uganda, Kenya and Tanzania Age 250 000–750 000 years	0°50′S,	32°50′E	(maps, Figs 2.2, 6.1)

Physical data

Area (km^2)	Volume (km^3)	Depth (m)			Drawdown (m)	Outflow/ vol. ratio
		Mean	Max.	Secchi		
68 000	2700	20	79	1.3–8.2[a]		

pH	Conductivity (μmho cm^{-1} at 20 °C)	Primary productivity		Temp. (°C)	O_2 ($mg\,l^{-1}$)
		$gC\,m^{-2}\,day^{-1}$	$gC\,m^{-2}\,year^{-1}$		
7.1–9.0	96	1.08–4.20[b]	950[b]	23.8–26.0	

Generalized water budget Rainfall on lake surface, 1260 mm; inflows, 330 mm; evaporation, 1310 mm; outflow (R. Nile), 280 mm.
Other physical information Stratified seasonally. Hypolimnion oxygenated. Now can become seasonally deoxygenated.

Fish

Total no. families	Total no. species	Non-cichlids*		Cichlids*	
		Genera	Species	Genera	Species
12	288	20 (1)	38 (16)	8 (4)	250 (247)

Introductions (1) *Lates niloticus*. First official stocking of Nile perch was 35 fish from Lake Albert planted off Entebbe Pier, Uganda, in May 1962. In September, 1963, 339 fish from Lake Turkana were put into the lake near Kisumu, Kenya. However, Nile perch were unofficially stocked before this date. In May 1960 a perch was caught at Bugungu just above the Ripon Falls and another in Waigala Hannington Bay in November 1960. (2) Tilapias: About four non-indigenous tilapiine species introduced, from 1951 onwards, from Lake Albert into Uganda, Kenya and Tanzania waters, including: *Oreochromis leucostictus*, *O. niloticus*, *Tilapia rendalli* and *T. zillii*. *O. leucostictus* does not appear to have competed with indigenous fish of more open water. It occupies pools and shallow areas behind papyrus beds. *T. zillii* and *O. niloticus* displaced the endemic *O. variabilis* and *O. esculentus* by 1970. *O. niloticus* is a superior competitor as it has a wider food spectrum, is more fecund, has a faster growth rate, grows to a larger size and lives longer than endemic *Oreochromis* spp. Possible hybridization between *O. niloticus* and *O. variabilis*.

Main commercial species 1971, *Oreochromis, Bagrus, Clarias, Haplochromis* and *Protopterus*; 1990, *Lates niloticus, Rastrineobola argentea* and *Oreochromis niloticus*.
Fishing gears used Gill nets, beach seines, longlines, trawling. Gill nets are mostly home-made and of no standardized materials and dimensions, making the establishment of an accurate index of fishing difficult. There is a traditional gill net fishery for 'tilapias', *Bagrus, Clarias* and *Protopterus*. In 1983, there were 54 000 fishers using about 11 000 boats (4000 in Kenya, 4000 in Tanzania and 3000 in Uganda). Trawl fishery started in the early 1970s in the Mwanza area (Tanzania) to exploit *Haplochromis* in the deeper parts of the lake. Beach seine fishery for *Rastrineobola* started in the Kenyan and Tanzanian sections but expanded to include a lamp fishery with ringnets and Danish seines and become a genuine pelagic fishery. Gill nets (mesh sizes from 6 to 16 inches, 153 to 405 mm) are used for *Lates niloticus*. Other minor, seasonal/local fisheries.
Total landings 110 000 t in 1968, 296 000 t in 1985.
Ecological considerations There is overfishing within both shallow inshore areas and river mouths, increased use of small-mesh gill nets and unrestricted use of beach and mosquito seines. The seasonal fisheries of the River Nzoia, Kenya, depend on the ascent of Lake Victoria fish (*Barbus* and *Labeo*). Growth: Nile perch 1984, $L_{\infty} = 257$ cm, $k = 0.09$ and $t_0 = -1$.

*Numbers in parentheses are endemic.
[a] Currently 1.5–2.5 m due to greatly increased phytoplankton biomass.
[b] Offshore.

LAKE VOLTA

Human-made reservoir,	Ghana,	7°10′N, 0°30′W	(location map, Fig. 1.1)
On River Volta,	Akosombo Dam closed in 1964		

Physical data

Area (km^2)	Volume (km^3)	Depth (m)			Drawdown (m)	Outflow/ vol. ratio
		Mean	Max.	Secchi		
8845	165	18.6	75		3	0.25:1[a]

pH	Conductivity (μmho cm^{-1} at 20 °C)	Primary productivity		Temp. (°C)	O_2 ($mg\,l^{-1}$)
		$gC\,m^{-2}\,day^{-1}$	$gC\,m^{-2}\,year^{-1}$		
		0.8–5.2	930		[b]

Other physical information Stratified: thermocline persistent but not permanent at about 17–20 m.

Fish

Total no. families	Total no. species	None-cichlids		Cichlids	
		Genera	Species	Genera	Species
?	?				

Introductions None.
Main commercial species *Alestes, Citharinus, Distichodus, Labeo, Lates niloticus, Sarotherodon galilaeus, Oreochromis niloticus, Tilapia zillii.*
Fishing gears used Gill nets.
Estimated sustainable fish yield 40 000 t $year^{-1}$, based on MEI 2.77 kg $ha^{-1}\,year^{-1}$.
Total landings 36 000 to 42 000 t $year^{-1}$ in 1971–1984, 60 000 t $year^{-1}$ in 1969.
Ecological considerations Pelagic clupeids (*Pellonula afzeliusi* and *Cynothrissa mento*) are underexploited and tilapias overexploited. Mormyrids virtually disappeared after filling.

[a] This value is low – cf. Lakes Nasser-Nubia, Cahora Bassa.
[b] Epilimnion, fully oxygenated (up to 140%); hypolimnion, 0–5%.

REFERENCES

Craig, J.F. (1992) Human-induced changes in the composition of fish communities in the African Great Lakes. *Rev. Fish. Biol. Fish.*, **2**, 125–43.

Vanden Bossche, J.P. and Bernacsek, G.M. (1990) *Source Book for the Inland Fishery Resources of Africa*, Vol. 1, FAO, Rome, *CIFA tech. Pap.*, **18**(1), 411 pp.

Vanden Bossche, J.P. and Bernacsek, G.M. (1991) *Source Book for the Inland Fishery Resources of Africa*, Vol. 2, FAO, Rome, *CIFA tech. Pap.*, **18**(2), 240 pp.

Author index

Abdel-Latif, A.-F. 523
Abdel-Malek, S.A. 318
Acere, T. 32, 56, 81, 106, 109, 135, 179, 180, 238, 485
Acheson, J.A. 238, 240
Achieng, A.P. 133, 211, 485
Adams, M. 491
Aebersold, P. 492
Akech, M.O. 41
Akersie, W.O. 238
Akiyama, T. 41
Alheit, J. 15
Ali, M. el-T. 360, 523
Alimoso, S.B. 393, 395
Allanson, B.R. 107, 134, 159, 487
Allen, K.F. 158
Allendorf, F.W. 485, 489
Amos, B. 485
Andersen, K.P. 158
Anderson, A.M. 80, 106, 211
Arthington, A.H. 485, 486, 490
Arunga, J.O. 106, 211
Asila, A. 30, 41, 80, 109, 133, 158, 179, 180, 212
Asowa-Okwe, C. 238
Avise, J.C. 485, 489

Babikwa, J. 238
Baccus, R. 491
Badjik, C.D. 523
Bagenal, T.B. 485
Bailey, C. 239
Bailey, R.G. 319
Baker, A.J. 485
Balarin, J.D. 239
Balirwa, J.S. 30, 55, 239
Balogun, J.K. 272, 332
Balon, E.K. 106, 272, 295, 332, 485–6, 542
Barel, C.D.N. 30, 41, 106, 108, 110, 133, 211, 486, 542
Barlow, C.G. 486
Barlow, J.C. 492
Barret, G. 239
Barrowsclough, G.F. 489
Bartell, S.M. 486
Barton, B.A. 490
Barton, N.H. 486
Basasibwaki, P. 394
Bashirwa, F. 296
Baskin, Y. 41
Baverstock, P.R. 491
Bax, N.J. 523
Bayley, P.B. 360
Bayona, J.D.R. 30, 523
Bazigos, G.P. 322, 394
Beadle, L.C. 30, 41, 158, 211, 295, 361, 542
Beauchamp, R.S.A. 30, 211
Beddington, J.R. 179, 361
Beeby, A. 318
Begg, G.W. 272–3, 542, 543
Bell-Cross, B. 272
Bell-Cross, C. 332
Bell-Cross, G. 272, 542
Benda, R.S. 106, 211
Bentzen, P. 486, 493
Berg, W.J. 488
Berger, M.R. 110
Bermingham, E. 485
Bernacsek, G.M. 332, 523, 525, 570
Bernatchez, L. 486

Berst, A.H. 486
Beverton, R.J.H. 30, 55, 133, 179
Bickham, J.W. 487
Billington, N. 486
Bishai, H.M. 361
Blake, B.F. 332
Bluhdorn, D.R. 486
Boivin, S. 486
Bonnell, M.L. 486
Booke, H.E. 489
Borgmann, U. 524
Bouton, N. 43, 110, 135, 180, 493, 525
Bowen, S.H. 318
Bowmaker, A.P. 272–3, 441, 542
Brabant, S. 239
Brassington, R.A. 486
Bravington, M.V. 394
Briand, F. 523
Brönmark, C. 107
Brookfield, J.F.Y. 486
Bruford, M.W. 486–7
Bruton, M.N. 15, 106, 107, 133, 134, 486, 487, 523, 542
Bugenyi, F.W.B. 31, 41, 133
Bundy, A. 180, 275, 524
Bunker, S. 239
Burgis, M. 158, 160
Burke, T. 486–7
Burne, R.H. 273
Butcher, D.A.P. 239
Butzer, K.W. 361
Bwathondi, P.O.J. 80, 107, 133, 179, 239

Caddy, J.F. 273
Cadwalladr, D.A. 31, 80, 107, 361
Cairns, E. 487
Calamari, D.A. 41
Cambridge Nabugabo Biological Expedition 80
Campton, D.E. 487
Carpenter, S.R. 133
Carvalho, G.R. 487, 491
Casselman, J.M. 489
Cavender, T.M. 491
Chakraborty, R. 490
Chande, A.I. 31, 180
Chapman, D.W. 295, 441, 542
Charlesworth, B. 486
Chilvers, R.M. 158
Chimanga, R. 275
Christensen, V. 15, 158
Christie, W.J. 523
CICS 239
CIFA 41, 107, 133, 179, 211
Clark, F. 318
Coche, A.G. 272, 295, 332
Cochran, W.G. 363
Cochrane, K.L. 273, 295, 542
Coenen, E.J. 180, 213
Coke, M. 273
Colaris, J.C.G. 239
Collart, A. 295–6, 487, 542
Colwell, R.K. 492
Cooke, J.G. 179, 361
Coopconsult-Propesca 332
Corbet, P.S. 31, 107
Cordone, A.J. 31, 56, 80, 108, 134, 159
Cornia, G.A. 239
Coulter, G.W. 15, 107, 134, 441, 453, 487, 523, 542
Cowx, I.G. 332
Coyne, J.A. 491
Craig, J.F. 15, 523, 570
Crapon de Caprona, M.D. 107
Crimmen, O. 108
Cronin, M.A. 487
Cross, D. 542
Crow, J.F. 487
Crul, R.C.M. 523
Csirke, J. 441
Cushing, D.H. 487
Cycon, D. 239

Dadswell, M.J. 490
Dadzie, S. 42, 80, 109, 160, 213, 332, 362
Dampha, N.F. 395
de Iongh, H.H. 296, 543, 544
DeAngelis, D.L. 361
Degubol, P. 15
Dehring, T.R. 489
Delacroix, J. 239
Derman, B. 239
Dhatemwa, C.M. 240

Diamond, J. 319
Dickie, L.M. 524
Diedrichsen, H. 296
Diggle, P.J. 134
Dixon, R.D. 239
Dodson, J.J. 486
Donda, S.J. 394, 395
Donnelly, B.G. 273
Dorit, R. 41, 106, 133, 211, 486, 542
Dowling, T.E. 487
Downing, J.A. 523
Dudley, R.G. 332
Dumont, H.J. 296, 542
Durand, J.R. 159, 453

EAFFRO 31, 211
EAHC 239
EASCO 239
Eccles, D.H. 394, 412, 542
Ecoutin, J.D. 159
Ecoutin, J.M. 453
Edge, T.A. 487
EIFAC 332
Elder, H.Y. 487
Ellis, C.M.A. 296, 441, 453, 542
Elton, C.S. 487
Emlen, J.M. 488
Entz, B.A.G. 332
EPADU 239
Erlich, H.A. 491
Evans, D.O. 523
Evans, M.S. 42
Evans, R.E. 543

Fahenensteil, G.L. 42
Falconer, D.S. 488
FAO 211, 394, 412
FAO/CIFA 488
FAO/UNEP 488
Fee, E.J. 296, 523
Ferguson, A. 239, 486, 488, 491–2
Ferguson, A.J.D. 361
Ferguson, M.M. 488
Fernando, C.H. 488
Ferris, S.D. 488
Fish, G.R. 41
Forney, J. 524
Franck, J.C.B. 488
Frank, V. 543, 544
Frank, V.G. 296
Frankel, O.H. 488
Franklin, I.R. 488
Freeman, D.M. 239
Frey, R.S. 239
Fritzsch, B. 107
Fryer, G. 41, 56, 80, 106, 107, 133–4, 159, 211–12, 319, 394, 412, 486, 488, 542
Fryer, G.E. 31

Gall, G.A.E. 488
Gambetta, D. 239
Garcia-Marin, J.L. 488
Garrod, D.J. 31, 56, 134, 487
Gaudet, J.J. 160, 319
Gayanilo, F.C. 159
Gee, J.M. 31, 56, 80, 158, 212, 361
Gelfand, D.R. 491
George, T.L. 319
Getabu, A. 31, 159
Gharret, A.J. 488, 489
Gideiri, Y.B.A. 361
Gjerstad, D. 361
Gliwicz, Z.M. 296, 543
Goeltenboth, F. 80
Goldschmidt, T. 30, 43, 81, 106, 107, 110, 133, 134, 135, 180, 493, 525
Goodman, D. 239
Gophen, M. 41, 42
Goudswaard, K. 43, 110, 135, 180, 493, 525
Goudswaard, P.C. 16, 30–1, 32, 56, 81, 106–8, 110, 133–5, 214
Graham, M. 31, 41, 56, 80, 134, 212
Gramling, R. 239
Grant, I.F. 81
Greboval, D.F. 42, 108, 109, 134, 135, 212, 213, 240, 241
Green, A. 489
Green, J. 543
Greenwood, P.H. 31, 41, 56, 80, 106, 107–8, 133, 159, 211, 486, 487, 542
Gregorius, H.-R. 488

Grevobal, D.F. 160
Grewe, P.M. 492
Grimm, A.S. 16, 395, 524
Grossman, Y.L. 492
Gruhn, I. 240
Gulland, J.A. 273, 361, 394, 412, 441

Hallerman, E.M. 489
Hamblyn, E.L. 31, 80, 212
Hancock, G. 240
Hannon, B. 159
Hanotte, O. 486–7
Harbott, B.J. 361
Harbott, B.S. 159
Harding, D. 394
Harper, D.M. 319
Harris, A.S. 486
Harris, C.K. 240
Harris, H. 488
Harrison, K. 108, 134, 488
Hart, P.J.B. 15, 180, 213, 275, 394
Hart, R.C. 107, 159, 487
Hastenrath, S. 41
Hauser, L. 487
Headrich, R.L. 524
Hebert, P.D.N. 486
Hecht, T. 362
Hecky, R.E. 41, 56, 80, 108, 134–5, 179, 241, 296, 490, 523
Henderson, B.A. 523
Henderson, H.F. 361, 453, 523
Herman, C. 296
Hickley, P. 319
Higuchi, R. 491
Hilborn, R. 15, 394
Hillis, D.M. 489
Hindar, K. 489
Hodgson, J.R. 133
Hodson, R.E. 492
Hoekstra, T.M. 212
Hoelzel, A.R. 489
Hoffman, G.L. 489
Holden, S. 394
Holt, S.J. 179
Hoogerhoud, R.J.C. 108, 110
Hopkinson, D. 488
Hopson, A.J. 31, 361
Horn, G.T. 491
Hughes, N. 41, 106, 211, 486, 542
Hughes, N.F. 56, 108, 134, 159, 179, 489
Hughes, R.N. 489
Humphreys, R. 490
Hutchinson, G.E. 31
Hutchinson, T.C. 213
Hutton, J.M. 273
Hyman, J. 524

Ihssen, P.E. 489
Iles, T. 543
Iles, T.D. 31, 56, 80, 107, 134, 319, 394, 412, 488, 523
Iles, T.G. 159
Iverson, I.L.K. 275

Jackson, P.B.N. 41, 80, 106, 107, 134, 211, 273, 394, 486, 487, 489, 542, 543
Jain, S.K. 489
Jansen, E.G. 240
Jarre, A. 159
Jeffreys, A.J. 489
Johannesson, K. 296
Joiris, C. 159
Jokonya, P. 275
Jolly, R. 239
Jones, H.L. 319
Jongman, R.H.G. 134
Jorde, P.E. 488
Joubert, C.S.W. 273
Jubb, R.A. 273
Jude, D.J. 42
Junor, F.J.R. 273–4, 543–4

Kajumulo, A.A. 41
Kallqvist, T. 361
Kanawabe, H. 106
Kaningini, M. 296
Kanyike, E.S. 212
Kapasa, C.K. 332
Kapetsky, J.M. 332
Kapuscinscy, A.R. 489
Karenge, L.P. 273
Karl, S.A. 489
Kasonde, J.M. 332

Kaufman, L. 108, 240
Kaufman, L.S. 41
Kawanabe, H. 41, 211, 486, 542
Kendall, R.L. 41
Kenmuir, D.H.S. 273, 361, 543
Kenya CBS 240
Kenya Fisheries Department 107, 134
Kerr, S.R. 524
Kessler, L.G. 485
Kibaara, D. 42
Kibaara, D.I. 32, 80, 109
Kilham, P. 42
Kilham, S.S. 42
Kimura, M. 487
Kinloch, B. 212
Kirby, P. 318
Kirema-Mukasa, C.T. 213, 240
Kirpichnikov, V.S. 489
Kiss, R. 296
Kitaka, G.E.B. 41
Kitakule, J.S. 213, 240, 242
Kitchell, J.F. 133
Kiyuku, A. 453
Kjaer, O. 361
Kling, G. 42
Kling, H. 296
Kling, H.J. 523
Knight, A.J. 489
Knudsen, K.L. 489
Kocher, T.D. 394
Koella, J.C. 109
Kolding, J. 273–4, 361–2, 363
Kongere, P. 240
Konings, A. 394
Kornfield, I. 489
Kricka, L.J. 489
Kriik, H. 108
Kristyanko, A. 80
Krueger, C.C. 489
Kruss, P.D. 41
Kudhongania, A.W. 31, 56, 80, 108, 134, 159, 180, 212, 523
Kuhry, B. 489

Laevastu, T. 159
Laing, Y. 523
Lalonde, S. 523
Lambert, J.D. 491
Lamboeuf, M. 296, 543
Lande, R. 489
Langerman, J.D. 274
Larkins, H.A. 159
Lauer, S. 81
Lauzanne, L. 159
Lays, J.F. 296
Lazzaro, X. 543
LBDA (Lake Basin Development Authority) 42
Leach, J.H. 524
Leary, R.F. 489
Leberg, P.L. 489
Lehman, J.T. 42
Lelek, A. 274, 332, 362
Lenski, R.E. 492
Levêque, C. 159, 453
Lewis, D.S.C. 274, 394, 395, 412, 524, 525
Lewis, W.M. 491
Lewontin, R.C. 491
Li, H.W. 490
Lien, L. 361
Ligtvoet, A. 31
Ligtvoet, W. 30–1, 80–1, 106–9, 133–5, 159–60, 180, 212
Lindem, T. 362
Lisle, A. 486
Liti, D. 361
Lock, J.M. 362
Loeschke, V. 489, 491
Loftus, K.H. 524
Lopes, S. 523
Lowe, R.H. 31, 56, 394
Lowe (McConnell), R.H. 56
Lowe-McConnell, R. 486
Lowe-McConnell, R.H. 32, 41, 106, 108, 211, 212, 274, 362, 524, 542
Lowery, R.C. 239
Luhmann, N. 240
Lymo, E. 212
Lysack, W. 524

Mabaye, A.B.E. 543
MacArthur, R.H. 319, 362
MacDonald, W.W. 109

Machena, C. 274, 543
Mack, R.N. 492
Maembe, T. 240
Maembe, T.W. 212
Magadza, C.H.D. 543
Magasa, J. 16
Magasa, J.H. 395, 412
Magnusson, K.G. 15
Magumba, K.M. 31
Magurran, A.E. 491
MAIF 241
Maikweki, J. 108
Makwinja, R.D. 543
Malvestuto, S. 319
Mangel, M. 319
Manini, P. 296
Mann, M.J. 32, 134, 296, 362
Mannini, P. 212, 240, 543–4
Marcus, L.F. 489
Margalef, R. 490
Markert, C.L. 490
Marsh, A.C. 394, 412, 491
Marsh, B.A. 394, 412, 491
Marshall, B.E. 15, 274, 332, 490, 524, 543
Marshall, T.R. 523
Marten, G.G. 42, 109, 134, 159, 212
Martin, A.P. 490
Maruyama, T. 490
Mather, P.B. 486, 490
Matthes, H. 296, 441, 544
May, B. 489
May, R.M. 490
Mayr, B. 394
Mbahinzireki, G.B. 109
McAllister, D.E. 487
McAndrew, B.J. 488
McCay, B. 240
McEvoy, A.F. 240
McGlade, J.M. 489
McIntyre, J.D. 491
McKaye, K.R. 543
McKenzie, A. 490
McLachlan, A.J. 274
McLeod, A.A.Q.R. 361
Mdaihli, M. 394–5
Melack, J.M. 42, 362, 523–4
Melvin, G.D. 490
Menyani, W.W. 543
Merona, B. 160, 362
Meyer, A. 394
MFEP 241
Mhone, O.K. 395, 543
Miller, D.J. 490
Miller, R.S. 487
Minev, J.G. 107
Minshull, J.L. 272, 542
Mitchell, D.S. 274, 544
Mitchell, S.A. 275, 544
Mkandawire, R. 239
Mkisi, M. 212
Mkoko, B.J. 394, 524
Mkumbo, O.C. 80, 108–9, 134, 180
Mngoma, S. 41
Moeed, A. 485
Moller, F. 490
Moran, P. 490
Moreau, J. 160, 362, 544
Moriarty, C.M. 160
Moriarty, D.J.W. 32, 56, 160
Moritz, C. 487, 489
Morris, M. 239
Mortimer, M.A.E. 332
Mosille, O. 179
Mosille, O.I.I.W. 32, 180
Mosille, O.W.I. 31
Moss, B. 275
Moyle, P.B. 490
Moyo, N.A.G. 275
MRAG 319
Mraja, A.H.S. 362
Mubanga, J. 441
Muchiri, S.M. 319
Mugidde, R. 42, 80
Mugyenyi, J.B. 241
Mulindabigwi, A. 543
Mullis, K. 491
Mumeka, A. 332
Munro, J.L. 544
Munyandorebo, J. 296
Munyandorero, J. 544
Mutamba, A. 543
Mutoro, D. 108
Muvuti, K.M. 319

Mwanyama, N.C. 395
Mwebaza-Mdawula, L. 241
Mwene-Beyanga, P. 213, 242

Naesje, T.F. 492
Nagoshi, M. 211, 486, 542
Nedelec, C. 296
Nei, M. 490
Nelson, K. 490
Nemec, T.F. 241
New Vision 241
Newell, B.S. 42
Nhwani, L.B. 212
Njuguna, S.G. 109
NORAD 362
N'Sibula, M. 544
Ntara, M. 242
Nyakageni, B. 160, 544
Nyeko, D. 180, 213

Obrdlik, P. 332
Ocenodongo, D. 32, 56, 81, 109
Ocenodongo, D.L. 135
Ochieng, E.O. 241
Ochumba, P.B.O. 32, 41, 42, 80, 109, 135
ODA 490
Odongkara, O.K. 241
Ogari, J. 30, 32, 41, 42, 80, 109, 158, 160, 179–80, 212–13, 241, 362
Ogari, J.T.N. 361
Oglesby, R.J. 523
Oglesby, R.T. 524
Ogunja, J.-A. 180
Ogutu-Ohwayo, R. 32, 42, 56, 80–1, 108–9, 135, 213, 241, 362, 490, 524, 544
Okaranon, J.O. 135
Okaronon, J. 32, 56, 81, 109
Okaronon, O. 241
Okedi, J. 32, 160, 241
Okedi, J.Y. 81
Okemwa, E.N. 109, 180, 213
Oldfield, M.L. 490
Oliphant, M.S. 275
Olseen, S. 41
Østvedt, O.J. 362
Ovenden, J.R. 490
Ow, M.D. 160

Paine, R.T. 16, 110, 493, 524
Palmer, J.D. 487
Palomares, M.L. 159–60
Palumbi, S.R. 490
Park, L.K. 490
Parkin, D.T. 486
Patton, J.C. 487
Pauly, D. 15, 158–60, 180, 394, 544
Payne, A.I. 160
Payne, I. 32, 135
Payne, N.R. 489
Pearce, M.J. 441
Petit, P. 453
Petr, T. 275, 332, 362
Peyton, J. 241
Philipp, D.P. 491
Pianka, E.R. 362
Pielou, E.C. 81
Pimm, S.L. 319
Pinkas, L. 275
Pitcher, T.J. 15, 16, 180, 213, 275, 394–5, 487, 524
Pla, C. 488
Plante, C. 523
Platteau, J.-P. 241
Pollingher, U. 41, 42
Polovina, J.J. 160
Popova, O.A. 544
Prodohl, P.A. 491
Puccia, C.J. 161

Qadri, S.U. 487
Quinn, T.W. 491
Quint, J.M. 239

Rabour, C.O. 135
Rabuor, C. 212
Ragin, C. 239
Rambiri, O. 212
Ramshaw, J.A.M. 491
Rand, A.S. 319
Rapport, D.J. 213
Reay, P. 16
Redclift, M. 239

Reeves, E. 241
Regal, P. 492
Regier, H.A. 213, 524
Reisenbichler, R.R. 491
Reusens, M. 544
Reuterwall, C. 491
Reynolds, J.E. 42, 109, 135, 160, 180, 213, 240, 241–2
Reyntjens, D. 296, 544
Rhodes, D.H. 363
Ribbink, A.C. 394, 412, 491
Ribbink, A.J. 107, 109, 135, 211, 213, 394, 412, 486, 487, 491, 524, 542, 544
Ricardo, C.K. 363
Richardson, A.E. 319
Richardson, B.J. 491
Richardson, J.L. 319
Ricker, W.E. 160, 319, 491
Rigler, F.H. 362
Rinne, J.N. 160
Robertson, A. 32
Robinson, R.L. 16, 395
Roby, D. 491
Roemer, M. 242
Roest, F. 296
Roest, F.C. 32, 441, 453, 544
Rogers, J.F. 56
Rohlf, F.J. 412
Ross, M.R. 491
Rudd, J.W.M. 296, 523
Rufli, H. 544
Ryder, R.A. 523, 524
Ryman, N. 485, 488–9, 491

Sabella, J.C. 239
Saiki, R.K. 491
St. Louis, V.L. 492
Sambilay, V.C. jun. 159
Sandlund, O.T. 492
Sanyanga, R.A. 274–5
Saunders, N.C. 485
Scavia, D. 32, 42
Schaefer, S.A. 491
Scharf, S.J. 491
Schnaiberg, A. 242
Schnieder, D.C. 523, 524
Schoener, T.W. 319
Scully, R.J. 332
Seghers, B.H. 491
Seisay, M.B.D. 393, 394–5
Seitz, A. 491
Selander, R.K. 486
Sevigny, J.M. 491
Shafland, P.L. 491
Sharp, B.J. 394, 412, 491
Sharp, G.D. 273
Shaw, P.W. 491
Shepherd, J.G. 441
Shields, W.M. 492
Shuter, B.J. 524
Siddiqui, A.Q. 319, 492
Simberloff, D. 492
Simmie, S. 32
Simon, R.C. 486
Siwo, J. 242
Smith, C. 16
Smith, M.H. 491
Snedecor, G.W. 363
Snoeks, J. 544
Sokal, R.R. 412
Solbrig, O.T. 492
Soriano, M. 159
Soriano, M.L. 159
Soulé, M. 490
Soulé, M.E. 488, 492
Sparre, P. 180
Spaturu, P. 319
Spearman, W.J. 487
Spiller, D.A. 319
Spliethoff, P.C. 296, 543, 544
Ssali, W.M. 180, 213
Ssentongo, G.W. 32, 160, 213, 395
Stearns, S.C. 109
Stewart, F. 239
Stewart, J.E. 492
Stoffel, S. 491
Stoneman, A.J. 80
Stoneman, J. 56
Symoens, J.J. 158, 160

Taggart, J.B. 491, 492
Tajima, F. 490
Talling, J.F. 43, 363

Tautz, D. 492
Taylor, E.B. 486, 492
TDRI 242
Temple, P.H. 56, 242
ter Braak, C.J.F. 134
Thein, S.L. 489
Thomason, M.A. 488
Thorpe, J.E. 488
Thorpe, J.P. 492
Tiedje, J.M. 492
Tier, C. 319
Tiews, K. 524
Tomiuk, J. 489
Toots, H. 544
Torres, F.S.B. 544
Townsend, C.R. 16
Tracey, C.R. 319
Trainer, T. 242
Travers, R. 108
Trewavas, E. 161, 211, 395, 412, 486, 492, 542
Tumwebaze, R. 213
Turner, G.F. 16, 394–5, 412, 524–5
Turner, G.T. 16
Turner, J.L. 16, 32, 109, 395, 412, 492, 525, 544
Tweddle, D. 16, 161, 394, 395, 412, 524, 525
Twongo, T. 31, 32, 56, 81, 108, 213, 242, 525

Uganda Fisheries Department Annual Report 1951 55
Ulanowicz, R.E. 161, 363
UNECA 242
UNICEF-Kenya 242
University of Leiden 242
Ursin, E. 158, 180
Utter, F. 488–9, 491, 492

van Lissa, J. 544
van Meel, L. 296
van Oijen, M. 43, 110, 135, 180, 493, 525
van Oijen, M.J.P. 32, 81, 109–10, 161
van Tongeren, O.F.R. 134
van Well, P. 441, 542
van Zahlinge, N.P. 393, 394–5
Vanden Bossche, J.P. 525, 570
Vanderpool, C.K. 242
Vanderpuye, C.J. 332
Vanni, M.J. 492
Venema, S.C. 180
Verbeke, J. 297, 545
Vetaas, O.R. 363
Visser, J. de 107
Vostradovsky, J. 545
Vuorinen, J. 492

Wadanya, J. 213
Wahabakomeye, J.B. 296
Wallace, R.K. jun. 319
Walters, C.J. 394
Wandera, S.B. 32, 81
Wang, J. 523
Wanink, J. 110, 134, 180, 493, 525
Wanink, J.H. 16, 81, 107, 109–10, 135, 214, 545
Wanjala, B. 161
Wannik, J. 43
Waples, R.S. 492
Ward, A.R. 213
Ward, R.D. 486, 489, 492
Waterhouse, J.C. 361
Watson, I.G. 81
Welcomme, R.L. 16, 32, 43, 57, 81, 110, 160–1, 213–14, 297, 361, 493, 523, 525, 545
Werimo, K.O. 180
White, B.N. 491
Whitehead, P.S.P. 32
Wiley, D.S. 242
Wilhm, J.T. 319
Willoughby, N.G. 81, 161, 395
Wilmot, R.L. 487
Wilson, A.C. 394
Wilson, D.C. 242
Wilson, E.O. 319, 362
Wilson, V. 489
Winans, G. 492–3
Winberg, G.G. 161
Witimani, J. 16, 395

Witte, F. 30, 32, 41, 43, 56, 81, 106, 107–8, 108–9, 110, 133–5, 159, 161, 180, 211, 214, 486, 493, 525, 542
Witte-Maas, E. 110, 135, 180, 493
Witte-Mass, E.L.M. 32, 109–10
Witte-Mas, E. 525
Witte-Mass, E. 43
World Resources 1990–91 542
Worthington, E.B. 43, 57, 81, 363, 493
Wright, J.M. 486, 488, 493

Yamaoka, K. 41, 106, 211, 486, 542
Yongo, E. 242

Zambia/Zimbabwe SADC Fisheries Project Report, no. 11, no. 26 272
Zaret, D. 319
Zaret, T.M. 16, 110, 493
Zwangenburg, K.C.T. 493

Species index

SCIENTIFIC NAMES

Aethiomastacembelus shiranus 421
Afromastacembelus frenatus 71
Alestes antinorii 301
Alestes baremose 73, 75, 280, 342, 345–7, 349–50, 357, 360, 548, 551, 563
Alestes brevirostris 405–6
Alestes dentex 75
Alestes ferox 351
Alestes 'furcicauda' 421
Alestes lateralis 323, 325–7, 329, 331
Alestes latifasciata 71
Alestes minutus 351
Alestes occidentalis 75
Alestes 'orange' 381
Alestes spp. 20, 73–4, 77, 114, 118–24, 140, 341, 350–1, 548, 566, 569
Aplocheilichthys spp. 280
Argulus japonicum 467
Aristochromis christyi 421
Astatoreochromis alluaudi 461
Astatotilapia calliptera 421
Astronotus ocellatus 461
Auchenoglanis spp. 74, 562
Aulonocara spp. 405

Bagrus altianalis 140
Bagrus amphigramma 301–2
Bagrus arcislongae 421
Bagrus barbus 461
Bagrus bayad 75, 340, 343–4, 346–7, 349, 352, 556
Bagrus bynni 75, 339, 341, 343–5, 347–50, 360
Bagrus docmac 20–1, 25, 29, 37, 45, 63–7, 69, 72, 75, 100, 146, 185–8, 554, 556
Bagrus kimberlevensis 461
Bagrus marequensis 251, 327
Bagrus meridionalis 417, 421, 461, 560
Bagrus natalensis 461
Bagrus niloticus 75
Bagrus paludinosus 327, 421
Bagrus poechii 327
Bagrus spp. 36, 73–4, 114, 118–24, 142, 148–9, 194–6, 230, 339, 341, 346, 548, 563, 568
Barbus paludinosus 552
Barbus spp. 20–1, 26, 63–7, 71–2, 74, 114, 118–24, 140, 186, 302, 324, 341–2, 358, 568
Barilius moorii 280
Barilius niloticus 75
Barotherodon galilaeus 548, 556, 563
Basilichthys bonariensis 461
Brycinus imberi 251, 253, 255, 259, 262, 267, 421, 532
Brycinus jacksonii 72
Brycinus lateralis 264, 529, 532, 535
Brycinus nurse 75, 78
Brycinus spp. 63
Buccochromis spp. 405, 409, 421

Carassius auratus 461
Carassius carassius 91, 461
Catla catla 461
Chaaoborus 537, 561, 579
Chilotilapia rhoadesii 405–6
Cichla ocellaris 9, 85, 461, 473

Citharinus citharus 73, 75, 338, 341–2, 344, 346–7, 349–50, 360, 548
Citharinus spp. 339, 343, 346, 348, 358, 563, 566, 569
Clarias gariepinus 37, 63–7, 69, 71–2, 75, 100, 251, 253, 255, 260–1, 263, 266, 327, 417, 421, 532, 536, 550, 555, 557
Clarias lazera 339, 341–2, 461, 554
Clarias mossambicus 20, 25, 552
Clarias ngamensis 555
Clarias spp. 36, 74, 114, 118–24, 142, 146, 148–9, 186, 230, 249, 327, 568
Copadichromis borleyi 387
Copadichromis cf. *chrysonotus* 421–2
Copadichromis cf. *eucinostomus* 405
Copadichromis cf. *prostoma* 387
Copadichromis cf. *virginalis* 381, 404–5, 421–2
Copadichromis chrysonotus 387
Copadichromis jacksoni 387
Copadichromis large spp. 405
Copadichromis pleurostigma 387
Copadichromis pleurostigmoides 387
Copadichromis quadrimaculatus 387
Copadichromis spp. 380, 387, 390
Copadichromis virginalis 387, 406
Coregonus albula 481
Corematodus taeniatus 405, 421
Ctenopharyngodon idella 460–1
Ctenopharynx intermedius 405, 421
Ctenopharynx nitidus 404–6
Ctenopharynx spp. 411
Cynothrissa 510
Cynothrissa menta 569
Cyprinus carpio 531, 461, 465

Dimidiochromis compressiceps 421
Dimidiochromis strigatus 421
Diplotaxodon limnothrissa 380–1, 519, 561
Diplotaxodon spp. 380, 390, 405, 519
Distichodus mossambicus 251
Distichodus niloticus 75, 342, 344–7, 349, 360
Distichodus schenga 251, 550
Distichodus spp. 74, 249, 339, 341, 343, 345, 348, 358, 563, 569

Engraulicypris minutus 280
Engraulicypris sardella 367, 417, 421, 519, 527, 530, 533, 561
Engraulicypris spp. 12, 280, 372, 387–8, 390
Engraulis sardella 539
Esox lucius 93, 461
Etroplus suratensis 461
Eutropius depressirostris 327, 329, 536
Eutropius niloticus 563

Gambusia affinis 460–1
Gambusia holbrooki 474
Gambusia spp. 301
Gnathonemus spp. 75, 562

Haplochromis angustifrons 554
Haplochromis discorhynchus 259–60, 267
Haplochromis ericotaenia 401
Haplochromis giardi 326
Haplochromis heusinkveldi 95, 97–8
Haplochromis laparogramma 93, 95, 97–8
Haplochromis mloto 404
Haplochromis mollis 401
Haplochromis nigripinnis 554
Haplochromis odoe 329
Haplochromis piceatus 92, 94–5, 98
Haplochromis pyrrhocephalus 95, 97–8
Haplochromis sphaerodon 401
Haplochromis spp. 19–20, 22–3, 27, 36–8, 40, 60, 64–9, 71–2, 75, 77, 79, 83–106, 111–12, 114, 118–24, 140–2, 146, 148–50, 185–8, 195–6, 205–6, 208, 232, 236–7, 324, 365, 370, 372–3, 380–7, 391–2, 416, 422–3, 457, 560, 568
Haplochromis squamipinnis 554
Hemitilapia oxyrhynchus 421
Hepstus odoe 555
Heterobranchus longifilis 251, 260, 536
Heterotis niloticus 339, 461

Hippopotamyrus discorhynchus 251–3, 532
Hydrocynus brevis 556
Hydrocynus forskalii 73, 342, 344–7, 349–50, 360, 513, 556, 563
Hydrocynus spp. 73–5, 77, 249, 339, 341–3, 351, 357–8, 548
Hydrocynus vittatus 251–5, 258–9, 262–4, 266–9, 529, 532, 535–6, 540, 550–1, 557
Hypophthalmichthys molitrix 461

Ictalurus punctatus 461

Labeo altivelis 251, 253, 255, 259–64, 268, 460, 532
Labeo congoro 251, 260
Labeo coubie 75
Labeo cylindricus 251, 260, 327, 421
Labeo horie 75, 339–41, 344–5, 347–9, 360
Labeo mesops 389, 417
Labeo molybdinus 327, 555
Labeo niloticus 563
Labeo rohita 461
Labeo spp. 74, 114, 118–24, 126, 188, 249, 329, 341–3, 348, 350, 358, 550, 568–9
Labeo victorianus 20–1, 29, 37, 45, 63–6, 71, 101, 460
Lates angustifrons 425–7, 518, 564
Lates longispinus 85, 347, 351, 566
Lates macrophthalmus 75, 85, 548
Lates mariae 425–7, 443, 518, 564
Lates microlepis 425–7, 517, 564
Lates niloticus 6, 11–12, 20, 28–30, 33, 36–7, 45, 49, 50–2, 54, 60, 64–6, 68–72, 75–6, 79, 85, 100, 111, 188, 206, 340, 346–7, 351, 461, 467, 473, 499, 551, 554, 556, 559, 566–7, 568–9
Lates spp. 8, 23, 25–9, 54, 73–4, 78, 138–9, 183, 186, 206, 208, 210, 339–47, 349, 351, 358, 360, 425–6, 432, 437–8, 440, 443, 445, 461, 506, 517–18, 520, 548, 560
Lates stappersi 425–7, 429–40, 443, 447–52, 513, 517–18, 564
Lebistes reticulata 301
Lepomis cyanellus 461
Lepomis gibbossus 461
Lepomis macrochirus 461
Lepomis microlophus 461
Lethrinops altus 405, 407
Lethrinops lethrinus 405, 421
Lethrinops longimanus 404–5
Lethrinops longipinnis 381, 405
Lethrinops macracanthus 404-6, 411
Lethrinops macrochir 421
Lethrinops microdon 381–2, 404, 407
Lethrinops mylodon 404–6, 410–11
Lethrinops 'oliveri' spp. 405–7
Lethrinops parvidens spp. 405
Lethrinops 'pink head' 421–2
Lethrinops small spp. 406–7
Lethrinops spp. 381
Lethrinops stridei/micrentodon 405
Limnothrissa 12, 269, 277–8, 281, 283, 285–6, 288–90, 295, 390, 426–7, 439, 447, 452, 510, 514, 517, 528–31, 533–8, 541
Limnothrissa miodon 6, 249, 251–2, 271, 277–8, 280–2, 294, 329–30, 382, 390, 410, 426, 428–9, 433, 438–40, 443, 451, 461, 477–8, 499, 510, 512, 528, 532–3, 541, 550, 557–8, 564

Malapterurus electricus 75, 251–2, 263–4
Maravichromis cf. *balteatus* 421
Maravichromis labidodon 421
Marcusenius macrolepidotus 251–3, 255
Micropterus dolomieu 461
Micropterus punctualatus 461
Micropterus salmoides 300–1, 303–4, 461
Mormyrops deliciosus 251, 260, 536
Mormyrops macrolepidotus 259–60, 267–8
Mormyrops spp. 75, 324
Mormyrus caschive 73

Mormyrus kannume 75
Mormyrus longirostris 251–5, 259–60, 262–4, 266, 268–9, 532
Mormyrus niloticus 75
Mormyrus spp. 114, 118–24, 186, 249, 562
Mormyrus victorie 75
Mylochromis anaphyrmus 410
Mylochromis spp. 405–6

Neobola bredoi 75
Nimbochromis livingstonii 421
Nimbochromis spp. 405

Oncorhynchus gorbuscha 476
Onorhynchus mykiss 301, 461, 466
Opsaridium microcephalum 421
Opsaridium microlepsis 421
Oreochromis andersonii 326, 461, 555
Oreochromis aureus 461
Oreochromis esculentus 8, 20–1, 23, 25, 29, 40, 45–6, 48–9, 51, 53–5, 59, 63, 67, 72, 101, 113–15, 118–24, 205, 460–1, 466, 517, 559, 566
Oreochromis hornorum 461, 465
Oreochromis karongae 372, 374–6, 379, 421–2
Oreochromis leucostictus 23, 48–9, 52–4, 60, 67, 69, 71–3, 75, 300–4, 306–8, 311–15, 317–18, 461, 466, 479, 548, 559, 566
Oreochromis lidole 7, 10, 372, 374–6, 379–80, 422, 514
Oreochromis macrochir 249, 251, 256, 461, 555
Oreochromis microchir 326–7
Oreochromis mortimeri 249, 251–6, 258, 260, 262–4, 266, 268–9, 532, 550
Oreochromis mossambicus 461, 465
Oreochromis niloticus 6, 8, 11, 23–4, 26, 28–9, 33, 37, 49–55, 60, 63, 68–9, 71–2, 75–6, 79, 99, 101, 103–4, 111-13, 115, 118–26, 140, 142, 147, 149–50, 155, 185–8, 206, 301, 353, 461, 465, 473, 499, 517, 548, 562, 566, 568–9
Oreochromis (*nyasalapia*) spp. 372
Oreochromis saka 422
Oreochromis shiranus 372, 417, 421, 461, 552
Oreochromis spilurus 461, 465–6
Oreochromis spilurus niger 301, 479
Oreochromis spp. 142, 147–8, 307, 387, 413, 416–17, 520, 559–60, 568
Oreochromis squamipinnis 372, 374–7, 379, 422
Oreochromis variabilis 20, 23–4, 37, 40, 46, 48–9, 51, 53, 54–5, 59, 63, 67, 72, 101, 460, 466, 559
Osphronemus gourami 461
Otopharynx 'argyrosoma b' 381
Otopharynx argyrosoma spp. 405–7, 421–2
Otopharynx auromarginatus spp. 405
Otopharynx cf. *tetraspilus* 421
Otopharynx speciosus 405
Otopharynx tetrastigma 421–2

Pellonula 512
Pellonula afzeliusi 556, 569
Pharyngochromis darlingi 251, 536
Placidochromis 'cf. *longimanus*' 381, 405
Placidochromis longimanus 414, 421
Placidochromis subocularis 405, 421
Poecilia reticulata 460–1
Poecilia spp. 301
Protomelas cf. *labridens* 421
Protopterus aethiopicus 20, 25–6, 37, 63–7, 69, 71–2, 100, 140, 142, 146, 148, 185–8, 554
Protopterus spp. 36, 74, 114–15, 118–24, 126, 194–6, 230, 517, 568
Pseudocrenilabrus philander 531, 536
Pseudotropheus elegans 405–6, 410, 421
Pseudotropheus livingstonii 405–6, 410, 421

Ramphochromis spp. 381–2, 405, 421
Rastrineobola 12, 155, 197, 209, 217, 226, 233, 237, 280, 516, 533

Rastrineobola argentea 28–30, 36, 50, 60, 62–7, 69, 71–2, 77, 103–4, 111–12, 114, 118–26, 131–2, 138, 140, 142, 147–9, 155, 183, 185, 187–9, 191, 206, 208, 531, 568
Rastrineobola sardella 519, 527
Rhamphochromis 519, 560

Salmo trutta 461
Salvelinus fontinalis 461
Schilbe intermedius 63–4, 66–7, 71–2, 532, 536
Schilbe mystus 20, 37, 69, 75, 251–3, 255, 262–3, 327, 555
Schilbe spp. 26, 115, 118–24, 341
Schilbe uranoscopus 342, 345–6, 349, 352, 360
Sciaenochomis spp. 405, 407
Serranochromis 557
Serranochromis angusticeps 555
Serranochromis codringtonii 249, 251–6, 258–60, 262–4, 266, 269, 536
Serranochromis macrocephalus 251, 253, 256, 262, 264, 267
Serranochromis robustus 251, 461
Sierrathrissa 512
Sierrathrissa leonensis 556
Stizostedion 502
Stolothrissa 8, 277, 447, 452, 510, 512–13, 517, 530–1
Stolothrissa tanganicae 277, 280, 425–30, 433–40, 443, 451, 461, 533, 557–8, 564
Synodontis 519
Synodontis afrofischeri 67, 69, 71–2
Synodontis kafuensis 327
Synodontis nebulosus 251, 260, 262
Synodontis njassae 405–7
Synodontis shall 73, 339–40, 342, 344–5, 347, 349, 352, 360
Synodontis spp. 20, 37, 63–6, 74–5, 115, 118–24, 140, 142, 146, 148, 186, 194, 341–3, 345, 352, 358
Synodontis victoriae 100
Synodontis zambezensis 251, 253, 255, 260–4, 267, 532, 536

Taeniochromis holotaenia 405
Taeniolethrinops furcicauda 404–5, 411
Taeniolethrinops praeorbitalis 404–5, 410–11
Tilapia esculentus 55
Tilapia melanopleura 23–4
Tilapia rendalli 48, 55, 60, 72, 251, 253–6, 263, 327, 372, 417, 421, 461, 536, 555, 557, 559, 566
Tilapia saka 372
Tilapia sparmani 555
Tilapia spp. 9, 19, 22–3, 36, 38, 40, 45, 47, 64–6, 69–71, 74, 76, 113–14, 127, 139, 144, 185–8, 191–2, 194–7, 210, 217, 221, 226–8, 232–4, 236–8, 299–319, 338–9, 355, 370, 372–80, 388, 392, 398, 418, 422, 463, 467, 479, 562
Tilapia variabilis 55
Tilapia zilli 6, 23–4, 48–9, 51–5, 60, 67, 69, 71–3, 75, 103, 300–4, 306–8, 312–15, 317–18, 461, 466, 556, 559, 566, 569
Tinca tinca 461
Tramitichromis lituris 421
Trematocranus placodon 421
Trematocranus spp. 405

Varicorhinus nasutus 251

Xiphophorus maculatus 461

COMMON NAMES

Asiatic branchiura 467

Bass, largemouth 300–2
Boolara carp 465
Branchiura, Asiatic 467
Bream
  benthivorous dwarf 142, 144, 148, 151–4
  Kafue 324
  planktivorous dwarf 142, 144, 148, 151–4

Bream *contd*
  predatory dwarf 142, 144, 148, 151–4
Brown squeaker 557
Butter catfish 539

Carp 465, 467, 531
  Chinese 460
Catfish 9, 142, 144, 148, 151–4, 155, 370–1, 373, 388, 398, 517, 536
Chambo 371–2, 374, 379, 413, 416–18, 420, 422–3, 539
  Malawi 7, 9, 12
Chinese carp 460
Chisawasawa 370–2
Cisco 462

Dagaa 111, 139, 142, 144, 148, 151–4, 193–6, 516, 532
  Lake Victoria 12, 111

Green tilapia 252
Guppy 302, 477

Haplochromine, rocky shore 14

Kafue bream 324
Kambuzi 287, 370–2, 416–20, 422–3
Kampango 417
Kapenta 267–8, 271, 294, 329, 425–6, 429, 434, 437, 481
Kariba tilapia 252
Kasawala 370
Kivu sardine 281

Lake salmon 425
Lake Victoria dagaa 12, 111
Lumpu 294
Lungfish 26, 142, 144, 148, 151–5

Malawi chambo 7, 9, 12, 514
Malawian usipa 12
Mbaba 370, 372, 417
Mbuna 391, 539
Mlamba 417
Mosquitofish 474
Mukeke 447, 449, 451

Ncheni 370–2
Nchila 417
Ndagala 294, 447, 451
Ndunduma 370–1, 380
Nile perch 6, 11–12, 20, 33–4, 37–40, 45, 60–1, 63, 67, 69, 71–3, 75–8, 84–9, 92, 95, 99–100, 103, 105, 111–13, 115–18, 120–4, 126–33, 138, 141–2, 144, 148, 151–4, 163–79, 181–214, 216–17, 219, 222–8, 232–4, 236–8, 351, 425, 428, 457, 460–1, 467, 473, 477, 495–7, 499, 505–7, 509–10, 512, 514–17, 520–2, 528, 532
Nile tilapia 33, 60, 67, 69–70, 72–3, 77, 111, 144, 148, 151–4, 176, 206–8, 473, 499–500, 514, 516
Ningu 20–1, 29, 37, 45, 63–6, 71, 101, 460
Nkacha 419–20, 423
Nvolo 425
Nyamu-nyamu 448

Omena 111

Pink salmon 476
Pupfish 462

Rainbow trout 301, 460
Rocky shore haplochromine 14, 391

Saguga 409
Salmon
  lake 425
  pink 476
Sardine
  clupeid 12, 282–6, 288–90, 292, 294–5, 329, 425–6, 443, 445
  freshwater 6, 11, 238, 267–9, 468, 495, 497, 499–500, 504–15, 528, 530, 535, 537
Sculpin 462
Snakehead 517
Snoutfish 142, 144, 148, 151–5
Squeaker 142, 144, 148, 155
  brown 557

Steelhead trout 466

Tigerfish 9, 14, 252, 255, 257, 267–9, 513, 529, 536
Tilapia, Kariba 252
Trout 460
  rainbow 301, 460
  steelhead 466

Usipa 367, 369–73, 387, 417, 519, 532
  Malawian 12
Utaka 369–73, 380, 387, 408, 539

Vendace 481

Walleye 502

Subject index

Page references in **bold** refer to tables; those in *italics* refer to figures.

Adaptability, long-term 472
African lakes
  cichlid preponderance 463
  endemic species *462*
  fish introductions 5–6, **461**
    genetic impacts 457–87
    reasons 5
  L. miodon establishment 538
  L. miodon growth parameters **282**
  L. miodon as successful species 529–35
  location 2–3, *3*
  population bottleneck examples 472–3
  species change and fisheries 1–15
  surface areas **4**
  total fish yields 521
Air cargo export 232
Algal biomass 78–9
Algal blooms 25, 103
Alleles, loss *470*, *475*
Allozyme electrophoresis 479–82
Animal feed plants 209
Anoxia 34, 37
Anoxic layer 79, 131
Aquarium fish trade 365, 496
Artificial lakes 496
  fish yield 512–15
  Limnothrissa introduction 541
  potential total yield 515
Artisanal light-fishing technique 290, *291*

Biodiversity
  and fish resources 458–9
  reduction 2
Biological control by fish predation 460
Biomass
  budget equations 137
  estimates 169, 402–4
Birds
  changes 104
  fish-eating 536
Boadzulu Island *375*, 384–5, 388–9, 392, 398–9, **405–7**, 411
Boat
  crew
    members 217
    size **219**
  managers 217
  operators 217–18
  owners 217
Boats, artisanal, analysis 446–8
Breeding status 52, *53*
Burundi
  industrial fishery CPUE *451*
  L. stappersi length *450*
  pelagic fisheries catches *451*
  rope length *450*
  stretched meshes *449*
Butimba Bay, *H. piceatus* female size **94**

Cambridge Nabugabo Biological Expedition 59
Cannibalism 8, 39, 533
Canoe density 164, *165*, 166

Cascade theory 13, 131
Catamarans 445, *446*, *448*
Catch
  assessment surveys (CAS) 415–16, 418, *420*
  changes in price and value 194, 197, 199
  composition in experimental gill nets **251**
  per unit effort (CPUE)
    Burundi 448–50
    Burundi artisanal fishery *452*
    calculation 169, 431
    data analysis 113
    experimental and commercial comparison **403**
    experimental fishing surveys *353–4*
    growth 189
    indicative change **190**
    Lake Kariba 254, *255*, 256, *257–8*, **259**, 260, *261*, 262–3
    Lake Kariba experimental fishery 250
    Lake Malawi tilapiine fisheries *376*, *378*
    Lake Tanganyika commercial fishery *432–3*
    mesh size 341, *342*, 343
    pair trawl fishery *385*
    relationship phases *386*
    seine fisheries *419*
    single-boat fishery *384*
    and stock size 178
Child health 226
Climatic changes 39
Collectors and marketers 217
Commercial catch and effort data 164, 166
Committee on Inland Fisheries of Africa 236
Competition 8, 535–6
  effects 267–8
  for food 54
  impact 23
Condition factor K *284*, 285
Conservation genetics 471–2
Cooperative societies 227–8
Cross correlation function (CCF) 117

Detritus as food 315, 318
Diet
  composition 141, **144**, 145
  studies 130
Diseases, introduction with fish 467
Distributional impacts 216

Echosounding surveys 289–90
Ecology
  changes in Mwanza Gulf 83–99
  of tilapias 307–14
Economy development 231–4
ECOPATH II
  balancing of model 145
  changes assessment 145
  input parameters **142–3**
  results 148–55
  routine 156–7
  species change analysis 137–61
ECOPATH model 7, 13, 156
  fish yield 500
Ecosystem
  alterations 216
  analysis 215
  damaged 138, 140
  deterioration factors 40
  fish introduction effects 457, 467–8
  fishery 238
  in stress 207
  structure and functioning differences 155
Ecotrophic efficiency 145
Electrophoretic mobility 480
Emin Pasha Gulf, haplochromine catch rates *88*
Employment
  benefits 201–2
  opportunities 225–6
Endemic
  fauna conservation 2
  species, threatened, mitigation measures 13–14

Environment, Lake Turkana
ecosystem 357
Environmental
change impact on Lake Victoria
fish stocks 19–32
degradation 78–9
issues 130–2
Experimental
fishing 62
investigation 12
Exploitation, long-term
sustainability 234
Export earnings 203–4
Exports in Africa 193

Fecundity, *L. miodon* **532**
Feeding behaviour, *Limnothrissa*
530–1
Feeding overlap 312, *313*
Fish
collectors 221
dealers 221
destination **221**
distribution 230
exports 236
and foreign exchange 234
filleting 204
frying 220
growth 55
household consumption
availability 226
increased value and fisher conflict
229
indigenous, and introduced
species 51–5
individual fitness reduction 465
introductions 5–6, 459–62
African lakes 462–3, 495–525
aquaculture opportunity 499
aquarium-trade benefits loss
498
costs of satisfying fisheries
497–8
decision making 497–9
ecological models 10
economic gain 499
establishment 6–9
fisheries economic value loss
497
genetic diversity loss 498
genetic impacts 457–63, *464*,
465–87
impacts 9–10, 461
on Lake Victoria 19–20,
28–9
indirect effects 467–73
instability and non-
sustainability of gains 498
international retribution for loss
of biodiversity 499
L. niloticus yields 509, **510**
Limnothrissa 510–12
Limnothrissa yields 510, **511**
loss of aquaculture potential of
endemic species 498
loss of cultural benefits from
endemic fauna 498
loss of tourism benefits from
endemic fauna 498
medicinal benefits loss 498
nutritional gain 499
reasons 5, 460
socio-economic gain in fishery
sector 499
socio-economic value loss 497
sport fishery income 499
successful 6
sustainable yields *507–8*
estimation 499–506
kills 34–5, 131
meal 27, 197, 232, 237
oils 224
processing 220
processors 219–20, 223
products, export 193
ranching 14
smoking 204
yield
plot per unit area *501*
relationship to primary
production 500–2
Fisheries
benefits competition 235
catches, ECOPATH II 141

Fisheries *contd*
  changes in African lakes 1–15
  commercialization 217
  development and assistance recommendations 14–15
  ease of entry 224
  ECOPATH use in management 156–7
  employment **202**
  fishing effort reduction 178
  haplochromine stock decline 129–30
  increased income 222
  L. niloticus assessment *165*
  management 232–3
  as shared resource 235–6
  wealth evaluation 2
  yield changes 60
Fishers
  activity *447*
  geographic mobility 225
  incomers 229
  increase 188–9
  interviews with 444
  quality of life 224–5
  socio-economic displacement 209
FISHIN' Project 191, 193, 194
Fishing
  effect 269–70
  effort 164, 187–92
    change **189**
    curve, equilibrium yield *174*
    stock impact 451–2
  gear
    analysis 446–8
    imported 223
    rising prices 218
  households, health of members 226–7
  methods, selective 208
  mortality 168, 439
  pressure 2, 127–30
    inshore 206
    Lake Kariba 246
    and tilapia persistence 307
  profitability 390
  regulations 232–3
  social organization 218
Floating fern 237, *238*, 264, 266–7, 300
  decline 538
  and fish abundance *540*
Food, *L. stappersi* 437, **438**
Food
  chains *139*, 140
  consumption 145, **146–7**
  habits 54
  security 233
  sources 7
  web 140, 313, *314*
    alteration 39
Foreign aid 236
Founder effects 474, *475*, 476–8
Fox surplus-production model 378

Gatun Lake, Panama 473
Gene flow changes 463
Genetic
  changes and fish resources 459
  diversity
    in bottlenecks 470
    determination 480
    level estimation 480
  impacts, groups 463
  resources conservation 484
  structure investigation 478–84
  variability 465
    loss 477, 484
Genotypic variability 484
Gill nets
  coarse 343
  fine 345–6
  ownership changes *419*
  see also Nets
Growth rate
  comparison **533**
  Limnothrissa 532
Guild structure 401
Gulland-Fox model *305*
Gulland's formula 169, 174, 356

Habitat
  changes assessment 2

detritus as buffer against variability 314–16
physical factors 7
variability and fish extinction rates 316 17
Haplochromis Ecology Survey Team (HEST) *84*, 98, 103, 128, 132
Herbivores impact 9–10
Heterozygosity
changes *476*
loss *471*, *475*
Hopson report 337, 343, **347**
Human consumption, fish predation 25–7
Hybrid fertility 481
Hybridization 55, 463–7
allozyme detection 481
impact 24
natural 481

Income, scale and patterns 222–4
Index of relative importance (IRI) 250
species changes *253*
Inland Fisheries Planning (IFIP) 188
Inland Fisheries Planning Project 236
Inshore netting of immature Nile perch 176
Introgression 463–5
allozyme detection 481

Kafue fishery system 324
Kafue Gorge, *see* Lake Itezhi-tezhi
Kafue River impoundment 322
Kenya
catch comparisons *114–15*
fishing community growth 227
introduced species national impact 231–4
Kenya Fisheries Department *86*, 113, 164
Kenyan Lake Basin Development Authority 230

Lake Albert
characteristics 548
commercial catch composition changes 73
fish diversity changes *76*
indigenous fish species 29
L. niloticus yields 510
Limnothrissa yields 511
taxa contribution **75**
taxa yield *74*
Lake Bangweulu
characteristics 549
L. niloticus yields 510
Limnothrissa yields 511
Lake Cahora Bassa
characteristics 550
fish yield model 513–14
L. niloticus yields 510
Limnothrissa yields 511
pelagic community biomass **535**
water transparency and plankton abundance *539*
zooplankton abundance **540**
Lake Chad
characteristics 551
L. niloticus yields 510
Limnothrissa yields 510
Lake Chilwa
characteristics 552
L. niloticus yields 510
Limnothrissa yields 511
Lake Edward
characteristics 553
L. niloticus yields 510
Limnothrissa yields 511
Lake George
characteristics 554
L. niloticus yields **510**
Limnothrissa yields **511**
Lake Idi Amin, *see* Lake Edward
Lake Itezhi-tezhi
characteristics 555
commercial catch *325*
fish production 329–30
fish yield 325–6
model 514
L. niloticus yields 511
Limnothrissa yields 511
map *322*

Lake Itezhi-tezhi *contd*
species
changes in fisheries 321–32
composition of catches 326–7
composition changes *326–7*
diversity **324**
water level variations *323*
Lake Kainji
characteristics 556
fish yield model 512
L. niloticus yields **510**
Limnothrissa yields 510
Lake Kariba
background information 247–9
biological development 245–6
Characidae 254–5
characteristics 557
Cichlidae 255–6
Clariidae 260
commercial fisheries 249–50
commercial species weight trends *262*
Cyprinidae 260
ECOPATH II assessment 270
fish yield model 512–13
fish-eating fish **536**
fishery 5
floating fern and fish abundance *540*
Limnothrissa
ecological effects 535
length–frequency distribution *534*
production **528**
spread *530*
yields 511
map *246*
Mochokidae 260, 262
Mormyridae 260
O. niloticus yields 510
pelagic community biomass **535**
Schilbeidae 262
species changes 245–73
species composition changes 252–4
species diversity and stabilized production 264
species length analysis **263**
water level fluctuation effects 265, **266**, 270–1
zooplankton abundance **540**
Lake Kariba Fisheries Research Institute (LKFRI) 250, **528**
Lake Kivu
artisanal fisheries development program 278
biological characteristics 280
characteristics 558
fish yield model 521
food availability and periodicity 286–8
geographical characteristics 278
L. niloticus yields 510
limnological characteristics 278
Limnothrissa
ecological impact 288–90
growth 281
introduction impact 277–97
length distributions *286*
length–frequency distribution *534*
length–weight relation 285
maturity 281, *282*, 283–4
migration and feeding 285–6
night feeding **288**
production 528
yields 511
Limnothrissa fishery 290, 292
socio-economic benefits 292–4
map *279*
organism abundance **287**
pelagic community biomass **535**
zooplankton abundance **540**
Lake Kossou
L. niloticus yields 510
Limnothrissa yields 511
Lake Kyoga
characteristics 559
commercial catch composition changes 67, 69
commercial fishery 50
diversity and stability of fish stocks 59–62, 67–72, 75–9

experimental catch composition 69, *70–1*, 72
fish species introduction impact on tilapias 45–57
fish yield *50*
fish yield model 515–16
introduced fish species 48–51
L. niloticus 51
introduction 60
yields 510
Limnothrissa yields 511
map of associated lakes and swamp *47*
species disappearance 466
tilapia distribution and fishery 46, 48
Lake Malawi
artisanal fishing fleet 393
catch statistical analysis 401
characteristics 560–1
cichlid communities changes 397–412
cichlid reduction 473
commercial importance of taxa **372**
commercial trawling 397–8
community structure alteration 406, **407**, 408
community structure changes 405–6
conservation of biological diversity 391
experimental trawl surveys 398, *399*
fish stocks management, conservation and species change 365–95
fish taxa from trawl catches **381**
fish yield model 518–20
fisheries
assessment problems 388–9
prospects 389–92
statistics 366–7
trends 370, **371**, **373**
fishing areas 366
limnology and exploitation **368–9**
food sources 7
future research priorities 391–2
haplochromine cichlid biomass *408*
haplochromine population conservation status 410–11
haplochromine trawl fisheries 380, 382–6
high biodiversity 2–3
L. niloticus yields 510
legal mesh size 411
Limnothrissa yields 511
map *367*
mechanized fisheries 380
possible introductions 390–1
seine fisheries 386–8
species composition changes 404, **405–7**
species identification 400–1
standing stock estimates and yields **404**
standing stock estimation 399
state of exploitation 408–9
stock identification 372, 374, 380
and management 370
tilapiine
breeding seasons *375*
catch species composition *376*
tilapiine fisheries 372, 374–80
assessment 374–9
trawling techniques 382
zooplankton abundance **540**
Lake Malawi National Park 391, 561
Lake Malombe 366, **368**, 374–5, 387, 389
catch estimates comparison *420*
catches *618*
fishing craft changes 416, *417*
map *414*
possible fishery collapse 424
sampling strata 415
seine net composition **422**
species composition **421**
change 413–24
species length distribution *423*
tilapiine catches *377*

Lake Mweru
- characteristics 562
- *L. niloticus* yields 510
- *Limnothrissa* yields 511

Lake Nabugabo
- experimental catches **72**
- *L. niloticus* introduction 60

Lake Naivasha
- ecosystem variability 300–1
- fish community **301**, 302
- fish introduction 466
- fish sample collection 307, 309
  - laboratory treatment 310–11
- introduced tilapia species persistence 299–319
- *L. leucostictus* diet 311–12
- map *308*
- natural hybrids 479
- sampling sites 309–10
- *T. zillii* diet 312
- tilapia fishery 302–7
  - catch *303–5*
- water levels, species biomass *306*

Lake Nasser-Nubia
- characteristics 563
- fish yield model 514–15
- *L. niloticus* yields 511
- *Limnothrissa* yields 511

Lake Nyasa, *see* Lake Malawi

Lake Rudolf, *see* Lake Turkana

Lake Tanganyika
- artisanal fishery
  - contribution 429
  - units 445–6
- characteristics 564–5
- commercial fishing grounds *430–1*
- energy anomaly 506
- exploitation effects on pelagic fish community 425–41
- fish biology summary 426, **427**, 428
- fish species 29
- fish yield model 517–18
- fishery investigations 443–5
- fishing gear study 445
- industrial catch and effort data 429–34
- *L. niloticus* yields 510
- *Limnothrissa*
  - length frequency distribution *534*
  - yields 511
- map *444*
- nature of fisheries 428
- pelagic community biomass **535**
- pelagic fisheries changes 443–53
- relationship between *S. tanganicae* and *L. stappersi* 434, **435–6**, 438–40
- stocks exploited 428–9

Lake Turkana
- catch rates versus lake levels 353–5
- Centropomids 351–2
- Characids 350–1
- characteristics 566
- Cichlids 353
- Citharinids 346, 348
- commercial yield and species composition **339**
- Cyprinids 348, 350
- experimental fishing surveys 340–6
- fishery history 337–8
- fishing effort *338*
- *L. niloticus* yields 510
- *Limnothrissa* yields 511
- local subsistence consumption *340*
- major taxa status 346, 348, 350–3
- map *336*
- Silurids 352–3
- species catch and mesh size *349*
- species changes and abundance 335–63
- species composition by mesh size *341*
- survey comparison 343
- water level and nutrients 354

Lake Turkana Fisheries Research Project 335, 340

Lake Victoria
anoxia 34, 37
biomass **148**
catch
composition by major species *185*
catch levels 183–5
characteristics 567–8
cichlid
reduction 473
species elimination 38
commercial catch composition changes 62–7
commercial fish species **20**
cross correlation analysis *122*
damaged ecosystem 138, 140
ECOPATH II model *149–50*
ECOPATH species change analysis 137–61
ecosystem disruptions 33
environmental change impact 24–5
eutrophication 37–8
experimental catch analysis **69**
exploitation impact 20–3
exploitation versus *Lates* predation 27
fauna and flora changes **100–2**
fish catch change *184*
fish diversity changes *76*
fish species introduction impact on tilapias 45–57
fish stock diversity and stability 59–81
fish theft 228
fish yield model 516–17
fisheries, pre-perch, post-perch 182–99
fishery role categories 217–22
fishing gear 20–1
fishing practices 29
food intake **148**
food web 131, *139*
future fisheries trend 207–8
haplochromine catch rate 128
haplochromine decline 88–91
haplochromine sampling areas *84*
impact of changes on fish stocks 19–32
introduced species
distribution and fishery 48–51
socio-economic impacts 215–41
Kenyan portion *35*
catch composition **36**
Kenyan region, fishery yield changes *64*
Kenyan sector, catch levels and species composition *187*
L. niloticus
annual catch *112*
annual yield *177*
distribution and increase 85, 87–8
fishery assessment 163–79
fishery development 181–214
fishery management 178
fishing mortality *172*
impact assessment 199–205
introduction 60, 181–2
MSY **175**
predation impact 25–7
weights *172*
yields 510
landings and landing value *199*
limnological changes 33–43
Limnothrissa yields 511
map *22*
mixed trophic impact matrices **153–4**
mixed trophic impacts *151–2*
net catch composition changes 67
new fishery 28–9
nutrient chemistry change 34
O. esculentis catch *21*
O. niloticus 49–50
Oreochromis leucostictus and *Tilapia zillii* 48–9
organism changes 103–4
PCA analysis results *117, 119*
post-Nile perch gains in catch *200*
predation on haplochromines 27
sediment changes 131

Lake Victoria *contd*
species
change analysis 111–35
composition 185–7
disappearance 466
diversity decline 19
introduction 2
per sampling season **90**
trawl catches **37**
stratification 34–5
Tanzanian sector
catch levels and species composition *186*
fishery yield changes *66*
fishing gear pattern *190*, 191
mesh size distribution *191*
tilapia distribution and fishery 46, 48
total annual catch *116*
trawl catch composition changes 67
trawl catches **26**
Ugandan region
trawl catch composition changes *68*
fishery yield changes *65*
Lake Victoria Exploitation Committee 233
Lake Volta
characteristics 569
L. niloticus yields 510
Limnothrissa yields 511
Lift nets 448, **449**
Lights for night fishing 237, 425, 447, *448*
Longlines 346
result comparison **347**

Macrophytes 103
effects 266–7
Malawi Fisheries Department 387, 397
Malnutrition
in Rwanda 293
in sub-Saharan Africa 527
Market expansion 192–3, 197
Maximum sustainable yield (MSY)
approximate estimators 173–4, **175**
demersal trawl fishery *383*
estimation 169–70
haplochromine catch *383*
L. niloticus 164
Lake Kivu sardines 292
Lake Malawi cichlids 400
Lake Malawi tilapiine fishery *379*
Lake Turkana prediction *356*, 357–9
Meristics 478–9
Mesh size, length and weight changes 263
Migrant labour 219, 225, 228, 329
Mitochondrial DNA analysis 482–3
Mixed trophic impact routine 150
Modelling investigation 13
Monkey Bay Fisheries Station 400
Monte Carlo bootstrapping 169, **175**
Morphoedaphic index (MEI) 500, 502
Morphometrics 478–9
Mortality rates, natural 358
MTF survey system 415–16, 418, *420*
Multidimensional scaling (MSD) 116, 118
Murchison Cataracts 366
Mwanza Gulf 27
commercial trawling 128
dwarfed *H. piceatus* 92, *94*
fecundity changes *98*
fish stocks abundance changes 99, 103
haplochromine
catch rates *86*, *88*
cichlid fauna changes 83–99
length–frequency distribution *91*, *93*
size reduction 91–4
surviving species 95, 97–9

Nets
catch comparison **344–5**
catch rate **71**
Lake Kariba sampling programme 263

large mesh 223
small mesh 21
see also Gill nets
Nile perch, *see Lates niloticus*
Nile tilapia, *see Oreochromis niloticus*
NORAD survey 337, 340, 343, **344–5**, **347**, *349*
Norwegian Agency for Development Cooperation, *see* NORAD
Nuclear DNA analysis 483–4
Nurseries, competition 53–4
Nutrient
cycling effects 537–8
inputs 33, 132
and fish production 265–6
Nutritional benefits 203
Nyanza Gulf, haplochromine catch rates *86*, *87*
Nyegezi Freshwater Fisheries Training Institute *86*

Outbreeding depression 465
Overexploitation 78
Overfishing 22, 33, 329
Overseas exports **193**

Pair trawl fisheries assessment 384–6
Pair-trawler units 218
Papyrus rafts 300
Papyrus swamp loss 24–5
Parasites, introduction with fish 467
Pearson product–moment correlation coefficient **266**
Pelagic Fish Resource Assessment Project 389
Phenotypic differences and genetic divergence 481
Phytobiomass 131
Phytoplankton 247, 311
Piscivores impact 9
Planktivores impact 9
Plankton
composition 436, **437**
changes **537**
after *Limnothrissa* 538
Pollution 40
Population
bottlenecks 469–71
growth, intrinsic rate 476–7
size, effective 468–9
size reductions 467–8, 473
stability and food web 315
structure models 388
Prawns 131
density changes 102
Predation 8–9
effects 267–8
L. niloticus 236–7
Lates impact 25–7
Predators 535–6
Prices, movement **195–6**, 197, **198**
Primary production rates 504
Principal components analysis (PCA) 113, 116, *117*, 118
Processing plants 220–1
Processors, local and industrial 217
Production and marketing benefits 199–200
Production/biomass ratio (P/B) 141
Protein supply by fish 365

Rainfall, monthly *283*
Random genetic drift 470
Reconnaissance dives 34–5, 37
Redistributional impacts 216
Relative trophic transfer efficiency (RTF) **505**
Reproductive potential/success 52–4
Reproductive strategy, *Limnothrissa* 531
Reservoirs
fishery development 330–1
surface area relationships **321**
water level and fish yield *328*
Rio Convention 1
Road improvement 232

Sardine, *see Limnothrissa*
Schaefer logistic model of population growth *305*, 306
Schaefer model, Gulland's modification 384

Schaefer surplus–production model 169, 171, **173**, 378, *379*, *383*
Scientists, interaction 157
Seine fisheries 386–8
Shannon–Weaver index of diversity 62
Shire River 413
  Upper 366, **368**, 375, 389
Sigulu Islands 231
Sino–Ugandan Joint Venture 233
Soil erosion 40
Spawning sites, competition 53–4
Spearman rank-correlation
  analysis **259**
  coefficient **266**
Species
  changes
    in African lakes 1–15
    fishery impacts 222
    from fishing 11–12
    from introductions 11
    impact evaluation 10–13
    in Lake Kariba 245–73
    mitigation of impacts 13–14
  composition
    dynamics 205–7
    Lake Malombe changes 413–24
  diversity
    changes 73, 75, 76–9
      analysis 60–2
    decline 331
    and ecological succession 264–5
    fish introduction effects 457
  elimination 468
  introduction
    biological factors that favour *Limnothrissa* 529–35
    community and organizational impacts 227–31
    individual and household impacts 217–27
    international and transnational impacts 234–7
    negative impact 51
    social and economic impacts 216
    in Lake Victoria fishery 28–9
    reduction in Lake Tanganyika 437–40
  transfer 460
Sport fishing 230, 234, 460
Ssese Islands
  fish smoking 227
  fish transport risk 229
Stress symptoms 360
Stunting 533
Surplus production, *L. niloticus* 169, 176
Surplus-production
  assessment 166–7
  model 171–2, **173**
Surplus-yield modelling 358
Sustainable yield prediction model 502–4
  baseline lake choice 506
  test 521–2
  uncertainties **505**
Sustainable yield values 504

Tagging experiments 163
Tanzania
  boat levels 188
  introduced species national impact 231–4
  see also Lake Victoria, Tanzanian sector
Tanzanian Fisheries Department 113
Tanzanian Fisheries Division 164
Tanzanian Freshwater Fisheries Training Institute 128
Third World pressures 496
Time series analysis 117, 118, 120–1, 125
Tourism, cash value 496
Transgenic fish 466–7
Trawl
  catch rate
    analysis of covariance **402**
    analysis of variance **403**
    in Kenyan and Ugandan waters *114–15*
  data, *L. niloticus 168*
  exploratory survey 23

fisheries, in Lake Malawi 392
fishing and catch rate decline 27
surveys, catch rates **402**
Turkana Fishermen's Cooperative Society 337

Uganda
boat levels 188
catch comparisons *114–15*
cross correlation analysis results *123–4*
fishing community growth 227
introduced species national impact 231
map *61*
PCA analysis results *120–1*
see also Lake Victoria, Ugandan region
Ugandan Fisheries Department 113, 164
UN Conference on Environment and Development 1

Vegetation loss 24–5
von Bertalanffy growth
equation **533**
for *L.miodon* 281
parameters 167
rate 168

Water
hyacinth 25, 77–8, 105, 207, 210
level
change 24–5
Lake Kariba 247, *248*
temperatures 38
transparency
Lake Kivu *283*
and plankton decline 537, *539*
Wood use for fish smoking 204, 208

Yield-per-recruit, *L. niloticus* 167–9
Yield-per-recruit model 170, *171*

Zambian Department of Fisheries **528**
Zambian Fisheries Department 323
Zooplankton 247
and *Limnothrissa* 541
production 283